T0282423

Myxomycetes: Biology, Systematics, Biogeography, and Ecology

Myxomycetes: Biology, Systematics, Biogeography, and Ecology

Edited by

Steven L. Stephenson
University of Arkansas, Fayetteville, AR
United States

Carlos Rojas
University of Costa Rica, San Pedro de
Montes de Oca, Costa Rica

Academic Press is an imprint of Elsevier
125 London Wall, London EC2Y 5AS, United Kingdom
525 B Street, Suite 1800, San Diego, CA 92101-4495, United States
50 Hampshire Street, 5th Floor, Cambridge, MA 02139, United States
The Boulevard, Langford Lane, Kidlington, Oxford OX5 1GB, United Kingdom

Library of Congress Cataloging-in-Publication Data
A catalog record for this book is available from the Library of Congress

British Library Cataloguing-in-Publication Data
A catalogue record for this book is available from the British Library

ISBN: 978-0-12-805089-7

For information on all Academic Press publications visit our website at
https://www.elsevier.com/books-and-journals

Publisher: Sara Tenney
Acquisition Editor: Kristi Gomez
Editorial Project Manager: Pat Gonzalez
Production Project Manager: Chris Wortley
Designer: Matthew Limbert

Typeset by Thomson Digital

Contents

5. Molecular Techniques and Current Research Approaches

Laura M. Walker, Thomas Hoppe, Margaret E. Silliker

6. Physiology and Biochemistry of Myxomycetes

Qi Wang, Yu Li, Pu Liu

10. Techniques for Recording and Isolating Myxomycetes

Diana Wrigley de Basanta, Arturo Estrada-Torres

11. Uses and Potential: Summary of the Biomedical and Engineering Applications of Myxomycetes in the 21st Century

Hanh T.M. Tran, Andrew Adamatzky

12. Myxomycetes in Education: The Use of These Organisms in Promoting Active and Engaged Learning

Katherine E. Winsett, Thomas Edison E. dela Cruz, Diana Wrigley de Basanta

13. Myxomycetes in the 21st Century

Carlos Rojas, Tetiana Kryvomaz

Contributors

Andrew Adamatzky The Unconventional Computing Centre, University of the West of England, Bristol, United Kingdom

Nikki Heherson A. Dagamac Institute of Botany and Landscape Ecology, Ernst Moritz Arndt University Greifswald, Greifswald, Germany

Thomas Edison E. dela Cruz University of Santo Tomas, Manila, Philippines

Uno Eliasson University of Gothenburg, Gothenburg, Sweden

Arturo Estrada-Torres Behavioural Biology Centre, The Autonomous University of Tlaxcala, Tlaxcala, Mexico

Sydney E. Everhart University of Nebraska, Lincoln, NE, United States of America

Thomas Hoppe Institute of Botany and Landscape Ecology, Ernst Moritz Arndt University, Greifswald, Germany

Bruce Ing Applied Science and Environmental Biology, University of Chester, Chester, United Kingdom

Harold W. Keller University of Central Missouri, Warrensburg, MO; Botanical Research Institute of Texas, Fort Worth, TX, United States of America

Courtney M. Kilgore Robeson Community College, Lumberton, NC, United States of America

Tetiana Kryvomaz Kyiv National Construction and Architecture University, Kyiv, Ukraine

Carlos Lado Royal Botanic Garden (CSIC), Madrid, Spain

Dmitry V. Leontyev H.S. Skovoroda Kharkiv National Pedagogical University, Kharkiv, Ukraine

Yu Li Engineering Research Center of Chinese Ministry of Education for Edible and Medicinal Fungi, Jilin Agricultural University, Changchun, Jilin, PR China

Pu Liu Engineering Research Center of Chinese Ministry of Education for Edible and Medicinal Fungi, Jilin Agricultural University, Changchun, Jilin, PR China

Dennis Miller University of Texas at Dallas, Richardson, TX, United States

Yuri K. Novozhilov Komarov Botanical Institute of the Russian Academy of Sciences, St. Petersburg, Russia

Ramesh Padmanabhan University of Texas at Dallas, Richardson, TX, United States

Carlos Rojas Engineering Research Institute, University of Costa Rica, San Pedro de Montes de Oca, Costa Rica

Adam W. Rollins Lincoln Memorial University, Harrogate, TN, United States

Subha N. Sarcar University of Texas at Dallas, Richardson, TX, United States

Martin Schnittler Institute of Botany and Landscape Ecology, Ernst Moritz Arndt University Greifswald, Greifswald, Germany

Margaret E. Silliker DePaul University, Chicago, IL, United States

Steven L. Stephenson University of Arkansas, Fayetteville, AR, United States

Hanh T.M. Tran Ho Chi Minh International University, Ho Chi Minh City, Vietnam

Laura M. Walker Smith College, Clark Science Center, Northampton, MA, United States

Qi Wang Engineering Research Center of Chinese Ministry of Education for Edible and Medicinal Fungi, Jilin Agricultural University, Changchun, Jilin, PR China

Katherine E. Winsett Wake Technical Community College, Raleigh, NC, United States

Diana Wrigley de Basanta Royal Botanic Garden (CSIC), Madrid, Spain

Preface

The myxomycetes (also called plasmodial slime molds or myxogastrids) have not been studied with the same intensity and geographical effort that have characterized many other groups of organisms. However, they are comparatively well known for being a group of microorganisms during much of their life cycle. The reproductive structures (fruiting bodies) produced by myxomycetes are often large enough be visible with the naked eye. Numerous examples are beautiful and their colorful miniature structures often capture the attention of both the trained and untrained eye.

It is commonplace to hear a "wow" or a similar expression of surprise when someone observes a specimen of myxomycete under a hand lens or microscope for the very first time. In fact, the same thing still happens to most "myxomycete people" when they see a new or unusual myxomycete, even after many years of studying these organisms. Edward O. Wilson popularized the concept of biophilia, which refers to the innate tendency of humans to seek connections with nature and other forms of life. Clearly, this applies to the microscopic world, including the myxomycetes, and is not likely to change any time soon. Hopefully, with this volume, more people will have the opportunity to develop an interest in both myxomycetes and other fascinating microscopic organisms as well.

It is our desire that this volume serve as a source of information that can help increase general bioliteracy by opening the door to the world of myxomycetes for the layperson, as well as representing a comprehensive reference work on the biology, systematics, biogeography, and ecology of the group for the expert. In the pages of this volume there is a considerable body of information on myxomycetes and their relationship with the world around them, as well as the modern scientific context within which they have been (and continue to be) studied. However, as is often the case with any comprehensive treatment of a group of organisms, these pages contain an even larger number of descriptions and accounts relating to laboratory and field stories, lifelong friendships and professional relationships, academic and nonacademic setbacks, financial stress, the problems of bureaucracy, and hundreds of thousands of miles traveled worldwide—using every possible means of transportation known to humans—all in the ongoing effort to learn more about myxomycetes. For this reason, the publication of this volume and all the work it represented was celebrated even before the first words were written.

This volume is the product of a collective effort of many individuals who have generated information on myxomycetes over a period of several decades. Of these, only the names of 29 coauthors from 11 countries appear. Consequently, we would like to acknowledge the extremely relevant support of field assistants, laboratory technicians, logistical staff members, and many others who have enabled the named individuals to generate the information presented in this volume. At the same time, we would like to acknowledge the support of multiple funding agencies over the years, without which much of the information that we now know about myxomycetes would not have been possible to obtain.

Our strong desire to communicate the scientific results that have been accumulated on myxomycetes has been the driving force in producing this volume. We have worked alongside numerous contributors, developing an updated and comprehensive treatment of a group of microorganisms for which the most comparable publication appeared almost 50 years ago. Since the present volume contains the work of many individuals from an earlier era, we would also like to acknowledge those now-historic figures who, in one way or another, were responsible for generating curiosity in others and thus provided the inspiration for modern researchers to carry out studies on this beautiful and challenging group of microorganisms.

At some point in our lives, there was someone who captured our attention and brought us into the world of myxomycetes. We all know the name of this person, different for each one, who influenced our careers more than any one of us could have known at the time. Hopefully, through this volume, the legacy of all of those who came before us can have a similar impact on someone, somewhere.

Steven L. Stephenson
Fayetteville, AR, United States

Carlos Rojas
Turrialba, Costa Rica
March 4, 2017

Introduction

Steven L. Stephenson*, Carlos Rojas**
**University of Arkansas, Fayetteville, AR, United States*
***Engineering Research Institute, University of Costa Rica, San Pedro de Montes de Oca, Costa Rica*

One of the earliest branches of the eukaryotic tree of life consists of an assemblage of amoeboid protists referred to as the supergroup Amoebozoa (Fiore-Donno et al., 2010). The most diverse members of the Amoeboza are the eumycetozoans, commonly referred to as slime molds, and the best known slime molds are the myxomycetes. The earliest published reference to a myxomycete (apparently *Lycogala epidendrum*) appears to have been made by the German botanist Thomas Panckow (1654), but some of the larger myxomycetes, such as *Fuligo septica*, have surely been noticed by humans for many thousands of years.

Since their "discovery" by scientists, the myxomycetes (also known as plasmodial slime molds, acellular slime molds, or myxogastrids) have been variously classified as plants, animals, or fungi. Because they produce aerial, spore-bearing structures which resemble those of certain fungi, and typically occur in some of the same ecological situations as fungi, myxomycetes traditionally have been studied almost exclusively by mycologists (Martin and Alexopoulos, 1969). Indeed, the name most closely associated with the group, first used by Link (1833), is derived from the Greek words *myxa* (which means slime) and *mycetes* (referring to fungi). However, abundant molecular evidence now confirms that they are amoebozoans and not fungi (Baldauf, 2008; Bapteste et al., 2002; Yoon et al., 2008). Interestingly, the fact that myxomycetes are protists was first pointed out by De Bary (1864), more than a century and a half ago, and he proposed the name Mycetozoa (literally meaning "fungus animal") for the group. However, myxomycetes continued to be considered as fungi by most mycologists until the latter half of the 20th century.

Approximately 1000 morphologically recognizable species of myxomycetes have been described. The traditional approach has been to assign these to six different taxonomic orders, but recent evidence derived from molecular studies suggests that these orders do not hold together as they have been circumscribed in the past. As such, the system of classification used for the myxomycetes needs to be revised.

Myxomycetes are free-living predators of other eukaryotic protists and bacteria and have been recorded in every terrestrial habitat investigated to date. The two trophic stages in the life cycle (amoeboflagellates and plasmodia) are usually cryptic, but the fruiting bodies are often large enough to be observed directly in nature. Fruiting bodies release spores that are dispersed by air or, more rarely, by animal vectors. Under favorable conditions, these spores germinate and give rise to amoeboflagellates, from which the plasmodium is ultimately derived. Myxomycetes are associated with a wide variety of different microhabitats, the most important of which are coarse woody debris, ground litter, aerial litter, and the bark surface of living trees. Specimens for study can be obtained as fruiting bodies that have developed in the field under natural conditions or cultured in the laboratory. A substantial body of data on the worldwide biodiversity and distribution of the reproductive stage of myxomycetes has been assembled over the past 200 years, but only limited information is available for their trophic stages. More recently, an appreciable body of data has become available on various aspects of the biology, phylogeny, and genetics of these organisms.

The primary objective of this volume is to provide an overview of the majority of what is currently known about this truly fascinating group of organisms. This volume represents what might be regarded as a greatly expanded treatment of the type of information provided by Gray and Alexopoulos (1968) in *Biology of the Myxomycetes* and Martin and Alexopoulos (1969) in *The Myxomycetes*, both compiled almost a half century ago.

Virtually all of the more recent field-based publications relating to myxomycetes have focused on the distributional aspects of the group or the description of species new to science. This is largely due to efforts to increase the information available on these organisms in regions of the world where studies had never been carried out. Based on the body of new information that has been generated, it seems appropriate to produce an updated, comprehensive treatment of the biology, ecology, and taxonomy of myxomycetes.

Furthermore, an increased focus on molecular biology in recent decades has greatly increased the understanding of the genetics, phylogenetics, and biochemical systems present in the different groups of myxomycetes, and this type of critical information calls for a reinterpretation of myxomycete biology within a modern framework.

This volume also represents an effort to bring what is currently known about the biology and potential applications of myxomycetes to as broad an audience as possible. The intent of the editors and the collaborators has been to reach out to students, naturalists, and academics equally. Although this is a difficult task to accomplish, all of the individuals who have contributed to the information presented herein have worked with this in mind. The present volume has been designed to include important, updated, and revised chapters on the basic biology, ecology, and taxonomy of this group of microorganisms, along with discrete sections on such subjects as myxomycete biogeography; the history of studies of the group; the results of molecular-based studies; their management and conservation; and their educational value and potential applicability in the context of modern learning objectives for students at every academic level.

The chapter authors and coauthors were drawn from a group of acknowledged authorities who work with myxomycetes. Such an effort to include so many coauthors and a wide range of content is remarkable in the sense that a comprehensive volume including cross-related information and various perspectives on myxomycetes has not previously been produced. In this sense, the present volume also has an appreciable historical value with respect to documenting the current state of the art with respect to investigations directed toward myxomycetes in the first part of the 21st century. From a comparative point of view, readers of this volume will have the opportunity to understand how studies of myxomycetes carried out over the past couple of centuries have often lagged behind comparable studies of many other groups of organisms. However, the relevance and extent of the accomplishments made by a number of past and current individuals interested in the group also are described.

In addition to chapters dealing with different aspects of myxomycete biology, such as biochemistry, phylogeny, taxonomy, ecology, and distribution, readers will find that contextual topics, such as history of research, isolation techniques, biophilic potential, uses and applications, and educational value, have also been included. The idea behind this strategy has been to incorporate different approaches that are normally not considered in "more difficult" scientific texts but which are very important if one is to understand the context within which myxomycetes have been studied. An advantage of this type of approach is that readers are provided with additional tools to generate new lines of study within a modern context.

We invite the readers of this volume to become immersed in the fascinating world of myxomycetes. We encourage naturalists, science professionals, and policy makers to include these microorganisms in their agendas. We hope this volume generates necessary insights for an integrated development of studies of myxomycetes and energizes more people to appreciate the role of microorganisms in nature and, perhaps more importantly, the hidden and yet simple beauty of the natural microscopic world that lies beyond the capability of the human eye.

ACKNOWLEDGMENTS

We are indebted to a number of individuals who were willing to review one or more chapters for us. These include Gražina Adamonytė, Charles Butterwell, Jim Clark, Uno Eliasson, Myriam de Haan, Eggehard Holler, Roland McHugh, Edward Haskins, Bruce Ing, Genaro Martinez, Wolfgang Marwin, Anna Ronikier, Wayne Rosing, Martin Schnittler, Frederick Spiegel, and Diana Wrigley de Basanta. We also thank Stephanie Somerville for the format revision of the complete volume.

REFERENCES

Baldauf, S., 2008. An overview of the phylogeny and diversity of eukaryotes. J. Syst. Evol. 46, 263–273.

Bapteste, E., Brinkmann, H., Lee, J.A., Moore, D.V., Sensen, C.W., Gordon, P., Durufle, L., Gaasterland, T., Lopez, P., Muller, M., Philippe, H., 2002. The analysis of 100 genes supports the grouping of three highly divergent amoebae: *Dictyostelium*, *Entamoeba*, and *Mastigamoeba*. Proc. Natl. Acad. Sci. USA 99, 1414–1419.

De Bary, A., 1864. Die Mycetozoen (Schleimpilze). Ein Beitrag zur Kenntnis der Niedersten Organismen, Leipzig, Germany.

Fiore-Donno, A.M., Nikolaev, S.I., Nelson, M., Pawlowski, J., Cavalier-Smith, T., Baldauf, S.L., 2010. Deep phylogeny and evolution of slime moulds (mycetozoa). Protist 161, 55–70.

Gray, W.D., Alexopoulos, C.J., 1968. Biology of Myxomycetes. Ronald Press, New York, NY.

Link, J.H.F., 1833. Handbuch zur Erkennung der nutzbarsten und am häufigsten vorkommenden Gewächse, vol. 3. Spenerschen Buchhandlung, Berlin, Germany, pp. 405–422, 432–433.

Martin, G.W., Alexopoulos, C.J., 1969. The Myxomycetes. University of Iowa Press, Iowa City.

Panckow, T., 1654. Herbarium Portabile. Berlin, Germany.

Yoon, H.S., Grant, J., Teckle, Y.I., Wu, M., Chaon, B.C., Cole, J.C., Logsdon, J.M., Patterson, D.J., Bhattacharya, D., Katz, L.A., 2008. Broadly sampled multigene trees of eukaryotes. BMC Evol. Biol. 8, 14.

Chapter 1

The Myxomycetes: Introduction, Basic Biology, Life Cycles, Genetics, and Reproduction

Harold W. Keller*,, Sydney E. Everhart†, Courtney M. Kilgore‡**
**University of Central Missouri, Warrensburg, MO, United States of America; **Botanical Research Institute of Texas, Fort Worth, TX, United States of America; †University of Nebraska, Lincoln, NE, United States of America; ‡Robeson Community College, Lumberton, NC, United States of America*

INTRODUCTION

What are myxomycetes? They are life forms that have baffled naturalists and scientists for more than 300 years. The first recorded observation in 1654 of the myxomycete *Lycogala epidendrum* fruiting bodies was by the German mycologist Thomas Pankow in his *Herbarium Portatile, oder behendes Kräuter- und Gewächsbuch* (see Chapter 2). Placement of the myxomycetes on the tree of life was controversial because different life cycle stages were emphasized over time by different researchers. A portion of the life cycle was considered animal-like and included in the Animal Kingdom; another portion was plant-like and included in the Plant Kingdom; and still another portion was considered fungal-like and included in the Fungi Kingdom. These organisms were also placed in the Protista Kingdom and more recently classified as Amoebozoans based on molecular evidence.

The historical name used for this group of organisms is myxomycetes (pronounced "mix-oh-my-seats"). This name was introduced by H.F. Link in 1833 in his *Handbuch zur Erkennung der nutzbarsten und am häufigsten vorkommenden Gewächse*. This treatment was used by the American myxomycologists Thomas H. Macbride, George W. Martin, and Constantine J. Alexopoulos. Another name for the group is Mycetozoa (literally fungus animals), introduced by the German mycologist Anton de Bary in 1859 and used mostly by European myxomycologists, especially the Listers (Arthur H. and daughter

Myxomycetes: Biology, Systematics, Biogeography, and Ecology. http://dx.doi.org/10.1016/B978-0-12-805089-7.00001-9

Gulielma) in three editions of their books published in 1894, 1911, and 1925 (Lister, 1894, 1911, 1925). The complete history of the taxonomy used for the myxomycetes and their placement in modern systems of classification are covered in detail in Chapters 3 and 7.

A review of myxomycete taxonomic history highlighted the past, present, and future (Keller, 2012), and encompassed the years from the 1700s and Linnaeus, through de Bary, Rostafińsky, the Listers, Macbride, Martin, Alexopoulos, and Farr, the latter four all being American myxomycologists of more recent time. This paper chronicles the history of the University of Iowa Myxomycete Collection and collectors, including eventual transfer of the collection to the National Fungus Collections (BPI) in Beltsville, MD, USA. The Myxomycete Collection at BPI numbers between 50,000 and 60,000 specimens, including many important type specimens. A current updated online inventory of myxomycete taxa, scientific binomials, and their authorities was consulted, and authority names will not be repeated here (Lado, 2005–16).

Myxomycetes are easy to find if you know what they look like and when and where to look for them. Generally, myxomycete fruiting bodies occur wherever there is sufficient decaying organic matter with adequate moisture and moderate temperatures. Some of the favorite collecting habitats on ground sites include forested areas with decaying logs, tree stumps, accumulated decaying leaves and needle litter, bark mulching around trees and shrubs, stacked wood piles, sawdust piles, compost heaps and garden mulching, hothouses with organic peat moss, basal dead leaves of ornamental flowers in flower beds, on piles of straw or baled hay in fields exposed to the weather elements, on bark of living trees and woody vines, on bark of living apple trees in apple orchards, on living lawn grass, and on herbivorous dung in pastures (Keller and Braun, 1999).

Many species have small fruiting bodies only a few millimeters in diameter and require a hand lens of 20× for detection. Weather-related conditions that favor the greatest diversity of species of myxomycetes are rainy periods that last several days with diurnal temperatures between 20 and 38°C; this coincides with the summer months of June–September in central and southeastern USA. Nivicolous (snowbank) myxomycetes occur in high-altitude mountainous regions of the world and develop under melting margins of snow cover in early springtime, usually in April and May in North America. In contrast, desert species of myxomycetes exhibit a surprising biodiversity in the Sonoran Desert of Arizona, occurring on the woody skeletons of cacti and dead parts of living plants in contact with the ground. Many new species have been discovered in arid habitats and desert regions throughout the world, whereas other species appear to be cosmopolitan and occur wherever decaying vegetation is present, for example, the tropics, temperate regions, wetlands (such as swamps, marshes, and bogs), grasslands, and even in the colder regions of the Arctic tundra and sub-Antarctic (Keller and Braun, 1999; Stephenson and Stempen, 1994).

MORPHOLOGY OF FRUITING BODIES

The spore-producing fruiting bodies of myxomycetes vary considerably in size. Some are tiny, less than 100 μm, and require microscopic examination; others are much larger, conspicuous, and visible to the naked eye, with dimensions up to 76 × 56 cm. The most common fruiting body type is a stalked sporangium, usually of definite size, shape, and color, with internal structural parts that are used to identify species. Stalked sporangia with spores propagate most species of myxomycetes, each with a combination of morphological characteristics either present or absent; externally a hypothallus and stalk, and internally a columella and a capillitium (system of threads) interspersed with spores and surrounded by an acellular peridium or wall (Fig. 1.1).

The fruiting body stage is used in dichotomous or synoptic keys to identify a particular species. Some shapes and colors are so distinctive that a species can be picture keyed or sight identified. However, most species require microscopic examination of spores and internal structures. Many species have "look-a-like" species that occur in similar habitats, with similar features, and require a careful

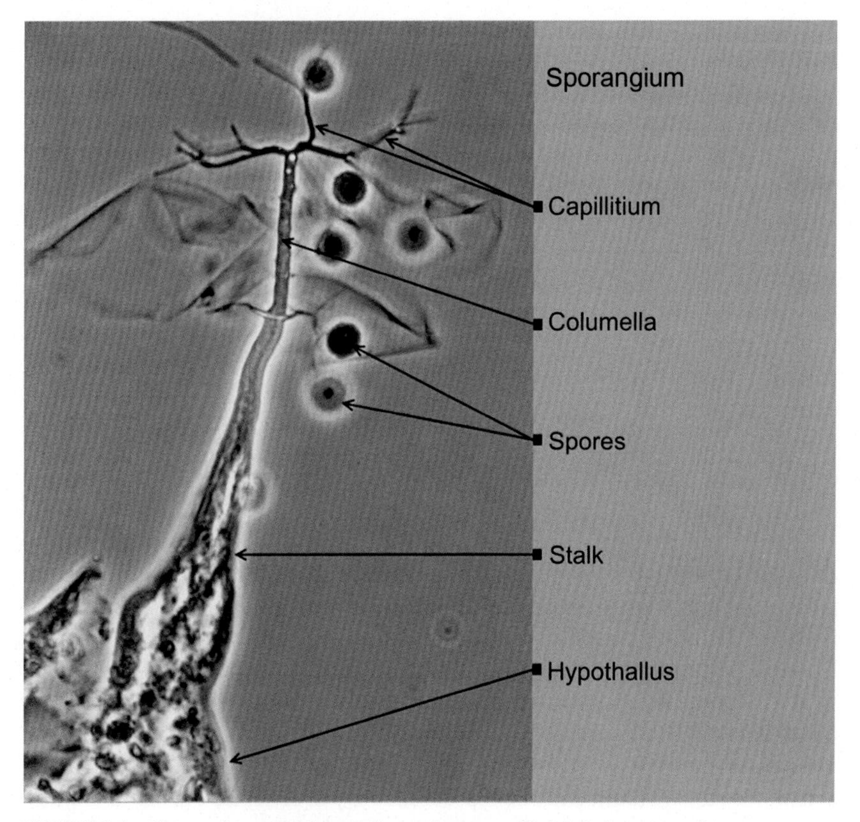

FIGURE 1.1 Sporangium of *Echinostelium arboreum* with labeled structural parts.

collecting and microscopic characterization using myxomycete monographs. The world monograph, *The Myxomycetes*, by Martin and Alexopoulos (1969) was published almost 50 years ago, and is still considered by many to be the most authoritative text on the taxonomy of the myxomycetes. It has many technical terms used in species descriptions, but lacks a glossary with definitions or illustrations of descriptive terms. This limits its use by beginners, where more recently published books include such resources—for example, Stephenson and Stempen (1994) and Keller and Braun (1999)—and are a better choice for beginning students of myxomycetes. None of these books include many new species described in the last 50 years. There are now approximately 980 described species, based on the latest count in Lado (2005–16). Therefore, a more comprehensive list of myxomycete-related terms is defined here, and in some cases illustrated, with species examples and literature sources. The reader should consult the life cycle and structural terminology definitions, to better understand the terms that follow in the topical sections.

FRUITING BODY TYPES

There are four fruiting body forms or types: sporangium, plasmodiocarp, aethalium, and pseudoaethalium. The sporangium may be stalked or sessile, often with a spherical-shaped spore case, and with a peridium that is acellular and surrounds the spores at some point during their development (Fig. 1.1). Spores are said to develop internally, and fruiting bodies are referred to as endosporous. Traditionally, there have been five recognized orders of endosporous myxomycetes: Echinosteliales, Liceales, Physarales, Stemonitales, and Trichiales. However, as outlined in Chapter 3, these taxa require revision based on recent molecular data. In certain minute species, a microscopic protoplasmodium produces only one sporangium, as in the genus *Echinostelium* (Martin and Alexopoulos, 1969) and a minute aphanoplasodium as in the genus *Macbrideola* (Martin and Alexopoulos, 1969). In contrast, phaneroplasmodial species, such as *Physarum polycephalum* (Fig. 1.2), produce a plasmodium that may grow to cover an area of a meter or more and form thousands of sporangia under ideal conditions (Keller and Braun, 1999; Martin and Alexopoulos, 1969; Stephenson and Stempen, 1994).

The plasmodiocarp is a sessile, elongated, worm-like, branched network or ring-shaped fruiting body, formed when the plasmodium concentrates protoplasm in situ in main veins during development. *Hemitrichia serpula* (Fig. 1.3) is one of the best examples of an entirely plasmodiocarpous habit (Keller and Braun, 1999; Martin and Alexopoulos, 1969) as is *Perichaena chrysosperma*, which is sporangiate to plasmdiocarpous (Martin and Alexopoulos, 1969). Several other species develop plasmodiocarps, with many in the Physarales. For example, *Physarum compressum* forms both stalked and sessile sporangia and plasmodiocarps (Martin and Alexopoulos, 1969) in the same cluster of fruiting bodies. *Physarum cinereum* forms sessile sporangia to short plasmodiocarps

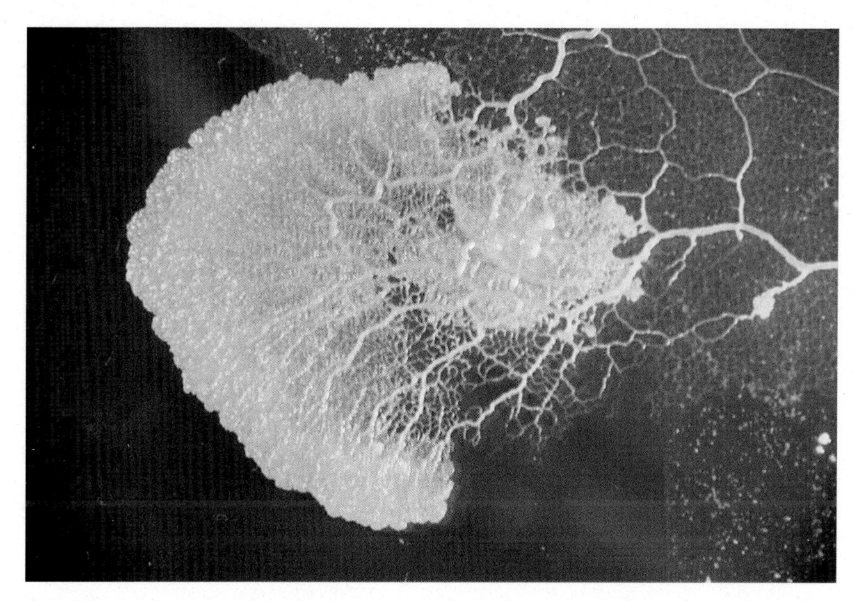

FIGURE 1.2 Bright yellow phaneroplasmodium of *Physarum polycephalum* on agar surface.

FIGURE 1.3 Plasmodiocarp of *Hemitrichia serpula* with peridium opened exposing yellow capillitial threads intermingled with yellow light-colored spores.

(Keller and Braun, 1999; Martin and Alexopoulos, 1969), as do *Physarum superbum* (Martin and Alexopoulos, 1969) and *P. serpula* (Martin and Alexopoulos, 1969). Laboratory experiments confirm that a single plasmodium can give rise to both stalked and sessile sporangia that intergrade into short or elongate plasmodiocarps of various lengths.

The aethalium is a large, sessile, round, or mound-shaped fruiting body formed from a single plasmodium, and is formed by all species in genera *Fuligo* and *Lycogala* (Figs. 1.4 and 1.5) (Martin and Alexopoulos, 1969; Stephenson and Stempen, 1994). These are the most widely known fruiting body types due to their relatively large size and frequent occurrence in urban landscapes. A world record aethalium of *F. septica* (76 × 56 cm) was discovered recently on the campus of the Botanical Research Institute of Texas on bark mulch (Keller et al., 2016).

Although similar in outer appearance to an aethalium, the pseudoaethalium represents the fusion of many sporangia packed together (Fig. 1.6). The degree to which sporangia retain internal side-walls varies. Individual sporangia are only partially fused in species of *Tubifera* (Fig. 1.7), with the side-walls still intact (i.e., *T. ferruginosa*; Stephenson and Stempen, 1994). In *Dictydiaethalium plumbeum*, the tops of sporangia are still discernable, but internal side-walls do not separate each sporangium, and only the angles of the walls remain as threads (Keller and Braun, 1999).

FIGURE 1.4 ***Lycogala epidendrum*** **aethalia broken open exposing pseudocapllitium and light colored spores covering piece of decayed wood.**

FIGURE 1.5 ***Lycogala epidendrum*** **aethalium closeup cut-away showing cortex warted surface and branching pseudocapillitial threads inside the spore chamber.**

FIGURE 1.6 ***Tubifera ferruginosa*** **pseudoaethalia top view showing closely packed individual sporangia with tops and side-walls intact.**

FIGURE 1.7 ***Tubifera ferruginosa*** **pseudoaethalia closeup lateral view showing side-walls and tops of intact individual sporangia.**

The stalked sporangium is the fundamental morphological unit found in most species of myxomycetes. However, these fruiting body types transition and intergrade into one another beginning with the sporangium that gradually undergoes transformation from individual discrete sporangia into worm-like plasmodiocarps. Then, through different degrees of sporangial fusion, it forms a pseudoaethalium where individual sporangia still retain their individual tops (in species of *Dictydiaethalium* and *Tubifera*), and finally, an aethalium may form, where the tops and side-walls become part of a much larger mound-shaped mass where the sporangia lose their individual identity. These fruiting bodies also intergrade from tiny, few-spored, stalked sporangia that grow on the bark surface of living trees and woody vines, to many thousands of sporangia producing many more spores, to a compound single, massive aethalium with thousands of spores on ground sites of decaying forest litter. These different fruiting body strategies for spore formation and release represent different evolutionary strategies for the dissemination of the maximum number of spores. A single phaneroplasmodium may form thousands of stalked sporangia with fewer spores per sporangium. In contrast, the plasmodiocarp and the aethalium form a single fruiting body to release thousands of spores. These fruiting body types occur in the different myxomycete orders, ensuring maximum spore formation and release.

MYXOMYCETE TERMINOLOGY

Myxomycete fruiting bodies are more conspicuous in natural habitats where they are collected for preservation in herbaria. This life cycle stage is used for the identification of species and is the basis for generic and species descriptions in monographs and other publications. A general review of the myxomycetes has updated the biodiversity inventory information, especially habitat descriptions and sampling (Spiegel et al., 2004), and in the same book a chapter on coprophilous fungi (Krug et al., 2004) has a limited section on myxomycete morphological terminology. Most technical terminology found in myxomycete publications is related to fruiting body occurrences and structural parts. Thus, a more comprehensive listing of defined morphological terms is included here, and examples found in myxomycete publications are provided with page numbers to facilitate location (Keller and Braun, 1999 = K & B; Martin and Alexopoulos, 1969 = M & A; Stephenson and Stempen, 1994 = S & S). Life cycle stages and reproductive and genetic terms are also included and defined (Figs. 1.8 and 1.9). Not every term found in literature sources is defined because some seem too vague and of questionable usefulness. Terms are arranged alphabetically.

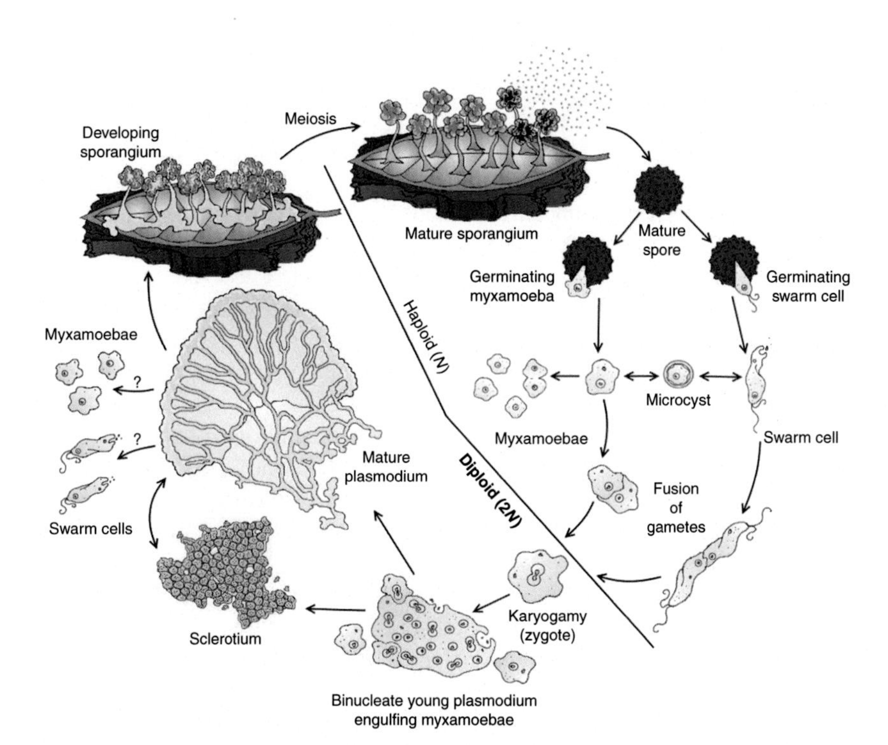

FIGURE 1.8 Heterothallic life cycle for *Physarum polycephalum*.

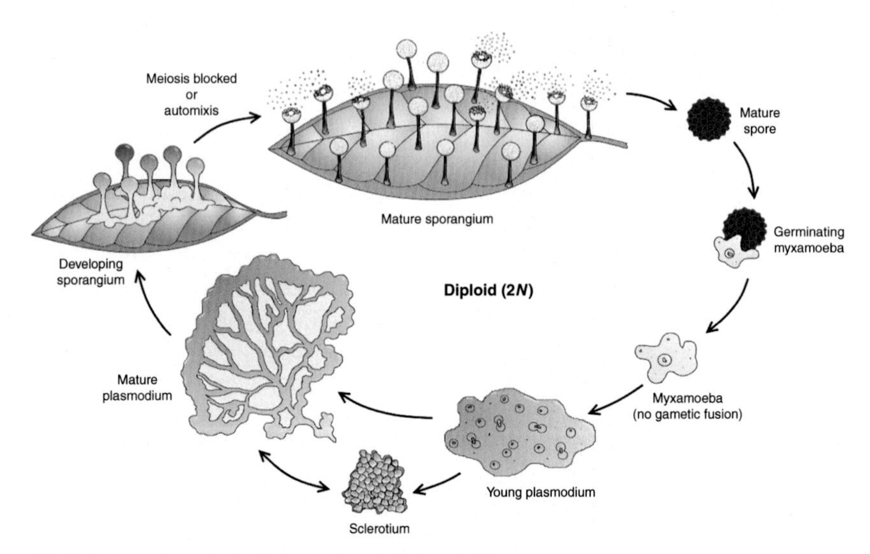

FIGURE 1.9 Apogamic life cycle for *Didymium iridis*.

Acuminate tips: Referring to the sharply pointed tips of capillitial elaters in species of *Trichia*, such as *T. botrytis* that has long, sharply tipped elaters (M & A: 138, 498–499; S & S: Plate 14).

Allantoid spores: These spores are hotdog or sausage-shaped and are unique in the myxomycetes (Keller et al., 1975). *Badhamia ovispora* is the sole example.

Anastomosing: Interconnecting branches, usually refers to the capillitium, especially used for species in the genera *Comatricha*, *Lamproderma* (Fig. 1.10) and *Stemonitis* (Fig. 1.11) (M & A: 191–240, 508–521; S & S: 28, 161).

Amoeboflagellates: A general term that refers to the haploid myxamoebae and haploid swarm cells (potential gametes) in the myxomycete life cycle, capable of interchanging developmentally based on the presence or absence of free water (Fig. 1.8).

Angular: Having angles or sharp corners, frequently used to describe the calcareous nodes in physaroid capillitium (Figs. 1.12 and 1.13). Angular calcareous nodes are described in *Physarum cinereum*, *P. contextum*, *P. leucopus*, and many other species of *Physarum* (M & A: 295, 307, 532–533, 536–537; S & S: 143–144).

Aphanoplasmodium: A plasmodial type characterized in its early stages of development by a network of flattened, thread-like, transparent, almost invisible strands that lack polarity and directional movement. These early stages lack distinct ectoplasmic and endoplasmic regions, and the streaming protoplasm is not coarsely granular. Free water favors early stages of development on agar cultures. Members of the Stemonitales and the genera *Comatricha*, *Lamproderma*, and *Stemonitis* are examples of this group.

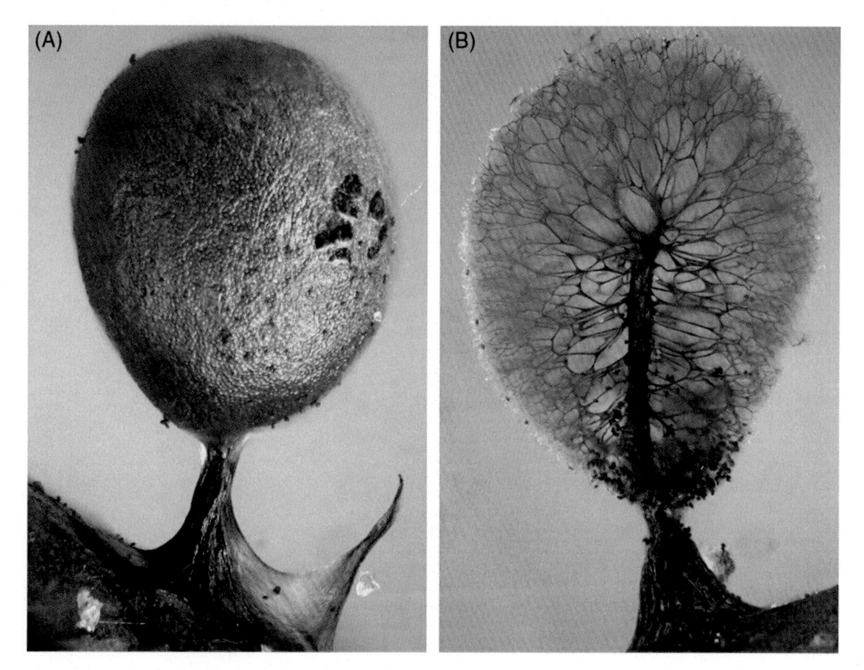

FIGURE 1.10 ***Lamproderma ovoideoechinulatum*** **stalked sporangium.** (A) Peridium with reflected bright iridescent colors. (B) Central profile showing columella rising to midpoint with attached capillitial threads branching and anastomosing with free tips at the periphery.

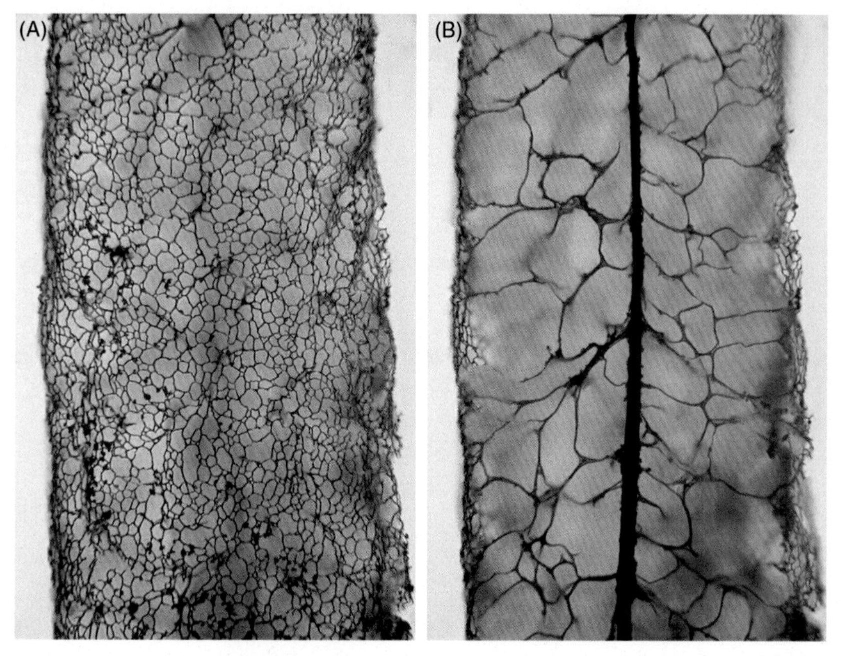

FIGURE 1.11 *Stemonitis smithii*: (A) surface view of sporangium showing capillitial network as a surface net with no free ends. (B) Sporangium in optical section with central columella giving rise to branching and anastomosing capillitium that unites at the surface as a surface net.

FIGURE 1.12 *Physarum bruneolum*: entire intact stalked sporangium (A) showing thickened calcareous peridium and hypothallus at base of stalk. (B) Image with cut-away interior showing physaroid capillitium of angular calcareous nodes connected to hyaline noncalcareous threads.

FIGURE 1.13 *Craterium leucocephalum*: (A) stalked sporangium with basal cup and white calcareous mostly angular nodes. (B) Centrally suspended calcareous pseudocollumella.

Apogamy (adj: apogamous): The condition of a myxomycete having a non-sexual life cycle without ploidy variation and therefore no fusion of haploid gametes (Fig. 1.9).

Apomixis (adj: apomictic): A life cycle where meiosis and subsequent fusion of gametes do not occur so that all stages are diploid (Fig. 1.9).

Arcuate: Curved like a bow, in reference to fruiting body shape. *P. chrysosperma* often has arcuate, sessile, plasmodiocarps (M & A: 110–111, 492–493).

Areolate: Having a pattern of block-like areas or polygons on the peridium, as in *Trichia floriformis* dehiscence (M & A: 161, 498–499).

Assimilative (trophic) phase: These are the myxamoebae, swarm cells, and plasmodia life cycle stages that are the feeding or nourishment phases (Fig. 1.8).

Asperulate: Appearing rough because of small warts or spines, as in *Physarum polycephalum* spore surface ornamentation (S & S: 148–149).

Attenuated: Gradually narrowed downward as the stalk of *Hemitrichia clavata* (M & A 148) and narrowed upward as the stalk of *Clastoderma* (K & B: 73–76; S & S: 95–96).

Axenic cultures: These cultures have only a single living species throughout the time-course of growth and development, with no other contaminating organisms present.

Badhamioid: A type of capillitium consisting of a network of calcareous tubules found in species of *Badhamia* (Fig. 1.14). This character is extremely

FIGURE 1.14 ***Badhamia panicea*** **showing badhamoid calcareous capillitial network.**

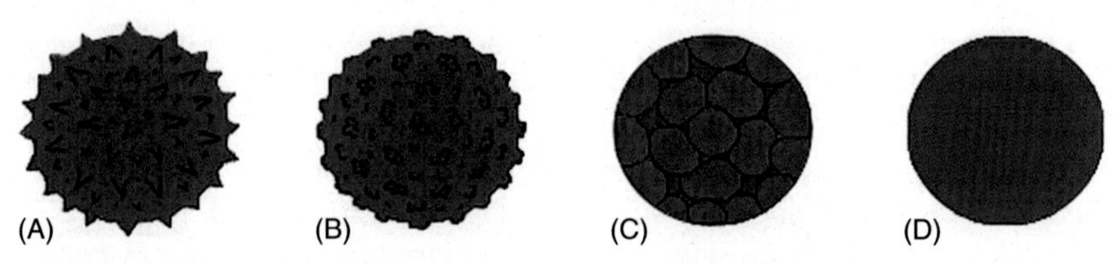

FIGURE 1.15 Spore ornamentation types: (A) spiny; (B) warted; (C) reticulate; and (D) smooth.

variable, sometimes appearing physaroid, with a network of threads lacking calcium carbonate in species that typically have a network of calcareous tubules (K & B: 28; M & A: 324–325; S & S: 31).

Bacteriovores: Any organism that feeds on bacteria as a source of food. Myxomycete myxamoebae and free-living plasmodia feed primarily on bacteria, and also on yeasts, fungal spores, algae, and possibly as opportunists engulfing other microorganisms, thus acting as "microbial predators."

Bordered reticulate: A distinctive spore ornamentation with a raised episporic network that appears to give the spore a border in optical section as seen with a compound microscope at 400–1000×. The size and numbers of the meshwork as seen in surface view are used in species descriptions and as key characters (Fig. 1.15C); *Comatricha rispaudii*, *Hemitrichia serpula*, and *Perichaena reticulospora* are a just few examples (K & B: 31; M & A: 152–153, 502–503; S & S: 34).

Bright (light)-spored: Having spores that are bright (light or pallid) colors in mass seen as hyaline, white, gray, yellow, orange, pink, and red. The orders Echinosteliales (*Echinostelium* and *Clastoderma*) and Trichiales with species in the genera *Arcyria* (Fig. 1.16B–C), *Hemitrichia* (Fig. 1.3), *Metatrichia*, *Perichaena*, and *Trichia* are members of this light-spored group (K & B: 40, 169–170; M & A: 110–165, 492–503; S & S: 35).

Bryophilous (musicolous) myxomycetes: Epiphytic species associated with and fruiting on liverworts and mosses. *Licea bryophila* and *L. hepatica* exclusively sporulate on liverworts, and *L. gleoderma* exclusively sporulates on mosses on the bark of living trees (Ing, 1994).

Bulbous: Referring to the bulb-like swollen tips of the elaters in some species of *Trichia*, for example, *T. lutescens* (M & A: 162).

Caespitose: Growing in clusters or dense tufts. *Comatricha caespitosa* takes its specific epithet from the dense tufts or clusters of stalked sporangia (M & A: 226–227, 512–513). Most *Stemonitis* species are caespitose, for example, *S. axifera*, *S. fusca*, and *S. splendens* among others (K & B: 171; M & A: 191–199, 508–511; S & S: 153, 154 Plate12).

Calcareous bodies (lime knots): Granular calcium carbonate, $CaCO_3$, in the Physaraceae is found in the capillitium as calcareous nodes, deposited as a covering on the peridium in varying thickness and layers, in the stalk and extending into the columella in the spore chamber. In the genus *Physarum*, expansions in the capillitial threads contain granular calcium carbonate (lime knots) interconnected by noncalcareous threads (Fig. 1.12). Use of the word "lime" (calcium oxide, CaO) in myxomycete literature is incorrect when it is substituted for calcium carbonate. Examples include many species of *Physarum* (K & B: 28).

Calyculus: A cup, referring to the persistent peridium forming a cup at the base of a sporangium. Most species in the genera *Arcyria* (Fig. 1.16C) and *Cribraria* have a calyculus. The depth and surface markings on the cup can be found in species descriptions (M & A: 91, 488–489; S & S: Plate 2, E–H).

Capillitium: A system of sterile threads usually attached to the columella, either simple or branching and anastomosing, to form a network intermingled with the spores inside the fruiting body (most species of myxomycetes). This definition refers to a "true" capillitium that occurs in the orders Echinosteliales (Fig. 1.1), Trichiales (Fig. 1.3), Physarales (Fig. 1.14), and Stemonitales (Fig. 1.11). These threads form from a system of preformed vacuoles that coalesce, giving rise to solid or hollow threads often with debris and of uniform diameter usually less than 6.0 μm. Threads in *Arcryia* species are more or less elastic and expand at maturity (Fig. 1.16C).

FIGURE 1.16 Fruiting body development of *Arcyria ferruginea*: (A) immature stalked sporangia in early stages of formation before final mature shape. (B) Sporangia freshly mature still in moistened state showing basal calyculus. (C) Gregarious mature orange sporangia with powdery light-colored spore mass and expanded elastic capillitial threads.

Centrioles (basal bodies): Cylindrical cell organelle of eukaryotes found in pairs in cytoplasm near the nucleus of flagellated cells (swarm cells) composed of the 9 + 2 arrangement of microtubules.

Cartilaginous: Consisting of a rather uniformly thickened peridial layer in some species of *Diderma*, especially *D. floriforme* and *D. trevelyani*, which have multiple calcareous layers. Peridial layers may be described as single, double, or triple and are often difficult to interpret accurately because of tight adhesion (M & A: 357–358, 371–372, 548–549, 552–553; S & S: 112–113, Plate 13A).

Cinereous: Resembling the color of ashes, bluish gray, represented by the fruiting bodies of *Physarum cinereum*, a common species frequently occurring on living St. Augustine lawn grass (K & B: 127, 174; M & A: 291–292, 532–533; S & S: 143–144).

Circumscissile: Refers to a dehiscence pattern where a special thin-walled area forms an encircling line usually around the apex of the sporangium. The best examples are *Perichaena corticalis* (M & A: 111–112, 492–493; S & S: Plate 15A) and *Metatrichia vesparium* (K & B: 169; M & A: 143–144, 502–503; S & S: 137–138).

Clavate: Club-shaped. Wider and thicker at the apex and narrower at the base. Generally applied to the shape of the sporangium as in the club or turbinate shape of *H. clavata* (M & A: 148, 502–503).

Clustered spores (spore balls): Spores adhering together in loose or tight clusters of 2–40, characteristic of some *Badhamia* species, including *B. bispora*, *B. capsulifera*, *B. crassipella*, *B. nitens*, *B. papaveracea*, *B. populina*, *B. utricularis*, and *B. versicolor*. This character also occurs scattered in other taxa, for example, *Dianema corticatum*, *Didymium synsporon*, *Macbrideola synsporos*, *Minakatella longifila*, *Perichaena syncarpon*, and *Trichia synsporon.* An interior cavity may be present or absent in spore clusters. The spore shape may be nearly spherical in loose clusters but distinctly ovate or pyriform in tight clusters. Spore clusters provide a diagnostic character useful in separating species in keys and species descriptions, for example, free spores versus clustered spores (K & B: 28; M & A: 251–261, 522–525; S & S: 34).

Coenocyte (syncytium): From multiple fusions and referring to the myxomycete plasmodium, a large amorphous mass of protoplasm not separated into individual cells, containing many nuclei and surrounded by a membrane (Figs. 1.2 and 1.8).

Cogs: Square-ended projections similar to cogs on a wheel, often accompanied by rings, half-rings, plates, reticulations, and minute spines. Capillitial threads of *Arcyria* species usually have ornamentation that includes cogs as part of the surface markings (M & A: 123–137, 494–497).

Coiled: A structure twisted around on itself, as in the capillitial threads of *Metatrichia vesparium* (M & A: 143–144, 502–503; S & S: 137–138, Plate 15).

Columella: A dome-shaped, spherical, or elongated central sterile structure within the sporangium that represents an extension of the stalk into the spore chamber (Figs. 1.10 and 1.11). It may be of various sizes and may serve as a supporting structure for the capillitium that may be attached in part or throughout its length. It may be a short protrusion or extend to the top of the sporangium. The columella, when present, is an important part of the species description, and is used in keys to identify the different species. This structure is absent in species of the Liceales and Trichiales.

Compressed: Flattened laterally as in the stalked sporangia of *Physarum compressum* that are compressed into a fan shape (M & A: 293, 532–533).

Conical: Cone-shaped, with a broad circular base and tapering to a point at the top. *Lycogala conicum* can be sight identified in the field because of its large conical fruiting body hence the specific epithet (M & A: 63–64, 482–483).

Coprophilous (fimicolous) myxomycetes: Species that grow and sporulate on mostly mammalian dung of herbivorous animals. A few species are obligate and are only known from dung, for example, *Kelleromyxa fimicola*, *Licea alexopouli*, and *Trichia brunnea*. These three species have evenly and greatly thickened spore walls without a thin-walled area or germination pore and appear to have adapted to passage through the intestinal tract of grazing herbivores (Eliasson and Keller, 1999; Krug et al., 2004).

Cortex: Thickened calcareous outer covering of the aethalium in the genus *Fuligo* (M & A: 266–267, 526–527; S & S: 123–124). This structure appears in all *Fuligo* species descriptions instead of the term peridium because it generally functions in the same way, protecting the black spore mass inside. *Fuligo septica* is a conspicuous example because of its large size (Keller et al., 2016) and well-developed cortex of calcareous granules. *Mucilago crustaea* also has a thickened cortex, but in this case, it is composed of crystalline not granular calcium carbonate. It is a look-alike species sometimes misidentified as *F. septica* since both are common species found in similar habitats.

Corticolous myxomycetes: This group of myxomycetes develops, grows, and sporulates on the bark surface of living trees and woody vines (Everhart et al., 2008, 2009; Snell and Keller, 2003). Most species have microscopic plasmodia (protoplasmodium or a reduced aphanoplasmodium) that exhibit rapid sporulation, usually in less than 24 h. These plasmodia produce single, tiny (less than 1 mm), stalked sporangia with an evanescent peridium and quick spore release. Life cycle stages, such as spores, microcysts, and sclerotia, are resting dormant stages capable of surviving unfavorable environmental conditions and reviving quickly when favorable conditions return, much like desert ephemeral plants. Most *Echinostelium* and *Macbrideola* species have tiny stalked sporangia that are the best examples of this short-lived life cycle strategy (Keller and Everhart, 2010).

Crowded: Refers to the species habit where fruiting bodies are massed or tightly packed together. Examples include several from the Trichiales: *Oligonema schweinitzii*, *Calonema aureum* (K & B: 163F), *Metatrichia vesparium*,

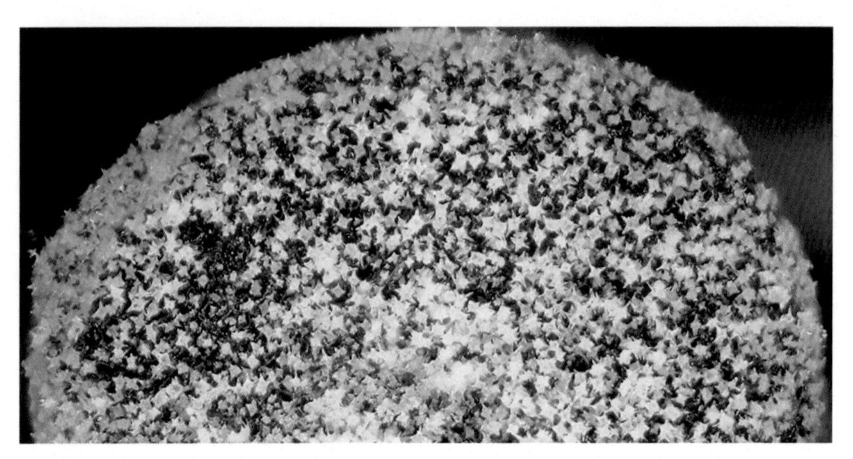

FIGURE 1.17 **Stellate crystals covering peridial surface typical of *Didymium iridis* sporangia.**

Trichia favoginea, and many *Arcyria* species, including *A. denudata*, *A. nutans*, and *A. versicolor* (M & A: 119–121, 126–127, 133–134, 136–137, 143–144, 160–161, 492–497; S & S: 137, 156–157, Plate 14A, Plate 15B).

Crystalline calcareous bodies: Loose crystals or scales scattered on the surface of the peridium (Fig. 1.16), or compacted to form a smooth eggshell-like outer crust, as in the genus *Didymium*. Two different groups of species are recognized: one based on the type of crystals such as the stellate or star-shaped crystals found on the peridium of *D. iridis* (Fig. 1.17) (S & S: 117–118) and *D. squamulosum*. The other group has an eggshell-like outer crust of compacted crystals that gives an outward appearance of *Diderma* species as in *Didymium difforme* and *D. quitense*. Members of the latter group are sometimes misidentified as *Diderma* species when there is failure to examine fractured peridial surfaces for protruding tips of crystals. A unique rhombohedral crystal was described from the stalk of *Diachea arboricola* using scanning electron microscopy (SEM) images (Keller et al., 2004).

Cylindrical: A structure that has the same diameter throughout its elongate length. Usually applied to the general shape of a sporangium, as in *Diachea leucopodia* (M & A: 178–179, 506–507; S & S: 107–108, Plate 4).

Dark-spored: Having spores in mass that are black or dark brown, such as the Physarales and Stemonitales (known as the dark-spored orders) (Figs. 1.10 and 1.18).

Dehiscent: Peridium splitting open at maturity along thinner lines like the petals of a flower as in *Diderma floriforme*, or like a star, as in *D. asteroides*, *D. radiatum*, or *D. trevelyani*.

Dextral: A term sometimes used to describe the winding in a right-handed direction of the spiral bands around the capillitial threads in *Hemitrichia* and *Trichia* species.

FIGURE 1.18 ***Lamproderma retirugisporum*** **stalked sporangia showing bluish iridescent peridia with black spores scattered on surface.**

Dichotomous: Refers to the branching pattern of the capillitium forking into two more or less equal parts as in *Lamproderma muscorum* and *Macbrideola cornea.*

Dictydine (plasmodic) granules: Microscopic (0.5–3.0 μm), usually dark-colored, strongly refractive spherical granules that are found only on fruiting bodies in the genera *Cribraria*, *Dictydium*, and *Lindbladia.* In species of *Cribraria* the dictydine granules are prominent on the peridial network and the calyculus (cup) when present. In *Lindbladia* these granules were characterized using SEM and X-ray microanalysis and shown to be hollow and contain calcium (Hatano et al., 1996). Earlier researchers (the Listers) assumed that the dark granules in the plasmodium were the source of the same granules on the fruiting bodies; hence they introduced the term plasmodic granules. However, since there has never been a developmental study to determine the origin and final deposition of these structures, dictydine granules is the preferred usage until more data from developmental studies are available. For a more detailed historical review of these terms, see Hatano et al. (1996).

Duplex: Refers to two types of capillitium as in *Leocarpus fragilis* (K & B: 173, Plate 13; S & S: 130), with a network of calcareous tubules connected to or distinct from a mostly noncalcareous system of capillitial threads, and as in *Willkommlangea reticulata* [M & A: 243–244, 522–523 (as *Cienkowskia*); S & S:

Plate 9B], with a capillitium of angular, flattened, calcareous nodes massed transversely as plates and slender, delicate, anastomosing threads forming a loose or dense net, mostly noncalcareous, or with a few calcareous nodes and numerous short, sharp-pointed branchlets.

Effused: A term applied to the habit of the fruiting body when thinly spread out and flattened on the surface of the substratum. The effused condition is found frequently in different species, but one special example is the broadly effused white fruiting bodies of *Diderma effusum* (M & A: 356–357, 546–547; S & S: 111) growing on decayed leaves.

Elaters: Free, usually unbranched and unattached, capillitial threads marked with spiral bands characteristic of species in the genus *Trichia*. These elastic threads are hygroscopic and undergo twitching movements that serve to disperse the spores in response to humidity changes in the surrounding environment (K & B: 29, 169; M & A: 160–161, 498–501; S & S: 155–159).

Endosporous: Having spores borne enclosed within a peridium (Fig. 1.10), such as in the orders Echinosteliales, Liceales, Trichiales, Physarales, Stemonitales. The order Ceratiomyxales has spores borne externally but this group is no longer included in the myxomycetes based on recent molecular evidence.

Epihypothallic development: A developmental type where stalked sporangia develop from the aphanoplasmodium through a series of individual primordial changes as the protoplasm condenses and concentrates into separate blebs that represent the sporangial initials destined to form the individual sporangia. The hypothallus forms on the lower side of the presporulating plasmodium and directly on the surface of the substratum. The stalk material is deposited at equidistant points on top of the hypothallus and elongates as more material (secreted internally from within the protoplasmic mass) is added to the tip of the stalk, carrying the prespore mass upward (Keller, 1982). This can be seen in time-lapse photography as the stalk forms first as a black line and the prespore mass appears to climb the stalk in dramatic fashion, and eventually forming the dark spore mass. This developmental type is characteristic of the Stemonitales as seen in several *Comatricha*, *Lamproderma*, and *Stemonitis* species.

Evanescent: Refers to a peridium that often disappears early in development in certain species of *Arcyria*, *Comatrichia*, *Macbrideola*, and *Stemonitis* (Fig. 1.11).

Extranuclear mitotic (open) division: Refers to mitotic nuclear division in myxamoebae wherein an extranuclear spindle forms and the nuclear membrane disappears during division.

Floricolous myxomycetes: An assemblage of species associated with the inflorescences of large neotropical living herbs, such as *Heliconia* and *Costus*, where rapidly decaying floral parts are enclosed by living bracts (Schnittler and Stephenson, 2002).

Foliicolous myxomycetes: Species restricted to growing and sporulating on decaying leaves on ground sites.

Fruiting body: A general term for the spore-bearing structure, also known as sporophore, sporocyst, spore case, sporotheca, or fructifications (Clark and Haskins, 2014).

Fugacious: Having a peridium that disappears soon after development, as in species of *Cribraria*, where portions are gone at maturity, resulting in a peridial network in the upper half of the spore case above the calyculus. Interstices of the peridial network are probably extremely thin and the membrane ephemeral because there is no evidence of its presence with scanning electron microscope observation.

Fungicolous myxomycetes: Species that feed and grow on fungi. *Physarum polycephalum* is a good example of a bright yellow phaneroplasmodium that feeds on and covers *Pleurotus ostreatus* (the oyster mushroom) and *Lentinus tigrinum* on decaying stumps, bark, logs, and the base of standing dead trees on ground sites (Keller et al., 2008). *Badhamia utricularis* also occurs on a wide variety of wood-rotting fungi, including *Stereum hirsutum* and *Phlebia radiata*, as a chrome-yellow plasmodium on decaying logs and forms extensive colonies of weakly stalked sporangia (Ing, 1994). These myxomycetes occur so frequently in this association with fungi it cannot be ascribed to chance.

Gregarious: A term used to describe the general habit when fruiting bodies grow in closely associated groups but not touching (as in crowded or heaped habits) or solitary. These terms are used in species descriptions to describe how fruiting bodies are spaced in a given habitat (Fig. 1.16A–C).

Helical (spiral) bands: These are thickened, raised bands that occur on the capillitial threads of *Hemitrichia* and *Metatrichia* species and on the elaters of *Trichia* species. This structural feature encircles the thread in a spiral pattern and is sometimes smooth, but often has an ornamentation of spines and warts. Helical may be a more appropriate term than spiral since helical refers to a three-dimensional structural arrangement around a cylinder, whereas spiral refers to a three-dimensional arrangement around a cone. The helical bands can be counted, and this is an important key character in the identification of *Trichia varia*, which has two or three bands, whereas most species have three to six bands.

Herbicolous myxomycetes: Species associated with a microhabitat of herbaceous, perennial, grassland plants with long floral stalks and dried fruits of capsules and composite heads. These myxomycetes apparently have adapted to higher summertime and colder wintertime temperatures and to drier and more alkaline conditions (Kilgore et al., 2009).

Hypothallus: A structure deposited by the plasmodium on the surface of the substratum, cementing the fruiting body in situ at the base. This structure is extremely variable and may be lacking; membranous and transparent; or tough, spongy, and conspicuous as in *Tubifera ferruginosa* (S & S:159, Plate 2A, 2B) and *T. microsperma*.

Heterokaryon: A multinucleate cell with genetically different nuclei, for example, *Didymium iridis* plasmodia.

Heterothallic: Describes a sexual reproductive system with mating types controlling amoeboflagellate fusion to produce a diploid plasmodium. This requires two genetically compatible sexual gametes controlled by a one-locus, multiple allelic mating system. These sexual systems are found in *Physarum polycephalum*, *Didymium iridis*, and *P. pusillum* (Fig. 1.8).

Nonheterothallic (homothallism): Describes a reproductive system where the sexual fusion of two amoebae carrying different mating types does not occur, and is replaced by either apogamy or homothallism. In apogamy (which can also be called apomixes = no meiosis) meiosis is partially blocked (automixis) and the resulting spore remains diploid and can thus produce diploid plasmodia without the need to mate with another amoeba (Fig. 1.9). In homothallism, which has not been confirmed in the myxomycetes, a sexual cycle with haploid amoebal fusion to produce a diploid plasmodium would occur between any two amoebae since there would be no mating types involved. Many species of myxomycetes have both heterothallic and nonheterothallic species, but several species are only nonheterothallic as in *Didymium difforme* and *D. saturnus* (Clark and Haskins, 2010).

Iridescence: Peridia of some fruiting bodies have brilliant combinations of metallic bronze, shining silver, and glittering gold, and beautiful bluish iridescent surfaces reminiscent of spectral rainbow colors, justifying the special phrase "biological jewels of nature." *Diachea* species are all iridescent, and most *Lamproderma* species are primarily bluish iridescent (Fig. 1.18). A species new to science, *D. arboricola*, has a spectacular array of iridescent colors and is only known from the bark of living trees in the upper tree canopy (Keller et al., 2004).

Intercalary swellings: It is difficult to know what the difference is between this term and vesicles, but in this case, they occur along the middle of the capillitial threads and the swellings are much smaller. Two species that consistently display this character, which aids in their identification include *Minakatella longifila* and *Perichaena minor* (Keller et al., 1973).

Intranuclear mitotic (closed) division: Nuclear division in the plasmodium does not involve centrioles and the nuclear membrane remains intact so spindle microtubules are confined within the nucleus.

Karyogamy: Fusion of two compatible nuclei.

Lignicolous myxomycetes: Those plasmodia that grow on or in decaying wood and form fruiting bodies on decaying tree logs, stumps, woody branches, and woody fragments on ground sites and standing dead trees. Species of *Lycogala* are good examples, especially *L. epidendrum* (Fig. 1.4).

Meiosis: A type of nuclear division in sexually reproducing organisms where chromosomes replicate once and divide twice, resulting in four chromosomal sets, each containing half the original chromosome number. Paired with cytokinesis, this typically results in the production of gametes (Fig. 1.8).

Microcyst: A round, tiny (4–7 μm in diameter), thin-walled, smooth, uninucleate structure that serves as a dormant resting stage in the life cycle of the myxomycetes. Forms when a myxamoeba encysts (Fig. 1.8).

Mitosis: Process of chromosomal replication and division and cytokinesis to produce two daughter nuclei genetically identical to the parent nuclei. Typically involves a series of stages consisting of prophase, metaphase, anaphase, and telophase (Fig. 1.8).

Myxamoeba: A nonflagellated, amoeboid, usually haploid cell that ultimately may function as a gamete in sexual life cycles (Fig. 1.8). The term "amoeboflagellate" is often used to encompass both this stage and the flagellated (swarm cell) stage in the life cycle.

Nivicolous (snowbank) myxomycetes: These myxomycetes occur in mountainous terrain at the snowline or underneath and along the margins of melting snowbanks. Fruiting occurs during spring on forest litter and rocks or sometimes on stems of living shrubs or trunks of trees. This is a distinct ecological group with species in the genera *Lamproderma* (Figs. 1.10 and 1.18), *Lepidoderma* (Fig. 1.19), *Comatricha*, and *Diderma* that are the most frequently collected.

Nodes: This is a general term used for a structure that serves as a common junction for connecting threads. In *Cribraria* the thickened nodes serve as the junction of several connecting threads that form the peridial network in the upper part of the sporangial wall. Sometimes used in the species descriptions of the genus *Physarum* for the calcareous connections of hyaline capillitial threads

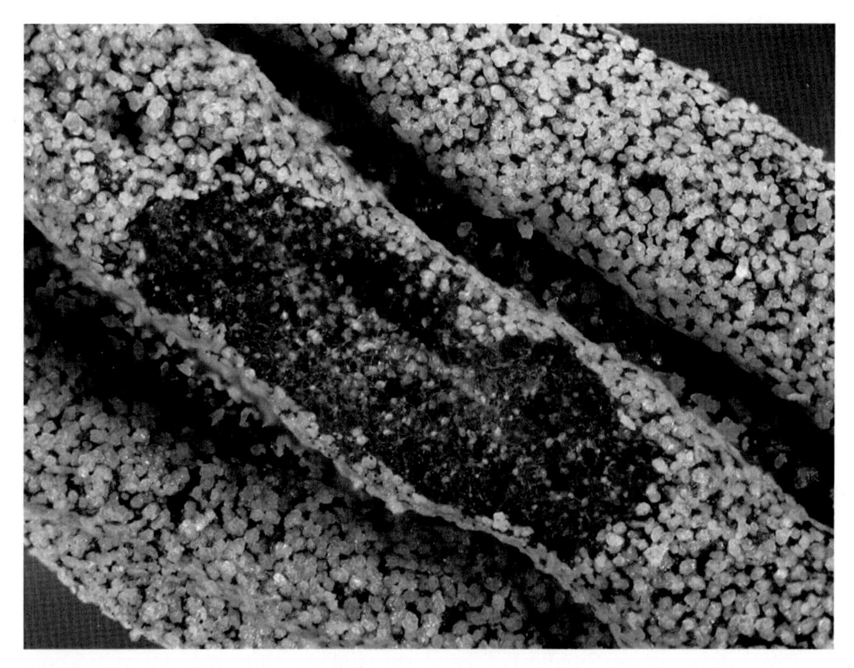

FIGURE 1.19 ***Lepidoderma granuliferum*** **sessile plasmodiocarps showing peridium covered with crystalline scales.**

(Fig. 1.12). Note that in *Cribraria* the node is part of the peridium, whereas the node in *Physarum* is part of the capillitial system.

Operculum: A lid-like structure found in some species of *Licea* where its apical position on a sporangium aids in the exposure and release of the spores. Some of the more common species that have an operculum are *L. kleistobolus* (M & A: 44–45, 480–481; S & S: 133B, 134), *L. operculata* (M & A: 480–481; S & S:133D, 135), and *L. parasitica* (M & A: 46–47, 480–481). The opercula are circular by way of a preformed thinner line of dehiscence that facilitates the lid popping open as the entire sporangium dries out at maturity.

Peridium: An acellular structural wall of varying thickness that encloses the spores in the endosporous myxomycetes (Fig. 1.18). At maturity, the peridium may be persistent as in *Perichaena* and either single-layered as in *P. microspora*, double as in *P. depressa*, or triple as in *Physarum bogoriense*. It may also be membranous and evanescent as in species of *Macbrideola* or *Stemonitis*.

Phagocytosis: The act of a plasmodium or myxamoeba feeding by a process of engulfment and ingestion of organic matter, mostly bacteria.

Phaneroplasmodium: This is the largest, most conspicuous, and colorful plasmodial type frequently seen in the field. It typically gives rise to many fruiting bodies that may cover an extensive area. At maturity, this plasmodium exhibits polarity and directional movement, terminating anteriorly in an advancing fan-shaped feeding edge, and posteriorly in a trailing network of veins. The veins deposit excreted matter along their margins, frequently appearing as "plasmodial tracks" on the substratum especially leaves. The entire plasmodium has a raised, three-dimensional aspect with definite margins. This type of plasmodium grows best under drier conditions where free water is absent. Members of the order Physarales are the best examples of this type of plasmodium (Figs. 1.2, 1.8, and 1.9).

Physaroid: A term applied to the capillitial system in species of *Physarum*. A network of slender, noncalcareous threads interconnected by enlarged calcareous nodes of various shapes (Fig. 1.12).

Pinocytosis: Type of plasmodial feeding whereby *Physarum polycephalum* engulfs liquid droplets from the surrounding medium.

Pitted (perforated): This term is applied here to the capillitial threads (tubules) of *Perichaena* species. SEM images clearly show capillitial surfaces with reticulations and pits, the latter as holes of various sizes. The first SEM images of a myxomycete new to science highlight the pits in *P. reticulospora* (Keller and Reynolds, 1971, Fig. 5). Additional SEM images of pitted capillitial threads were documented for *P. quadrata* (Keller and Eliasson, 1992, Figs. 24–28) and *P. calongei* showing a conspicuous pitted reticulum (Lado et al., 2009, Figs. 19–21).

Plasmogamy: Fusion of protoplasts of two cells preceding nuclear fusion.

Protoplasmodium: Smallest of the plasmodial types that remains microscopic throughout its existence. The highly granular and homogeneous protoplasm has a plate-like shape that fails to develop vein-like reticulate strands and advancing fans typical of other plasmodial types. Each plasmodium gives rise to a single, tiny sporangium, as in species of *Echinostelium*.

Pseudocapillitium: A capillitium assumed to develop differently than a true capillitium, though there is little evidence to support this notion. Occurring as thread-like, membrane-like, or perforated structures intermingled with the spore mass in *Reticularia* species. Appears as a system of irregular tubes in aethalia of *Lycogala* species, such as *L. epidendrum* (M & A: 63, 482–483; S & S: 136), where margins are wrinkled and much wider than 6 μm in diameter (Figs. 1.4 and 1.5).

Pseudocolumella: A calcareous mass freely suspended in the center of the sporangium, resembling a true columella but disconnected from the tip and the base of the columella or to the internal surface of the peridium. Three examples are *Craterium leucocephalum* (Fig. 1.13), *Physarella oblonga*, and *Physarum stellatum* (M & A: 330–331, 542–543).

Pulvinate: Cushion or mound-shaped. Used in connection with the aethalium of *Fuligo septica* (M & A: 526–527; S & S: Plate 2D).

Reniform: Kidney-shaped. Referring to sporangia, as in *Physarum compressum*.

Reversible protoplasmic streaming: This term is also referred to as cytoplasmic streaming, rhythmic streaming, or shuttle streaming. Movement is directional first toward the anterior feeding edge, then slows and pauses, and then reverses direction, flowing toward the posterior end. In time-lapse photography this results in pulsating movements in the plasmodia and also in the developing sporangia. This streaming occurs in both the aphanoplasmodium and phaneroplasmodium type, but the latter best example is the yellow plasmodium of *P. polycephalum* that is used in most teaching laboratories (Fig. 1.2).

Rugose: Wrinkled, as in the peridial surface of *Diderma rugosum*, which is reticulately wrinkled, marking the lines of dehiscence into polyhedral segments.

Sclerotium: A dormant resting structure in the myxomycete life cycle formed from the plasmodium under unfavorable conditions, such as lower temperatures, decreased moisture, depleted nutrients, and aging factors (Figs. 1.8 and 1.9).

Sessile: Without stalks; in many species of myxomycetes (Figs. 1.3, 1.14, and 1.19).

Shapes of fruiting bodies: Discoid (disc- or plate-shaped); fusiform (spindle-shaped, wider in the middle and tapering at both ends); globose (spherical or baseball-shaped); subglobose (nearly globose); infundibuliform (funnel-shaped); lenticular (having the shape of a double convex lens); obovate (ovate but tapering to base); obpyriform (pear-shaped, broader above and tapering to base); ovate (egg-shaped, wider at base and tapering toward apex); pyriform (pear-shaped); turbinate (top-shaped or the shape of an inverted cone); umbilicate (having a small depression, usually on the underside of the spore case above the stalk).

Sinistral: With spiral bands winding in a left-handed direction, opposite of dextral.

Spherules (macrocysts): Multinucleate, walled, plasmodial segments of various sizes that form through encystment of the plasmodium into a sclerotium. These spherules regenerate the plasmodial stage once favorable environmental conditions return (Gray and Alexopoulos, 1968). These structures have also been denoted as macrocysts, but this may create confusion since the same term is used in the life cycle stage of cellular slime molds (Figs. 1.8, 1.9).

Spores: Microscopic walled haploid reproductive units formed inside the fruiting body. Spore color, shape, size, and ornamentation are important characteristics used in species descriptions. Spores usually are spherical in shape and free as single units or aggregated into loose or tight clusters. Size ranges from 5 to 22 μm in diameter, and surfaces may be spiny, warty, or reticulate (partially, evenly, or banded), smooth or variations of these basic markings (Fig. 1.15A–D).

Spore mass: A group of spores. The color of spores en masse is a common descriptive phrase used to determine the dark (black or brown, Fig. 1.12) versus light (red, yellow, or white spore, Fig. 1.16) character. This can be determined either by opening the mature sporangium and noting the color of spores collectively or observing the color of the spore dust on the bottom of a collecting box. Specimens should be in a fresh state because color changes can occur, especially fading, when sporangia are exposed to the weather elements for longer periods of time.

Sporulation: The series of developmental stages from plasmodium to mature fruiting body with spores (Fig. 1.16A–C).

Stalk: A structure that elevates the spore case with spores above the substratum. Sometimes referred to as a stipe (adj: stipitate) in species descriptions. Stalks may be broader at the base and tapered upward and described as subulate or furrowed with grooves, as in *D. rugosum.*

Subhypothallic development: A type of development seen in species that develop from phaneroplasmodia (e.g., the order Physarales) wherein the protoplasm becomes concentrated into hemispherical mounds with the hypothallus forming on the surface of the plasmodium from the outer slime sheath. The fluid protoplasm migrates from underneath the hypothallus, pushing upward so that each mound eventually elongates into a columnar structure (the future stalk). Development proceeds so that the hypothallus, stalk, and sporangial wall form from, and are continuous with, the upper external surface of the plasmodium. Time-lapse photography shows the pulsating protoplasm actually blowing through the stalk much like air in a balloon and the early immature sporangial stages still pulsating as more protoplasmic material enters through the stalk, eventually forming the spore-forming apex. The granular debris in the plasmodium sometimes includes fungal spores, green algae, and other particulate matter that ends up forming the central core of the stalk. Species in the genera *Physarum* and *Didymium* are examples of this type of development (Keller, 1982).

Succulenticolous myxomycetes: A group of species that occur in deserts and xeric habitats primarily on species of *Opuntia* and other cacti (Lado et al., 1999).

These myxomycetes grow and fruit on rotting cladodia (cactus pads) and dried stems of cacti and arborescent giant succulents (*Agave* and *Carnegia*) in southwestern USA and similar regions of the world.

Surface net: In *Stemonitis* species, the system of branching and anastomosing capillitial threads which arise from the columella and eventually fuse at the surface into a complete net (Fig. 1.11) (M & A: 191–202, 508–510; S & S: 83).

Sutures or plates: Structures involved in a special mode of dehiscence that results in splitting open along preformed definite lines of demarcation in mature sporangia, releasing spores. Found especially in certain species in the genus *Licea*, for example, *L. minima* (M & A: 45, 480–481; S & S: 133C) and *L. pusilla*.

Swarm cell: A comma-shaped, usually biflagellate, uninucleate, haploid cell capable of functioning as a sexual gamete. This structure has a corkscrew swimming motion that distinguishes it from other aquatic microorganisms (Fig. 1.8). As already noted, the term "amoeboflagellate" is often used to encompass both this stage and the nonflagellated (myxamoeba) stage in the life cycle.

Synchronous nuclear division: Simultaneous or synchronous nuclear division occurs in the phaneroplasmodium of *Physarum polycephalum* and *P. gyrosum*. This has been shown in a series of real time movie film images produced by James Koevenig at the University of Iowa (Koevenig, 1961).

Syngamy: Fusion of sexual gametes (myamoebae and swarm cells) (Fig. 1.8).

Trabecula: An unbranched calcareous column or pillar attached above and below, bridging the vertical space in the fruiting body as in *Badhamiopsis ainoae* (Keller and Brooks, 1976, p. 836, Figs. 1–4) and *Didymium sturgisii* (M & A: 398, 558–559).

Trophic stages: Life cycle stages relating to nutrition or feeding activity (myxamoeba and plasmodium).

Tubular (hollow) versus solid: This refers to the capillitial threads generally in Trichiales taxa. *Minakatella longifila* has tubular threads (once thought to be solid), which resulted in reclassification of this species from the Dianemataceae (solid threads) to Trichiaceae (hollow or tubular threads; Keller et al., 1973).

Vesicles: This is a general term that refers to enlargements, usually in the capillitial threads. Two of the more noteworthy examples are *Didymium serpula* (M & A: 396, 558–559), which has large yellow vesicular bodies, and *Brefeldia maxima*, which has nodes bearing multicellular vesicles (M & A: 170–171, 504–505; S & S: 85).

Warted (verrucose): Having small rounded protuberances, usually referring to surface spore ornamentation (Fig. 1.15B).

Waxy (oil droplets): This term is applied to the apparent waxy or oily inclusions in the two species of *Elaeomyxa*: *E. cerifera* and *E. miyazakiensis*. These species have oil or wax in the stalk, the columella, peridium, or the capillitium as granules, globules, or inclusions. This long-standing interpretation of waxy or oily inclusions is based on direct microscopic observation, but convincing evidence from chemical tests is lacking (M & A: 175–176, 504–505).

MYXOMYCETE LIFE CYCLE STAGES

The biology of the myxomycetes narrative follows the stages in the life cycle beginning with the spore (Clark and Haskins, 2010, 2013, 2016; Everhart and Keller, 2008; Gray and Alexopoulos, 1968). The spore germinates giving rise to either a myxamoeba or a swarm cell that may form a microcyst under adverse conditions, or given optimal conditions, may fuse with genetically compatible types, eventually developing into a plasmodium. The plasmodium is the stage that develops into the fruiting bodies where spores are formed. Each of these stages is described in detail.

The spore—The typical myxomycete spore is spherical in shape and can be as small as 4 μm in diameter (as in *Stemonitis smithii*) or as large as 22 μm (as in *Fuligo megaspora*). Most of the species have spores in the size range of 7–12 μm. Spore size, surface ornamentation, shape, and color are important characters used in all species descriptions and identification keys. For proper identification, sporangia should be in a mature state with a powdery spore mass to avoid premature drying that may result in aberrant spore development. It is tempting to collect premature, brightly colored fruiting bodies in the latter stages of sporulation (Fig. 1.16A), but this may, and often does, result in specimens that are hardened, aberrant, and have agglutinated spores that cannot be identified. Furthermore, even specimens that appear to be normally mature sometimes have "giant spores" present in microscopic mounts that lack ornamentation and should not be measured in making spore counts, used to determine size, or used as a descriptive character for ornamentation.

Great care must be taken in selecting a field specimen in good condition and to avoid badly weathered, moldy, or insect-damaged specimens. Internal structures are sometimes obscured by a spore-filled sporangium, and therefore spores should be removed without damaging the branching patterns and attachments of capillitial threads when present. This is especially true of many species in the Physarales and Stemonitales. This can be done in the field by carefully preselecting wind-blown sporangia or by gently blowing away spores free of the spore case. This also can be facilitated by firmly grasping the sporangium with needle-nose forceps and gently blowing on the sporangium to remove most of the spores. Another equally effective method is to create an air stream by blowing through a flexible rubber tubing with a glass pipette which has a very small terminal opening. Sometimes it may take more effort by placing broken sporangia in a water-filled test tube and vigorously shaking. Additional procedures for handling and preparing specimens for study and instructions for making homemade dissecting tools are available in Sundberg and Keller (1996) and Keller and Braun (1999).

How high do myxomycete spores fly? Dry spore dispersal is mainly by wind currents over long distances. Evidence from high altitude atmospheric sites, such as airplanes or mountaintops, is lacking. Most airborne sample sites are at or near ground level up to 25 m (Brown et al., 1964). One study at the

University of Texas at Austin obtained rooftop samples using a Rotorod and isolation on agar culture medium. More than 16 different species of myxomycetes, all belonging to the Physarales, were grown to maturity; however, none were identified to species. Tesmer and Schnittler (2007) measured the sedimentation velocity of myxomycete spores based on experimental protocols that followed Stokes' Law for the terminal velocity of small spherical bodies in air. These aerodynamic experiments were performed under controlled laboratory conditions using airtight and earthed glass cylinders that determined the terminal velocity of myxomycete spores represented by seven different species with spore sizes from 5 to 15 μm in diameter. Sedimentation velocities depended on size. Their Table 1 clearly shows that *Stemonitis smithii* and *S. fusca* have the smallest spore diameters and the lowest sedimentation velocity, and therefore potentially longer dispersal distances over time, whereas *Lamproderma atrosporum* and *L. sauteri* have the largest spore diameters and the highest sedimentation velocity, and therefore shorter dispersal distances over time. The authors make the case that species of myxomycetes with stalked spore cases (about 58% of all described species) increase the rate of airborne spore dispersal, which is selection pressure for stalked habit morphology. In addition, the forest canopy seems to be a favorable habitat for high species diversity presumably because of the launching points at various heights above ground level (Tesmer and Schnittler, 2007).

Tree canopy studies by Keller et al. (2009) of living *Juniperus virginiana* trees (Eastern Red Cedar) in central and southeastern USA over a period of 35 years were based on bark samples from approximately 250 trees. This study documented the highest species diversity of myxomycetes (54 species) ever recorded for a single tree species. These were mostly field-collected corticolous myxomycetes on bark samples above 2 m from living trees in open fields, along fence lines, or in cemeteries. Species of myxomycetes harvested from moist chamber bark cultures were also included in this biodiversity list. The bark of *J. virginiana* is fibrous, spongy, and highly water absorbent, with a nearly neutral pH, so it hosts many more species of myxomycetes that are pH generalists. It appears that this tree species also has launching and landing sites for myxomycete spores because of the tree's habit of 20–30 m in height, exposed trunks without branches, and spatially separated horizontal branches. Such architecture may account in part for the higher species biodiversity.

Insects may also act as vectors of spore dispersal, especially beetles in the family Leiodidae and species of *Anisotoma* and *Agathidium* (Stephenson and Stempen, 1994) and various species of flies in the family Mycetophilidae. Many different organisms were directly observed either ingesting and/or feeding on spores (Keller and Smith, 1978), or incidentally scattering spores with their movements, including mites, isopods, springtails, wood lice, tardigrades, collembolans, psocoptera, dipterans, and millipedes (Ing, 1994). Slugs *Philomycus carolinianus* and *P. flexuolaris* were reported to feed on immature sporangia of the myxomycete *Stemonitis axifera* (Keller and Snell, 2002).

SEM has resulted in more accurate species descriptions of spore ornamentation and the discovery of fine structure not possible with a compound microscope oil immersion optical lens (Keller et al., 1975; Keller and Schoknecht, 1989). *Badhamia ovispora and Fuligo megaspora* represent a case study in earlier inaccurate descriptions of spore ornamentation. *Badhamia ovispora* was misdescribed in the holotype description as smooth and oval-shaped when SEM showed an allantoid shape and surface markings as minute, raised, plaque-like areas. *Fuligo megaspora* was inaccurately described as having coarsely tuberculate markings arranged in irregular lines, or rough-tuberculate, surface markings often united into irregular and incomplete reticulation, or tuberculate to subreticulate, or closely spinulose. Comparisons using SEM, however, highlight the presence of an incomplete episporic reticulum with a serrated upper edge that lacks the circular "halo" seen in bordered reticulate spores. The serrated edge was undoubtedly the source of "closely spinulose" and the pattern of the tips of the serrations the source of the "tuberculate" references.

Most species of myxomycetes have spores with some kind of ornamentation, spiny, warted, reticulate, or variations of these markings (Fig. 1.15). The presence of a stalk to support the spore case promotes higher spore dispersal rates than sessile sporangia or plasmodiocarps. Furthermore, smooth spores appear to be rare in the myxomycetes (only one species has smooth spores in the Physarales, *Badhamia apiculospora*, and there are none in the Stemonitales and Trichiales). Interestingly, in contrast, the genus *Licea* (Lado, 2005–16) has 29 species with smooth spores out of a total of 59 species, most species in the genus being sessile and not stalked. There is no experimental evidence to explain this disproportionately high number of smooth spores in *Licea*.

Myxomycete spores float on the surface of water presumably because of surface tension and small size. Wetting agents, such as solutions of ethyl alcohol, are used to submerge spores in water mounts. Spore ornamentation also accounts for hydrophobic effects as recently described by Hoppe and Schwippert (2014). Their study grouped spore ornamentation into three types: reticulate, spiny, and smooth. Spores of 17 species of myxomycetes were examined by SEM to characterize spore ornamentation type and then the spores were photographed on the water surface to determine contact angle. Calculations were made to determine the surface energy of *Metatrichia floriformis* (reticulate spore type), *Fuligo septica* var. *candida* (spiny spore type), and *Licea parasitica* (smooth spore type). Their Fig. 3 shows the reticulate spore type floating on the water surface, the spiny spore type half sunken in the water, and the smooth spore type totally submerged in the water. Their conclusions were that the hydrophobicity effect was caused by the spore ornamentation type and hydrophobines present on the spore surface.

Spores have been purported to remain viable and germinate after 75 years, as in *Hemitrichia clavata* (Gray and Alexopoulos, 1968), although it is possible specimens that old may have been contaminated with the passage of time. Spore viability decreases with age, but some species have spores that germinate within 30 min, especially in freshly matured specimens of *F. septica* that can be used in teaching laboratories (Keller and Braun, 1999). More than 120 species of myxomycetes are recorded as having spores that germinate (Gray and Alexopoulos, 1968), and this occurs either by the split or pore method. Typically, members of Physarales germinate by a wedge or V-shaped split and members of the Stemonitales (*Stemonitis*) and Liceales (*Licea*) by the pore method.

Three different life cycle types were proposed by Ross (1957) based on spore germination events and the duration of the flagellate stage: (1) the briefly flagellate type (protoplasts emerge from the spore as a myxamoeba which produce flagella shortly thereafter); (2) the flagellate type (wherein germinating spores give rise only to flagellate swarm cells that remain active 100–130 h); and (3) the completely flagellate type (develops flagella immediately after leaving the spore and remains flagellated for longer periods of time—more than 130 h—and syngamy only occurs between two swarm cells).

The myxamoeba—Myxamoebae are colorless, uninucleate, and microscopic (about 10 μm) and divide mitotically by extranuclear division. It is this stage that is essential to complete the myxomycete life cycle since the swarm cell and microcyst stages are not generally necessary. The myxamoeba and swarm cell are referred to as amoeboflagellates since each stage is interconvertible and condition-dependent: free water in the case of the swarm cell and a drier condition in the case of the myxamoeba. It is the myxamoebal stage that is capable of multiple divisions that results in large populations of cells that are potentially immortal if a heterothallic strain is maintained. In contrast, the swarm cell does not divide but is capable of feeding during periods when excessive water precludes myxamoebal feeding and division. These life cycle stages enable survival during unfavorable environmental periods, as well as take advantage of more favorable growth conditions and at the same time maintain a population reservoir of reproductive cells, especially in soil and forest litter. Furthermore, *Lycogala epidendrum*, *Trichia decipiens*, and *Tubulifera arachnoidea* myxamoebae were observed with bacterial prey food microorganisms, and test measurements were made to determine the highest rate of cell movement. Results showed that *T. decipiens* had the highest rate of cell movement, followed by *T. arachnoidea*, and *L. epidendrum*, showing species-specific patterns of cell motility. In addition, cell size of myxamoebae also differed significantly, with *L. epidendrum* having considerably larger myxamoebae (Hoppe and Kutschera, 2015).

Myxomycete myxamoebae that carry only one sexual mating type locus represent sexual gametes that can be cultured in perpetuity and can build up populations of gametes through growth and division (Fig. 1.8). Feeding behavior is by lobose pseudopodia that engulf primarily bacteria. Myxamoebae can be transferred from culture to culture and through mating and genetic crosses it was determined that mating types operated as a single locus-multiple allelic system. Ray Collins and his students popularized *Didymium iridis* and myxomycetes as the system of choice to study mating types, apogamy, and plasmodial incompatibility, and this system still stands today as one of the great scientific advancements in myxomycete research (Keller and Everhart, 2010).

The swarm cell—In the presence of free water, the swarm cell becomes a biflagellate, comma-shaped swimming cell. Typically, two whiplash flagella are attached to the anterior cone-shaped end: one short, recurved, and sometimes difficult to see as it is appressed to the cell membrane, and the other much longer, propelling the swarm cell through the water. The corkscrew swimming motion is due to rotational body movement combined with flagellar beating that is unique to the myxomycetes (Fig. 1.8). There appears to be little directional movement, but more experimentation is needed to determine how these cells react to chemical gradients in the water (Everhart and Keller, 2008; Keller and Everhart, 2010).

Swarm cells are sexual gametes that may mate given genetically compatible pairs. Observation of many cultures of swarm cells in hanging drop slides and in microchambers suggests that the swarm cell stage does not undergo cell division, sometimes referred to as binary fission. Mating of two genetically compatible swarm cells occurs through attachment at their posterior ends. This is sometimes comical to watch since as mating progresses the swimming motion becomes more labored as the flagella at their free anterior end thrash about making very little directional headway. Eventually the swarm cells fuse and initiate the diploid phase. This type of mating behavior is unusual especially when compared to the green alga *Chlamydomonas reinhardtii* wherein mating occurs through the pairing and fusion of two equally biflagellate cells by the tips of their flagella.

The microcyst—These structures are tiny (4–7 μm in diameter), highly refractive, thin-walled, uninucleate, and a highly resistant stage. Encystment of myxamoebae is induced by a combination of adverse environmental conditions including lack of food, overcrowding, accumulation of toxic metabolic byproducts, drought or drier conditions, wide temperature fluctuations, or too much water. Excystment occurs by the pore method as a portion of the wall is dissolved and the protoplast exits. Microcysts give myxomycetes the ability to survive environmental extremes, and provide a potential reservoir of additional propagating units that can shorten the time required to complete sporulation (Gray and Alexopoulos, 1968).

The plasmodium—Once mating has occurred with the fusion of genetically compatible haploid amoeboflagellates, the protoplasm comingles (plasmogamy), and then the two nuclei fuse (karyogamy) to form a diploid zygote in heterothallic strains. Early formation zygotes begin a series of synchronous intranuclear divisions without forming cell walls, thus resulting in a multinucleate plasmodium. Zygote coalescence was observed that also increases the size of the plasmodium. Over time the plasmodium feeds on bacteria, yeast, and other microorganisms, including myxamoebae (cannibalism), by phagocytosis and fusions with other plasmodia (Gray and Alexopoulos, 1968). This results in growth and an increase in size.

There are three major types of plasmodia (previously defined) that are associated with certain orders, as in the aphanoplasmodium with the Stemonitales

(*Stemonitis* spp.), phaneroplasmodium with the Physarales (*Physarum* spp.) (Fig. 1.2), and protoplasmodium with the Echinosteliales (*Echinostelium* spp.) and with some Liceales (*Licea* spp.). Perhaps a fourth type is a combination of the first two as in the Trichiales (*Perichaena* spp.) (Gray and Alexopoulos, 1968). Of these, the phaneroplasmodial type is usually the largest and often is brightly colored yellow (the most commonly encountered color in the field). Other colors are often seen, including white (very common), black, gray, lead-colored, red, rose, pink, orange, purplish, bluish, greenish hues, and colorless. In many cases the plasmodium is unknown (Martin and Alexopoulos, 1969). Plasmodial color is extremely variable in the same species and changes occur with changes in pH or when incorporating the color of a pigmented bacterial food source, or as fruiting progresses, thereby suggesting color is not a very dependable taxonomic character. Plasmodial color when observed is recorded in species descriptions (Gray and Alexopoulos, 1968; Martin and Alexopoulos, 1969).

Some properties of the myxomycete plasmodium deserve special mention. The regeneration properties are well documented since small sections of the plasmodium can be transferred and subcultured on agar in petri dishes. Cultures can thus be maintained over many generations as long as a food source, often *Escherichia coli* bacteria, and sterile old-fashioned oat flakes, is provided (Gray and Alexopoulos, 1968).

Synchronous mitotic nuclear divisions occurring in the plasmodium were filmed using time-lapse photography and showed the chromosomes all dividing at the same stage, for example, all nuclei in prophase at the same time, metaphase at the same time, and so on (Koevenig, 1961). Due to this unique property of synchronization, the plasmodial stage of *Physarum polycephalum* has been used experimentally as a model organism in cancer research. Plasmodia of *P. polycephalum* can be cultivated under controlled axenic laboratory conditions, and form a highly regulated amorphous, simple, plasma membrane-bound "cell," compared with cancerous cells that have lost their internal controls and divide out of control to form a cancerous tumor (Keller and Everhart, 2010).

The plasmodium is membrane-bound, acellular, and naked. However, the veins are surrounded by a hyaline sheath with enclosed refuse matter that leaves a "plasmodial track" on the agar surface. Inside the sheath is a plasma membrane that encloses the living protoplasm consisting of an outer, more viscous plasmagel and an inner, more liquid plasmasol (Gray and Alexopoulos, 1968).

What are the protoplasmic organelles and inclusions found in the plasmodia of myxomycetes? Nuclei vary in size (3–8 μm), appear plastic and capable of changing shape, have a single nucleolus, and have tiny chromosomes that are difficult to count but vary in $2N$ number from 50 to 100 based on light microscope counts. Mitochrondria are present as small granular inclusions along with pigment granules in colored plasmodia, food vacuoles, and contractile vacuoles. Food vacuoles are involved in the ingestion of solid food particles by a series of events that occur at the advancing edge of plasmodium. Engulfment results when a depression and advancing sides surround a bacterium, eventually enclosing it, digesting the contents, and finally egestion of waste products (Gray and Alexopoulos, 1968).

Plasmodial membranes and pseudopods (lobopods) are also involved in the formation of diploid myxamoebae and swarm cells (Fig. 1.8), but life cycle illustrations usually fail to show these interesting and controversial morphological events. The observations of Indira (1964, 1969) merit special consideration in this regard as they include detailed descriptions and line drawing illustrations of *Physarum* species, *Arcyria cinerea*, and *Stemonitis herbatica* plasmodial development, and the direct formation of myxamoebae and swarm cells. The latter occurs in *A. cinerea* when the leading edge of advancing plasmodial fan cuts off a small bit of protoplasm from a pseudopod as a myxamoeba that eventually develops flagella and swims away. A series of similar events occurred along the veinlets. These observations were repeated many times with many different cultures, and this also occurred in *S. herbatica* and thus perhaps represented a method of asexual reproduction. Extra precautions were taken to ensure no spores were present in the cultures. The first author can confirm that *Stemonitis* and *Comatricha* species grown in culture often have swarm cells that remain abundant in the culture for prolonged periods of time. Swarm cells that were transferred to agar plates eventually developed into plasmodia even though mating swarm cells were never observed. Diploid swarm cells produced from mature diploid plasmodia were never seen being ingested (Indira, 1969).

Phase contrast microscopic observations and photographic evidence were obtained from microcultures that showed myxamoebae formation from a diploid, multinucleate plasmodium (Ross, 1967). The culture was tentatively identified as *Physarum pusillum*, and although a rare event (and the ultimate fate of the amoeboid cells could not be determined), the plasmodium had the potential to produce diploid myxamoebae. These are interesting implications for the myxomycete life cycle, and additional questions arise: Are these myxamoebae capable of mating or engulfed by the plasmodium and ingested or do they add to the ploidy level? Do they encyst and play no role in the growth or development of the plasmodium? Much more research is needed to answer these questions.

A special property of the myxomycete plasmodium is protoplasmic streaming, also called shuttle streaming, reversible or rhythmic streaming, or pulsating streaming. This plasmodial internal movement can be observed easily in teaching laboratories even at lower magnifications of 10–20× and sometimes with the naked eye. Students are spellbound observing protoplasm moving forward toward the feeding end, then interrupted by a pause, reversing directional flow, and moving toward the posterior network end. Indeed, the streaming protoplasm within a plasmodial strand can reach speeds of up to 1350 μm/s, which is the fastest recorded rate for any microorganism (Gray and Alexopoulos, 1968). The rate of protoplasmic streaming can be timed with a stopwatch to determine speed of flow, duration of directional flow forward, time of pause, and time of reversible flow posteriorly. Students can do a simple experiment using

P. polycephalum plasmodia to answer the questions: Where is the longest duration flow? Progressive toward the anterior end? Or regressive toward the posterior end?

The sclerotium—Plasmodia are capable of forming a dormant, resting, or resistant stage and thereby surviving adverse environmental conditions of starvation, excessive cold or hot temperatures, extremely dry conditions, and overwintering in temperate climates. The sclerotial stage of *Physarum polycephalum* is used to start cultures in teaching laboratories. Sclerotia go through a process of dehydration, forming large spherical isodiametric spherules. These spherules range from 24 to 40 μm and are multinucleate with up to 14 nuclei. Sclerotia may form on the bark of living trees as a thin sheet of horny material and accounts in part for the rather short periods of time (12–24 h) required to complete the life cycle with fruiting bodies. Physaraceous plasmodia may form sclerotia on filter paper that encyst as thin flattened sheets of hardened matter, often in the shape of the plasmodium (Fig. 1.8). Filter paper can be wetted, and plasmodia can be revived on agar in 30–60 min as migrating and feeding plasmodia (Gray and Alexopoulos, 1968).

Sporulation—Once the plasmodium has been triggered to develop into a fruiting body, the process cannot be reversed. Fruiting body formation usually occurs at night or in early morning hours (Fig. 1.15A–C). Experiments with different light regimes showed that light triggers sporulation in species with yellow-pigmented plasmodia, whereas nonpigmented plasmodia sporulated equally well when exposed to light or in total darkness (Gray and Alexopoulos, 1968).

Other factors involved with the induction of sporulation include a combination of optimal growth age at a time when nutrients are exhausted, dark incubation, and presence of niacin. Importance of nutrient depletion can be demonstrated by transferring an actively growing plasmodium to nutrient-poor water agar that stimulates sporulation. It appears that an optimal temperature of 20–25°C and more acid pH between 4.0 and 5.0 supports the growth and development of most species of myxomycetes, but this can vary somewhat with individual species. Most of this experimental evidence is based on *P. polycephalum*, and more research is required before broad generalizations can be made for the myxomycetes (Gray and Alexopoulos, 1968). The fruiting body sequence of events was described for epihypothallic and subhypothallic development in the terminology section and will not be repeated here.

GENETICS AND REPRODUCTION

Mating types are determined through a genetically regulated system that controls fusion of myxamoebae and swarm cells. Compatible mating types will result in the formation of a diploid zygote. Heterothallic species of myxomycetes produce myxamoebae that require a different and compatible mating type for production of plasmodia and fruiting bodies. Consequently, single-spore isolates of heterothallic species never form zygotes, and do not produce plasmodia or fruiting bodies in culture or in nature. Strains that can form fruiting bodies from single-spore cultures are considered nonheterothallic, with two potential mechanisms through which this may arise: apogamy or homothallism. In general, it is thought that true homothallism does not exist, and heterothallism is considered most common for myxomycetes.

In apogamic (without gametes) isolates, the spores develop directly into plasmodia due to their diploid (or polyploid) nuclei having two (or more) sets of chromosomes, which carry two different mating type alleles. This diploid state is maintained by an incomplete meiosis (automixis) during sporulation, which results in the retention of the diploid state in the spore. An alternate term for this system is apomixes (without meiosis) since the lack of meiosis produces the diploid amoebal state.

Many myxomycete isolates grown in culture often are nonheterothallic (sometimes called homothallic), which would be considered a truly asexual (apomictic; derived from a single spore) mechanism. Under this scenario, germination of an apomictic spore occurs without undergoing meiosis, thus producing a diploid or polyploid myxamoeabe that is able to convert to the plasmodial stage without mating. These nonheterothallic isolates are able to complete the entire life cycle as 1*N* or 2*N*, without mating.

Homothallism, which has never been demonstrated in the myxomycetes, would involve a sexual reproductive system with gamete fusion and meiosis and an alternation between haploid amoeba and diploid plasmodia. However, this system would require the absence of a mating type system, so that any two haploid gametes (including identical sister amoeba derived from a single spore) could fuse to produce the diploid plasmodium. A special type of homothallism (called selfing) can occur in some normally heterothallic isolates. Haploid amoeboid clones (which can undergo normal sexual fusion) may also produce haploid plasmodia by themselves without crossing; this behavior is apparently due to a mutation in the locus or loci governing heterothallic mating.

Homothallic mating, also called selfing or cryptic sex, may also exist, wherein there has been a breakdown in the locus or loci governing heterothallic control. Clark and Haskins (2010) provides a table of heterothallic and nonheterothallic reports for 50 different species of myxomycetes.

Mating type alleles—The mating system relies on a single gene locus (*mt*) with multiple alleles, where some species have up to 18 different alleles. These genes govern sexual fusion, where heterothallism (requiring different *mt* alleles) is known from systems like *Didymium iridis* as being controlled by a single-locus multiallelic system or in *Physarum polycephalum* two loci (matA, matB), each with multiple alleles (Betterley and Collins, 1983; Clark, 1995; Collins, 1975; Kawano et al., 1987). In general, the evolutionary role of heterothallic mating systems is to ensure populations of closely related individuals are unable to self, thus avoiding inbreeding depression or the accumulation of deleterious alleles that negatively affect fitness.

Genetics of the plasmodium—There are genes known to regulate plasmodial fusion (plasmogamy), which is a system of compatibility determination analogous to vegetative compatibility in true fungi. In total, there have been more than 10 fusion genes identified that function to enable self- and nonself-recognition. Between plasmodia that have compatible vegetative compatibility groups (VCGs), fusion of plasmodia with different genomes can result in a heterokaryotic ($2N + 2N$) nuclear state. Inside the same plasmodium, fusion of genetically different nuclei may occur to form a diploid ($2N$), following which, meiosis of recombined heterokaryotic nuclei returns the plasmodium to a haploid (N) state. Such recombination without formation of fruiting bodies is called the parasexual cycle, although there is no published experimental evidence to show that this can occur.

ACKNOWLEDGMENTS

We wish to thank Renato Cainelli from Trieste, Italy, who provided most of the beautiful fruiting body photographic images. He used focal plane merging (z-stacking) which combines multiple closeup images taken at different focal distances with a camera system to create greater depth of field. Computer software was used to stack and blend the images so that the spherical spore case and stalk are both in focus. Most of the structures shown here would have portions of the final image out-of-focus with a standard camera macro lens. Harold W. Keller provided images for Figs. 1.1 and 1.2. Courtney Kilgore provided the artwork for the life cycles and other illustrations that were newly prepared for this chapter. Brooke Best at the Botanical Research Institute of Texas provided editorial assistance with the manuscript and Jim Clark gave advice on the genetics and life cycles.

REFERENCES

Betterley, D.A., Collins, O.N.R., 1983. Reproductive systems, morphology, and genetical diversity in *Didymium iridis* (Myxomycetes). Mycologia 75, 1044–1063.

Brown, R.M., Larson, D.A., Bold, H.C., 1964. Airborne alga: their abundance and heterogeneity. Science 143, 583–585.

Clark, J., 1995. Myxomycete reproductive systems: additional information. Mycologia 87, 779–786.

Clark, J., Haskins, E.F., 2010. Reproductive systems in the myxomycetes: a review. Mycosphere 1 (4), 337–353.

Clark, J., Haskins, E.F., 2013. The nuclear reproductive cycle in the myxomycetes: a review. Mycosphere 4 (2), 233–248.

Clark, J., Haskins, E.F., 2014. Sporophore morphology and development in the myxomycetes: a review. Mycosphere 5 (1), 153–170.

Clark, J., Haskins, E.F., 2016. Mycosphere Essays 3. Myxomycete spore and amoeboflagellate biology: a review. Mycosphere 7 (2), 86–101.

Collins, O.N.R., 1975. Mating types in five isolates of *Physarum polycephalum*. Mycologia 67, 98–107.

de Bary, A., 1859. Die mycetozoan: Ein beitrag zur kenntnis der niedersten thiere. Zeitschrift für wissenschaftliche Zoologie 10, 88–175.

Eliasson, U.H., Keller, H.W., 1999. Coprophilous myxomycetes: updated summary, key to species, and taxonomic observations on *Trichia brunnea*, *Arcyria elaterensis*, and *Arcryia stipata*. Karstenia 39, 1–10.

Everhart, S.E., Ely, J.S., Keller, H.W., 2009. Evaluation of tree canopy epiphytes and bark characteristics associated with the presence of corticolous myxomycetes. Botany 87, 509–517.

Everhart, S.E., Keller, H.W., 2008. Life history strategies of corticolous myxomycetes: the life cycle, plasmodial types, fruiting bodies, and taxonomic orders. Fungal Divers. 29, 1–16.

Everhart, S.E., Keller, H.W., Ely, J.S., 2008. Influence of bark pH on the occurrence and distribution of tree canopy myxomycete species. Mycologia 100, 191–204.

Gray, W.D., Alexopoulos, C.J., 1968. Biology of the Myxomycetes. Ronald Press Co., New York, NY.

Hatano, T., Arnott, H.J., Keller, H.W., 1996. The genus *Lindbladia*. Mycologia 88, 316–327.

Hoppe, T., Kutschera, U., 2015. Species-specific cell mobility of bacteria-feeding myxamoebae in plasmodial slime molds. Plant Signal. Behav. 10 (9), e1074368.

Hoppe, T., Schwippert, W.W., 2014. Hydrophobicity of myxomycete spores: an undescribed aspect of spore ornamentation. Mycosphere 5 (4), 601–606.

Indira, P.U., 1964. Swarmer formation from plasmodia of myxomycetes. Trans. Brit. Mycol. Soc. 47, 531–533.

Indira, P.U., 1969. On the plasmodium of myxomycetes. Univ. Iowa Stud. Nat. Hist. 21 (2), 1–36.

Ing, B., 1994. The phytosociology of myxomycetes. New Phytol. 126 (2), 175–201.

Kawano, S., Kuroiwa, T., Anderson, R.W., 1987. A third multiallelic mating-type locus in *Physarum polycephalum*. Microbiology 133 (9), 2539–2546.

Keller, H.W., 1982. The myxomycetes. Parker, S.P. (Ed.), Synopsis and Classification of Living Organisms, vol. 1, McGraw Hill, New York, NY, pp. 165–172.

Keller, H.W., 2012. Myxomycete history and taxonomy: highlights from the past, present, and future. Mycotaxon 122, 369–387.

Keller, H.W., Aldrich, H.C., Brooks, T.E., 1973. Corticolous myxomycetes II: notes on *Minakatella longifila* with ultrastructural evidence for its transfer to the Trichiaceae. Mycologia 62, 768–778.

Keller, H.W., Aldrich, H.C., Brooks, T.E., Schoknecht, J.D., 1975. The taxonomic status *of Badhamia ovispora*: a myxomycete with unique spores. Mycologia 67, 1001–1011.

Keller, H.W., Braun, K.L., 1999. Myxomycetes of Ohio: Their Systematics, Biology, and Use in Teaching (Ohio Biological Survey Bulletin New Series, Vol. 13, No. 2). Ohio Biological Survey, Columbus, OH.

Keller, H.W., Brooks, T.E., 1976. Corticolous myxomycetes IV: *Badhamiopsis*, a new genus for *Badhamia ainoae*. Mycologia 68, 834–841.

Keller, H.W., Eliasson, U.H., 1992. Taxonomic evaluation of *Perichaena depressa* and *Perichaena quadrata* (myxomycetes) based on controlled cultivation, with additional observations on the genus. Mycol. Res. 96, 1085–1097.

Keller, H.W., Everhart, S.E., 2010. Importance of myxomycetes in biological research and teaching. Fungi 3 (1), 29–43.

Keller, H.W., Everhart, S.E., Skrabal, M., Kilgore, C.M., 2009. Tree canopy biodiversity in temperate forests: exploring islands in the sky. S.E. Biol. 56 (1), 52–74.

Keller, H.W., Kilgore, C.M., Everhart, S.E., Carmack, G.J., Crabtree, C.D., Scarborough, A.R., 2008. Myxomycete plasmodia and fruiting bodies: unusual occurrences and user friendly study techniques. Fungi 1 (1), 24–37.

Keller, H.W., O'Kennon, R., Gunn, G., 2016. World record myxomycete *Fuligo septica* fruiting body (aethalium). Fungi 9 (2), 6–11.

Keller, H.W., Reynolds, D.R., 1971. A new *Perichaena* with reticulate spores. Mycologia 63, 405–410.

Keller, H.W., Schoknecht, J.D., 1989. *Fuligo megaspora*, a myxomycete with unique spore ornamentation. Mycologia 81, 454–458.

Keller, H.W., Skrabal, M., Eliasson, U.H., Gaither, T.W., 2004. Tree canopy biodiversity in the Great Smoky Mountains National Park: ecological and developmental observations of a new myxomycete species of *Diachea*. Mycologia 96, 537–547.

Keller, H.W., Smith, D.W., 1978. Dissemination of myxomycete spores through the feeding activities (ingestion-defecation) of an acarid mite. Mycologia 70, 1239–1241.

Keller, H.W., Snell, K., 2002. Feeding activities of slugs on myxomycetes and macrofungi. Mycologia 94, 757–760.

Kilgore, C.M., Keller, H.W., Ely, J.S., 2009. Aerial reproductive structures on vascular plants as a microhabitat for myxomycetes. Mycologia 101, 303–317.

Koevenig, J.L., 1961. Slime Molds I: Life Cycle. U5518, 30 min. sd. Color. Film. Bureau of Audio Visual Instruction, Extension Division, University of Iowa, Iowa City.

Krug, J.C., Benny, G.L., Keller, H.W., 2004. Coprophilous fungi. In: Mueller, G.M., Bills, G.F., Foster, M.S. (Eds.), Biodiversity of Fungi: Inventory and Monitoring Methods. Elsevier Academic Press, Burlington, pp. 467–499.

Lado, C., 2005–2016. An online nomenclatural information system of Eumycetozoa. Real Jardín Botánico de Madrid, CSIC, Madrid, Spain. Available from: http://www.nomen.eumycetozoa.com

Lado, C., Mosquera, J., Beltrán-Tejera, E., 1999. *Cribraria zonatispora*, development of a new myxomycete with unique spores. Mycologia 91, 157–165.

Lado, C., Wrigley de Basanta, D., Estrada-Torres, A., Carvajal, E.G., Aguilar, M., Hernández-Crespo, J.C., 2009. Description of a new species of Perichaena (myxomycetes) from arid areas of Argentina. Anales del Jardín Botánico de Madrid, 63–70, 66S1.

Lister, A., 1894. A monograph of the mycetozoa. Printed by order of the Trustees, London.

Lister, A., 1911. A monograph of the mycetozoa. Revised by G. Lister. Printed by order of the Trustees of the British Museum, London.

Lister, A., 1925. A monograph of the mycetozoa. Revised by G. Lister. Printed by order of the Trustees of the British Museum, London.

Martin, G.W., Alexopoulos, C.J., 1969. The Myxomycetes. University of Iowa Press, Iowa City.

Ross, I.K., 1957. Syngamy and plasmodium formation in the myxogastres. Am. J. Bot. 44, 843–850.

Ross, I.K., 1967. Formation of amoeboid cells from the plasmodium of a myxomycete. Mycologia 59, 725–732.

Schnittler, M., Stephenson, S.L., 2002. Inflorescences of neotropical herbs as a newly discovered microhabitat for myxomycetes. Mycologia 94, 6–20.

Snell, K.L., Keller, H., 2003. Vertical distribution and assemblages of corticolous myxomycetes on five tree species in the Great Smoky Mountains National Park. Mycologia 95, 565–576.

Spiegel, F.W., Stephenson, S.L., Keller, H.W., Moore, D.L., Cavender, J.C., 2004. Sampling the biodiversity of mycetozoans. In: Mueller, G.M., Bills, G., Foster, M.S. (Eds.), Biodiversity of Fungi: Inventory and Monitoring Methods. Elsevier Academic Press, Burlington, pp. 547–576.

Stephenson, S.L., Stempen, H., 1994. Myxomycetes: A Handbook of Slime Molds. Timber Press, Portland.

Sundberg, W.J., Keller, H.W., 1996. Myxomycetes: some tools and tips on collection, care, and use of specimens. Inoculum 47 (4), 12–14.

Tesmer, J., Schnittler, M., 2007. Sedimentation velocity of myxomycete spores. Mycol. Prog. 6, 229–234.

FURTHER READING

Bundschuh, R., Altmüller, J., Becker, C., Nürnberg, P., Gott, J.M., 2011. Complete characterization of the edited transcriptome of the mitochondrion of *Physarum polycephalum* using deep sequencing of RNA. Nucleic Acids Res. 39, 6044–6055.

Kopp, D., 2012. Assembly of the *Didymium iridis* Mitochondrial Genome by Genome Walking. College of Science and Health Theses and Dissertations. Paper 20. Depaul University, Chicago. Available from: http://via.library.depaul.edu/csh_etd/20

Chapter 2

The History of the Study of Myxomycetes

Bruce Ing*, Steven L. Stephenson**

**Applied Science and Environmental Biology, University of Chester, Chester, United Kingdom;*

***University of Arkansas, Fayetteville, AR, United States*

THE EARLY PERIOD

The earliest reference to a myxomycete is probably that of the German botanist Thomas Panckow, whose *Herbarium Portatile* (Panckow, 1654) contained a brief description and a woodcut depicting what is likely to have been a species of *Lycogala* (Fig. 2.1). We know nothing more about the writer. During the 17th century, myxomycetes, if they were noticed at all, were grouped with the fungi, which in turn were regarded as plants. The first real attempt to classify plants was made by the Frenchman Joseph Pitton de Tournefort (1656–1708), who in his publication *Institutiones rei Herbariae* (de Tournefort, 1700) described *Lycoperdon sanguineum sphericum* as a puffball, which is almost certainly *Lycogala epidendrum*. It is not surprising that this genus should be the first to be noticed, since it is very common and rather conspicuous. The next species to be described is also common and conspicuous. The Frenchman Jean Marchant (1659–1738) gave an account of *Les Fleurs de la Tannée* (Marchant, 1727), with "pale yellow foam-like masses on oak bark tan, which develops a golden crust beneath which is a fine black powder." This clearly refers to what we still call "Flowers of tan," namely *Fuligo septica.* Marchant thought it was a sponge.

The first scientific treatment of fungi, including myxomycetes, was by the Italian Pier Antonio Micheli (1679–1737). Born of a poor family, he was self-taught but developed into such a good botanist that he became superintendent of public gardens in Florence, a royal appointment. Regarded by later mycologists as the founder of their discipline, he was interested in all groups of plants as well as fungi and lichens. The great luminary Elias Magnus Fries (1794–1878) credited him with creating order out of the chaos in which fungal classification found itself at that time. His work *New Genera of Plants, Arranged After the Method of Tournefort* (Micheli, 1729) included, within the division Fungi, four genera of myxomycetes—*Clathroides*, *Clathroidastrum*, *Lycogala*, and

Myxomycetes: Biology, Systematics, Biogeography, and Ecology. http://dx.doi.org/10.1016/B978-0-12-805089-7.00002-0

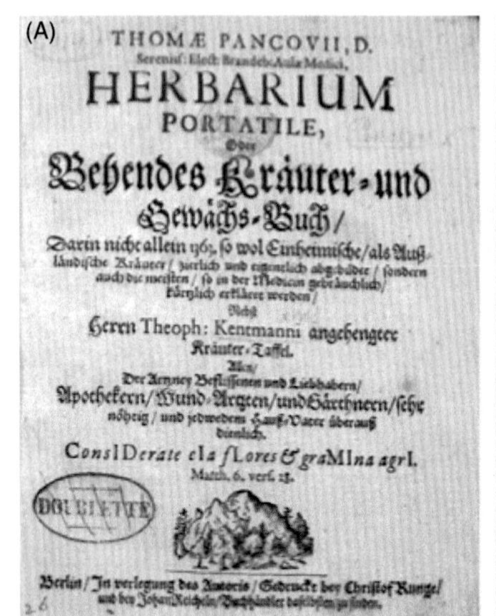
(A)
THOMÆ PANCOVII, D.
HERBARIUM
PORTATILE,
Behendes Kräuter- und
Gewächs-Buch/
Herren Theoph: Kentmanni angehengter
Kräuter-Taffel.
Apotheckern/ Wund-Artzten/ und Gärthnern/ sehr
ConsIDerate eLa fLores & graMIna agrI.

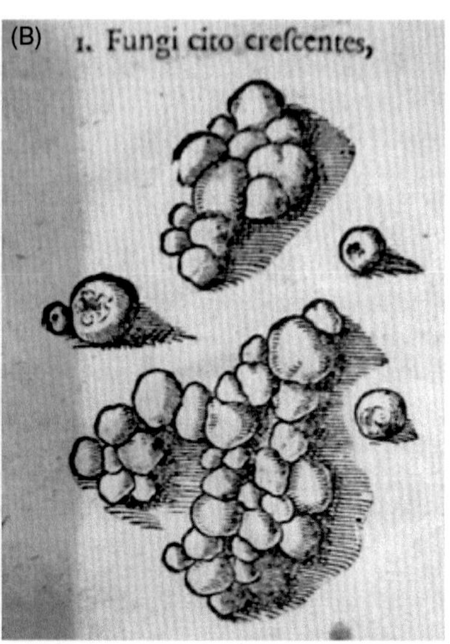

FIGURE 2.1 (A) Title page from Thomas Panchkow's Herbarium Portatile, published in 1654. (B) Woodcut which accompanied what is apparently the first known description of a myxomycete. *(Images used with permission from the Natural History Museum Library and Archives.)*

Mucilago. The first two contain species of *Arcyria* and *Stemonitis*, and the others encompass members of genera, such as *Reticularia*, *Fuligo*, *Comatricha*, *Physarum*, *Didymium*, and *Trichia*. *Ceratiomyxa* is recognizable under the rust genus *Puccinia*. Micheli appears to be the first person to have used a microscope in the study of fungi and to realize the role of spores.

The Swiss Albrecht von Haller (1708–77) published *A Systematic and Descriptive List of Plants Indigenous to Switzerland* (von Haller, 1742). The myxomycete genera are those of Micheli with two additions—*Embolus* and *Sphaerocephalus*. As circumscribed by Haller, these contained an odd mixture of species now placed in *Physarum*, *Comatricha*, and *Trichia*. Haller was not sure about *Clathroides* but included five species, one of which is definitely *Metatrichia vesparia*. A more comprehensive account of Swiss plants was published in 1762, including the description of the genus *Trichia*, although not quite as we know it today. Haller was not impressed with the binomial system of nomenclature as proposed by Linnaeus and did not adopt it. The Englishman John Hill (1716–75) published *A History of Plants* (Hill, 1751), which added little to our knowledge but gave us the name *Arcyria*. He was convinced that there were male and female flowers in the fruiting bodies. As was the case in those days, his names were often long and flowery sentences in Latin.

Eventually, we come to Carl Linnaeus (1701–78), who was born in rural Sweden. Although he trained in medicine, his interests were in the natural world. He set about putting this world in order and in several seminal works he largely achieved this goal. His *Systema Plantarum* (Linnaeus, 1763) has 1515 pages of higher plants but only 17 pages of fungi, including 7 species of myxomycetes in the fungal genera *Lycoperdon*, *Clathrus*, and *Mucor*. He was clearly not very interested in fungi. Perhaps his greatest gift was the binomial system of nomenclature. As was the custom at that time, the Latin name was a long descriptive phrase, beginning with a word that we would recognize as a generic name. Linnaeus' innovation was that he placed in the margin a single epithet appropriate to that species. Whether this was deliberate we cannot know, but we can be grateful for it. His few specimens are in the herbarium of the Linnean Society of London. The impact of Linnaeus was that he inspired botanists in the second half of the 18th century to add to our knowledge. Among these were the Austrian Giovanni Antonio Scopoli (1723–88), who published *Flora Carniolica* (Scopoli, 1772), which included several myxomycetes, and the German Jakob Christian Schäffer (1718–90), whose *Fungi Bavarici* (1762–74) listed six myxomycetes in the genus *Mucor* (Schäffer, 1762).

August Johann Georg Carl Batsch (1761–1802) was born in Jena, Germany, and published his *Elenchus Fungorum* (Batsch, 1783) with descriptions in German and Latin and color engravings of eight species of myxomycetes, of which those of *Cribraria cancellata* (as *Mucor*) and *Metatrichia vesparia* (as *Lycoperdon*) are very good. Jean Baptiste François Bulliard (often mistakenly called Pierre) (1752–93) was born in France and is best known for his *Histoire des Champignons de la France* (Bulliard, 1791). The myxomycetes were placed with puffballs under four genera—*Lycoperdon*, *Sphaerocarpus*, *Reticularia*, and *Trichia*. The arrangement of the 30 or so species is confusing, but the illustrations are so good that the species can be identified with confidence. Due to this, this work became an early standard reference for the myxomycetes.

Heinrich Adolph Schrader (1767–1836) was Professor of Botany at Göttingen in Germany and was primarily interested in flowering plants. In 1797, he published his *Nova Genera Plantarum*, and all of the organisms considered were myxomycetes (Schrader, 1797). Among these were *Cribraria*, *Dictydium*, *Licea*, and *Didymium*, which we still recognize. He included 27 species, which are well described. This is especially the case for *Cribraria*. He noted that although the organisms he considered were regarded as fungi, they developed from a mucilaginous state. Heinrich Tode (1733–97) seems to be the first worker to describe the movement of a plasmodium, in his *Fungi of Mecklenburg* (Tode, 1790). He called it *Mesenterica tremelloides* but clearly did not understand what was involved. Charles Persoon (1761–1836) was born in South Africa of a Dutch father. On his parents' death, he came to Europe and visited many universities, learning as much as possible about fungi. In 1801 his *Synopsis Methodica Fungorum* appeared (Persoon, 1801). The myxomycetes were still classified with the puffballs and *Mucor* but were now placed in 11 genera, all still current, and most of the 80 species he considered are still accepted. Persoon renamed *Mesenterica* as *Phlebomorpha* in 1823 but had no more idea of its significance than Tode. His collections are in Leiden.

The Scot James Dickson (1738–1822) described and illustrated *Leocarpus fragilis* (as *Lycoperdon*) in 1785 (Dixon, 1785). During the same period James Sowerby of London (1757–1822) was producing his *Coloured Figures of English Fungi and Mushrooms*, which appeared between 1797 and 1809 (Sowerby, 1809). These contained some of the best drawings of myxomycetes produced thus far. In 1801 and 1803, Heinrich Christian Friedrich Schumacher (1757–1830) published two accounts of the plants of Saeland, Denmark (Schumacher, 1803). One hundred species of myxomycetes were described, of which 72 are ascribed to Schumacher himself. There are no illustrations. However, of the 72 species, only 2 are still accepted today.

A far more important partnership was formed at the same time as Schumacher's poor effort. Ludwig David von Schweinitz (1780–1834) was born at Bethlehem, Pennsylvania, a Moravian colony. In 1798, he came to Germany to study and there met Johannes Baptiste von Albertini (1769–1831), and they became friends. The two worked together for some years and produced their *Conspectus Fungorum* in 1805 (Albertini and Schweinitz, 1805), with its color illustrations and full descriptions. Seventy-three species of myxomycetes were included but still considered as fungi. Nine of the species were new to science, and six of these are still accepted today. Schweinitz returned to America and published some of the first accounts of North American myxomycetes (Schweinitz, 1822, 1832). During the first part of the 19th century, a number of European botanists described myxomycetes. These included Johann Heinrich Friedrich Link (1767–1851), L.P.F. Ditmar and Christian Gottfried Ehrenberg (1795–1876) from Germany, the Norwegian Søren Christian Sommerfelt (1794–1838), and the Swiss Augustin Pyramus de Candolle (1778–1841) (Ditmar, 1813; Ehrenberg, 1818; Link, 1809; Sommerfelt, 1826). Their new species are still recognized.

The most eminent and influential mycologist of the period was Elias Magnus Fries. Born at Femsjö in rural Sweden, he grew up surrounded by nature and became devoted to the study of fungi. Beginning in 1815 and continuing for some 60 years, he published many seminal works that laid the foundations of modern mycology. In his *Systema Mycologicum* (Fries, 1829), he produced the best work thus far on myxomycetes. Although Fries still included them in the Gasteromycetes, he erected a separate suborder—the Myxogastres (or slime stomachs)—by virtue of the mucilaginous early stage, the plasmodium. Thus began the great debate of the 19th century as to how to classify the myxomycetes, a problem finally resolved only in the late 20th century. Fries investigated the plasmodium of several species, now aware that this was part of the life cycle. The main gap in his work was the lack of microscopic detail of spore ornamentation, calcareous inclusions, or, for example, the spiral bands on the elaters of *Trichia*. These had been observed and commented on by the microscopist R.A. Hedwig in 1802.

In his *Flora Cryptogamica Germaniae* (Wallroth, 1833), Carl Friedrich Wilhelm Wallroth (1792–1857) coined the term myxomycetes, or slime fungi, for what had long been called the Myxogastres, and so the debate continued. One admirer of Fries was the Reverend Miles Joseph Berkeley (1803–89). A country parson with a large family, he found time to become known as the father of British mycology. However, his interests were actually worldwide and his correspondents came from all corners of the globe. He was sent specimens from Australia, North America, Cuba, Sri Lanka, as well as many parts of Europe and all parts of Great Britain. He is best known for his work on the larger fungi, in which he followed the Friesian line, but he was also active in studies of the myxomycetes. His book *British Fungi* (Berkeley, 1836) rapidly became the standard work on the subject. Under the Myxogastres, he recorded 63 species from Britain, and 4 of these were species new to science. In addition to his own work, Berkeley was an enthusiastic collaborator, notably with Christopher Edmund Broome of Batheaston, Somerset (1802–86), with whom he described the beautiful *Diderma lucidum* from a ravine at Dolgarrog in North Wales. For many years, they published a long series of *Notices of British fungi* in the *Annals and Magazine of Natural History*, an influential journal of its day. He also collaborated with Moses Ashley Curtis (1801–72) and Henry William Ravenel (1814–87) on American species. Like Fries, Berkeley did not refer to microscopic characters but preferred to use features visible with a hand lens. Perhaps less well known is the fact that he was the first to recognize that the potato famine which occurred in the Irish and West Highlands of Scotland during the 1840s was due to a blight caused by the fungus-like organism *Phytophthora infestans* and not to bad air as had been suggested by many others. As to the debate on the taxonomic affinities of the myxomycetes, Berkeley came down on the side of the fungal relationship because the spores were enclosed in a sac. The nature of the plasmodium and the earlier amoeboid stage (the myxamoebae) in the life cycle had been used by the next scientist (de Bary) to make a major contribution to the study of myxomycetes as evidence of their animal, or at least protozoan, affinities. However, Berkeley disagreed.

THE FIRST SCIENTIFIC STUDIES

Heinrich Anton de Bary (1831–88) trained as a medical practitioner but soon turned his attention to botany and held senior posts at several German and Austrian universities. He was especially interested in the physiology, morphology, and developmental biology of fungi, and his descriptions of the structure and life cycle of myxomycetes were the most complete to date. In his 1859 book *Die Mycetozoa* (or fungus animals), he brought into use a term which has remained with us in one-way or another ever since (de Bary, 1859). He germinated spores and cultivated the resultant "swarm cells" and suggested that the plasmodium was formed by the union of myxamoebae. He also demonstrated that the sporocysts (or fruiting bodies) arose from the plasmodium. His later book, *The Comparative Morphology and Biology of the Fungi, Mycetozoa and Bacteria* (de Bary, 1884), is a classic of its type.

During this most productive era, Leo Cienkowski (1822–87) from Poland observed the union of myxamoebae and introduced the term "plasmodium" in a paper published in 1863 (Cienkowsky, 1863). He also observed the ingestion of food particles, thus strengthening the idea of a protozoan affinity for the myxomycetes. Also active at this time were Oscar Brefeld (1839–1925), who added to our knowledge of the biology of the group, and the Russians Mikhail Stepanovich Voronin (1833–1903) and Andrei Sergeevich Famintsin (1835–1918), who worked out the life history of *Ceratiomyxa*, which was outlined in a paper published in 1873 (Famintsin and Woronin, 1873).

As the century drew nearer to its close, several more stars began to shine in the myxomycete galaxy. The Pole Józef Tomasz Rostafinski (1850–1928) compiled the first monograph of the group. His *Śluzowce Monografia* was written with the guidance of de Bary and published in 1873 (Rostafinski, 1873). It was beautifully illustrated and rapidly became the standard work on the myxomycetes. A large number of Rostafinki's names are still in use. Unfortunately, *Śluzowce Monografia* was never translated into English in full, but a partial translation was made by Mordecai Cubitt Cooke (1825–1914), who was Head of Mycology at the Royal Botanic Gardens, Kew. Cooke's *The Myxomycetes of Great Britain* was published in 1877 (Cooke, 1877). Cooke also wrote on some American collections and collaborated with several other workers. In addition, he produced books of a more general kind in an effort to popularize fungi.

Cooke's successor at Kew was George Massee (1847–1917), who was one of the founders of the British Mycological Society and served as its first president. He also wrote handbooks on British fungi but is best remembered for his *Monograph of the Myxogastres*, published in 1892 (Massee, 1892). The title is instructive, since to Massee myxomycetes were fungi and he argued against the idea that they might be related to protozoans. The book has useful descriptions and clear, color illustrations. Numerous species were introduced as new, but few of these have survived. It is possible that he was aware of another book on its way and was in a hurry to publish, and there is some evidence of carelessness. Massee was an argumentative Yorkshireman who would not take "yes" for an answer. His relationship with the following worker was stormy.

ARTHUR AND GULIELMA LISTER

Arthur Lister F.R.S. (1830–1908) (Fig. 2.2) was the son of Joseph Jackson Lister F.R.S., a distinguished microscopist, whose brother was Joseph Lister (1827–1912), the pioneer in antiseptic surgery. Members of his family were Quakers, and they loved all things natural. He was a keen sportsman and a skilled artist. His profession was ultimately that of wine merchant, and he and his family lived comfortably in a town house at Leytonstone in east London, close to Epping Forest. The family also owned a house at Highcliffe, near Lyme Regis in Dorset, and both areas benefited from his fieldwork. He became interested in myxomycetes in 1877 and quickly became immersed in the collections

FIGURE 2.2 **Arthur Lister.** *(From* Mycological Notes, *1919.)*

at Kew, the British Museum, Strasbourg and Paris. He also began to experiment with plasmodia and in 1877 demonstrated protoplasmic streaming in *Badhamia utricularis* to the Linnean Society (Lister, 1888). He also showed clearly the ingestion of food particles by plasmodia. From the onset of his work on myxomycetes, he was accompanied by his daughter, Gulielma, who acted as his secretary, field companion and fellow artist. (More information on Miss Lister, as she was universally known, is provided later.) Lister had a wide range of correspondents from all parts of the world, and specimens were constantly arriving, examined, described, and deposited in the Botany Department at the British Museum (Natural History) in South Kensington. Papers on the myxomycetes of Japan, Antigua, and many other countries, but especially from Britain, filled the pages of the *Journal of Botany* for many years, initially by himself and then in collaboration with Gulielma, who continued the tradition after her father's death. The assiduous fieldwork and intuitive understanding of what constituted a species culminated in the appearance in 1894 of *A Monograph of the Mycetozoa*, published by the museum (Lister, 1894). This was immediately seen to be superior to the book recently produced by Massee, and if they had not been on bad terms before, they certainly were after Lister's treatment of the myxomycetes was published. Massee edited the *Naturalist*, an influential journal published in Yorkshire, and many of his editorials were full of barbed criticism, some very spiteful, of his rival's work. Lister replied only when he felt suitably irritated, and always in more moderate tones. It was a classic case of a professional (Massee) being jealous of a better amateur (Lister). However, Lister had the

FIGURE 2.3 **Gulielma Lister.** *(From* Transactions of the British Mycological Society, *1950.)*

last laugh, as he became a Fellow of the Royal Society and Massee did not! He continued to write papers with Gulielma until his death.

Meanwhile, across the Atlantic a new star was rising. Thomas Houston Macbride (1848–1934) was impressed with Lister's *A Monograph of the Mycetozoa* and became determined to add more information on the American range of species. He accomplished this in 1899 with his *North American Slime Moulds* (Macbride, 1899), with a second edition in 1922 (Macbride, 1922). These two works are of particular importance because they were the basis of yet another work, *The Myxomycetes* (Macbride and Martin, 1934), which Macbride coauthored with George W. Martin (1886–1971). This aspect of the history of the study of myxomycetes will be picked up later in this chapter.

Gulielma Lister (1860–1949) (Fig. 2.3) continued her father's work and also inherited his status as *the* world authority on myxomycetes. She prepared the second edition of *A Monograph of the Mycetozoa* (Lister and Lister, 1911) and included beautiful color paintings. The third edition (Lister, 1925) was much enlarged through her own work. As a skilled artist, she produced some of the finest drawings ever made of myxomycetes, and we are fortunate that the originals of both Listers' illustrations have been carefully preserved in the Botany Library of the Natural History Museum. In her will, she bequeathed half to the Museum and half to the British Mycological Society, of which she had been president on two different occasions. In 1904, she became one of the first women to be admitted as a Fellow of the Linnean Society, hitherto an all-male preserve. In 1929, she was elected as vice-president. She continued to describe species, produce valuable lists, and communicate with workers everywhere, but with advancing years she began to confine her activities to Epping Forest. Here she trained a new generation of British students of her beloved "myxies."

Her artistic skills were not confined to myxomycetes. She provided the figures for Dallimore and Jackson's *Handbook of the Coniferae* (Dallimore and Jackson, 1923, with editions in 1931 and 1948) and the color plates for F.J. Hanbury's *Illustrated Monograph of the British Hieracia* (1889–98) (Hanbury, 1889). With her father, she had recorded their observations on the Mycetozoa in a series of notebooks, 74 in all, covering 60 years. The Lister notebooks contain a wealth of information—descriptions, sketches, drawing and paintings, notes on particular specimens, lists from field trips, letters from correspondents, and draft replies. The entries began in 1887 and the last one appeared in 1947. On her death, she bequeathed all the notebooks to the British Mycological Society, and they are kept in the Cryptogamic Botany Department of the Natural History Museum in South Kensington. Fuller biographies can be found in Scott (1908), Wakefield (1950), and Ainsworth and Balfour-Browne (1960). Joseph Jackson Lister F.R.S. (1857–1927) was the son of Arthur and brother of Gulielma. He was primarily a zoologist and spent his entire career at Cambridge University. He contributed the article on the Mycetozoa to the 11th edition of *Encyclopaedia Britannica* (1911).

As was the case for her father, Gulielma collaborated with workers around the world. Of interest was her association with William Cran (1854–1933), who had sent specimens to Arthur from Antigua, where he had been a lecturer at a theological college. On his return to Scotland, he became a Congregational minister at Skene, near Aberdeen. Gifted with amazing eyesight, he started to research the minute myxomycetes found on the bark of living trees. With G. Lister he described what we now call *Macbrideola cornea* and *Paradiacheopsis fimbriata* (Lister, 1917, 1938). His notebooks and collections are in the herbarium of the Botany Department of Aberdeen University.

Gulielma Lister could also be credited with publishing the first broad account of myxomycete ecology. In a paper presented to the Essex Field Club, she outlined the range of ecosystems and niches in which these organisms occur. Later work on this aspect of the myxomycetes can be regarded as an extension of what was provided in this paper (Lister, 1918). After her death, the work was continued by several excellent amateurs, including W.D. Graddon, J. Ross, and H.J. Howard. The latter was based in Norwich and in addition to his efforts to continue to record British species, he also collected material in Switzerland and elsewhere. His collections are in the Castle Museum, Norwich, with duplicates at Kew and the Natural History Museum in London. Howard also wrote on myxomycete ecology and was the first to comment on the differences that existed from place to place in the assemblages of myxomycetes present (Howard, 1948).

MORE RECENT STUDIES IN EUROPE

The new generation began work in 1957 when Bruce Ing (b. 1937) became a student at the University of Cambridge. In his first week, while looking for bryophytes, he encountered *L. fragilis*. After searching in second hand bookshops, he discovered a small booklet illustrating the models of myxomycetes in the Natural History Museum in London. The authors were the Listers, and there was a drawing of *Leocarpus*. On referring to the 1925 *Monograph* in the library with its beautiful paintings, he was captivated. As there was no other worker specializing in myxomycetes at that time in Britain, he decided to devote as much time as possible to determine which species had been recorded as British and Irish. This culminated in the first checklist of British species in 1968 (Ing, 1968), which has been revised twice (Ing, 1980, 1985, 2000b). He also communicated with other workers around the world, notably George Martin and Constantine Alexopoulos in America and especially with Elly Nannenga-Bremekamp (born as Neeltje Elizabeth Bremekamp) in the Netherlands, with whom he developed a strong working relationship until her untimely death in 1996.

Early on in his career he was fortunate to be able to work on material brought back from Nigeria as a result of a student expedition to that country, and in 1964 he published a new genus, *Metatrichia*, which is essentially tropical apart from the common *Metatrichia floriformis* (Ing, 1964). A second expedition to Nigeria allowed another life-long collaboration (this time with Roland McHugh) to develop, especially in regard to Irish myxomycetes (see later). However, his most satisfying achievement was to organize the first International Congress on the Systematics and Ecology of Myxomycetes, which was held in Chester, England, in 1993. The congress has been held every 3 years since then. His work on ecology was summarized in 1994 in a paper on the phytosociology of myxomycetes (Ing, 1994), and it is likely that the paper redirected some of the energy of other workers to this aspect of their biology. His work on the British and Irish species resulted (Ing, 1999b) in the first comprehensive account of the species of these islands, although this is now in need of updating and revision. Ing's collections and myxomycete library and records will be eventually deposited at Kew.

During this same period, interest in the group increased, and Ing was joined by Peter Holland, Roland McHugh, and David Mitchell in studies which covered a wide geographical area and triggered further collaboration with workers overseas. Some of these studies will be discussed later. In Europe, the most prominent worker was Nannenga-Bremekamp. She became the person to consult after both Martin and Alexopoulos had died. She enjoyed a fruitful collaboration with virtually all the leading workers of the day and especially with Yukinori Yamamoto from Japan. Her extensive collections and notebooks are now housed in the Botanical Department of the National Museum in Brussels.

Elsewhere in Europe, the pace of study was quickening as a new generation of postwar scientists and naturalists took up the study of myxomycetes. In Greece, there was little interest until Constantine Alexopoulos, on a sabbatical from his university in America, spent a year collecting material from all over the country. This was the stimulus needed, and workers from many parts of Europe began to add more data. This will be summarized in a pending publication by Ing and Zervakis. Visits to Greece by Ing in 1988 and 1998 have yielded many more species, including nivicolous examples, such as *Dianema aggregatum*, new to Europe. Schnittler and Novozhilov (1993) added *Lepidoderma crustaceum* from Crete. A few corticolous species from Cyprus were reported by Ing (1987a), and a foray of the British Mycological Society in 2011 yielded over 100 species, all but one being new records for the island. The first reports from Bulgaria were by Hinkova (1951), followed by Hinkova and Draganov (1965). Markov (1962) added more data, while Reid and Vanev (1984) and Vanev and Reid (1986) included four species of myxomycetes. A modern checklist of the myxomycetes of the former Yugoslavia was provided by Ing and Ivancevic (2000).

The Mediterranean region, with its preponderance of limestone ecosystems, has been well studied, and a valuable synthesis of the occurrence of myxomycetes was provided by Lado (1994). The first checklist from Malta (Briffa, 1998) included 73 species, an indication of the favorable conditions in most parts of the Mediterranean region. Briffa et al. (2000) discussed some of the rarer species on the list. Surprisingly, there has been no overview of the myxomycetes of Italy. A first list was published by Pirola (1968) and expanded by Pirola and Credaro (1971), but no systematic studies of the entire country have been carried out. The species found in the alpine region of the Alto-Adige in the Sud Tirol area of the north were studied by Ing (2003), and a British Mycological Society foray to Sardinia in 2006 produced over 100 species. A recent development has been the transfer of annual "days of study" of nivicolous myxomycetes from the Savoie Alps, in France, to an area to the south of Turin, in Italy. This concentration of interest has been a factor in recruiting many more students of myxomycetes as well as generating far more appreciation among other mycologists and naturalists.

Spain can claim to be one of the most active and productive countries in terms of producing a modern generation of myxomycologists. The main centers are the National Botanic Garden in Madrid, where Carlos Lado and his team are based; the University of Alcala de Henares, with Gabriel Moreno and his team; and the University of Barcelona, with Enric Gràcia. Their combined influence is now worldwide, especially in Latin America, and their contribution to our knowledge cannot be overestimated. The development of other centers in Spain has been encouraged by them. Their home activities embrace the Balearic Islands and the Canary Islands, both groups being provinces of Spain. Research based at the University of La Laguna on Tenerife, in cooperation with the various National Parks in the islands, has meant that the archipelago is well documented and several new species have been found in the unique environments of the islands (Beltrán Tejera, 2001). The nature of the Spanish legacy is shown in a brief selection of their publications.

Nomenmyx, a nomenclatural database, was published by Lado (2001) and has become a most valuable tool for taxonomists. Lado and Pando (1997) produced the first part of a treatment of myxomycetes in *Flora Mycologica Iberica*. Lado and Siquier (2014) published an illustrated catalogue of the species found in the Balearic Islands, and Moreno et al. (2001) made extensive use of electron micrographs in the account of the species occurring in the province of Extremadura. The body of new information that the current generation of Spanish myxomycologists has generated is indeed remarkable.

Portugal has a long history of studies of myxomycetes, with the earliest list being that of Torrend (1908), who reported about 100 species. In more recent times, M.G. Almeida, at the University of Lisbon, began a series of studies which lasted more than 20 years and extended to the myxomycetes of Portugese territories in Africa. The Azores archipelago is a province of Portugal, and an interesting range of species, including *Perichaena microspora* on mule dung, was reported by Ing.

France has a long and distinguished history for the study of myxomycetes, but like so many other countries, this was interrupted by two wars. In the 20th century, as with much of Europe, the new peace brought with it a renewed interest in all branches of natural history, and France was no exception. With the oldest mycological society in the world—the French society was founded in 1895 (1 year before the British!)—there was a firm foundation of mycology and a tradition to foray intensively as both a scientific and a social activity. There had been several earlier national and regional accounts, but the first modern list was provided by Cochet (1977), with additions and correction in 1980 and 1984 by Cochet and Bozonnet (Cochet and Bozonnet, 1980, 1984). Bozonnet himself was a pioneer in the studies of nivicolous myxomycetes in France and, together with other members of the Fédération Mycologique et Botanique Dauphine-Savoie, built up a great tradition of alpine mycology in the French Alps. One of his colleagues is Marianne Meyer, who became the doyenne of alpine myxomycology. One of her lasting achievements was the inauguration of the "study days" on nivicolous myxomycetes each spring in the Savoie Mountains where she lives. These meetings were heavily attended by enthusiastic French naturalists and also attracted myxomycologists from all over Europe to the little village of St. Paul sur Isère. Moreover, her collaboration with Père Bozonnet and Michel Poulain, the latter known for his brilliant photographs of myxomycetes, resulted in the publication in 2011 of *Les Myxomycètes* in two volumes. This was not just an account of the French species; it also contains a high proportion of the global total and will be a major reference source for many years to come.

The Netherlands have a long history of mycology, and myxomycetes have figured strongly. During the second half of the 20th century, Elly Nannenga-Bremekamp patiently collected, illustrated, and described numerous species from all over the world and collaborated in print with virtually all the leading workers of the period. Her many papers, mostly in English, were full of important details relating to particular species. In addition to exhaustively surveying the Dutch species, she visited several European countries, notably France, Germany, Switzerland, and Great Britain. Her major work was *De Nederlandse Myxomyceten*, published (in Dutch) in 1974 (Nannenga-Bremekamp, 1974), with separate additions in 1979 and 1983. It has been translated into English, with the rather pretentious title of *A Guide to Temperate Myxomycetes*, by Feest and Burggraaf (1991). In spite of some features relating to poor production and a number of omissions in the work, it is a very useful guide to a major portion of the assemblage of myxomycete found in Europe but is of less use elsewhere.

Belgium's first note seems to be that of Marchal (1895), in which he described *Trichia varia* var. *fimicola* (now recognized as *T. fimicola*) along with a range of coprophilous fungi. Based at the National Botanic Garden at Meise, Jean Rammeloo carried out systematic studies on many critical genera (e.g., Rammeloo, 1978a) but perhaps is better known for his work in the former Belgian colonies in Africa. Inspired by him, Myriam de Haan published an annual series of additions to the Belgian list since 1996 (Haan, 1996). Wishing to encourage children to enjoy myxomycetes, she has produced the delightfully illustrated *The Adventures of Mike the Myxo*, which introduces the reader to the structure, life cycle, and ecology painlessly!

One of the earliest references from Austria is that of Sauter (1841), after whom *Diderma sauteri* was named, who listed myxomycetes from the vicinity of Salzburg. He was followed by Wettstein (1886), who recorded the fungus flora of Steiermark (the former Styria). A major work to include Austrian species was Schinz's volume in Rabenhorst's *Kryptogamenflora*, which covered Austria, Germany, and Switzerland and was very much based on the second edition of Lister's monograph (Schinz, 1920). This can be regarded as the first comprehensive list from all three countries. Gottsberger (1966) expanded on the previous studies in a detailed account from Steiermark. In recent years, the leading worker in Austria has been Wolfgang Nowotny, who has written on the species in Upper Austria but is best known for his coauthorship of *Die Myxomyceten* (see later). Germany, as in so many areas of science, has produced a wealth of information on myxomycete systematics, distribution and ecology, and this has been synthesized into one of the key publications of recent years. Hermann Neubert, Wolfgang Nowotny, Karlheniz Baumann, and Heidi Marx (Neubert et al., 1993, 1995, 2000) produced a detailed and lavishly illustrated three-volume account of the myxomycetes of Germany, Austria, with references to Switzerland, in *Die Myxomyceten*. Among the younger generation must be mentioned Martin Schnittler, based at the University of Greiswald, who is taking the study of myxomycete ecology to new levels, especially in collaboration with several other workers (Schnittler, 2000).

Switzerland has a special place in the history of myxomycete ecology as many of the earliest studies of nivicolous species took place in the Alps and the Jura. The Listers visited Arolla in the early part of the 20th century and collected and described several species (Lister, 1913; Lister and Lister, 1908). Although a number of mycologists reported myxomycetes from Switzerland in the 19th century, the first systematic work was carried out by Charles Meylan, a schoolteacher from Ste. Croix in the Vaud Jura. Between 1908 and 1937, he produced a series of papers describing new species from the mountains of the Jura and the Bernese Oberland (Meylan, 1908, 1935, 1937). He was well placed in Ste. Croix, which sits below the Chasseron, one of the highest hills in the Swiss Jura and associated with rich limestone grassland. This became the type locality for many of his taxonomic novelties. He corresponded with Gulielma Lister and contributed a paper on Japanese species (Meylan, 1935). His herbarium is housed at the Lausanne Botanical Garden and has been visited by workers from many countries. Since 1992, Bruce Ing has collected material and collated records of the Swiss species. He has published two lists of corticolous myxomycetes, including such rarities as *Paradiacheopsis erythropodia* (Ing, 1997, 1999a).

The former Czechoslovakia, especially Bohemia, has received some attention. Celakowsky (Celakovsky, 1893) published a long list, but little was added until Cejp (1962) made several additions and Wichansky (1964) discussed some rare species. Current work on the impact of pollution in the forest regions should prove to be interesting. An important contribution on Romanian species was that of Brandza (1928), which included a wealth of ecological data on more than 170 species. Later contributions come from Eftimie (1965) and Forstner (1969, 1971). Hungary has been less well served, with the earliest reference being that of Moesz (1933). More recently, Toth (1963), Ubrizsy (1967a,b), and Zeller and Toth (1971) have added additional records of myxomycetes from the country.

Raciborski (1884) described new species from around Kraków in Poland and Skupienski (1926) made a number of additions to what was known about the myxomycetes of this region of Europe. Jarocki (1931) was especially interested in montane species from the Carpathian Mountains, but it was Krzemieniewska (1960) who brought together previous knowledge of the Polish species in her monograph. Stojanowska (1982) dealt with the species from the Sudetan region, with more than 100 different species being recorded. The latest checklist is that of Drozdowicz et al. (2003).

The former USSR was largely neglected in the past, but many of the new states in Europe, as well as those in Asia, are being actively studied. Yuri Novozhilov from St. Petersburg produced a detailed account of the Russian species in 1993 (published in Russian) and a more accessible list from the Leningrad region (Novozhilov, 1999) as well as a wealth of papers on the more remote parts of Russia and Asia. A long list of Ukrainian species is found in Minter and Dudka (1996), the latter a result of cooperation between the University of Kiev and the International Mycological Institute under the auspices of the Darwin Project. The myxomycetes were largely collected and identified by Tanya Krimovaz and Bruce Ing. The sixth International Congress on the Systematics and Ecology of Myxomycetes was held in Yalta in 2008, and an international foray for nivicolous myxomycetes was also held. A list of myxomycetes from Belarus was published by Moroz and Novozhilov (1994).

The three Baltic states which border Russia have been quite well studied. Lithuanian species were treated in the first volume of *Mycota Lithuaniae* by Mazelaitis and Stanevičienė (1995), and Gražina Adamonytė is currently combining floristics with ecology (e.g., Adamonytė, 2005). Vimba and Adamonytė

(2003) discussed some Latvian species. A list of Estonian fungi (Raitvir, 1980) included some 50 species of myxomycetes, and Ing (1990) added another 30 species (most of them corticolous), while Adamonytė (2000) listed 34 species from the hitherto understudied south-eastern part of the country.

Northwest Europe has always been very active in mycology, and myxomycetes are well studied in the region. In Denmark, Raunkiaer produced the first list in 1888 (Raunkiaer, 1888), and W.T. Elliott (Elliott, 1926) examined the Danish material in the University of Copenhagen herbarium. A monograph on the Danish species was produced by Bjørnkaer and Klinge (Bjørnekaer and Klinge, 1964), and Onsberg (1970) added five species to the list. In 1991, a foray by the British Mycological Society added several and three more species were recorded from the Faeroes by Möller (1958).

R.E. Fries published an early list of Swedish myxomycetes in 1899 and greatly expanded it in 1912 (Fries, 1899, 1912). Rolf Santesson, who had recently added *Listerella paradoxa* to the country's records, added a number of other species in 1964 (Santesson, 1964). However, the most important recent worker is Uno Eliasson, who contributed additional records (Eliasson, 1975; Eliasson and Sunhede, 1972). In 1983, Schinner, working in Swedish Lappland, added 13 more to the Swedish checklist, 8 of which were new records for Scandinavia (Schinner, 1983). Yet more additions came from Eliasson and Gilert (2007), with 20 species new to Sweden and 9 which are globally rare. Norway is less well studied, but an early record was reported by Karlsen (1934), and Johannesen (1984) added 49 species to the Norwegian checklist.

Finland has produced several respected workers, and many of their publications are listed in this account. Prominent among these is Marja Härkönen, based at the University of Helsinki but active not only in Finland but also in Africa. However, an earlier list of Finnish species was published by Hintikka (1963). Härkönen was particularly interested in species of corticolous myxomycetes (Härkönen, 1977, 1978), but she covered other species as well and produced a detailed checklist in 1979 (Härkönen, 1979). Her latest work, *Suomen Limasienet* (*The Myxomycetes of Finland*), is a collaborative work with Elina Varis (Härkönen and Varis, 2012). It contains valuable details on the distribution and ecology of myxomycetes and is beautifully illustrated.

Iceland has not been neglected by myxomycologists. Rostrup (1903) listed 6 species and Gøtzsche (1984, 1990) increased the total to 46, 3 of which are nivicolous species. Trees are sparse on the island, with *Betula*, *Pinus*, and *Salix* providing the only available substrates for corticolous myxomycetes, but six species in this group have been reported.

The island of Ireland has a climate well suited to myxomycetes, and both home-grown and visiting mycologists have recorded their finds. Several amateur naturalists communicated their finds to Gulielma Lister and were encouraged to publish lists. The early records were summarized by Adams and Pethybridge (1910), but the first lists from specialists were by Gunn (1918, 1919, 1920) and Stelfox (1915). A visit to Dublin by Bruce Ing in 1965 led to frequent collecting trips that have continued to the present day, and David Mitchell also found Ireland to be irresistible from the standpoint of collecting myxomycetes. Roland McHugh came to live and work in the Dublin area, and the combined efforts of this trio have produced a healthy catalogue of 224 species (Ing, 1966; Ing and McHugh, 1988, 2012; Ing and Mitchell, 1980).

STUDIES IN THE UNITED STATES

In the United States, the first extensive collections and reports of myxomycetes were by Schweinitz in the early part of the 19th century. Schweinitz, who was mentioned earlier in this chapter, is often considered at the "father" of North American mycology. His publication entitled *Synopsis Fungorum Carolinae Superioris* (Schweinitz, 1822) represented the first truly significant contribution to mycology in the United States. Other early mycologists who collected and studied myxomycetes to at least some extent were Moses Ashley Curtis (1808–72), Job Bicknell Ellis (1829–1905), Charles Horton Peck (1833–1917), Andrew Price Morgan (1836–1907), and William Codman Sturgis (1862–1942). Most of their efforts were limited to eastern North America, particularly the northeastern states. However, Thomas Nutall (1786–1859), an English botanist who spent more than 30 years in the United States, reported three species of *Fuligo* and three of *Trichia* from the south central United States in his paper "*Collections toward a flora of the Territory of Arkansas*" (Nuttall, 1837). Later, Sturgis collected extensively in Colorado. Sturgis had been born in Boston and earned three degrees at Harvard University. However, he became dean of the School of Forestry at Colorado College in 1904, and this provided him with an opportunity to collect and study myxomycetes in a portion of the United States where little information was available on these organisms. At some point, just after 1900, Sturgis visited the Listers in England, and he maintained contact with Gulielma Lister after her father had died. During his studies of myxomycetes, Sturgis also established contact with Robert L. Hagelstein (1870–1945), who became one of the leading authorities on the group in the United States during the 1920s and 1930s.

As noted earlier, Thomas Macbride and George Martin (Fig. 2.4) collaborated on *The Myxomycetes* (Macbride and Martin, 1934). Several decades later, Martin collaborated with Constantine J. Alexopoulos (1907–86) (Fig. 2.5) to produce their comprehensive world monograph, also entitled *The Myxomycetes*. The Martin and Alexopoulos (1969) monograph, published by the University of Iowa Press, is now long out of print. However, it remains the single most definitive treatment on the myxomycetes, literally representing a "bible" for those individuals engaged in studies of these organisms. The illustrations used in the monograph, which were done by Ruth McVaugh Allen (1913–84), a gifted and extremely talented artist, are an almost indispensable resource. These same illustrations were used in *The Genera of Myxomycetes*

FIGURE 2.4 George Martin. *(Courtesy of the University of Iowa Herbarium.)*

FIGURE 2.5 Constantine Alexopoulos. *(From Transactions of the British Mycological Society, 1988.)*

(Martin et al., 1983), an abbreviated version of *The Myxomycetes* produced by Marie L. Farr (b. 1927).

During the interval between the publication dates of the two different works entitled *The Myxomycetes*, Hagelstein published his *Mycetozoa of North America* (Hagelstein, 1944), which was based largely on collections of myxomycetes in the New York Botanical Garden. This book, self-published by the author and not widely distributed worldwide, is actually quite valuable for the ecological observations that Hagelstein recorded for particular species of myxomycetes. Hagelstein had an interesting career. After graduating from high school in Brooklyn, Hagelstein joined J. and D. Lehman Company, a glove manufacturer located in New York City. As such, he did have the scientific background one typically associates with the individuals who have studied myxomycetes. Hagelstein retired as a manager of the company in 1925 and dedicated himself full-time to his scientific studies. His attention was soon captured by the myxomycetes, although at first he studied them concurrently with the diatoms. Hagelstein's first publication on myxomycetes appeared in 1927 (Hagelstein, 1927), and this was followed by a series of other publications on the group. He was appointed as honorary curator of myxomycetes at the New York Botanical Garden in 1930 and held this position until his death in 1945. Although best known for the studies he carried out in the northeastern United States (particularly Pennsylvania and the Long Island area of New York), Hagelstein collected from Canada to Florida, as well as in Puerto Rico. His constant collecting partner on many of his trips was Joseph H. Rispaud, after whom he named the relatively uncommon *Comatricha rispaudii* (now known as *Paradiachea rispaudii*).

As well as the primarily taxonomic publications of the period, one of the most influential and indeed iconic items to appear was the account by William Crowder of the myxomycetes of Long Island, accompanied by a fine series of color paintings (Crowder, 1926). This edition of the *National Geographic* is now a collectors' item.

Alexopoulos was unquestionably one of the leading authorities on the myxomycetes in the United States from late 1950s until his death in 1986. He taught at several universities, but his most productive period was when he was at the University of Texas at Austin. Although widely known for myxomycetes, his research interests also encompassed fungi and his *Introductory Mycology* (Alexopoulos, 1952) became the "standard" textbook on the subject.

In 1958, Marie Farr, who was born in Vienna, Austria, but received her PhD in mycology from the University of Iowa, came to the US National Fungus Collections. Although perhaps best known for her studies of myxomycetes, she also worked extensively with black mildews and other fungi found in the tropics. Farr's *How to Know the True Slime Molds* (Farr, 1981) represented one of the first examples of a true field guide to these organisms, and her earlier work *Flora Neotropica, Volume 16: Myxomycetes* (Farr, 1976) remains the most comprehensive treatment of these organisms for this region of the world.

Almost 2 decades later, Steve Stephenson collaborated with Henry Stempen to produce *Myxomycetes: A Handbook of Slime Molds*, published by Timber Press in 1994 and currently still in print (Stephenson and Stempen, 1994).

Other workers studying myxomycetes during the latter part of the 20th century but not already mentioned include Harold Keller, who worked extensively with the corticolous myxomycetes and published a series of papers on this group (e.g., Keller and Brooks, 1977). He was the first individual to direct his studies to those species of corticolous myxomycetes that occur in the canopy of trees (Keller et al., 2004). Another important individual is Ed Haskins, who has worked on the developmental aspects of myxomycetes and carried out detailed cytological and genetic studies of selected species. Haskins collaborated with Jim Clark on several papers (e.g., Clark and Haskins, 2014; Haskins and Clark, 2016) dealing with various aspects of the biology of myxomycetes. During his career, Clark worked extensively on mating types in myxomycetes and was able to bring into culture a number of species for the first time. Donald Kowalski at Chico State University in California contributed much of what is known about the nivicolous myxomycetes associated with snowbanks in alpine areas of the western United States (e.g., Kowalski, 1967, 1970).

STUDIES THROUGHOUT THE WORLD

The study of myxomycetes worldwide received a tremendous boost in 2003 when a project entitled "Global Biodiversity of Eumycetozoans" was approved for funding by the National Science Foundation of the United States. The project, based at the University of Arkansas, was directed by Stephenson. The financial support (>2 million dollars) provided by the National Science Foundation supported a totally unprecedented effort to survey the eumycetozoans (but with the myxomycetes representing the primary target group) in regions of the world (e.g., Australia, Africa, southern South America, Central Asia, and Madagascar) where these organisms were understudied. These surveys generated more than 100 papers for publication (the vast majority of which considered myxomycetes) in addition to support the graduate training of several students who earned their PhD degrees studying myxomycetes.

Lado and Wrigley de Basanta (2008) provided a comprehensive review of the myxomycetes of the Neotropics (defined as including all of North, Central, and South America between the Tropic of Cancer and the Tropic of Capricorn). They indicated that almost half of all described species had been reported from this region of the world. However, the number of species recorded from certain countries varied rather widely. Mexico, from which 323 species had been reported at the time the data in the review were compiled, was the country with the highest total. This reflects in large part the efforts of individuals, such as Arturo Estrada-Torres at the University of Taxcala. However, there were no known records from El Salvador and fewer than 25 species had been reported for Honduras, the Bahamas, Haiti, Guyana, Paraguay, and Surinam. In contrast, the totals for Brazil and Costa Rica were 206 and 143 species, respectively. It should be noted that the total number of species of myxomycetes known from Costa Rica has since increased to at least 225 (Rojas et al., 2010, 2015). Mexico was also represented by the largest number of publications (138) on myxomycetes, with Brazil second with 114. All of these are listed by Lado and Wrigley de Basanta (2008).

Since the review referred to earlier appeared, there has been a major effort, largely carried out by Carlos Lado, Arturo Estrada-Torres, and Diana Wrigley de Basanta, to document the myxomycetes of the arid regions of Chile and Peru. This has yielded a series of important papers, including those by Lado et al. (2011, 2013, 2014), with others still yet to appear. These surveys, along with other surveys carried out in the deserts of Mexico, indicated that there is a distinct ecological group of myxomycetes associated with decaying portions of succulent plans, such as various cacti. A major paper on the myxomycetes of the southernmost portion of South America was published by Wrigley de Basanta et al. (2010).

Although most studies (particularly early examples) of myxomycetes took place in Europe and North America, there have been several other countries around the world where many individuals have made significant contributions to our knowledge of this group of organisms. As will be noted, in many instances, these individuals either sent specimens to and/or collaborated with specialists on the myxomycetes who resided in either Europe or the United States. One such country is India, which was granted independence from Britain in 1948.

The first collection of a myxomycete in India apparently occurred about 1830, with a few additional records being reported during the remainder of the 19th century (Thind, 1973). However, the first serious work on the group was carried out by a Mrs. A. Drake, who collected myxomycetes from several different regions of India during the period of 1911–27. Her collections were mostly identified by Gulielma Lister and were deposited in Kew or the herbarium of the British Museum. Drake's collections represented about 74 different species. In 1924, 36 of these were reported in a paper by Lister which was entitled *Myxomycetes from North India* (Lister, 1924). This was the very first publication on the myxomycetes of India. In 1934, S.A. Lodhi published a description of Mrs Blake's collections (Lodhi, 1934). Almost 2 decades later, two Indian mycologists, K.S. Thind (1917–91) in northern India and V. Agnihothrudu (1930–99) in southern India, began their studies of myxomycetes. These studies, often carried out in collaboration with their studies, yielded a whole series of papers, some of which described species new to science. During the 1950s, George Martin worked with Thind and in the 1970s Nannenga-Bremekamp described several new species from the Indian Himalaya with S.S. Dhillon (Dhillon and Nannenga-Bremekamp, 1977, 1978). In 1977, Thind's *Myxomycetes of India* was published (Thind, 1977). In this monograph, 182 species were described and illustrated, which represented the total known from all of India prior to about 1974. T.N. Lakhanpal began his studies of the myxomycetes of north-western

India (primarily around Kulu and Shimla) in 1965 but soon extended his efforts to other parts of the state of Himachal Pradesh. Lakhanpal earned his PhD at the University of Delhi under the direction of R.G. Mukerji. Lakhanpal and Murkerji collaborated on a series of papers that reported new records and an appreciable number of new species from India. Their efforts culminated in the publication of the book *Taxonomy of the Indian Myxomycetes* (Lakhanpal and Mukerji, 1981). Ing reported on some corticolous species from the Mumbai region in 1981, and the same topic was addressed by Chopra et al. (1992). In 1987, Stephenson spent 3 months in India working with Lakhanpal, and this ultimately led to the publication what is perhaps the first paper to consider global patterns of biodiversity in myxomycetes (Stephenson et al., 1993).

In contrast to India, Pakistan has been far less studied, although it is possible that before partition some Indian reports may refer to what we now know as Pakistan. The earliest records are by Lodhi (1934), followed by further accounts in 1956, 1960, 1969, and 1970, some of which were coauthored with other individuals. Visiting Japanese workers have added considerably to the number of species recorded from the country (Yamamoto, 1998; Yamamoto et al., 1992, 1993). There appears to be no reports from Bangladesh.

The mountainous country of Nepal has also received some attention. Two early accounts were by J. Poelt (1965) and S.C. Singh (1971). Visits by Japanese workers have added many more species (Hagiwara and Bhandary, 1982; Nannenga-Bremekamp and Yamamoto, 1988, 1990; Yamamoto and Hagiwara, 1990). A visit by Martin Gregory in 1995 yielded the rare corticolous species *Licea atricapillata* at high-elevations; he also visited Sikkim in 1997 and again *L. atricapillata* grew in the cultures prepared back in England, together with *Licea erectoides*. Sharma and Lakhanpal (1999) reported 75 species from Bhutan.

The island of Sri Lanka is well known for its biodiversity, but relatively little is known of the myxomycetes. The earliest records are by Berkeley and Broome (1873, 1876) and there are papers by Petch (1909, 1910), but this fascinating island still needs to be surveyed more completely. The first reports of myxomycetes from Myanmar (Burma) were by Reynolds and Alexopoulos (1971), but Ko Ko et al. (2013) also reported several additional records. The neighboring country of Thailand has been better studied, with the same paper by Reynolds and Alexopoulos listing 42 species, mostly from the lowlands. Siwasin and Ing (1982), Ing et al. (1987), and Tran et al. (2006, 2008) added many additional species, and a comprehensive checklist of all the species known from the country was provided by Ko Ko et al. (2010). There were apparently no records from Cambodia until the paper published by Ko Ko et al. (2015), who also reported the first records from Laos (Ko Ko et al., 2012). Tran et al. (2014) published on the myxomycetes of Vietnam.

Moving south into Malaya, the few accounts include those of Sanderson (1922), Lister (1931a), and Emoto (1931c). Kuthubutheen (1981) included four species of myxomycetes in an account of fungi from the Langkawi Islands, and a single species was included in the report of a Kew expedition to Pahang and Negri Sembilan in 1985 (Oldridge et al., 1986). It is clear from these papers that much remains to be done in this biologically diverse country.

Indonesia, despite its significance as an evolutionary hotspot, has not been well studied in recent years but was an attraction to early workers. Raciborski published a list of Javanese species in 1898 (Raciborski, 1898) and in the same year Penzig (Penzig, 1898) produced a detailed account of 82 species of myxomycetes collected in the Botanic Garden at Buitenzorg, Java. Emoto (1931a) added more species to the Java list. Boedijn (1927) recorded a small number of myxomycetes from Sumatra and in 1940 added more from the Krakatoa Island group.

An expedition in 1992 to Brunei Darussalam yielded 26 species, with *Erionema aureum* being the most interesting record (Ing and Spooner, 1998). Only three species had been recorded previously from Boneo, all regarded as plant pathogens. The Philippines were visited by D.R. Reynolds, who listed a large number of species in 1981 (Reynolds, 1981). Since 2007, Thomas Edison E. dela Cruz and his students at the University of Santo Tomas have carried out surveys for myxomycetes in several different regions throughout the country.

In writings from the 9th century attributed to the Chinese scholar Twang Ching-Shih, there is reference to a certain substance *kwei hi* (literally "demon droppings") that is of a pale yellowish color and grows in shady damp conditions. It is quite possible that this is the earliest known report of a myxomycete in the literature, and the species involved could have been the common and conspicuous *F. septica*. However, the first definitive records from China (from what is now the province of Manchuria) were published by Emoto between 1931 and 1938 (Emoto, 1931b, 1933, 1934, 1938), and also by Skvortzow (1931). However, Shuqun Teng and his wife were the individuals most responsible for beginning serious studies of the myxomycetes of China. They were most active during the period between 1932 and 1947. Buchet (1939) added a few records from the northern portion of the country, Champion and Mitchell (1980) listed several species from Hong Kong, and Ing (1987b) added several other records. A more comprehensive account of the myxomycetes of Hong Kong species was provided by Chao-hsuan Chung (1997). Since the 1980s, Yu Li at Jilin Agricultural University and some of his students, such as Qi Wang, Shuanglin Chen, Shuyan Liu, Xiaoli Wang, and Pu Liu, have carried out studies of the taxonomy, life cycle, molecular biology, and biochemistry of myxomycetes (Liu, 1980, 1981, 1982, 1983). Jilin Agricultural University hosted the 8th International Congress on the Systematics and Ecology of Myxomycetes in 2014.

The island of Formosa, now called Taiwan, attracted students from early times and the first record of myxomycetes came from Nakazawa (1929, 1931). It was half a century before workers from the University of Taipei began a systematic study of the group, including some most interesting work on structural development. In terms of floristics and distribution studies, the field was led by Liu between 1980 and 1983. Fimicolous myxomycetes were reported for the first time by Chung and Liu (1995). Little is known about the myxomycetes of Korea other than those reported in a paper by Nakagawa (1934).

Japan has a long tradition of studies of myxomycetes, which has been supported at the highest level. The Listers received material from Japanese workers and published many papers, the earliest being in 1904 (Lister and Lister, 1904). They described several new species, one of which, *Hemitrichia imperialis*, was named in honour of Emperor Hirohito (Lister, 1929). Unfortunately, this was later synonymized with *Arcyria stipata*. However, a second species (*Diderma imperiale*) was subsequently named after the emperor by Emoto in the same year, with the emperor himself having collected it. The emperor was a keen naturalist and an expert on marine invertebrates, and he eventually became an honorary member of the Linnean Society of London. He was so interested in natural history that on the palace grounds in Tokyo there was a laboratory devoted to the study of myxomycetes, with Yoshikado Emoto as the director. The latter has already been mentioned several times as the author of papers on Asian myxomycetes, and he was a greatly respected scientist. However, there were several other contributors before him, notably Kumagusu Minakata, who was an avid correspondent with the Listers (as was Emoto). The genus *Minakatella* was named for Minakata, and the latter published a number of lists of Japanese species between 1908 and 1927 (Minakata, 1908, 1927). Charles Meylan, from Switzerland, was sent material from Japan in 1935 and published four new varieties, only one of which survives today. Inevitably, there was a long gap caused by the Second World War, during which little or no activity was possible. However, although Emoto had retired, he was anxious to see a series of paintings of Japanese myxomycetes published and this became *The Myxomycetes of Japan* (Emoto, 1977). This sparked the modern age of myxomycology and mycetozoology in Japan. Two men have led this resurgence, one (Hiromitsu Hagiwara) a professional biologist and the other (Yukinori Yamamoto) an amateur. From 1981 to the present they have produced a wide range of publications, with Hagiwara specializing in dictyostelids but also contributing to delightful books of beautiful photographs designed to encourage naturalists to look at myxomycetes. Yamamoto has collaborated with several other specialists, notably Nannenga-Bremekamp, and has described and distributed exsiccatae of many new species. Two works should be mentioned, the first a joint production of Hagiwara and Yamamoto entitled *Myxomycetes of Japan* (Hagiwara and Yamamoto, 1995) and Yamamoto's *The Myxomycete Biota of Japan* (Yamamoto, 1998). Today, Japan has its own Society of Myxomycetology, with a dedicated journal. Japan and China rival America and Europe in their enthusiasm for myxomycetes!

The remainder of Asia has been less well studied until recently. The great expanses of Mongolia have yielded interesting assemblages of myxomycetes (Novozhilov and Golubeva, 1986) and there are a few records from Turkmenistan (Annalieu, 1960). Reports of myxomycetes from Kazakhstan have been published by Golovenko (1957, 1960), Glustchenko et al. (2002)

published an illustrated book on myxomycetes. Siberian species were listed by Lavrov (1926, 1932) and those from the Kamtchatka Peninsula were included in a report by Transhel (1914). The remaining areas of the former USSR were discussed in the European section given earlier.

Three countries in western Asia have been studied to various degrees. Turkey is another hotspot for biodiversity and has been visited by many biologists, although few mycologists. The first records were reported by Gobelez (1963). Härkönen and Uotila added many corticolous species in their paper of 1983 (Härkönen and Uotila, 1983), and Härkönen followed this with a number of additional species in 1988. Ergul and Gugin (1993) and Ergul and Dülger (2000) listed even more species, and Ing (2000a) added some corticolous examples. Al-Doory (1959) recorded three species from Iraq. Ramon (1968) produced the first list from Israel, and this was followed by two contributions from Binyamini (1986, 1987). Twenty species were reported from Saudi Arabia by Yamamoto and Hagiwara (2003).

The continent of Africa has been much neglected and considering the range of ecosystems available for study, this needs to be rectified. The earliest reference to African species is by Durieu (1848), as part of *Flore d'Algerie*, with references to many species associated with desert plants. A full account of the species known from Algeria, Morocco, and Tunisia was given by Maire et al. (1926), in which 58 species are recorded, including the description of the nivicolous *Lepidoderma peyerimhoffii*. The same three countries were covered to an even greater extent by Faurel et al. (1965), with 113 species being listed, including several nivicolous species. Morocco was the subject of a report by Rammeloo (1973a), in which more species were added, and additions were also made to the Tunisian list by Mitchell and Kylin (1984).

Central Africa was the region adopted by mycologists from Belgium, which resulted in the beautifully illustrated series the *Flore Illustrée des Champignons d'Afrique Centrale* (e.g., Rammeloo, 1981a, 1983) and Buyck (1983a). Rwanda was especially studied and yielded six new species in the genera *Arcyria*, *Diderma*, *Perichaena*, and *Trichia* (Rammeloo, 1981b; Buyck, 1983b). West Africa has been visited by a number of workers and also by European mycologists and naturalists working in various areas of agriculture, especially in oil palm plantations, and they have thoughtfully left a record of the myxomycetes they observed.

The Congo basin was studied in the late 19th century, and some species were listed by Bresadola and Saccardo (1899). Rammeloo (1973b) described *Trichia arundinariae* (now placed in *Metatrichia*) from Zaire. Kranz (1964) reported *Arcyria insignis* and *Physarum cinereum* from Equatorial Guinea, whereas Lado and Teyssiere (1998) added 40 species, including the rare *Physarum plicatum*. Dixon (1959) listed eight species from Ghana, and Härkönen (1981), working in the Gambia, recorded 22 species which developed in moist chamber culture, 5 of which were new records for Africa. Gracia (1986) listed 3 species from the Ivory Coast. Farr (1959) worked on the collections made by

O.F. Cook in the 1890s in Liberia and listed 45 species, including the description of *Physarum tesselatum.*

The myxomycetes of Nigeria are better known than in many parts of West Africa, partly as a result of the comprehensive account by Farquason and Lister (1916) of the species found in the southern portion of the country, especially in oil palm plantations and other cultivated areas. In 1963, a student expedition from Imperial College in London brought back many samples, enabling Ing (1964) to add to the list, including the description of *Metatrichia horrida.* A second expedition in 1966 yielded *Arcyria major* and the white form of *Physarella oblonga*, both new to Africa (Ing and McHugh, 1968). Some myxomycetes from Sierra Leone were collected by F.C. Deighton when he was a government plant pathologist in the 1940s. The material is in Herbarium IMI, which has now been incorporated into the Kew collections. Ing (1967) worked on the collection, which included *A. stipata* and *Perichaena pulcherrima*, both new to Africa. Rammeloo described *Hemitrichia rubrobrunnea* from Sierra Leone in 1978 (Rammeloo, 1978b), a species which is morphologically rather similar to *P. pulcherrima.* More recently, Angela Ejale of the University of Benin in Nigeria has produced a number of publications on the myxomycetes of the country, including a comprehensive taxonomic treatment of the species known from southern Nigeria (Ejale, 2011).

East Africa has fared a little better in terms of research on myxomycetes. Farquason and Lister (1916) referred to several species as having been found in East Africa, but no further details are available. There is circumstantial evidence from the presence of some Kenyan specimens at Kew that this was the source, and these may have been collected in the 1890s. This is also suggested by Ebbels (1972, 1973), who collected a few species in Tanzania and Uganda. Martin and Alexopoulos (1969) also listed "East Africa" without details, for a few species. Rammeloo and Mitchell (1994) listed myxomycetes from Malawi and Zambia, including over 50 species, many of them rare. Almeida (1974b) listed 13 myxomycetes from Mozambique, at the time a colony of Portugal. Ndiritu et al. (2009) provided a report of myxomycetes from Kenya.

Tanzania is by far the best studied country in East Africa, if not the entire continent. This is largely due to the activities of Marja Härkönen and her colleagues from Helsinki. Between 1980 and 1999, the mycoflora was catalogued, especially in relation to its importance to the indigenous people. The emphasis on myxomycetes was a secondary goal but resulted in a wealth of published data (Härkönen and Saarimäki, 1991, 1992; Ukkola, 1998a,b,c; Ukkola et al., 1996). More than 120 species are recorded, including the description of *Licea poculiformis.*

The earliest reference to myxomycetes in southern Africa is by Fries (1848) in his *Fungi Natalenses*, where a few species are listed from Natal. This was followed by Kalchbrenner (1882), who added several species. However, the first attempt at compiling a checklist were those of Duthie (1917, 1918), who recorded some 80 species, many of them identified by Gulielma Lister, to whom the material had been sent. Doidge (1950) listed about 90 species, most of them repeating Duthie's records. The important ecosystems of the Cape area should be a priority for further research. Almeida (1973, 1974a) listed 45 species in the first paper and added 22 in the second.

The African islands in the Indian Ocean have been little visited. The atoll of Aldabra, famous for its giant tortoises, was studied by a Royal Geographical Society expedition in 1980, and a surprisingly wide range of myxomycetes was recorded (Ing and Hnatiuk, 1981). Kryvomaz et al. (2015) reported the results of the first survey of an island in the Seychelles as part of a project that is still ongoing. Fourteen species were listed from Mauritius by Wiehe (1948), whereas Adamonyte et al. (2011) reported on the myxomycetes of La Réunion Island. The first report of myxomycetes from Madagascar was by Patouillard (Patouillard, 1928), and little other information was available until the results of a recent expedition to the island was published by Wrigley de Basanta et al. (2013).

The first report of myxomycetes from the continent of Australia was by Berkeley (1839), who listed two species from Tasmania (then known as Van Diemen's Land) which had been collected by R.W. Lawrence and Ronald Gunn and then sent to William Hooker. The two species—*Aethalium septicum* (now known as *F. septica*) and *Stemonitis fusca*—are both relatively common and form relatively conspicuous fruiting bodies. Later, Berkeley (1845) mentioned eight species (including *S. fusca*) identified from specimens collected by James Drummond in Western Australia and sent to Hooker. In another early report, Berkeley (1859) reported 16 species identified from specimens collected by Ronald Gunn and William Archer in Tasmania during the period of 1839–43. Interestingly, two of these (the first described originally as *Stemonitis echinulata* but renamed *Lamproderma echinulatum* and the second described originally as *Trichia metallica* but now known as *Prototrichia metallica*) were new to science. Some years later, Berkeley (1881) also reported nine species of myxomycetes from Queensland. Cooke added more records in 1888 (Cooke, 1888a,b) and McAlpine (1895) listed the known Australian species, but there appears to have been no additional reports during the remainder of the 19th century.

Cheel (1918) added more records and was followed by Cleland (1927) and Fraser (1933). In 1933 Gulielma Lister described *Dictydium rutilum* but does appear to have otherwise considered Australian species (Lister, 1933). Flentje and Jeffery (1952) reported on species from South Australia, Hnatiuk (1978) on those from Western Australia, and Cribb (1986) added two species from Queensland. As part of the Bicentennial Celebrations, an Anglo–Australian expedition visited the Kimberley Region of Western Australia in 1988. The myxomycetes were reported by Ing and Spooner (1994) and included some rare tropical species with disjunct distributions. There was also a strong similarity between the assemblages in limestone areas with similar sites in the Mediterranean region. David Mitchell (1995) reviewed all records of Australian myxomycetes, listing 146 species, but not including those from the Kimberley

Expedition. Thus, the total known from Australia at that time was 150. This number was increased to 177 by McHugh et al. (2003). Later Lloyd (2014) produced an illustrated booklet, entitled *Where the Slime Mould Creeps*, on the Tasmanian species. It is clear from other branches of natural history that there is much more to be discovered among the myxomycetes of Australia.

New Zealand has enjoyed the attention of both visiting and resident naturalists, and a full list of previous works is given in Stephenson (2003). The earliest records were those of Berkeley (1855) and Cooke (1879). Lister and Lister (1905) also contributed an account of 35 species, including one (*Physarum dictyospermum*) new to science. Cheesman and Lister (1915) added more species, and Rawson (1937) provided the first comprehensive list. The next such list was produced by Mitchell (1992), with 60 species reported as new to the country, thus increasing the total known from the country to 155 species. Stephenson (2003) provided the first monographic treatment of the group in New Zealand in his *Myxomycetes of New Zealand*. More than 180 species are described and illustrated and a full historical background is given.

The island groups in the Pacific Ocean have received little attention in spite of their rich diversity of habitats. However, New Caledonia (Huguenin and Kohler, 1969; Lister, 1922), the Marshall Islands (Rogers, 1947), Cook Islands (Whitney and Olive, 1983), Fiji and Western Samoa (Kylin and Mitchell, 1987), Fiji and Vanuatu (Yamamoto and Matsumoto, 1992), and Vanuatu (Yamamoto and Hagiwara, 2005) have demonstrated the potential for further work in the region. Of special importance are the Galapagos and Hawaiian groups, with their critical role in speciation; valuable studies on their myxomycetes have been made by Eliasson and Nannenga-Bremekamp (1983) and Eliasson (1991), respectively.

The South Polar region poses difficulties for the biologist, but a few expeditions have reported on myxomycetes, with some species, especially those associated with long-persistent snow packs, being the same as those encountered on mountains in both temperate and tropical regions. Horak (1966) described *Diderma antarcticola* from material collected on the Antarctic Peninsula, and Ing and Smith (1980, 1983) recorded a few species from the Antarctic Peninsula and South Georgia. The most intensive survey for myxomycetes in any high-latitude region of the Southern Hemisphere was carried out on subantarctic Macquarie Island by Stephenson in 1995 (Stephenson et al., 2007). The results of this survey and surveys of subantarctic Campbell Island and the Auckland Islands were provided by Stephenson (2011a).

FUTURE STUDIES OF MYXOMYCETES

The account provided herein has necessarily concentrated on the history of collecting and recording myxomycetes around the world, and this was the major activity until the second half of the 20th century. In the early days, considerable attention was also directed to the structure and life cycle of myxomycetes, and toward the end of the 19th century, this was extended to physiology. As biological science progressed in the second half of the 20th century, the physiology, biochemistry and genetics of myxomycetes began to attract experimental biologists (e.g., Alexopoulos, 1953; Hawker, 1952; Sobels, 1950). A valuable overview was provided by Gray and Alexopoulos (1968) with their *Biology of the Myxomycetes*. Another useful introduction was that of Ashworth and Dee (1975), which placed some emphasis on genetics. The field of biosystematics was highlighted by the review by O'Neil Ray Collins (1979) and this was followed by years of dedicated research by Jim Clark and his students at the University of Kentucky. Clark's experience in disentangling morphospecies and biological species has led to some powerful discussions at international meetings and is well reviewed in his paper of 2000. A more recent review of the species concept in myxomycetes was provided by Walker and Stephenson (2016), whereas the progress that has been made in the studies of myxomycetes since the publication of the Martin and Alexopoulos monograph was described by Stephenson (2011b).

During this same period, some heavyweight publications were produced, such as those by Dove and Rusch (1980) on the growth and differentiation of *Physarum polycephalum* in the laboratory and the highly influential two-volume *Cell Biology of Physarum and Didymium* by Aldrich and Daniel (1982). In the 21st century we have seen the rise of molecular taxonomy, which to date has largely reinforced our traditional system of classification, but we cannot be complacent! We have entered a bizarre period in which myxomycetes are being used to design transport systems, operate computers and even play the piano (e.g., Adamatzky, 2010). All of these disciplines will be examined in later chapters. However, as is the case for all experimental biology, repeatability of experimentation must be based on a sound knowledge of precisely which organism is actually being studied!

REFERENCES

Adamatzky, A., 2010. Physarum Machines: Computers from Slime MouldWorld Scientific, Singapore.

Adamonytė, G., 2000. New data on Estonian myxomycete biota. Folia Cryptog. Estonica 36, 7–9.

Adamonytė, G., 2005. Slime molds on *Heracleum sosnowskii* in Lithuania. Mikol. Fitopatol. 39, 1–5.

Adamonyte, G., Stephenson, S.L., Michaud, A., Seraoui, E.H., Meyer, M., Novozhilov, Y.K., Krivomaz, T.I., 2011. Myxomycete species diversity on the island of La Réunion (Indian Ocean). Nova Hedwigia 92, 523–549.

Adams, J., Pethybridge, G.H., 1910. A census catalogue of Irish fungi. Proc. R. Ir. Acad. B 28, 120–166.

Ainsworth, G.C., Balfour-Browne, F.L., 1960. Gulielma Lister centenary. Nature 188, 362–363.

Albertini, J.B., Schweinitz, L.D. von., 1805. Conspectus Fungorum in Lusatiae Superioris. Kummer, Leipzig.

Al-Doory, Y., 1959. Myxomycetes from Iraq. Mycologia 51, 299–300.

Aldrich, H.C., Daniel, J.W., 1982. Cell Biology of *Physarum* and *Didymium*. Academic Press, New York, NY.

Alexopoulos, C.J., 1952. Introductory Mycology. Wiley, New York.

Alexopoulos, C.J., 1953. Myxomycetes developed in moist chamber culture on bark from living Florida trees, with notes on an undescribed species of Comatricha. Q. J. Florida Acad. Sci. 16, 254–262.

Almeida, M.G., 1973. Contribução para o conhecimento dos myxomycetes de Angola I. Bol. Soc. Brot. 47, 277–297.

Almeida, M.G., 1974a. Contribução para o conhecimento dos myxomycetes de Angola II. Bol. Soc. Brot. 48, 187–203.

Almeida, M.G., 1974b. Contribução para o conhecimento dos myxomycetes de Moçambique. Bol. Soc. Brot. 48, 205–210.

Annalieu, S.A., 1960. New data on the mycoflora of Turkmenia. Izvestia Akademie Nank Turkmen S. S. R. Biological Sciences 1960, 25–34.

Ashworth, J.M., Dee, J., 1975. The Biology of Slime Moulds. Edward Arnold Ltd., London.

Batsch, J.G.C., 1783. Elenchus fungorum. Halle.

Beltrán Tejera, E., 2001. Fungi. In: Izquierdo, I., Martín, J.L., Zurita, N., Arechavaleta, M. (Eds.), Lista de especies silvestres de Canarias (hongos, plantas y animales terrestres). Consejería de Política Territorial y Medio Ambiente del Gobierno de Canarias, La Laguna, pp. 29–62.

Berkeley, M.J., 1836. British fungi. In: Smith, J.E. (Ed.), The English Flora 2. London.

Berkeley, M.J., 1839. Notices on British fungi. Ann. Mag. Nat. History 1, 198–208.

Berkeley, M.J., 1845. Decades of fungi. Decades III–VII. Australian fungi. London J. Bot. 4, 42–56, 298–315.

Berkeley, M.J., 1855. Fungi. In: Hooker, J.D. (Ed.), The Botany of the Antarctic Voyage 2. Flora Novae-Zelandiae Pt. 2, pp. 172–210.

Berkeley, M.J., 1859. Fungi. In: Hooker, J.D. (Ed.), Flora of Australia 2. The Botany of the Antarctic Voyage of H. M. Discovery Ships Erebus and Terror, in the Years 1839–43. Flora Tasmaniae, Vol. II. Monocotyledones and Acotyledones. Novell Reeve, London. (Myxomycetes, pp. 266–269)

Berkeley, M.J., 1881. Australian fungi II. J. Linn. Soc. London Bot. 18, 383–389.

Berkeley, M.J., Broome, C.E., 1873. Enumeration of the fungi of Ceylon. J. Linn. Soc. London Bot. 14, 29–140.

Berkeley, M.J., Broome, C.E., 1876. Supplement to the enumeration of the fungi of Ceylon. J. Linn. Soc. London Bot. 15, 82–86.

Binyamini, N., 1986. Myxomycetes from Israel I. Nova Hedwigia 42, 379–386.

Binyamini, N., 1987. Myxomycetes from Israel II. Nova Hedwigia 44, 351–364.

Bjørnekaer, K., Klinge, A.B., 1964. Die Danischen Schleimpilze. Friesia 7, 149–280.

Boedijn, K.B., 1927. Mycetozoa von Sumatra. Miscellania Zoolog. Sumatra 24, 1–4.

Brandza, M., 1928. Les Myxomycètes de Neamtz (Moldavie.). Bull. Soc. Myc. Fr. 44, 249–300.

Bresadola, J., Saccardo, P.A., 1899. Fungi congoenses. Bull. Soc. Roy. Belg. 38, 163.

Briffa, M., 1998. First checklist of the myxomycetes of Malta. Xjenza 2, 28–34.

Briffa, M., Moreno, G., Illana, C., 2000. Some rare myxomycetes from Malta. Stapfia 73, 151–158.

Buchet, T.S., 1939. Contribution à la flore mycologique de la Chine septentrionale 1. Bull. Soc. Myc. France 55, 220–225.

Bulliard, J.B., 1791. Histoire des champignons de la France. Paris.

Buyck, B., 1983a. Flore illustrée des champignons d'Afrique Centrale. Fasc.11. Diderma. Meise, Belgium.

Buyck, B., 1983b. *Diderma petaloides* Buyck, a new myxomycete species from Rwanda. Bull. Jardin Bot. Natl. Belg. 53, 293.

Cejp, K., 1962. A contribution to the mycoflora of slime fungi (Myxomycetes) in Bohemia, especially in West Bohemia. Acta Musei Nationalis Prague 18B, 61–80.

Celakovsky, L., 1893. Die Myxomyceten Böhmens. Arkiven Natur Lands Böhmens 7, 1–88.

Champion, C.L., Mitchell, D.W., 1980. Some myxomycetes collected in Hong Kong. Bull. Brit. Mycol. Soc. 14, 135–137.

Cheel, E.C., 1918. Notes on Australian Myxies (Mycetozoa, Myxomycetes). The Australian Naturalist 4, pp. 6-12.

Cheesman, W.N., Lister, G., 1915. Mycetozoa of Australia and New Zealand. London J. Bot. 53, 203–212.

Chopra, R.K., Nannenga-Bremelamp, N.E., Lakhanpal, T.N., 1992. Some new taxa of corticolous myxomycetes from the N.W. Himalayas, India and a note on Cribraria from Japan. Proc. K. Ned. Akad. Wetensc. 95, 41–50.

Chung, C., 1997. Slime Moulds of Hong Kong. Yi Jsien Publishing Co., Taipei.

Chung, C., Liu, C., 1995. First report of fimicolous myxomycetes from Taiwan. Fungal Sci. 10, 33–35.

Cienkowsky, L., 1863. Das Plasmodium. Pringsheim Jahrbücher 3, 400–411.

Clark, J., Haskins, E.F., 2014. Sporophore morphology and development in the myxomycetes: a review. Mycosphere 5, 153–170.

Cleland, J.B., 1927. Notes on a collection of Australian myxomycetes. Transactions of the Royal Society of South Australia 51, pp. 62–64.

Cochet, S., 1977. Les myxomycetes de France. Bull. Soc. Myc. Fr. 93, 159–200.

Cochet, S., Bozonnet, J., 1980. Les myxomycetes de France: nouveautés et complements. Bull. Soc. Myc. Fr. 96, 115–120.

Cochet, S., Bozonnet, J., 1984. Les myxomycetes de France: nouveautés et complements. Bull. Soc. Myc. Fr. 100, 39–64.

Collins, O.R., 1979. Myxomycete biosystematics: some recent developments and future research opportunities. Bot. Rev. 45, 145–201.

Cooke, M.C., 1877. The Myxomycetes of Great Britain. Williams and Norgate, London.

Cooke, M.C., 1879. New Zealand fungi. Grevillea 8, 54–68.

Cooke, M.C., 1888a. Australian fungi. Grevillea 16, 72–76.

Cooke, M.C., 1888b. Australian fungi. Grevillea 17, 7–8.

Cribb, A.B., 1986. Two slime fungi new to Queensland from Krombit Tops. Queensland Nat. 27, 22–23.

Crowder, W., 1926. Marvels of mycetozoa. Natl. Geogr. Mag. 49, 421–443.

Dallimore, W., Jackson, A.B., 1923. A Handbook of Coniferae, Including Ginkgoaceae. Edward Arnold, London.

de Bary, A., 1859. Die Mycetozoen. Zeitschrift fur wissenschaftwehe Zoologie 10, pp. 88–175.

de Bary, A., 1884. The Comparative Morphology and Biology of the Fungi, Mycetozoa and Bacteria. Clarendon Press, Oxford, English translation, 1887.

de Tournefort, J.P., 1700. Institutiones rei Herbariae. Parisiis, E Typographia Regia. France.

Dhillon, S.S., Nannenga-Bremekamp, N.E., 1977. Notes on some myxomycetes from the northwestern part of the Himalaya. Proc. K. Ned. Akad. Wetensc. C 80, 257–266.

Dhillon, S.S., Nannenga-Bremekamp, N.E., 1978. Notes on some myxomycetes from the northwestern part of the Himalaya. Proc. K. Ned. Akad. Wetensc. C 81, 141–149.

Ditmar, L.P.F., 1813. Die Pilze Deutschlands, In: Sturm, J. (Ed.), Deutschlands Flora. in Abbildungen nach der Natur mit Beschreibungen, 2, pp.55–56.

Dixon, J., 1785. Fasciculus cryptogrammaticae Britannicae. London.

Dixon, P.A., 1959. Myxomycetes of Ghana: 1: *Stemonitis* and *Comatricha*. J. W. Afr. Sci. Assoc. 5, 101–104.

Doidge, E.M., 1950. The South African fungi and lichens. Bothalia 5, 1–1094.

Dove, W.F., Rusch, H.P., 1980. Growth and Differentiation in *Physarum polycephalum*. Princeton University Press, Princeton, NJ.

Drozdowicz, A., Ronikier, A., Stojanowska, W., Panek, E., 2003. Myxomycetes of Poland: a checklist. Krytyczna lista śluzowców Polski, 10.

Durieu, M.C., 1848. Flore d'Algerie. Explor. Sci. Bot. Algerienne 1, 400–423.

Duthie, A.V., 1917. African myxomycetes. Trans. R. Soc. S. Afr. 6, 297–310.

Duthie, A.V., 1918. South African myxomycetes. S. Afr. J. Sci. 15, 456–460.

Ebbels, D.B., 1972. Additions to the mycoflora of south-west Uganda. J. E. Afr. Nat. Hist. Soc. 133, 1–6.

Ebbels, D.B., 1973. Myxomycetes in East Africa. Bull. E. Afr. Nat. Hist. Soc. 1973, 14–18.

Eftimie, E., 1965. Contributu la cunoastreen myxomyceteen di Romania. Anali Stiint. Univ. Cuza Ser. 2, 355–358, 11.

Ehrenberg, C.G., 1818. Fungorum nova genera. Jahrbüche die Gewachskunde, Berlin.

Ejale, A., 2011. Myxomycetes from Southern Nigeria. Lambert Academic Publishing, Saarbrücken.

Eliasson, U., 1975. Myxomycetes in the nature reserve of the Gothenburg Botanical Garden. Svensk Bot. Tidskrift 69, 105–112.

Eliasson, U., 1991. The myxomycete biota of the Hawaiian Islands. Mycol. Res. 95, 257–267.

Eliasson, U., Gilert, E., 2007. Additions to the Swedish myxomycete biota. Karstenia 47, 29–36.

Eliasson, U., Nannenga-Bremekamp, N.E., 1983. Myxomycetes of the *Scalesia* forest, Galapagos Islands. Proc. K. Ned. Akad. Wetensc. C 86, 143–153.

Eliasson, U., Sunhede, S., 1972. Some Swedish records of myxomycetes. Svensk Bot. Tidskrift 66, 18–24.

Elliott, W.T., 1926. Danish myxomycetes contained in the Botanical Museum of the University of Copenhagen. Bot. Tidsskrift 39, 357–367.

Emoto, Y., 1931a. Javanische Myxomyceten. Bull. Buitenzorg Bot. Garden III 11, 161–164.

Emoto, Y., 1931b. Die Myxomyceten der Südmanschurien. Bot. Mag. Tokyo 45, 229–234.

Emoto, Y., 1931c. The Malayan myxomycetes. London J. Bot. 69, 38–42.

Emoto, Y., 1933. Die Myxomyceten der Südmanschurien 2. Bot. Mag. Tokyo 47, 200–202.

Emoto, Y., 1934. On the Manchurian myxomycetes. Trans. Nat. Hist. Soc. Manchuria 4, 7–12.

Emoto, Y., 1938. Myxomycetes in Manchuria. Proc. Jap. Assoc. Adv. Sci. 13, 63–69.

Emoto, Y., 1977. The Myxomycetes of Japan. Sangyo Tosho Publishing Co., Tokyo.

Ergul, G., Dülger, B., 2000. Myxomycetes of Turkey. Karstenia 40, 189–194.

Ergul, C., Gugin, F., 1993. Turkye icni yen ihi myxomycetes tatescu. Turk. J. Bot. 17, 267–271.

Famintsin, A., Woronin, M.S., 1873. Über zwei neue Formen von Schleimpilzen. Mem. Acad. St. Petersburg VII 20, 1–16.

Farquason, C.O., Lister, G., 1916. Notes on South Nigerian mycetozoa. London J. Bot. 54, 121–133.

Farr, M.L., 1959. O. F. Cook's myxomycete collection from Liberia and the Canary Islands. Lloydia 22, 295–301.

Farr, M.L., 1976. Flora Neotropica. Monograph No. 16. Myxomycetes. The New York Botanical Garden, New York.

Farr, M.L., 1981. How to Know the True Slime Molds. McGraw-Hill Science/Engineering/Math, Dubuque.

Faurel, L., Feldman, J., Schotter, G., 1965. Catalogue des myxomycètes de l'Afrique du Nord. Bull. Soc. Hist. Nat. Afrique N. 55, 7–39.

Feest, A., Burggraaf, Y., 1991. A Guide to Temperate Myxomycetes, English translation of Nannenga-Bremekamp, N.E., De Nederlandse Myxomyceten. Bio Press Ltd., Bristol.
Flentje, N.T., Jeffery, M.W., 1952. A note on some slime moulds from South Australia. J. S. Aust. Dept. Agric. 55, 297–300.
Forstner, S., 1969. Second contribution to the study of myxomycetes in Rumania; Third. Communle Bot. 9, 77–88.
Forstner, S., 1971. Second contribution to the study of myxomycetes in Rumania; Third. Communle Bot. 12, 211–220.
Fraser, L., 1933. The mycetozoa of New South Wales. Proc. Linn. Soc. NSW 58, 431–436.
Fries, E.M., 1829. Systema mycologicum. Upsala, Sweden.
Fries, E.M., 1848. Fungi Natalenses. Stockholm, Sweden.
Fries, R.E., 1899. Sveriges myxomyceter. K. Vetensk. Akad. Förh. Ofaren 56, 215–246.
Fries, R.E., 1912. Den svenska myxomycet-floran. Svensk Bot. Tidskrift 6, 721–802.
Glustchenko, V.I., Leontyev, D.V., Akulov, A.Y., 2002. The slime moulds. Kharkov.
Gobelez, M., 1963. The mycoflora of Turkey. Mycopathologia 19, 296–314.
Golovenko, I.N., 1957. Myxomycetes on Tienshan Fir. Urchen. Zap. Kazakh Univ. 29, 95–106.
Golovenko, I.N., 1960. K flore Mikromitsetov Kazakhstana. Frunze Izvestdata Akademie Kirghiz S.S.R., 122–139.
Gottsberger, G., 1966. Die Myxomyceten der Steiermark. Nova Hedwigia 12, 203–296.
Gøtzsche, H.F., 1984. Contributions to the myxomycete flora of Iceland. Acta Bot. Islandica 7, 15–26.
Gøtzsche, H.F., 1990. Notes on Icelandic myxomycetes. Acta Bot. Islandica 10, 3–21.
Gracia, E., 1986. Tres mixomicetes de la Costa d'Ivori. Folia Bot. Miscellanica 5, 141–147.
Gray, W.D., Alexopoulos, C.J., 1968. Biology of the Myxomycetes. Ronald Press Co., New York, NY.
Gunn, W.F., 1918. Irish myxomycetes. Ir. Nat. J. 27, 174.
Gunn, W.F., 1919. Irish myxomycetes. Ir. Nat. J. 28, 45–48.
Gunn, W.F., 1920. Irish myxomycetes. Ir. Nat. J. 29, 76.
Haan, M. de, 1996. Twee nieuwe Myxomyceten voor Belgie. Steerbeckia 17, 7–10.
Hagelstein, R., 1927. Mycetozoa from Puerto Rico. Mycologia 19, 35–37.
Hagelstein, R., 1944. The Mycetozoa of North America: Based Upon the Speciments in the Herbarium of the New York Botanical Garden. Hafner Publishing Co., New York, NY.
Hagiwara, H., Bhandary, H.R., 1982. Myxomycetes from Central Nepal I. In: Otani, Y. (Ed.), Reports on the Cryptogamic Study in Nepal. Miscellaneous Publication of the National Science Museum, Tokyo, pp. 119–124.
Hagiwara, H., Yamamoto, Y., 1995. Myxomycetes of Japan. Heibonsha, Tokyo.
Hanbury, F.J., 1889–1898. An Illustrated Monograph of the British Hieracia. The Author, London.
Härkönen, M., 1977. Corticolous myxomycetes in three different habitats in southern Finland. Karstenia 17, 19–32.
Härkönen, M., 1978. On corticolous myxomycetes in northern Finland and Norway. Ann. Bot. Fennici 15, 32–37.
Härkönen, M., 1979. A checklist of Finnish myxomycetes. Karstenia 19, 8–18.
Härkönen, M., 1981. Gambian myxomycetes developed in moist chamber culture. Karstenia 21, 21–25.
Härkönen, M., Saarimäki, T., 1991. Tanzanian myxomycetes; first survey. Karstenia 31, 31–54.
Härkönen, M., Saarimäki, T., 1992. Tanzanian myxomycetes; first survey (addition). Karstenia 32, 6.
Härkönen, M., Uotila, P., 1983. Turkish myxomycetes developed in moist chamber culture. Karstenia 23, 1–9.
Härkönen, M., Varis, E., 2012. Suomen limasienet. Luonnontieteellinen keskusmuseo, Helsinki.

Haskins, E.F., Clark, J., 2016. A guide to the biology and taxonomy of the Echinosteliales. Mycosphere 7, 473–491.

Hawker, L.E., 1952. The physiology of myxomycetes. The Lister Memorial Lecture. Trans. Brit. Mycol. Soc. 35, 177–187.

Hill, J., 1751. A History of Plants. Thomas Osborne, London.

Hinkova, T., 1951. On Bulgarian myxomycetes. Bull. Bot. Inst. Sofia 2, 265–268.

Hinkova, T.S., Draganov, S., 1965. On Bulgarian myxomycetes. God. Sofia Univ. Biol. Faculty 58, 163–167.

Hintikka, V., 1963. Notes on Finnish myxomycetes. Karstenia 6–7, 110.

Hnatiuk, R.J., 1978. Records of myxomycetes in Western Australia. W. Aust. Herbarium Res. Notes 1, 17–18.

Horak, E., 1966. Sobre dos nuevas especies de hongos recolectadas en el Antárctico. Contrib. Antarctic Inst. Argentina 104, 1–13.

Howard, H.J., 1948. The mycetozoa of sand-dunes and marshland. Southeast Nat. 5, 26–30.

Huguenin, B., Kohler, F., 1969. Qulelques myxomycètes de Nouvelle-Caledonie. Bull. Soc. Myc. Fr. 85, 381–384.

Ing, B., 1964. Myxomycetes from Nigeria. Trans. Brit. Mycol. Soc. 47, 49–55.

Ing, B., 1966. New Irish myxomycetes. Ir. Nat. J. 15, 225–226.

Ing, B., 1967. Myxomycetes from Sierra Leone. Trans. Brit. Mycol. Soc. 50, 549–553.

Ing, B., 1968. A Census Catalogue of British Myxomycetes. The Foray Committee of the British Mycological Society, London.

Ing, B., 1980. A revised census catalogue of British myxomycetes. Bull. Brit. Mycol. Soc. 14, 97–111.

Ing, B., 1985. A revised census catalogue of British myxomycetes. Bull. Brit. Mycol. Soc. 19, 109–115.

Ing, B., 1987a. Corticolous myxomycetes from Cyprus. Mycotaxon 30, 195.

Ing, B., 1987b. Myxomycetes from Hong Kong and southern China. Mycotaxon 30, 199–201.

Ing, B., 1990. New records of myxomycetes in Estonia. Proc. Estonian Acad. Sci. Biol. 39, 271–276.

Ing, B., 1994. Tansley Review No. 62. The phytosociology of myxomycetes. New Phytol. 126, 175–201.

Ing, B., 1997. Corticolous myxomycetes from Switzerland 1. Mycol. Helvetica 9, 3–19.

Ing, B., 1999a. Corticolous myxomycetes from Switzerland 2. Mycol. Helvetica 10, 25–40.

Ing, B., 1999b. The Myxomycetes of Britain and Ireland: An Identification Handbook. Richmond Publishing Co., Slough.

Ing, B., 2000a. Corticolous myxomycetes from Turkey. Karstenia 40, 63–64.

Ing, B., 2000b. A Census Catalogue of the Myxomycetes of Great Britain and Ireland, second ed. Miscellaneous Publication of the British Mycological Society, Kew.

Ing, B., 2003. The myxomycetes of Alto Adige—Trentino. Il Micologo. Cunino 13, 22–30.

Ing, B., Hnatiuk, R.J., 1981. Myxomycetes of Aldabra Atoll. Atoll Res. Bull. 249, 1–10.

Ing, B., Ivancevic, B., 2000. A checklist of myxomycetes from former Yugoslavia. Stapfia 73, 135–150.

Ing, B., McHugh, R., 1968. Myxomycetes from Nigeria II. Trans. Brit. Mycol. Soc. 51, 215–220.

Ing, B., McHugh, R., 1988. A revision of Irish myxomycetes. Proc. R. Ir. Acad. B 88, 99–117.

Ing, B., McHugh, R., 2012. A second revision of Irish myxomycetes. Proc. R. Ir. Acad. B 112, 1–20.

Ing, B., Mitchell, D.W., 1980. Irish myxomycetes. Proc. R. Ir. Acad. B 80, 277–304.

Ing, B., Siwasin, J., Samaranpan, J., 1987. Myxomycetes from Thailand II. Mycotaxon 30, 197.

Ing, B., Smith, R.I.L., 1980. Two myxomycetes from South Georgia. Brit. Antarctic Survey Bull. 50, 118–120.

Ing, B., Smith, R.I.L., 1983. Further myxomycete records from South Georgia and the Antarctic Peninsula. Brit. Antarctic Survey Bull. 59, 80–81.

Ing, B., Spooner, B.M., 1994. Myxomycetes from the Kimberley Region, Western Australia. Bot. J. Linn. Soc. 116, 71–76.

Ing, B., Spooner, B.M., 1998. Myxomycetes from Brunei Darussalam. Kew Bull. 53, 829–834.

Jarocki, J., 1931. Mycetozoa from the Czarnohora Mountains in the Polish eastern Carpathians. Bull. Acad. Polonica B 11, 447–464.

Johannesen, E.W., 1984. New and interesting myxomycetes from Norway. Nordic J. Bot. 4, 513–520.

Kalchbrenner, K., 1882. Fungi Macowaniani. Grevillea 10, 143–147.

Karlsen, A., 1934. Studies on myxomycetes I: new records for Norway. Bergens Museum Årbok, Naturvidensk rekke 1, 1–8.

Keller, H.W., Brooks, T.E., 1977. Corticolous myxomycetes VII: contributions towards a monograph of *Licea*, five new species. Mycologia 69, 667–684.

Keller, H.W., Skrabal, M., Eliasson, U.H., Gaither, T.W., 2004. Tree canopy biodiversity in the Great Smoky Mountains National Park: ecological and developmental observations of a new species of *Diachea*. Mycologia 96, 537–547.

Ko Ko, T.W., Hanh, T.T.M., Stephenson, S.L., Mitchell, D.W., Rojas, C., Hyde, K.D., Lumyong, S., Bahkali, A.H., 2010. Myxomycetes of Thailand. Sydowia 62, 243–260.

Ko Ko, T.W., Ko Ko, Z.Z.W., Rosing, W.C., Stephenson, S.L., 2013. Myxomycetes from Myanmar. Sydowia 65, 267–276.

Ko Ko, T.W., Rosing, W.C., Stephenson, S.L., 2015. First records of myxomycetes from Cambodia. Österreichische Zeitschrift für Pilzkunde 24, 31–37.

Ko Ko, T.W., Tran, H.T.M., Clayton, M.E., Stephenson, S.L., 2012. First records of myxomycetes from Laos. Nova Hedwigia 96, 73–81.

Kowalski, D.T., 1967. Observations on the Dianemaceae. Mycologia 59, 1075–1084.

Kowalski, D.T., 1970. The species of *Lamproderma*. Mycologia 62, 621–672.

Kranz, J., 1964. Fungi collected in the Republic of Guinea I. Sydowia 17, 132–138.

Kryvomaz, T., Michaud, A., Stephenson, S.L., 2015. First survey for myxomycetes on Mahe Island in the Seychelles. Nova Hedwigia 102, 3–4.

Krzemieniewska, H. 1960. Sluzowce Polski na tle sluzowców europejskich. PWN, Warsaw.

Kuthubutheen, A.J., 1981. Notes on the macrofungi of Langkawi. Malayan Nat. J. 34, 123–130.

Kylin, J.H., Mitchell, D.W., 1987. Contribution to the myxomycete flora of Fiji and Western Samoa. Nova Hedwigia 45, 375–381.

Lado, C., 1994. A checklist of myxomycetes in Mediterranean countries. Mycotaxon 52, 117–185.

Lado, C., 2001. Nomenmyx. Cuadernos de Trabajo de Flora micológica Ibérica 16. Editorial CSIC, Madrid.

Lado, C., Pando, F., 1997. Myxomycetes I: ceratiomyxales, echinosteliales, liceales, trichiales. Flora Mycol. Iberica 2, 1–323.

Lado, C., Siquier, J.S., 2014. Myxomycetes de las Islas Baleares. Real Jardin Botánico CSIC, Madrid.

Lado, C., Teyssiere, M., 1998. Myxomycetes from Equatorial Guinea. Nova Hedwigia 67, 421–441.

Lado, C., Wrigley de Basanta, D., 2008. A review of Neotropical Myxomycetes (1828–2008). Anales Jard. Bot. Madrid 65, 211–254.

Lado, C., Wrigley de Basanta, D., Estrada-Torres, A., 2011. Biodiversity of Myxomycetes from the Monte Desert of Argentina. Anales Jard. Bot. Madrid 68, 61–95.

Lado, C., Wrigley de Basanta, D., Estrada-Torres, A., García-Carvajal, E., 2014. Myxomycete diversity of the Patagonian Steppe and bordering areas in Argentina. Anales Jard. Bot. Madrid 71, 1–35.

Lado, C., Wrigley de Basanta, D., Estrada-Torres, A., Stephenson, S.L., 2013. The biodiversity of myxomycetes in central Chile. Fungal Divers. 59, 3–32.

Lakhanpal, T.N., Mukerji, K.G., 1981. Taxonomy of the Indian myxomycetes. Bibliotheca Mycol. 78, 1–531.

Lavrov, N.N., 1926. Material k mikoflore nizov'ev reki Eniseya i ostrovov Eniseiskogo zaliva (data on the Mycoflora of the lower reaches of the Enisei River and Islands of Enisei Bay). Izvestiya Tomskogo Gosudarstvennogo Universiteta, 76 (2), 158-177.

Lavrov, N.N., 1932. Opredelitel' rastitel'nykh parazitov kul'turnykh i dikoras- tushchikh poleznykh rastenii Sibiri. Vypusk I. Polevye, ogorodnye, bakhchevye i tekhnicheskie kul'tury (key to parasites of cultivated and useful wild plants of Siberia. Part I. Field, Melon Field, Garden and Industrial Crops). Tomsk, Russia.

Link, J.H.F., 1809. Observationes in ordines plantarum naturales. Magazin der Gesellschaft naturforschenden Freundes zu Berlin iii, 3–42.

Linnaeus, C., 1763. Systema plantarum, ed. 2, vol. 2. Holmiae.

Lister, A., 1888. Notes on the plasmodium of *Badhamia utricularis* and *Brefeldia maxima*. Ann. Bot. 2 (5), 1–23.

Lister, A., 1894. Monograph of the Mycetozoa. British Museum, London.

Lister, G., 1913. Notes on Swiss mycetozoa, 1912. London J. Bot. 51, 95–109.

Lister, G., 1917. Two new British species of *Comatricha*. London J. Bot. 55, 121–123.

Lister, G., 1918. The mycetozoa: a short history of their study in Britain; an account of their habitats generally, and a list of species recorded from Essex. Essex Field Club Special Memoir 6. London.

Lister, G., 1922. Mycetozoa from New Caledonia. Bot. J. Linn. Soc. 46, 94–96.

Lister, G., 1924. Mycetozoa from North India. London J. Bot. 62, 16–20.

Lister, G., 1925. Monograph of the Mycetozoa, third ed. British Museum, London.

Lister, G., 1929. A new species of *Hemitrichia* from Japan. Trans. Brit. Mycol. Soc. 14, 225–227.

Lister, G., 1931a. Notes on Malayan mycetozoa. London J. Bot. 69, 42–43.

Lister, G., 1933. A new species of D*ictydium* from Australia. London J. Bot.

Lister, G., 1938. The Rev. William Cran and his scientific work. London J. Bot. 76, 319–327.

Lister, A., Lister, G., 1904. Notes on mycetozoa from Japan. London J. Bot. 42, 97–99.

Lister, A., Lister, G., 1905. Mycetozoa from New Zealand. London J. Bot. 43, 111–114.

Lister, A., Lister, G., 1908. Notes on Swiss mycetozoa. London J. Bot. 46, 216–219.

Lister, A., Lister, G., 1911. Monograph of the Mycetozoa, second ed. British Museum, London.

Liu, C.H., 1980. Myxomycetes of Taiwan I. Taiwania 25, 141–151.

Liu, C.H., 1981. Myxomycetes of Taiwan II. Taiwania 26, 58–67.

Liu, C.H., 1982. Myxomycetes of Taiwan III. Taiwania 27, 64–85.

Liu, C.H., 1983. Myxomycetes of Taiwan IV. Taiwania 28, 89–116.

Lloyd, S., 2014. Where the Slime Mould Creeps. Devonport, Tasmania.

Lodhi, S.A., 1934. Indian Slime Moulds. Publication Bureau University of Punjab, Lahore.

Macbride, T.H., 1899. North American Slime Moulds. Macmillan, New York, NY.

Macbride, T.H., 1922. North American Slime Moulds, second ed. Macmillan, New York, NY.

Macbride, T.H., Martin, G.W., 1934. The Myxomycetes. Macmillan, New York, NY.

Maire, R., Patouillard, N., Pinoy, E., 1926. Myxomycètes de l'Afrique du Nord. Bull. Soc. Hist. Nat. Afrique N. 17, 38–43.

Marchal, E., 1895. Champignons coprophiles de Belgique. Bull. Soc. Roy. Bot. Belg. 34, 125–149.

Marchant, J., 1727. Les fleurs de la tannée. Memoirs of the Academy of Science, Paris, 1727, 335–339.

Markov, M., 1962. A contribution to the fungus flora of Bulgaria. Izvestaia Bot. Inst. Sofiya 10, 185–193.

Martin, G.W., Alexopoulos, C.J., 1969. The Myxomycetes. University of Iowa Press, Iowa City.

Martin, G.W., Alexopoulos, C.J., Farr, M.L., 1983. The Genera of Myxomycetes. University of Iowa Press, Iowa City.

Massee, G., 1892. Monograph of the Myxogastres. Methuen & Co., London.
Mazelaitis, J., Stanevičienė, S., 1995. Mycota Lithuaniae I. Mokslo ir enciklopedijų leidykla, Vilnius.
McAlpine, D., 1895. Systematic Arrangement of Australian Fungi, Myxomycetes. Agriculture Department, Victoria, pp 193–199.
McHugh, R., Stephenson, S.L., Mitchell, D.W., Brims, M.H., 2003. New records of Australian Myxomycota. N. Z. J. Bot. 23, 487–500.
Meylan, C., 1908. Connaisance des myxomycetes du Jura. Bull. Soc. Vaud. Sci. Nat. 44, 285–302.
Meylan, C., 1935. Myxomycètes japonais. Bull. Soc. Vaud. Sci. Nat. 58, 321–324.
Meylan, C., 1937. Nouvelle contribution à la connaissance des myxomycetes du Jura et des Alpes. Bull. Soc. Vaud. Sci. Nat. 59, 479–486.
Micheli, P.A., 1729. New Genera of Plants, Arranged After the Method of Tournefort. The Autrhor, Florence.
Minakata, K., 1908. A list of Japanese myxomycetes. Bot. Mag. Tokyo 22, 317–323.
Minakata, K., 1927. A list of the Japanese species of mycetozoa. Bot. Mag. Tokyo 41, 41–47.
Minter, D.W., Dudka, I.O., 1996. Fungi of Ukraine. CAB International, Egham.
Mitchell, D.W., 1992. The myxomycetes of New Zealand and its island territories. Nova Hedwigia 55, 231–256.
Mitchell, D.W., 1995. The Myxomycota of Australia. Nova Hedwigia 60, 269–295.
Mitchell, D.W., Kylin, H., 1984. Some Tunisian myxomycetes. Bull. Brit. Mycol. Soc. 18, 64–65.
Moesz, G., 1924–1933. Fungi Hungariae. Folia Cryptog. 1, 111.
Möller, F.H., 1958. Fungi of the Faeroes, Part II. Munksgaard, Copenhagen.
Moreno, G., Illana, C., Castillo, A., García, J.R., 2001. Myxomycetes de Extremadura. Impresos Postalx, Alcala.
Moroz, E.L., Novozhilov, Y., 1994. Novya i redkordy mixomiceti Belarus. Mikologie Fitopatologie 28, 21–27.
Nakagawa, K., 1934. A list of mycetozoa from Korea. J. Chosen Nat. Hist. Soc. 17, 17–33.
Nakazawa, R., 1929. A list of Formosan mycetozoa. Trans. Nat. Hist. Soc. Formosa 19, 16–20.
Nakazawa, R., 1931. The rare mycetozoan *Minakatella longifila* G. Lister was found in Formosa. Trans. Nat. Hist. Soc. Formosa 21, 191–192.
Nannenga-Bremekamp, N.E., 1974. De Nederlandse Myxomyceten. Koninklijke Nederlandse Natuurhistorische Vereniging, Zutphen.
Nannenga-Bremekamp, N.E., Yamamoto, Y., 1988. *Stemonitis laxifila* (Myxomycetes), a new species from Nepal. In: Watanabe, M., Malla, S.B. (Eds.), The Cryptogams of the Himalayas I. National Science Museum, Japan, Tsukuba, pp. 29–30.
Nannenga-Bremekamp, N.E., Yamamoto, Y., 1990. Two new species and a new variety of Myxomycetes from Nepal. Proc. K. Ned. Akad. Wetensc. 93, 281–286.
Ndiritu, G.G., Spiegel, F.W., Stephenson, S.L., 2009. Rapid biodiversity assessment of myxomycetes in two regions of Kenya. Sydowia 61, 287–319.
Neubert, H., Nowotny, W., Baumann, K., 1993. Die Myxomyceten Deutschlands und des angrenzenden Alpenraumes unter besonderer Berücksichtigung Österreichs, Band 1 Ceratiomyxales, Echinosteliales, Liceales, Trichiales. Baumann, Gomaringen.
Neubert, H., Nowotny, W., Baumann, K., Marx, H., 1995. Die Myxomyceten und des angrenzenden Alpenraumes unter besonderer Berücksichtigung Österreichs, Band 2 Physarales. Baumann, Gomaringen.
Neubert, H., Nowotny, W., Baumann, K., Marx, H., 2000. Die Myxomyceten und des angrenzenden Alpenraumes unter besonderer Berücksichtigung Österreichs, Band 3 Stemonitales. Baumann, Gomaringen.

Novozhilov, Y.K., 1999. Myxomycetes of the Leningrad region. In: Balashova, N.B., Zavarzin, A.A. (Eds.), In: Proceedings of St. Petersburg Society of Naturalists, Ser. 6. St. Petersburg University Press, St. Petersburg, pp. 197–204.

Novozhilov, Y.K., Golubeva, O.E., 1986. Epiphytic myxomycetes from the Mongolian Altai and the Gobi Desert. Mikologyi Fitopatologyi 20, 368–374.

Nuttall, T., 1837. Collections toward a flora of the Territory of Arkansas. Transact. Am. Philos. Soc. 5, 140–203.

Oldridge, S.G., Pegler, D.N., Reid, D.A., Spooner, B.M., 1986. A collection of fungi from Pahang and Negri Sembilan (Malaysia). Kew Bull. 41, 855–872.

Onsberg, P., 1970. Five new myxomycetes recorded in Denmark. Friesia 9, 344–347.

Panckow, T., 1654. Herbarium portabile. Berlin, Germany.

Patouillard, N., 1928. Contribution à l'étude des chasmpignons de Madagascar. Mém. Acad. Malg. 6, 1–49.

Penzig, O., 1898. Die Myxomyceten der Flora von Buitenzorg. The Author, Leiden.

Persoon, C., 1801. Synopsis Methodica Fungorum. The Author, Göttingen.

Petch, T., 1909. New Ceylon fungi. Ann. R. Bot. Garden Peradeniya 4, 299–307.

Petch, T., 1910. A list of the mycetozoa of Ceylon. Ann. R. Bot. Garden Peradeniya 4, 309–371.

Pirola, A., 1968. Lista di mixomiceti Italiani. Giornale Bot. Italiano 102, 21–32.

Pirola, A., Credaro, V., 1971. Contributo alla flora mixomicetologica Italiano. Giornale Bot. Italiano 105, 157–165.

Poelt, J., 1965. Myxomyceten aus Nepal. Khumbu Himalay 2, 59–70.

Raciborski, M., 1884. Myxomycetum agri Cracoviensis, genera, speces et varietates novae. Rozpr. Akad. Umiej. 12, 69–86.

Raciborski, M., 1898. Über die javanischen Schleimpilze. Hedwigia 37, 50–55.

Raitvir, A. (Ed.), 1980. Eesti seente. Tartu.

Rammeloo, J., 1973a. Contribution a la connaisance des myxomycètes du Maroc. Bull. Soc. Sci. Nat. Phys. Maroc. 53, 31–35.

Rammeloo, J., 1973b. *Trichia arundinariae* sp. nov. (Myxomycetes-Trichiaceae) from the National Kahinu Park (Zaire.). Bull. Jardin Nationale Belgique 43, 349–352.

Rammeloo, J., 1978a. Systematische studie van de Trichiales en de Stemonitales (Myxomycetes) van Belgie. Verhandelingen van de Koninklijke Academie voor Wetenschappen, Letteren en schone Kujnsten van Belgie No. 146.

Rammeloo, J., 1978b. *Hemitrichia rubrobrunnea*, a new myxomycete from Sierra Leone. Bull. Jardin Nationale Belgique 48, 383–386.

Rammeloo, J., 1981a. Flore illustrée des champignons d'Afrique Centrale. Fasc. 8-9. Jardin Botanique National de Belgique, Meise.

Rammeloo, J., 1981b. Five new myxomycete species (Trichiales) from Rwanda. Bull. Jardin Nationale Belgique 51, 229–230.

Rammeloo, J., 1983. Flore illustrée des champignons d'Afrique Centrale. Fasc.11. Meise.

Rammeloo J., Mitchell, D.W., 1994. Contribution towards the knowledge of the myxomycetes of Malawi and Zambia. In Proceedings of XIIIth Plenary Meeting AETFAT, Malawi, 1, pp. 785–793.

Ramon, E., 1968. Myxomycetes of Israel. Israel J. Bot. 17, 207–211.

Raunkiaer, L., 1888. Myxomycetes Daniae. Bot. Tidsskrift 17, 20.

Rawson, S.H., 1937. A list of mycetozoa collected in the vicinity of Dunedin, New Zealand. Trans. Proc. R. Soc. N. Z. 66, 351–353.

Reid, D.A., Vanev, S., 1984. New or interesting fungi from Bulgaria. Trans. Brit. Mycol. Soc. 83, 415–421.

Reynolds, D.R., 1981. South-east Asian myxomycetes II: Philippines. Kalikasan 10, 127–150.
Reynolds, D.R., Alexopoulos, C.J., 1971. South-east Asian myxomycetes I: Thailand and Burma. Pac. Sci. 25, 33–38.
Rogers, D.P., 1947. Fungi of the Marshall Islands, Central Pacific Ocean. Pac. Sci. 1, 92–107.
Rojas, C., Schnittler, M., Stephenson, S.L., 2010. A review of the Costa Rican myxomycetes (Amebozoa). Brenesia 73–74, 39–57.
Rojas, C., Valverde, R., Stephenson, S.L., 2015. New additions to the myxobiota of Costa Rica. Mycosphere 6, 709–715.
Rostafinski, J.T., 1873. Versuch Eines Systems der Mycetozoen. Inaugural Dissertation. University of Strassberg, Germany.
Rostrup, E., 1903. Islands Svampe. Botan. Tidsskr. 25, 281–335.
Sanderson, A.R., 1922. Notes on Malayan mycetozoa. Trans. Brit. Mycol. Soc. 7, 239–256.
Santesson, R., 1964. Swedish myxomycetes. Svensk Bot. Tidskrift 58, 113–124.
Sauter, A.E., 1841. Beiträge zur Kenntnis der Pilz-Vegetation des Ober-Pizgaues, im Herzogthune Salzburg. Flora 24, 305–320.
Schäffer, J.C., 1762–1774. Fungi bavarica. Ratsibon.
Schinner, F., 1983. Myxomycetes aus dem Gebiet des Torne Träsk (Abisko) in Schwedisch Lappland. Sydowia 36, 269–276.
Schinz, H., 1920. Die Pilze, X. Myxogasteres, In: Rabenhorst, L., Grunow, A. (Eds.), Dr. L. Kryptogamen-Flora Von Deutschland, Oesterreichs und der Schweiz, E. Kummer, Leipzig.
Schnittler, M., 2000. Ecology and biogeography of myxomycetes. Doctoral dissertation, Freidrich-Schiller-Universität Jena.
Schnittler, M., Novozhilov, Y.K., 1993. *Lepidoderma crustaceum*, a nivicolous myxomycete, found on the island of Crete. Mycotaxon 71, 387–391.
Schrader, H.A., 1797. Nova Genera Plantarum. Lipsiae pp. 32.
Schumacher, H.C.F., 1803. Enumeratio plantarum in partibus Sallandiae septentrionalis et orientalis crescentia. Havniae.
Schweinitz, L.D. von., 1822. Synopsis fungorum Carolinae superioris. Schriften Naturforschenden Gesellschaft zu Leipzig 1, 20–131.
Schweinitz, L.D. von., 1832. Synopsis fungorum in America Borealis et Media degentum. Trans. Am. Philos. Soc. Philadelphia, ser. 2 IV, 141–316.
Scopoli, G.A., 1772. Flora Carniolica. Editio Secunda 1, 1–448.
Scott, D.H., 1908. Arthur Lister, F. R. S. Lond J. Bot. 46, 331–334.
Sharma, R., Lakhanpal, T.N., 1999. Systematics and ecology of myxomycetes in the Royal Kingdom of Bhutan. ICSEM3, Abstracts, 54.
Singh, S.C., 1971. Some myxomycetes from Kathmandu Valley (Nepal). Indian Phytopathol. 24, 715–721.
Siwasin, J., Ing, B., 1982. Myxomycetes from Thailand. Nordic J. Bot. 2, 369–370.
Skupienski, F.X., 1926. Contribution à l'étude des myxomycetes de Pologne. Bull. Soc. Myc. Fr. 42, 142–169.
Skvortzow, B.W., 1931. Mycetozoa from North Manchuria, China. Phillipp. J. Sci. 46, 85–93.
Sobels, J.C., 1950. Nutrition de quelques myxomycètes en culture pures et associées et leurs propriétes antibiotiques. Antonie Van Leeuwenhoek 16 (3), 123–243.
Sommerfelt, S.C., 1826. Supplementum florae Lapponicae quam editit Dr. Georgius Wahlenburg pp.1–331.
Sowerby, J., 1797–1809. Coloured Figures of English Fungi. London.
Stelfox, M.D., 1915. Myxomycetes from the Dingle promontory. Ir. Nat. J. 24, 37–39.
Stephenson, S.L., 2003. The Fungi of New Zealand, Vol. 3: Myxomycetes of New Zealand. Fungal Diversity Press, Hong Kong.

Stephenson, S.L., 2011a. Myxomycetes of the New Zealand subantarctic islands. Sydowia 63, 215–236.

Stephenson, S.L., 2011b. From morphological to molecular: studies of myxomycetes since the publication of the Martin and Alexopoulos monograph. Fungal Divers. 50, 21–34.

Stephenson, S.L., Kalyanasundaram, I., Lakhanpal, T.N., 1993. A comparative biogeographical study of myxomycetes in the mid-Appalachians of eastern North America and two regions of India. J. Biogeogr. 20, 645–657.

Stephenson, S.L., Laursen, G.A., Seppelt, R.D., 2007. Myxomycetes of subantarctic Macquarie Island. Aust. J. Bot. 55, 439–449.

Stephenson, S.L., Stempen, H., 1994. Myxomycetes: A Handbook of Slime Molds. Timber Press Inc., Portland.

Stojanowska, W., 1982. Myxomycetes Sudetow I. Acta Micol. 19, 207–243.

Thind, K.S., 1973. The myxomycetes of India. Proc. Natl. Acad. Sci. India 1972, 47–64.

Thind, K.S., 1977. The Myxomycetes of India. Indian Council of Agricultural Research, New Delhi.

Tode, H.I., 1790. Fungi of Mecklenburg, Lüneberg

Torrend, C., 1908. Catalogue raisonné des myxomycetes du Portugal. Bol. Soc. Portug. Ci. Nat. 2, 55–73.

Toth, S., 1963. Data to the knowledge on the coprophilous microscopic fungi in Hungary I. Ann. Nat. Hist. Museum Hungary 55, 181–185.

Tran, D.Q., Nguyen, H.T.N., Tran, H.T.M., Stephenson, S.L., 2014. Myxomycetes from three lowland tropical forests in Vietnam. Mycosphere 5, 662–672.

Tran, H.T.M., Stephenson, S.L., Hyde, K.D., Mongkolporn, O., 2006. Distribution and occurrence of myxomycetes in tropical forests of northern Thailand. Fungal Divers. 22, 227–242.

Tran, H.T.M., Stephenson, S.L., Hyde, K.D., Mongkolporn, O., 2008. Distribution and occurrence of myxomycetes on agricultural ground litter and forest floor litter in Thailand. Mycologia 100, 181–190.

Transhel, V., 1914. Die Pilze und Myxomyceten Kamtschatkans. Expedition à Kamtchatka 2, 535–576.

Ubrizsy, G., 1967a. Review of the mycoflora of Hungary I. Acta Phytopathol. Acad. Sci. Hung. 2, 153–172.

Ubrizsy, G., 1967b. Review of the mycoflora of Hungary II. Acta Phytopathol. Acad. Sci. Hung. 2, 267–285.

Ukkola, T., 1998a. Some myxomycetes from Dar es Salaam (Tanzania) developed in moist chamber cultures. Karstenia 38, 27–36.

Ukkola, T., 1998b. Myxomycetes of the Usambara Mountains, NE Tanzania. Acta Bot. Fenn. 160, 1–37.

Ukkola, T., 1998c. Tanzanian myxomycetes to the end of 1995. Publications in Botany from the University of Helsinki No. 27. Helsinki.

Ukkola, T., Härkönen, M., Saarimäki, T., 1996. Tanzanian myxomycetes: second survey. Karstenia 36, 51–77.

Vanev, S., Reid, D.A., 1986. New taxa and chorologic data for the Bulgarian fungus flora. Fitologija 21, 63–70.

Vimba, E., Adamonytė, G., 2003. Additional data on Latvian myxomycetes. Folia Cryptog. Estonica 40, 57–61.

von Haller, A., 1742. A Systematic and Descriptive List of Plants Indigenous to Switzerland. The Author, Bern.

Wakefield, E.M., 1950. Miss Gulielma Lister. Trans. Brit. Mycol. Soc. 33, 165–166.

Walker, L.W., Stephenson, S.L., 2016. The species problem in the myxomycetes revisited. Protist 167, 319–338.

Wallroth, F.W., 1833. Flora cryptogramica Germaniae.In: Compendium Florae Germanicae, Sectio II, Plantae Cryptogamicae. Flora cryptogamia s. Cellulosae Scripserunt Math. Jos. Bluff et Carol. Aug. Fingerhuth, Tomus IV, pp. 923.

Wettstein, R. von, 1886. Vorarbeiten zu einer Pilzflora der Steiermark. Verhandelingen zur zoologische-botanische Gesellschaft, Wien 35, 529–618.

Whitney, K.D., Olive, L.S., 1983. A new *Didymium* from Rarotonga, Cook Islands. Mycologia 75, 628–633.

Wichansky, E., 1964. Rare and less known myxomycete species in Bohemia and Moravia. Ceska Mykologia 18, 55–59.

Wiehe. P.O., 1948. The plant diseases and fungi recorded from Mauritius, Mycological Papers 24. Commonwealth Mycological Institute, Kew.

Wrigley de Basanta, D., Lado, C., Estrada-Torres, A., Stephenson, S.L., 2010. Biodiversity of myxomycetes in subantarctic forests of Patagonia and Tierra del Fuego, Argentina. Nova Hedwigia 90, 45–79.

Wrigley de Basanta, D., Lado, C., Estrada-Torres, A., Stephenson, S.L., 2013. Biodiversity studies of myxomycetes in Madagascar. Fungal Divers. 59, 55–83.

Yamamoto, Y., 1998. The Myxomycete Biota of Japan. Tokyo.

Yamamoto, Y., Hagiwara, H., 1990. Myxomycetes from Central Nepal II. In: Watanabe, M., Malla, S.B. (Eds.), Cryptogams of the Himalayas 2: Central and Eastern Nepal. National Science Museum, Tsukuba, pp. 33–40.

Yamamoto, Y., Hagiwara, H., 2003. Myxomycetes from the Asir Mountains, Saudi Arabia. Bull. Natl. Sci. Museum Tokyo B 29, 23–29.

Yamamoto, Y., Hagiwara, H., 2005. Myxomycetes of the Yatsugatake Mts., Central Japan. Bull. Natl. Sci. Museum Tokyo 31 (3), 79–88.

Yamamoto, Y., Hagiwara, H., Sultana, K., 1992. Myxomycetes from Northern Pakistan I. In: Nakaike, T., Malik, S. (Eds.), Cryptogamic Flora of Pakistan Vol. 1. National Science Museum, Tokyo, pp. 109–117.

Yamamoto, Y., Hagiwara, H., Sultana, K., 1993. Myxomycetes from Northern Pakistan II. In: Nakaike, T., Malik, S. (Eds.), Cryptogamic Flora of Pakistan Vol. 2. National Science Museum, Tokyo, pp. 25–42.

Yamamoto, Y., Matsumoto, J., 1992. Three myxomycetes collected in Fiji and Vanuatu. Hikobia 11, 217–218.

Zeller, L., Toth, D., 1971. Myxomycete data from Hungary. Ann. Sci. Univ. Budapest Biol. Sect. 13, 269–278.

FURTHER READING

Ainsworth, G.C., 1952. The Lister notebooks. Trans. Brit. Mycol. Soc. 35, 188–189.

Alexopoulos, C.J., 1959. Myxomycetes from Greece. Brittonia 11, 25–40.

Alexopoulos, C.J., 1963. The myxomycetes II. Bot. Rev. 29, 1–78.

Almeida, M.G., 1964. Contribução para o estudo dos myxomycetes de Portugal. Boletim da Sociedade Portuguesa ded Ciências Naturals. Ser. 2, 172–185, 10.

Boedijn, K.B., 1941. The mycetozoa, fungi and lichens of the Krakatoa group. Bull. Buitenzborg Bot. Garden 16, 358–429.

Clark, J., 2000. The species problem in myxomycetes. Stapfia 73, 39–53.

de Haan, M., no date. The adventure of Mike the myxo. Antwerp.

Dhillon, S.S., 1978. Bark culture myxomycetes new to India. Botaniske Notiser 131, 6.

Gracia, E., 1983. Mixomicetes nuevos o interessantes para la Flora Iberica y Balear. Collectanea Bot. 14, 281–284.

Härkönen, M., 1988. Some additions to the knowledge of Turkish myxomycetes. Karstenia 27, 1–7.

Ing, B., 1982. A revised census catalogue of British myxomycetes. Bull. Brit. Mycol. Soc. 16, 26–35.

Ing, B., 1983. Some corticolous myxomycetes from Bombay. Mycotaxon 17, 299–300.

Lawrow, N.N., 1927–1931. Beiträge zur Schleimpilzflora Sibirens I,II. Mitteilung Tomske Botanische Gesellschaft 2, 10–21.

Li, Y., Li, H.-Z., 1989. Myxomycetes of China. Mycotaxon 35, 429–436.

Lister, A., 1898. Mycetozoa of Antigua and Dominica. London J. Bot. 36, 113–122.

Lister, A., Lister, G., 1906. Mycetozoa from Japan. London J. Bot. 44, 227–230.

Lister, G., 1931b. New species of mycetozoa from Japan. London J. Bot. 69, 297–298.

Lister, G., 1915. Japanese mycetozoa. Trans. Brit. Mycol. Soc. 5, 67–84.

Lodhi, S.A., 1951. Some myxomycetes from western Pakistan. Sydowia 5, 375–383.

Lodhi, S.A. (as Ahmad, S.), 1956. Fungi of West Pakistan. Biological Society of Pakistan, Monograph 1. Lahore.

Lodhi, S.A. (as Ahmad, S.), 1969. Fungi of West Pakistan, Supplement. Biologia Monograph 5. Lahore.

Lodhi, S.A., (as Ahmad, S.) Asad, F., 1970. Coprophilous fungi from West Pakistan, pt. 3. Pakistan Journal of Scientific and Industrial Research 12, 239–243.

Martin, G.W., Thind, K.S., Sohi, H.S., 1957. The myxomycetes of the Mussoorie Hills (India) IV. Mycologia 49, 128–133.

Martin, G.W., Thind, K.S., Rehill, P.S., 1959. The myxomycetes of the Mussoorie Hills (India) X. Mycologia 51, 159–162.

Martin, G.W., Lodhi, S.A., Khan, N.A., 1960. Additional myxomycetes from West Pakistan. Sydowia 14, 281–284.

Novozhilov, Y.K., 1993. Myxomycetes of Russia. St. Petersburg.

Poulin, M., Meyer, M., Bozonnet, J., 2011. Les Myxomycètes, 2 vols. Fédération mycologique et botanique Dauphiné-Savoie, Sevrier.

Yamamoto, Y., Hagiwara, H., 2000. Myxomycetes collected in Vanuatu. Ann. Tsukuba Bot. Garden 21, 127–133.

Yamamoto, Y., Hagiwara, H., Sultana, K., 1995. Myxomycetes from Northern Pakistan III. In: Watanabe, M., Malla, S.B. (Eds.), Cryptogams of the Himalayas 3. National Science Museum, Tokyo, pp. 37–44.

Chapter 3

The Phylogeny of Myxomycetes

Dmitry V. Leontyev*, Martin Schnittler**
**H.S. Skovoroda Kharkiv National Pedagogical University, Kharkiv, Ukraine;*
***Institute of Botany and Landscape Ecology, Ernst Moritz Arndt University Greifswald, Greifswald, Germany*

INTRODUCTION

To the extent that science organizes knowledge in the form of testable statements, we may only call a classification "scientific" if two different experts using the same data can produce the same classification. However, taxonomy has only approached the task of fulfilling of this requirement in the 21st century. Prior to the onset of molecular research, different "phenotypic" or "morphological" classifications were used, known in the East European scientific tradition under the names *ecomorphema* (Kusakin and Drozdov, 1994) and *epimorphema* (Zmitrovich, 2010). These classifications were based on morphological, anatomical, physiological, biochemical, and many other criteria available through direct observation. In order to develop these classifications, scientists were forced (1) to choose the most important characters among the hundreds of available data, and (2) to bind each type of characters to the certain level of classification, for instance, using the shape for delimiting classes and the pigmentation for delimiting orders, or vice versa.. In both cases, decisions depended strongly on the views of individual researchers; therefore the number of classifications was sometimes comparable to the number of taxonomists working on the particular group in question. For example, in classifications of myxomycetes, higher taxa were delimitated on the basis of spore color (Corda, 1837; Hagelstein, 1944; Lister, 1894, 1911, 1925; Rostafinski, 1875), presence or absence of a capillitium (Raunkiær, 1888; Torrend, 1907), occurrence of lime deposits (Yatchevskiy, 1907), or these three characteristics considered simultaneously (Macbride, 1922; Martin and Alexopoulos, 1969; Nannenga-Bremekamp, 1991). By the middle of the 20th century, practically all possible combinations of easily observable morphological characters had been used to develop a classification of the group. The end of discussions in the 1960s was triggered more by general fatigue than by finding of the proper answer.

Myxomycetes: Biology, Systematics, Biogeography, and Ecology. http://dx.doi.org/10.1016/B978-0-12-805089-7.00003-2

The principal alternative to the epimorphema is represented by the phylogenetic classification or *phylema* (Farris, 1990), based on the reconstruction of the *phylogeny* (i.e., the sequence of evolutionary events). Phylema recodes the evolutionary tree in the form of a taxonomical hierarchy. In such a classification, closely related species are assigned to one genus, while more distant ones will not be, even if they share a number of morphological characters with each other. Unlike the epimorphema, which is based on arbitrary selection and ranging of characters, a phylema operates by a single criterion, the structure of genealogical relations. This approach contradicts the famous principle of Linnaeus, who postulated that the "natural system" must take into consideration all of the essential characters of organisms (Linnaeus, 1758). Based on this, opponents of phylogenetic classifications often emphasize that the phylogeny is only one among hundreds of criteria, which have to be considered by taxonomists (Mayr, 1963). At the same time, an example of the most successful natural classification, the periodic table of chemical elements, shows that one essential character (the atomic weight) can provide a fundamental basis for understanding most of the other properties of the objects being classified. In the case of biological systems, the best choice for such a criterion would be a phylogeny. The near-identical reproduction of nucleotide sequences together with the accumulation of changes (mutations) in these sequences form the basis of life as a continuous process. All of the diversity expressed in taxa is a direct result of this process and may be ordered in an optimum way on the basis of phylogeny.

However, this does not mean that phylogenetic classifications are absolutely free of subjectivism. The same data set will yield the same classification only if exactly the same analytical methods are used. Such a level of uniformity seems unattainable and even undesirable. However, taking into consideration that there existed only one consequence of evolutionary events, it may be expected that more extensive data sets, processed by more powerful analytical methods, will give more and more congruent results.

A second element of subjectivism is added when phylogenies are expressed in the context of hierarchical classifications. Evolution is a continuous process, and thus the resulting phylogenies indicate gradual changes, whereas ranks in a hierarchical classification are discrete steps, and their number is limited by practical reasons. The number of nodes of the evolutionary tree is obviously larger than the number of acceptable levels in a practically applicable biological classification. In addition, different ranks may be assigned to the same consequence of branches. For instance, if the clade represented by a certain class includes two subclades, these clades may be assigned to subclasses, orders of the same subclass, or even suborders of the same order. This means that any formal classification will simplify the structure of genealogical relationships, but this simplification strongly depends on the concepts adopted by a particular scientist. One of the possible ways to resolve this problem is to establish a consensus about basic nodes of phylogenies providing the starting points for a classification. For instance, if we accept for conservative reasons that the highest level of the classification is called a domain, while the level, which unifies all the Chordata has to be called a phylum, this gives us the limits within which taxa, such as the Opisthokonta, Metazoa, Bilateria, or Deuterostomia may be categorized.

PHYLOGENETIC REVOLUTION IN TAXONOMY

The concept of phylogenetic classification has existed for more than 150 years and was initially proposed by Haeckel (1866) on the basis of the ideas put forward by Charles Darwin. However, at the time of Haeckel, as well as for more than 100 years thereafter, any assumptions about the origin and relationships between taxa were made on the basis of the indirect data represented by comparative morphology. More effective were fossil records, but these are not available for most prokaryotes and lower eukaryotes. In the absence of well-defined evidence of genealogical relations, classifications of the 19th and 20th centuries were declared to be phylogenetic, but in fact they were not. In the same way as for morphological classifications, these early phylogenies were based on arbitrary selection and ranking of phenotypic characters, and therefore did not comply with the principle of independent reproduction of data. Solving this problem required a fundamentally different approach.

The first real step in the revolution in taxonomy was made by Willi Hennig and his followers, who developed formal procedures for the recoding of evolutionary trees into hierarchical classifications. One of the most important ideas proposed by them was the principle of monophyly, according to which any valid taxon must consist of an ancestral species and all its descendants, with no exceptions (Hennig, 1966). Other groups recognized by Hennig were: (1) *polyphyletic* taxa, such as the Algae, Plantae, or Fungi *sensu lato*, which consist of selected descendants of the common ancestor of all eukaryotes, and (2) *paraphyletic* taxa, such as the Green Algae, Gymnosperms, Pisces, or Reptilia, which include most of the descendants of the same ancestor, except for one "advanced" group (the vascular plants, angiosperms, tetrapods, and birds, respectively). Both paraphyletic and polyphyletic groups, according to Hennig, cannot be used in classification. This rule questioned the justification for the existence of many traditional taxa, and this caused a long-standing debate.

Crick (1958) postulated that the evolution of organisms can be reduced to the evolution of their nucleic acids. This statement was soon confirmed by the comparison of amino acid sequences of proteins in vertebrates; more related groups have been shown to have more similar sequences (Kendrew, 1963). This meant that a comparison of biopolymer sequences can be used to determine the relationships between taxa. But could the gene similarity be the result of convergence, as is the case for many morphological characters of organisms? In most cases, the answer is "no," because the vast majority of gene mutations are neutral with respect to natural selection, having no influence on the conformation of proteins or structural ribonucleic acid (RNA) (Kimura, 1968). The independence of these gene mutations from the pressure of natural selection makes them independent from adaptations. As such, similar genotypes, in contrast to similar phenotypes, may reliably indicate evolutionary relationships.

The step from this theory to practice required the selection of genes, which could serve as a marker of phylogeny. To become useful in phylogenetic studies, the studied gene must (1) originate from a single ancestral gene, and (2) be present in all the organisms being studied. A suitable gene corresponding to these requirements was found in 1974 by Carl Woese and appeared to encode the structure of ribosomal RNA (rRNA). Ribosomes probably arose even before the cells (Bokov and Steinberg, 2009; Root-Bernstein and Root-Bernstein, 2015) and did not change their function for nearly 3 billion years. A comparison of the structure of the 16/18S rRNA gene was used as the basis for the first molecular phylogeny of living organisms (Woese, 1990). The modern phylogenetic classifications (deep tree of life) appearing over the next 2 decades showed the potential of this approach.

MARKER GENES FOR MYXOMYCETES

Among tens of thousands of genes encoded in eukaryotic genomes, only five groups have been widely applied thus far to reconstruct the phylogeny of myxomycetes, and three of these are associated with the ribosomes. These are (1) coding rRNA genes (Vogt and Braun, 1976), (2) the internal transcribed spacers, ITS1 and ITS2, located between rRNA genes (Martin et al., 2003; Fiore-Donno et al., 2011), and (3) the self-splicing group I introns located within the coding sequences of the rRNA genes (Wikmark et al., 2007). As independently inherited alternative markers, the gene of the eukaryotic elongation factor, EF-1α (Fiore-Donno et al., 2011; Feng et al., 2016), and the gene of mitochondrial cytochrome *c* subunit I, COI (Rundquist and Gott 1995; Feng and Schnittler, 2015), have been used in certain studies. Schnittler et al. (2017) provided an overview of the available markers used to barcode myxomycete species (which are not necessarily most suitable for phylogenetic studies).

rRNA forms the skeleton of ribosomes. In 80S ribosomes, which function in the cytoplasm of eukaryotes, the small subunit (SSU) contains 18S rRNA, while the large one comprises three RNA molecules (5S, 5.8S, and 28S). Genes 18S, 5.8S, and 28S are located within one transcription unit, the nuclear ribosomal operon, which includes additional noncoding sections, the nontranscribed spacer (NTS), the external transcribed spacer (ETS), and two internal transcribed spacers (ITS1 and ITS2). Their sequence within the transcription unit is NTS—ETS—18S—ITS1—5.8S—ITS2—26/28S (Steitz, 2008).

rRNA constitutes nearly 80% of the total cellular RNA. Such an amount requires an intensive transcription. This may be achieved by the production of multiple extrachromosomal copies of rRNA genes. These copies, concentrated primarily in the nucleolus, are called mini-chromosomes (Torres-Machorro et al., 2010) or ribosomal DNA (rDNA) (Johansen et al., 1992). The existence of multiple copies greatly helps to obtain sequences from scanty or limited material, making rDNA the best accessible marker for phylogeny studies. Moreover, with rare exceptions, the SSU sequences are homogeneous, since one of the parental genotypes is usually eliminated in crosses during plasmodial development (Ferris et al., 1983; Feng and Schnittler, 2015) and this strongly enhances the readability of a sequence. Finally, in contrast to protein-coding genes, where the variation is relatively evenly distributed (and found mostly among the third base of each triplet), rDNA shows a pattern of alternating conservative and extremely variable sections according to the secondary structure of the rRNA (Smit et al., 2007). This greatly helps in the identification of target regions for primers and sequence alignment (Schnittler et al., 2017).

A disadvantage of rDNA as a phylogenetic marker is that universal primers, able to amplify this gene in all species of myxomycetes, are still unknown (Feng and Schnittler, 2017). Another problem is created by the frequent occurrence of group I introns in this part of genome (Fiore-Donno et al., 2013). The evolution of these parasitic genes was most likely shaped by horizontal gene transfer and thus their presence does not occur in parallel with the evolution of the host organism (Feng and Schnittler, 2015). Therefore, complete SSU sequences become unpredictable in size and are difficult to obtain. Only the first ca. 600 bases of rDNA are free of introns. The information from such a small section is usually sufficient for barcoding of species but is often inadequate to resolve deeper evolutionary relationships.

Despite these and other problems, rDNA remains the most widely applied phylogenetic marker in studies of myxomycetes. The overwhelming majority of the phylogenetic information currently available for these organisms was obtained using different portions of rDNA (Fiore-Donno et al., 2005, 2008; Leontyev et al., 2014a,b, 2015; Walker et al., 2015). The most important for taxonomy is the 5′-domain of 18S rDNA, which consists of 18 helices and ends before the first intron insertion site. It contains four polymorphic helices called E8-1, E10, E10-1, and E11, which vary, especially in their paired (stem) regions, even in closely related species (Fiore-Donno et al., 2012).

Other markers appear less useful in studies of myxomycete phylogeny. For example, the group I introns in myxomycetes are too polymorphic to distinguish taxa at a species or higher level. Most importantly, their evolutionary history is different from that of the host (the myxomycete) (Feng and Schnittler, 2015). It has been found that even different populations of the same species may differ in the presence of a particular intron. However, group I introns are valuable tools to study population genetics and variations in life cycle of myxomycetes (Wikmark et al., 2007). The internal transcribed spacers of rDNA have the same disadvantage. The ITS2 sequences of the related species *Didymium iridis* and *Physarum polycephalum* display only a 53% match (Martin et al., 2003); similar relationships were found for two species of *Lamproderma* (Fiore-Donno et al., 2011). In addition, ITS regions are linked with coding ribosomal genes, and therefore cannot serve as an independent control marker for them. A potentially useful marker for myxomycete phylogenies is the mitochondrial COI gene (Schnittler et al., 2017), for which near-universal primers are available (Feng and Schnittler, 2015; Rundquist and Gott, 1995). In addition, this marker displays insertional RNA editing in myxomycetes (Schnittler et al., 2017).

One of the most appropriate alternatives to the ribosomal genes is represented by the EF-1α gene. In contrast to all other markers discussed so far, this nuclear gene shows Mendelian inheritance (Feng et al., 2016). Like COI, this marker is a protein-coding gene, thus variation is largely limited to the third base of a triplet and as a result is rather low. Splicesomal introns (Feng et al. 2016) prove this rule. As a single-copy gene, this marker is slightly more difficult to amplify. However, it has been used successfully for dark-spored myxomycetes (Feng et al., 2016; Fiore-Donno et al., 2011). The EF-1α gene, together with other protein-coding genes, was also used for disclosing the evolutionary relationships among major groups of Amoebozoa (Fiz-Palacios et al., 2013).

OBTAINING SEQUENCES OF MYXOMYCETE DNA

To obtain DNA sequences, a researcher should take the following steps: (1) select material from which DNA can be obtained, (2) extract DNA from this material, (3) amplify selected portions of the genome, and (4) carry out sequencing of received amplicons using one of the existing techniques. Clean axenic cultures are obviously the best source of DNA for phylogenetic studies. However, this source is often not suitable for myxomycetes because only a few selected species, mostly from the families Physaraceae and Didymiaceae, have been cultured on agar media (and even these cultures often are not axenic). Therefore, the main sources of DNA for phylogenetic research of myxomycetes are fruiting bodies, kept in personal collections and herbaria. Among the elements of a fruiting body, only mature spores usually contain genomic DNA; all other parts, with rare exceptions, have a noncellular structure and do not contain nuclei (Fiore-Donno et al., 2008, 2010).

Specimens of myxomycetes used in molecular studies must meet the following requirements (Leontyev, 2016a; Postel, 2014): (1) have no visible evidence of contamination by filamentous fungi and invertebrates, (2) have an intact peridium or cortex (this reduces the risk of contamination with the spores of another species), (3) be placed in an individual paper box in the field, with this box kept in a zip-lock plastic bag to avoid cross-contamination of specimens, and (4) should not be older than about 10 years, otherwise successful DNA extraction becomes much less likely.

A time-efficient approach for DNA extraction is provided by commercially available kits, which include a set of reagents and filters (see examples in Anonymous, 2013a, 2015), but extraction works as well with a set of reagents that are available in any laboratory. In the first stage of extraction, the fruiting bodies have to be homogenized to break down the spore walls. To do this, fruiting bodies should be placed in plastic tubes along with steel balls or glass beads, kept in freezer for ca. 30 min and then shaken on the laboratory mill for 5 min. The next step is the lysis of cellular and nuclear membranes with lysis buffers, which contain detergents. Simultaneously, or at the next stage, proteins and RNA can be destroyed with the use of proteinase K and RNAse, respectively. Next, the suspension is treated with a concentrated salt solution to precipitate debris (broken cell walls, proteins, lipids, and RNA) and then this debris is filtered to remove it. Finally, DNA is purified from the reagents used in the previous steps. To do this, sodium acetate and ethanol or isopropanol is added to the solution, causing the DNA to immobilize on the membrane of the spin column. Immobilized DNA is then washed 2–3 times by alcohol, which seeps through the filter during centrifugation. After this has taken place, the alcohol is removed by centrifugation, and the DNA pellet is redissolved in a slightly alkaline buffer (Tris/EDTA, Tris/Borate/EDTA, etc.) or in ultrapure water.

The extract obtained in this manner contains the total DNA of the studied species of myxomycete. For Sanger sequencing of the selected gene (e.g., the 18S rRNA), a strong increase in its concentration in needed. This amplification step is provided by a polymerase chain reaction (PCR). This method is based on repeatedly copying the DNA region using artificially synthesized primers, short DNA chains which are complementary to both ends of the studied fragment of DNA (Lodish et al., 2012; Sambrook and Green, 2012). The success of PCR strongly depends upon the proper choice of primers. In contrast to higher animals and plants, universal primers able to amplify any informative portion of the genome in all species of myxomycetes are difficult to find. Therefore, different specific primers have been developed for different groups, using sequences, received by universal primers as a template. Primers used for sequencing the DNA of the dark- and bright-spored myxomycetes are mostly different (Fiore-Donno et al., 2005, 2008, 2013).

To obtain evidence that the targeted portion of genome has been successfully amplified, a gel electrophoresis may be carried out. For this, a fraction of the obtained solution of PCR-products (amplicons) is transferred to the surface of agarose gel and run for 20–30 min on the electrophoresis apparatus at a voltage of 95–100 V/cm. Under the influence of the electromagnetic field, amplicons move within a gel from the negative to the positive electrode. The speed of their movement depends on the length of the DNA fragment (short fragments move quicker). Thus, the position of amplicons in the gel after electrophoresis allows us to determine their relative lengths and indirectly indicates the success of amplification. Visualization of amplicons is provided by different intercalating or fluorescent dyes, added before or after electrophoresis.

If electrophoresis shows that PCR resulted in amplification of the selected gene, PCR products are usually purified using commercial kits (see example in Anonymous, 2013b) or chelating agents, such as Chelex 100 (Walsh et al., 1991). Purified fragments are ready to be sequenced, if diluted to the standard concentration for the particular device being used.

Due to the still unprecedented low error rate of this method, studies of selected genes of myxomycetes for taxonomic purposes are usually carried out by Sanger sequencing with the dye-terminator method (Sambrook and Green, 2012; Smith et al., 1986). The PCR product, containing the amplified fragments of the studied gene, is subjected to a second sequencing PCR where only one of the two primers, DNA polymerase, a set of four nucleotide triphosphates (dNTP), and four fluorescent dideoxynucleotide triphosphates (ddNTP) is added. The ddNTPs lack the 3′-hydroxyl group, which stops further elongation after their inclusion in the new DNA chain. As a result of the reaction, a set of DNA fragments with different lengths and ends labeled with different fluorescent dyes is produced. The reaction mixture is exposed to capillary electrophoresis, during which the fragments of DNA move in the electromagnetic field. In a detection window the four different dyes are excited by a laser, and the fluorescence is detected. Since the speed of fragment movement depends on its length, the sequence of registered fragments reproduces the sequence of nucleotides in the studied gene (Lodish et al., 2012). A resulting chromatogram is usually recorded in an *.ab1 format file, with the nucleotide sequence automatically reconstructed by the computer. Obtained sequence may further corrected manually by visual inspection of chromatograms, for example, using the software ChromasLite (Technelysium Pty Ltd).

The sequence obtained then has to be checked to ensure that it actually represents (1) the selected species and not a contaminant, and (2) the selected portion of the genome. To achieve this, the search of similar sequences in online databases is carried out. The largest database of sequences, the NCBI GenBank (https://www.ncbi.nlm.nih.gov/genbank/), is equipped with a powerful sequence search tool, BLAST (https://blast.ncbi.nlm.nih.gov/Blast.cgi), which automatically compares the sequence in question against all sequences maintained in the database and calculates the percentage of overlap.

HOW TO BUILD A TREE?

The first stage in sequence analysis is to build an alignment, where sequences obtained from related species are stacked together in such a way that a maximum number of columns with identical bases can be achieved with a minimum number of gaps in between. This procedure is based on an assumption that all the variants of a certain gene, represented in related species, have evolved from one hypothetical sequence, which was present in the last common ancestor of the group. Therefore, the purpose of alignment is to identify the mutations that occurred in the studied gene of this ancestor during its evolution. If one finds that a nucleotide is present in sequence A but is absent in sequence B, this indicates that either the insertion of a nucleotide has happened in A or a deletion of a nucleotide has occurred in B (it is initially unknown which of these two hypotheses is more probable). To align the sequences A and B, one has to enter a gap in sequence B instead of the lacking nucleotide. This shifts the following part of sequence B one step forward, bringing it in line with those of sequence A, in which the loss of nucleotide is not observed. If number of nucleotides in compared sequences does not change but some of them are replaced by others, the purpose of alignment is to find a position where replacement has occurred (Graur and Li, 2000; Li, 1997; Nei and Kumar, 2000).

The quality of the alignment critically affects the reliability of the subsequent phylogenetic hypotheses, because all algorithms of phylogenetic analyses use an alignment as primary information. The most obvious way to obtain an accurate alignment is to do it manually, by consequent checking of all the positions of compared sequences. Contrary to expectations, this method is of limited use. Manual alignment intuitively aims to establish the most complete match between the sequences, and this can be achieved only by introducing the maximum number of gaps. However, it has been proven that insertions and deletions occur during the course of evolution considerably less often than replacement of nucleotides, since in protein-coding genes they will lead to a reading frame shift. Therefore, a more desirable (and more comprehensible) approach to the comparison of sequences is the use of formal algorithms, in which the formation of insertions and deletions (opening of the gap) is charged by penalties. This allows to obtain a less "perfect" but more realistic evolutionary scenario (Hall, 2011; Loytynoja and Goldman, 2008).

Among the alignment algorithms used for automatic alignment, the most popular are ClustalW, ClustalX, MUSCLE (Multiple Sequence Alignment) and MAFFT (Multiple Alignment using Fast Fourier Transform). The latter algorithm is often considered to be most effective when aligning sequences with long nonhomologous introns or evolutionarily distant homologous sections with low similarity (Katoh et al., 2002).

The final result of a phylogenetic study is the phylogram, which can be seen as a hypothesis of evolutionary divergences. Phylograms are created on the basis of sequence alignment using one of more than ten algorithms, which are divided into two groups, distance algorithms and probability algorithms. The first group of algorithms is based on the measurement of evolutionary distances (i.e., the number of mismatches) between sequences. Using Ockham's razor (the simplest of several hypotheses explaining the phylogeny is considered the most likely; principle of parsimony), distance algorithms build a tree based on minimal total distance. This method of phylogenetic reconstruction is used by such algorithms as the Unweighted Pair Group Method with Arithmetic Mean, Minimum Evolution, and Neighbor Joining. In contrast, probability algorithms consider differences between the sequences at specific sites separately. The goal of these methods is the reconstruction of an evolutionary scenario that leads to the existing sequences with the highest probability. This group of algorithms includes methods, such as Maximum Parsimony, Maximum Likelihood, and Bayesian Inference. The last two methods are most popular in current phylogenetic studies in myxomycetes (Fiore-Donno et al., 2005, 2012, 2013; Kretzschmar et al., 2016; Leontyev et al., 2015; Nandipati et al., 2012).

The *Maximum Likelihood* method is based on the use of formal models of genome evolution, in which the probability of each mutation (e.g., A → T substitution) is previously deducted. The respective programs calculate the likelihood of different scenarios that could lead to a certain sequence if evolution took place according to the specified model. The *Bayesian Inference* is used to update the probability for an evolutionary hypothesis as more evidence or information becomes available. Primary analysis of data makes it possible to create a list of most probable phylogenetic models without involving the full amount of data. The full data set is used to calculate the likelihood of these initial models and find the most likely tree explaining the full alignment (Graur and Li, 2000; Li, 1997; Nei and Kumar, 2000).

It is impossible to know to what extent a phylogram corresponds to the real sequence of evolutionary events. However, we can estimate the statistical reliability that the proposed phylogram was not obtained under the influence of accidental events. This probability is estimated through the use repetitive sampling methods, with bootstrapping as the most effective. In this method, the positions of analyzed alignment are placed in random order, and each position may be evaluated one to several times or completely ignored. As a result, 1000 or more so-called pseudo-alignments are generated, each containing a randomly transformed sequence. Then, each pseudo-alignment is used to build a tree. The whole set of these pseudo-trees is compared with the true phylogram. The percentage of pseudo-trees, which contain the same node as resulting phylogram is regarded as the confidential assessment of this node (Hall, 2011).

With the methods outlined earlier, during the last decade the still very fragmentary data has provided insights into the position of myxomycetes within eukaryotic organisms and the main steps in myxomycete evolution, although the phylogenetic tree of myxomycetes is still scattered with numerous blank spots. The following paragraphs summarize the knowledge currently available.

WHERE ARE MYXOMYCETES IN THE TREE OF LIFE?

In the 1980s, it was shown that cellular organisms split into three basal domains: Archaea, Bacteria, and Eukaryota (Woese, 1990). However, studies led to the conclusion that the Archaea are paraphyletic with the Eukaryotes and therefore cannot be considered as a separate domain. Together with the Eukaryotes, Archaea may form a joint domain Arkarya (Arkaryota, Eocyta, and Karyota), as opposed to the second domain called Bacteria (Spang et al., 2015).

Treatments (Adl et al., 2005, 2012; Burki, 2014; Burki et al., 2016; Cavalier-Smith, 2010; Forterre, 2015; Ruggiero et al., 2015) have recognized three supergroups within the eukaryotes, which have been named the Excavata, Diaphoretickes, and Amorphea (Fig. 3.1). The subdomain Excavata is composed of ancient, mainly heterotrophic unicellulates and includes two large taxa: the Discoba, having mitochondria with discoidal cristae (e.g., Jakobida, Kinetoplastida, Euglenida, and Heterolobosea) and the Metamonada, possessing

FIGURE 3.1 The position of the myxomycetes (arrow) within the eukaryotic tree of life (author of the picture: D.V. Leontyev). 1. Collodictyonida; 2. Oxymonadida; 3. Trichomonadida; 4. Hypermastigida; 5. Diplomonadida; 6. Retortamonadida; 7. Malawimonadida; 8. Jakobida; 9. Kinetoplastida; 10. Euglenida; 11. Heterolobosea; 12. Flabellinea; 13. Euamoebida; 14. Arcellida; 15. Copromyxida; 16. Variosea; 17. Pelobiontida; 18. Mastigamoebida; 19. Entamoebidae; 20. Protosporangiida; 21. Dictyosteliomycetes; 22. Myxomycetes; 23. Apusomonadida; 24. Ancyromonadida; 25. Breviatida; 26. Nuclearida; 27. Fonticulida; 28. Aphelidea; 29. Microsporidia; 30. Cryptomycota; 31. Chytridiomycota; 32. Amastigomycota; 33. Filasterea; 34. Corallochytrida; 35. Mesomycetozoea; 36. Choanoflagellata; 37. Placozoa; 38. Parazoa; 39. Radiata; 40. Xenacoelomorpha; 41. Protostomia; 42. Deuterostomia; 43. Mesostigmatophyceae; 44. Klebsormidiophyceae; 45. Zygnematophyceae; 46. Charophyceae; 47. Coleochaetophyceae; 48. Nephroselmidophyceae; 49. Pyramimonadophyceae; 50. Chlorodendrophyceae; 51. Chlorophyceae; 52. Ulvophyceae; 53. Marchantiophyta; 54. Bryophyta; 55. Anthocerotophyta; 56. Tracheophyta; 57. Cyanidiophytina; 58. Rhodophytina; 59. Glaucophyta; 60. Katablepharida; 61. Goniomonas; 62. Cryptophyceae; 63. Centrohelida; 64. Haptophyta; 65. Ciliophora; 66. Apicomplexa; 67. Protaveolata; 68. Dinoflagellata; 69. Bicosoecida; 70. Labyrinthulomycetes; 71. Opalinida; 72. Bolidophyceae; 73. Bacillariophyceae; 74. Synurophyceae; 75. Xanthophyceae; 76. Phaeophyceae; 77. Chrysophyceae; 78. Raphidophyceae; 79. Actinophryida; 80. Oomycota; 81. Cercomonadida; 82. Euglyphida; 83. Plasmodiophoromycetes; 84. Haplosporidia; 85. Gromiida; 86. Chlorarachniophyceae; 87. Foraminifera; 88. Radiolaria.

hydrogenosomes or mitosomes instead of mitochondria (e.g., Oxymonadida, Trichomonadida, and Diplomonadida). The second subdomain, the Diaphoretickes, unites unicellular and multicellular, often highly organized forms, a considerable portion of which are photosynthetic. Within this group, the superkingdom Archaeplastida is composed of the three main kingdoms of primarily photosynthetic organisms: the glaucophytes, red algae, and green plants (including the green algae and terrestrial plants). Another superkingdom, called Sar (an acronym for Stramenopiles + Alveolate + Rhizaria), is composed of three kingdoms with a heterotrophic or secondarily photosynthetic type of feeding. It is made up of the Chromista (e.g., ochrophyte algae, Bicosoecida, Oomycota, and Opalinida), Alveolata (e.g., Dinoflagellata, Ciliata, and Apicomplexa) and Rhizaria (e.g., Cercozoa, Foraminifera, Chlorarachniophyta, and Radiolaria). The groups called Cryptista (Cryptophyta) and Haptista (Haptophyta and centrohelid Heliozoa) form separate branches within the Diaphoretickes (Burki et al., 2016). Finally, the third subdomain, the Amorphea (Adl et al., 2012), consists of two main groups, which can be considered as superkingdoms. These are the Obazoa (Brown et al., 2013) and the Amoebozoa (Adl et al., 2012). The Obazoa includes unicellular Apusozoa and the large group Opisthokonta, which, in turn, includes the Holozoa (choanoflagellates, animals and related groups) and Holomycota, also known under the name Nucletmycea (true fungi and related groups).

The second superkingdom, called Amoebozoa, is characterized by broad blunt pseudopodia, tubular mitochondrial cristae, and a flagellar apparatus with one or two strongly unequal flagella (Cavalier-Smith, 2009). Studies of the morphology and ultrastructure of myxomycetes, as well their life cycle and sexual processes, has led to the conclusion that the myxomycetes unambiguously belong to this group of eukaryotes (Olive, 1975; Spiegel, 1990, 1991). This assumption is strongly supported by molecular phylogenies (Baldauf and Doolittle, 1997; Baldauf et al., 2000).

Amoebozoans are divided on two large clades. These are the Lobosa, usually having several finger-like pseudopodia, and the Conosa, for which a single, broadly hemispherical pseudopodium is typical. The amoeboid cells of myxomycetes demonstrate this feature of the Conosa, and indeed, the molecular phylogeny supports that they belong to this branch of the Amoebozoa (Fiz-Palacios et al., 2013; Shadwick et al., 2009).

The group Conosa consists of three main branches. These are (1) the aerobic Eumycetozoa (or Mycetozoa *sensu stricto*), (2) anaerobic Archamoebae (Cavalier-Smith, 2009), and (3) the Variosea, which is represented by several basal branches of the Conosa (Cavalier-Smith et al., 2015). The taxon Eumycetozoa was initially proposed by Olive (1975) for three groups capable to form fruiting bodies: the Protostelia (the fruiting body usually consists of a single spore and a noncellular stalk, the feeding stage is a plasmodium), the Dictyostelia (the fruiting body with numerous spores and a cellular stalk in most instances, the feeding stage is a pseudoplasmodium), and the Myxomycetes (the fruiting body with numerous spores and noncellular stalks, the feeding stage is a plasmodium).

However, as it has been shown by both morphological and molecular studies that the Eumycetozoa include a number of nonfruiting amoebal groups (Baldauf et al., 2000). Moreover, the protostelioid type of fruiting body formation, usually considered as ancestral for all of the Eumycetozoa, has appeared in the evolution of the Amoebozoa on at least eight different occasions (Shadwick et al., 2009). Only five of these eight lineages involved belong to the Eumycetozoa. These are the protosporangiids, protosteliids, soliformoviids, cavosteliids, and schizoplasmodiids (Shadwick et al., 2009). Among these five, the protosporangiid clade (*Protosporangium*, *Clastostelium*, and the macroscopic genus *Ceratiomyxa*) show the closest affinity to myxomycetes in the structure of the fruiting body (with a fruiting body producing multiple spores in *Protosporangium*), the stalk formation and the morphology of sexual processes. The obligate amoebae of the protosporangiids and especially the protoplasmodium of *Ceratiomyxa* possess morphological and ultrastructural similarities to the plasmodia of myxomycetes (Shadwick, 2011). This led to the conclusion, that the fruiting bodies of myxomycetes are homologous with the fruiting bodies of protosporangiids and the last common ancestor of myxomycetes certainly was characterized by a protosporangiid-like fruiting body, the further evolution of which included the differentiation of a peridium and the formation of capillitial structures.

In summary, the position of myxomycetes in the hierarchy of higher-ranking taxa can be shown as outlined below [with examples of genera derived from the last common ancestor in brackets]:

Domain: Eukarya Woese (1977) [*Euglena*, *Rosa*, *Laminaria*]
  Subdomain: Amorphea Adl et al. (2012) [*Homo*, *Agaricus*]
    Superkingdom: Amoebozoa Lühe (1913) emend. Cavalier-Smith (1998) [*Amoeba*, *Arcella*]
      Kingdom: Conosa Cavalier-Smith (1998) [*Pelomyxa*, *Entamoeba*]
        Phylum: Eumycetozoa L.S. Olive (1975); = Mycetozoa de Bary (1859) sensu strictu: [*Protostelium*, *Cavostelium*, *Soliformovum*]
          Subphylum: Macromycetozoa Fiore-Donno et al. (2010) [*Dictyostelium*, *Ceratiomyxa*, *Protosporangium*, *Clastostelium*]
            Class: Myxomycetes G. Winter (1880) p.p.; = Myxogastria Fr. (1829) emend. Cavalier-Smith (2013).

The latter taxon unites only the plasmodial slime molds, forming multisporous fruiting bodies with endogenous spores. The separation of the myxomycete branch from the rest of the Eumycetozoa was recently dated as having taken place some 700 million to 1 billion years ago (Fiz-Palacios et al., 2013).

MAJOR BRANCHES OF THE MYXOMYCETE TREE

The first efforts to elucidate the phylogeny of the major branches of the myxomycetes was made by Fiore-Donno et al. (2005), using complete 18S rDNA and EF1α sequences for 11 species of myxomycetes which represented all five classical orders (i.e., the Echinosteliales, Liceales, Trichiales, Stemonitales, and

Physarales). The results from this study indicated that the myxomycetes are a monophyletic group which splits into two basal clades. The first clade unites members of the traditional orders Stemonitales and Physarales, which have dark (different tints of brown or black) spores pigmented by abundant melanin. The second clade includes the Liceales and Trichiales, which possess bright spores (red, orange, pink, yellow, or olive) colored by various organic pigments, while melanin, if present at all, occurs in their spores in very low concentrations (Kalyanasundaram, 1994). The fifth order, the Echinosteliales, was shown to occupy a basal position in the phylogeny but appears to have a strong affinity to the dark-spored group. Except for the small genera *Barbeyella* and *Clastoderma*, members of the Echinosteliales have mostly hyaline spores but are connected to the dark-spored group by the presence of a columella, a structure that represents an extension of the stalk inside the sporotheca (Fiore-Donno et al., 2005).

Based on these data, Cavalier-Smith (2013), using zoological nomenclature, separated the class Myxogastrea into two superorders. The first of these, the Columellidia (those having a columella), united the dark-spored myxomycetes, or the Fuscisporidia (consisting of the traditional orders Stemonitales and Physarales) and the related Echinosteliales. Interestingly, this classification generally corresponds to the systems of Rostafinski (1875) and his followers, including Lister (1894, 1911, 1925) and Hagelstein (1944) but contrasts the classifications, which recognize four or five orders (e.g., Martin and Alexopoulos, 1969; Nannenga-Bremekamp, 1991; Poulain et al., 2011). Results from molecular studies refuted as well the assumption that the separation of the main branches of the myxomycetes corresponds to the presence of a capillitium or lime deposits (Raunkiær, 1888; Torrend, 1907; Yatchevskiy, 1907). In a similar fashion, the separation of the Stemonitales into the subclass Stemonitomycetidae on the basis of an epihypothallic fruiting body development (Ross, 1973) was not supported by molecular phylogenies.

Subsequent studies of both the Columellidia (Fiore-Donno et al., 2008, 2012) have supported that the basic position within Columellidia is occupied by the Echinosteliales, which include several studied species of *Echinostelium* (Fiore-Donno et al., 2008, 2012), the monotypic genus *Barbeyella* (Fiore-Donno et al., 2012), and the enigmatic stalkless myxomycete *Semimorula* (Fiore-Donno et al., 2009). The position of the genus *Clastoderma* remains unclear; however, there are reasons to believe that it also belongs to the Echinosteliales (Kretschmar et al., 2016), showing all of the features of this order, including a stalk composed of granular material, a thin columella and filamentous capillitium (Poulain et al., 2011).

Apart from the echinosteliid superclade (Table 3.1, Fig. 3.2), dark-spored myxomycetes were found to split into three clades, two of which partially correspond to the classical Stemonitales and Physarales, but the third and most separated one is formed by the genus *Meriderma*, with most of its species known before 2011 as the *Lamproderma atrosporum* group (Fiore-Donno et al., 2008, 2012; Poulain et al., 2011). This Meridermid clade is united by a peridium splitting

TABLE 3.1 Differences Between the Traditional Classification of the Myxomycetes (Five Orders Plus *Ceratiomyxa*) Found in Most Published Monographs, and Informal Groups (Named After the Genus Unambiguously Assignable to the Group, Which is Listed First) Emerging From Molecular Phylogenies

Informal groups based on phylogenies	Traditional classification
Macromycetozoa	
Protosporangiida (*Ceratiomyxa*)	Order Ceratiomyxales
Myxogastria	Class Myxogastria (Myxomycetes)
Dark-spored basal clade (Collumellidia)	
Echinosteliid superclade (*Echinostelium*)	Order Echinosteliales
Fuscisporoid superclade	
Meridermid clade (*Meriderma*)	Order Stemonitales p.p.
Stemonitid clade (*Stemonitis, Comatricha*)	Order Stemonitales p.p.
Lamprodermid clade (*Lamproderma, Didymium, Kelleromyxa, Physarum*)	Orders Physarales, Stemonitales p.p.
Bright-spored basal clade (Lucisporidia)	
Cribrarioid superclade (*Cribraria*)	Order Liceales p.p.
Trichioid superclade	
Reticularioid clade (*Reticularia, Lycogala, Tubifera*)	Order Liceales p.p.
Liceoid clade (*Licea*)	Order Liceales p.p.
Trichoid clade (*Trichia, Dianema, Dictydiaethalium*)	Order Trichiales, Order Liceales p.p.

For orientation, myxomycete genera with an isolated position or rich in species are listed.
Source: Modified after Stephenson, S.L., Schnittler, M., 2017. Myxomycetes. In : Archibald, J.M., Simpson, A.G.B., Slamovits, C.H., (Eds.) Handbook of the Protists. Springer, New York.

into fragments attached to the free ends of capillitial threads. It unites at least all known species of *Meriderma* and the *Collaria rubens* (=*Comatricha rubens*) (Fiore-Donno et al., 2008, 2012). This connection of peridium and capillitium seems to be an ancient character of the dark-spored myxomycetes and is also displayed by some Echinosteliales (Schnittler et al., 2000).

The second (Stemonitid) clade of the Fuscisporidia includes most of the Stemonitales, namely, those having a fugacious peridium, which completely disappears in mature fruiting bodies. The third (Lamprodermid) clade of dark-spored myxomycetes unites, with no exceptions, all of the members of the Physarales, and the remaining part of Stemonitales, which possess a membranaceous peridium that forms large fragments and is not attached to the capillitium. A close

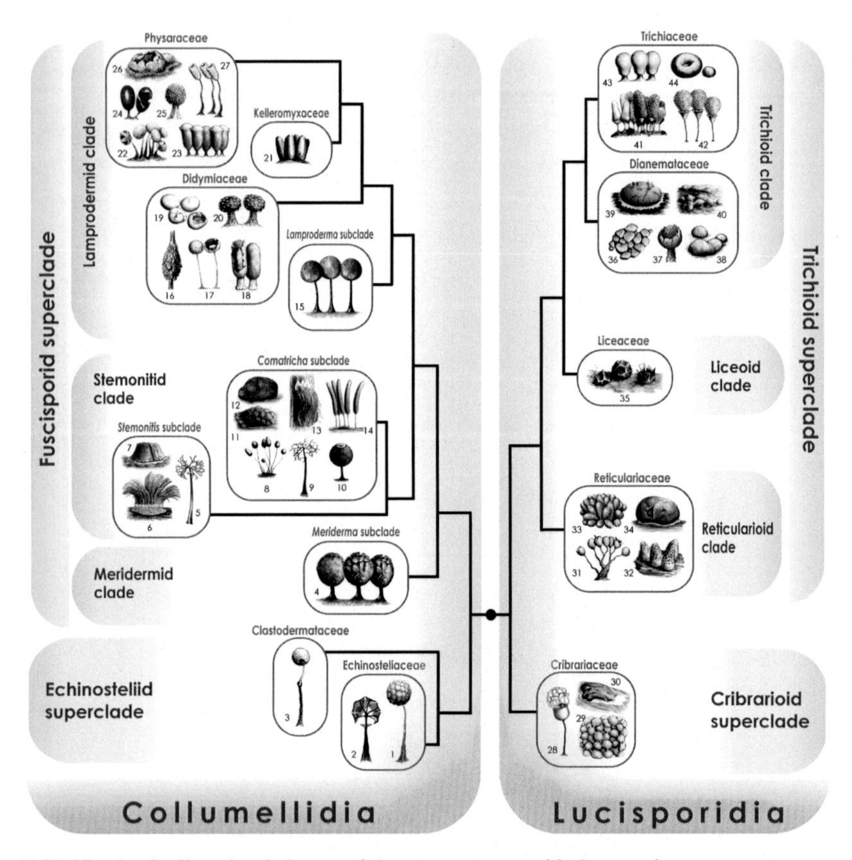

FIGURE 3.2 Outline of a phylogeny of the myxomycetes, with the most important genera assigned to major clades (author of the picture: D.V. Leontyev) . 1. *Echinostelium*; 2. *Barbeyella;* 3. *Clastoderma*; 4. *Meriderma;* 5. *Macbrideola*; 6. *Stemonitis*; 7. *Symphytocarpus*; 8. *Comatricha*; 9. *Paradiacheopsis*; 10. *Enerthenema*; 11. *Brefeldia;* 12. *Amaurochaete*; 13. *Stemonaria*; 14. *Stemonitopsis*; 15. *Lamproderma*; 16. *Mucilago*; 17. *Didymium*; 18. *Diachea*; 19. *Diderma*; 20. *Lepidoderma*; 21. *Kelleromyxa*; 22.*Physarum*; 23. *Craterium*; 24. *Leocarpus*; 25. *Physarina*; 26. *Fuligo*; 27. *Physarella*; 28. *Cribraria*; 29. *Lindbladia*; 30. *Licaethalium*; 31. *Alwisia*; 32. *Lycogala*; 33. *Tubifera*; 34. *Reticularia*; 35. *Licea*; 36. *Dianema*; 37. *Prototrichia*; 38. *Calomyxa*; 39. *Dictydiaethalium*; 40. *Licea variabilis* (shows affinities to Dianemataceae); 41. *Arcyria;* 42. *Hemitrichia*; 43. *Trichia*; 44. *Perichaena.*

relationship of *Lamproderma* with members of the traditional order Physarales may seem strange, because it strongly contradicts the traditional classification. However, it is mainly the absence of lime in fruiting bodies that separates *Lamproderma* and related genera from the Physarales, whereas the true peridium in all species is thin and separated from the capillitium.

The position (as well as the monophyly) of many genera in the dark-spored myxomycetes remains problematic, since only a small proportion of species has been involved in phylogenetic studies thus far. Examples include the large genera *Physarum* and *Badhamia*, which are most likely to be paraphyletic, as indicated by the partly independent markers SSU and LSU rDNA, EF-1α, and ITS1 (Fiore-Donno et al., 2012; Nandipati et al., 2012), or the position of the monotypic genus *Kelleromyxa*, described originally as a species of *Licea* (*Licea fimicola*). The latter seems to assume an isolated position among the dark-spored myxomycetes, which inspired the description of the separate family Kelleromyxaceae for this genus (Erastova et al., 2013).

The phylogeny of bright-spored myxomycetes have been investigated by Fiore-Donno et al. (2013). Only a fraction of all species in this group (35) were involved, but these represent 18 genera (about four fifths of the described genera). Data from two molecular markers (18SrDNA and EF-1α), unambiguously support a basal position of the family Cribrariaceae (Cribrarioid superclade, Table 3.1, Fig. 3.2), which does not correspond to the traditional position of the Cribrariaceae in the order Liceales (Poulain et al., 2011). This discovery provided a strong argument for the revision of the order Liceales, which has been believed to be artificial for a long time (Eliasson, 1977, 2017).

A second part of the traditional Liceales (called Reticularioid clade in Table 3.1, Fig. 3.2) is composed of the genera *Lycogala*, *Reticularia*, and *Tubifera,* which form a monophyletic clade in the phylogeny of Fiore-Donno et al. (2013), whereas *Dictydiaethalium* shows more affinities with the traditional Trichiales (Leontyev et al., 2014a). Contrary to the suggestion of several taxonomists (e.g., Lister, 1894; Macbride, 1922; Torrend, 1907), the genus *Lycogala* is not deeply separated from *Tubifera* and *Reticularia*.

The history of the genus *Alwisia*, with its single known species *Alwisia bombarda* later synonymized with *Tubifera*, is a good example of a situation in which molecular phylogenies may as well support the reerection of a genus. Within the Reticulariaceae at least four subclades have been identified, and these correspond to the genera *Alwisia*, *Lycogala*, *Reticularia*, and *Tubifera* (Leontyev et al., 2014a,b, 2015). 18S rDNA sequences indicate that *Alwisia* should be separated from *Tubifera* and occupies a basal position within the Reticulariaceae (Leontyev et al., 2014a). Three new species have been described within *Alwisia*, forming together with the well-known *A. bombarda* a monophyletic branch (Leontyev et al., 2014a,b).

A third (Liceoid) clade was represented in the study by Fiore-Donno et al. (2013) by four species in the genus *Licea* (*Licea castanea*, *Licea marginata*, *Licea parasitica*, and *Licea variabilis*). However, only three of these support the clade, whereas one (*L. variabilis*) appears to have close affinities with the Trichoid clade (Table 3.1, Fig. 3.2). This example shows that the monophyly of a genus can only be established only if all described species are involved in a study.

The Trichoid clade includes all of the bright-spored myxomycetes which do not belong to the Cribrariaceae, Liceaceae, and Reticulariaceae and have been traditionally considered as the members of the order Trichiales, which at the current stage of knowledge seems to be much more a natural group than the undoubtedly paraphyletic Liceales. However, the sequence of divergences in the evolution of this group is not yet clear (Fiore-Donno et al., 2013); it is more or less clear only, that the traditional Dianemataceae (*Calomyxa*, *Dianema*) plus the genera *Prototrichia* and *Dictydiaethalium* seem to form a basal clade. A number of genera (examples are *Arcyriatella*, *Calonema*, and *Minakatella*) have never been investigated with molecular markers.

Perhaps due to their lack of any economic importance, myxomycetes are probably one of the last major groups of eukaryotes whose diversity and phylogeny have been studied by molecular methods. The available molecular phylogenies directed to disentangle major branches within the group (Fiore-Donno et al., 2005, 2008, 2009, 2010, 2012, 2013; Nandipati et al., 2012) have provided evidence that the traditional five-order classification of myxomycetes is nonnatural. A growing number of case studies add data on the position and verify or deny the monophyly of single families and genera, including *Kelleromyxa* (Erastova et al., 2013), Reticulariaceae (Leontyev et al., 2014a,b, 2015), *Perichaena* (Walker et al., 2015), and Echinosteliaceae (Kretzschmar et al., 2016). With more molecular studies, a revised classification of myxomycetes, as outlined in Fig. 3.2, may soon be within reach.

MORPHOLOGY VERSUS PHYLOGENY

Even the limited phylogenetic data currently available support the conclusion that some very conspicuous characters traditionally considered as important for the classification of myxomycetes seem to have evolved several times independently. The most prominent example is the trend toward compound fruiting bodies, including the following stages: solitary sporocarps as fruiting bodies → fascicled fruiting bodies (sharing a common stalk) → pseudoaethalia (individual fruiting bodies still discernible) → aethalia (individual fruiting bodies indiscernible). A good example outlined earlier provide the Reticulariaceae (excluding *Dictydiaethalium*) with the sequence *Alwisia* → *Tubifera* → *Reticularia*. It may be expected, that at least some of the genera that have been erected for species with compound fruiting bodies may not be justified in the light of molecular investigations, a likely example will be *Cribraria* (solitary sporocarps as fruiting bodies) and *Lindbladia* (pseudoaethalia). Detailed investigations which include as many species as possible need to be carried out to determine in which cases these genera represent distinctive clades in phylogenetic trees. The transformation of the stalked forms to the sessile ones, followed by the formation of pseudoaethalia and aethalia, along with a simultaneous change of the spore dissemination type from the active mode (with the help of a capillitium) to the passive mode (through rain and insects) may be a general tendency of morphological evolution in the myxomycetes (Leontyev et al., 2014c; Leontyev, 2016b).

In contrast, we can expect that more inconspicuous characters, even including some not yet discovered, may support a system based on molecular data. An example is the structure of the peridium and its connection with the capillitium, which supports the separation of *Meriderma* from *Lamproderma* (Poulain et al., 2011). Similarly, hollow stalks filled with spore-like cells represent another, hitherto overlooked, character that seems to support the inclusion of *Trichia decipiens* into a modified genus *Hemitrichia* (Fiore-Donno et al., 2013). Such inconspicuous characters may escape selective pressure, thus becoming somewhat independent from variation in ecological conditions, whereas characters of the fruiting body that affect the dispersal abilities of spores depend strongly upon the environment. For these reasons, some of the most conspicuous characters in the myxomycetes (such as type of fruiting body or the degree of stalk and capillitium development) may be less suitable for supporting a natural system. Due to the plasticity of characters that are under high selective pressure, it can be expected that systematic revisions of classical genera, which include more than just a few species will often result in new generic affiliations for some of these species.

Currently, myxomycete taxonomy seems to have reached the same point at which research on green plants and multicellular animals had arrived several decades ago. Molecular data have shattered the traditional system, which had been accepted for more than a century. However, these results are not yet sufficient to construct a fully comprehensive new system down to the species level. Of the apparently 1000 currently accepted species of myxomycetes, virtually all of which were described on the basis of a morphospecies concept, about 150 dark-spored and 70 bright-spored are represented by partial 18S rDNA sequences (Feng and Schnittler, 2017), and complete SSU sequences are known for a much smaller number of species. For this reason, new combinations for species names should be postponed until the respective genus or family has been treated in a monographic way by a combination of morphological and molecular methods. It will require time and continuous research efforts to develop a comprehensive natural system of classification for the myxomycetes. The studies reviewed in this chapter provide a starting point for such an effort.

REFERENCES

Adl, S.M., Simpson, A.G., Farmer, M.A., Andersen, R.A., Andersen, O.R., Barta, J.R., Bowser, S.S., Brugerolle, G., Fensome, R.A., Fredericq, S., James, T.Y., Karpov, S., Kugrens, P., Krug, J., Lane, C.E., Lewis, L.A., Lodge, J., Lynn, D.H., Mann, D.G., McCourt, R.M., Mendoza, L., Moestrup, Ø., Mozley-Standridge, S.E., Nerad, T.A., Shearer, C.A., Smirnov, A.V., Spiegel, F.W., Taylor, M.F.J.R., 2005. The new higher level classification of eukaryotes with emphasis on the taxonomy of Protists. J. Eukaryot. Microbiol. 52, 399–451.

Adl, S.M., Simpson, A.G., Lane, C.E., Lukeš, J.L., Bass, D., Bowser, S.S., Brown, M.W., Burki, F., Dunthorn, M., Hampl, V., Heiss, A., Hoppenrath, M., Lara, E., Le Gall, L., Lynn, D.H., McManus, H., Mitchell, E.A.D., Sharon, L., Mozley-Stanridge, E., Parfrey, L.W., Pawlowski, J., Rueckert, S., Shadwick, L., Schoch, C.L., Smirnov, A., Spiegel, F.W., 2012. The revised classification of Eukaryotes. J. Eukaryot. Microbiol. 59, 429–493.

Anonymous, 2013a. Invisorb Spin Plant Mini Kit. User Manual. Stratec Molecular GmbH, Berlin.

Anonymous, 2013b. MSB Spin PCRapace. User manual. Stratec Molecular GmbH, Berlin.

Anonymous, 2015. DNeasy Plant Handbook. Quiagen N.V, Venlo.

Baldauf, S.L., Doolittle, W.F., 1997. Origin and evolution of the slime molds (Mycetozoa). Proc. Natl. Acad. Sci. U.S.A. 94, 12007–12012.

Baldauf, S.L., Roger, A.J., Wenk-Siefert, I., Doolittle, W.F., 2000. A kingdom-level phylogeny of eukaryotes based on combined protein data. Science 290, 972–977.

Bokov, K., Steinberg, S.V., 2009. A hierarchical model for evolution of 23S ribosomal RNA. Nature 457, 977–980.

Brown, M.W., Sharpe, S.C., Silberman, J.D., Heiss, A.A., Lang, B.F., Simpson, A.G., Roger, A.J., 2013. Phylogenomics demonstrates that breviate flagellates are related to opisthokonts and apusomonads. Proc. R. Soc. Lond. 280 (1769), 20131755.

Burki, F., 2014. The Eukaryotic Tree of Life from a global phylogenomic perspective. Cold Spring Harb. Perspect. Biol. 6, a016147.

Burki, F., Kaplan, M., Tikhonenkov, D.V., Zlatogursky, V., Minh, B.Q., Radaykina, L.V., Smirnov, A., Mylnikov, A.P., Keeling, P.J., 2016. Untangling the early diversification of eukaryotes: a phylogenomic study of the evolutionary origins of Centrohelida, Haptophyta and Cryptista. Proc. R. Soc. Lond. 283 (1823), (pii: 20152802).

Cavalier-Smith, T., 2009. Megaphylogeny, cell body plans, adaptive zones: causes and timing of eukaryote basal radiations. J. Eukaryot. Microbiol. 56, 26–33.

Cavalier-Smith, T., 2010. Deep phylogeny, ancestral groups and the four ages of life. Philos. Trans. R. Soc. Lond. 365, 111–132.

Cavalier-Smith, T., 2013. Early evolution of eukaryote feeding modes, cell structural diversity, and classification of the protozoan phyla Loukozoa, Sulcozoa, and Choanozoa. Eur. J. Protistol. 49 (2), 115–178.

Cavalier-Smith, T., Fiore-Donno, A.M., Chao, E., Kudryavtsev, A., Berney, C., Snell, E.A., Lewis, R., 2015. Multigene phylogeny resolves deep branching of Amoebozoa. Mol. Phylogenet. Evol. 83, 293–304.

Crick, F.H., 1958. On protein synthesis. Symp. Soc. Exp. Biol. 12, 138–163.

Corda, A.C.J., 1837. Icones Fungorum hucusque cognitorum. J. G. Calve, Praga.

Eliasson, U., 1977. Recent advances in the taxonomy of myxomycetes. Botanical Notes 130, 483–492.

Eliasson, U., 2017. Review and remarks on current generic delimitations in the myxomycetes, with special emphasis on *Licea*, *Listerella* and *Perichaena*. Nova Hedwigia 104, 343–350.

Erastova, D.A., Okun, M.V., Novozhilov, Y.K., Schnittler, M., 2013. Phylogenetic position of the enigmatic myxomycete genus *Kelleromyxa* revealed by SSU rDNA sequences. Mycol. Prog. 12, 599–608.

Farris, J.L., 1990. Haeckel, history and Hull. Syst. Zool. 39, 81–88.

Feng, Y., Klahr, A., Janik, P., Ronikier, A., Hoppe, T., Novozhilov, Y.K., Schnittler, M., 2016. What an intron may tell: several sexual biospecies coexist in *Meriderma* spp. (Myxomycetes). Protist 167 (3), 234–253.

Feng, Y., Schnittler, M., 2015. Sex or no sex? Independent marker genes and group I introns reveal the existence of three sexual but reproductively isolated biospecies in *Trichia varia* (Myxomycetes). Org. Divers. Evol. 15, 631–650.

Feng, Y., Schnittler, M., 2017. Molecular or morphological species? Myxomycete diversity in a deciduous forest in northeastern Germany. Nova Hedwigia 104, 359–380.

Ferris, P.J., Vogt, V.M., Truitt, C.L., 1983. Inheritance of extrachromosomal rDNA in *Physarum polycephalum*. Mol. Cell. Biol. 3 (4), 635–642.

Fiore-Donno, A.M., Berney, C.J., Pawlowski, J., Baldauf, S.L., 2005. Higher-order phylogeny of plasmodial slime molds (Myxogastria) based on elongation factor 1-A and small subunit rRNA gene sequences. J. Eukaryot. Microbiol. 52, 201–210.

Fiore-Donno, A.M., Clissmann, F., Meyer, M., Schnittler, M., Cavalier-Smith, T., 2013. Two-gene phylogeny of bright-spored Myxomycetes (slime moulds, superorder Lucisporidia). PLoS ONE 8, e62586.

Fiore-Donno, A.M., Haskins, E.F., Pawlowski, J., Cavalier-Smith, T., 2009. *Semimorula liquescens* is a modified echinostelid myxomycete (Mycetozoa). Mycologia 101, 773–776.

Fiore-Donno, A.M., Kamono, A., Meyer, M., Schnittler, M., Fukui, M., Cavalier-Smith, T., 2012. 18S rDNA phylogeny of *Lamproderma* and allied genera (Stemonitales, Myxomycetes, Amoebozoa). PLoS ONE 7, e35359.

Fiore-Donno, A.M., Meyer, M., Baldauf, S.L., Pawlowski, J., 2008. Evolution of dark-spored myxomycetes (slime-molds): molecules versus morphology. Mol. Phylogenet. Evol. 46, 878–889.

Fiore-Donno, A.M., Nikolaev, S.I., Nelson, M., Pawlowski, J., Cavalier-Smith, T., Baldauf, S., 2010. Deep phylogeny and evolution of slime moulds (Mycetozoa). Protist 161, 55–70.

Fiore-Donno, A.M., Novozhilov, Y.K., Meyer, M., Schnittler, M., 2011. Genetic structure of two protist species (Myxogastria, Amoebozoa) suggests asexual reproduction in sexual amoebae. PLoS ONE 6, e22872.

Fiz-Palacios, O., Romeralo, M., Ahmadzadeh, A., Weststrand, S., Ahlberg, P.E., Baldauf, S., 2013. Did terrestrial diversification of amoebas (Amoebozoa) occur in synchrony with land plants? PLoS ONE 8, e74374.

Forterre, P., 2015. The universal tree of life: an update. Front. Microbiol. 6, 717.

Graur, D., Li, W.H., 2000. Fundamentals of Molecular Evolution, second ed. Sinauer Associates,, Sunderland MA.

Haeckel, E., 1866. Generelle Morphologie der Organismen: allgemeine Grundzüge der organischen Formen-Wissenschaft, mechanisch begründet durch die von Charles Darwin reformirte Descendenz-Theorie, Berlin.

Hall, B.G., 2011. Phylogenetic Trees Made Easy. A How-to Manual, fourth ed. Sinauer Associates, Sunderland MA.

Hagelstein, R., 1944. The Mycetozoa of North America. Mineola, New York.

Hennig, W., 1966. Phylogenetic Systematics. University of Illinois Press, Urbana.

Johansen, S., Johansen, T., Haugli, F., 1992. Extrachromosomal ribosomal DNA of *Didymium iridis*: sequence analysis of the large subunit ribosomal RNA gene and sub-telomeric region. Curr. Genet. 22, 305–312.

Kalyanasundaram, I., 1994. Occurrence of melanin in bright-spored myxomycetes. Cryptogam. Mycol. 15 (4), 229–237.

Katoh, K., Misawa, K., Kuma, K., Miyata, T., 2002. MAFFT: a novel method for rapid multiple sequence alignment based on fast Fourier transform. Nucleic Acids Res. 30 (14), 3059–3066.

Kendrew, J.C., 1963. Mioglobin and the structure of proteins. Science 139 (2), 1259–1266.

Kimura, M., 1968. Evolutionary rate at the molecular level. Nature 217 (5129), 624–626.

Kretschmar, M., Kuhnt, A., Bonkowski, M., Fiore-Donno, A.M., 2016. Phylogeny of the highly divergent Echinosteliales (Amoebozoa). J. Eukaryot. Microbiol. 63 (4), 453–459.

Kusakin, O.G., Drozdov, A.L., 1994. Phylema of the Organic World. Nauka, Sankt-Petersburg.

Li, W.-H., 1997. Molecular Evolution. Sinauer Associates, Sunderland MA.

Leontyev, D.V., 2016a. Myxomycetes of the family Reticulariaceae: molecular phylogeny, morphology and systematics. Thesis for Doctor of Science degree in Biology (Dr. Sci. Biol.). M.G. Kholodny Institute of Botany of the NAS of Ukraine, Kyiv (in Ukrainian).

Leontyev, D.V., 2016b. The evolution of sporophore in reticulariaceae (myxomycetes). Ukr. Bot. J. 73 (2), 178–184.

Leontyev, D.V., Schnittler, M., Moreno, G., Stephenson, S.L., Mitchell, D.W., Rojas, C., 2014a. The genus *Alwisia* (Myxomycetes) revalidated, with two species new to science. Mycologia 106, 936–948.

Leontyev, D.V., Schnittler, M., Stephenson, S.L., 2014b. A new species of *Alwisia* (Myxomycetes) from New South Wales and Tasmania. Mycologia 106, 1212–1219.

Leontyev, D.V., Schnittler, M., Stephenson, S.L., 2014c. Pseudocapillitium or true capillitium? A study of capillitial structures in *Alwisia bombarda* (myxomycetes). Nova Hedwigia 99 (3–4), 441–451.

Leontyev, D.V., Schnittler, M., Stephenson, S.L., 2015. A critical revision of the *Tubifera ferruginosa* complex. Mycologia 107, 959–985.

Linnaeus, C., 1758. Systema naturæ per regna tria naturæ, secundum classes, ordines, genera, species, cum characteribus, differentiis, synonymis, locis. Laurentius Salvius, Stockholm.

Lister, A., 1894. A Monograph of the Mycetozoa. British Museum, London.

Lister, A., 1911. A Monograph of the Mycetozoa, second ed. (revised by G. Lister). British Museum, London.

Lister, A., 1925. A Monograph of the Mycetozoa, third ed. (revised by G. Lister). British Museum, London.

Lodish, H., Berk, A., Kaiser, Ch.A., Krieger, M., Bretscher, A., Ploegh, H., Amon, A., Scott, M.P., 2012. Molecular Biology of the Cell, seventh ed. Routledge, New York.

Loytynoja, A., Goldman, N., 2008. Phylogeny-aware gap placement prevents errors in sequence alignment and evolutionary analysis. Science 320, 1632–1635.

Macbride, T.H., 1922. The North-American Slime Molds. McMillan Co, New York.

Martin, G.W., Alexopoulos, C.J., 1969. The Myxomycetes. Iowa Univ. Press, Iowa City.

Martin, M.P., Lado, C., Johansen, S., 2003. Primers are designed for amplification and direct sequencing of ITS region of rDNA from myxomycetes. Mycologia 95, 474–479.

Mayr, E., 1963. Animal Species and Evolution. Belknap Press of Harvard University Press, Cambridge.

Nandipati, S.C., Haugli, K., Coucheron, D.H., Haskins, E.F., Johansen, S.D., 2012. Polyphyletic origin of the genus *Physarum* (Physarales, Myxomycetes) revealed by nuclear rDNA minichromosome analysis and group I intron synapomorphy. BMC Evol. Biol. 12, 166.

Nannenga-Bremekamp, N.E., 1991. A Guide to Temperate Myxomycetes (Feest A., E. Burgraff. De Nederlandse Myxomyceten, Engl. transl.). Biopress Ltd., Bristol.

Nei, M., Kumar, S., 2000. Molecular Evolution and Phylogenetics. Oxford University Press, New York.

Olive, L.S., 1975. The Mycetozoans. Academic Press, New York.

Postel, E., 2014. Two gene phylogeny of the genus *Tubifera* (Myxomycetes): Phylogeny of the protein elongation factor EF1α. Bachelor thesis. Ernst Moritz Arndt University, Department of Biology, Greifswald.

Poulain, M., Meyer, M., Bozonnet, J., 2011. Les myxomycetes. Fédération Mycologique et Botanique Dauphiné-Savoie.

Raunkiær, C., 1888–89. Myxomycetes Daniae eller Danmarks Slimsvampe. Botanisk Tidsskrift 17, 20–105.

Root-Bernstein, M., Root-Bernstein, R., 2015. The ribosome as a missing link in the evolution of life. J. Theor. Biol. 367, 130–158.

Ross, I.K., 1973. The Stemonitomycetidae, a new subclass of Myxomycetes. Mycologia 65, 477–485.

Rostafinski, J.H., 1875. Sluzowce (Mycetozoa). Monografia, Paris.

Ruggiero, M.A., Gordon, D.P., Orrell, T.M., Bailly, N., Bourgoin, T., Brusca, R.C., Cavalier-Smith, T., Guiry, M.D., Kirk, P.M., 2015. A higher level classification of all living organisms. PLoS ONE 10 (4), e0119248.

Rundquist, B.A., Gott, J.M., 1995. RNA editing of the coI mRNA throughout the life cycle of *Physarum polycephalum*. Mol. Genet. 247 (3), 306–311.

Sambrook, M.R., Green, J., 2012. Molecular Cloning, fourth ed. Cold Spring Harbor, New York.

Schnittler, M., Shchepin, O., Dagamac, N.H.A., Borg Dahl, M., Novozhilov, Y.K., 2017. Barcoding myxomycetes with molecular markers: challenges and opportunities. Nova Hedwigia 104, 183–209.

Schnittler, M., Stephenson, S.L., Novozhilov, Y.K., 2000. Ultrastructure of *Barbeyella minutissima* (Myxomycetes). Karstenia 40, 159–166.

Shadwick, L.L., 2011. Systematics of Protosteloid amoebae. Theses and Dissertations. 221.

Shadwick, L.L., Spiegel, F.W., Shadwick, J.D., Brown, M.W., Silberman, J.D., 2009. Eumycetozoa = Amoebozoa? SSUrDNA phylogeny of protosteloid slime molds and its significance for the amoebozoan supergroup. PLoS ONE 4, e6754.

Smit, S., Widmann, J., Knight, R., 2007. Evolutionary rates vary among rRNA structural elements. Nucleic Acids Res. 35 (10), 3339–3354.

Smith, L.M., Sanders, J.Z., Kaiser, R.J., Hughes, P., Dodd, C., Connell, C.R., Heiner, C., Kent, S.B., Hood, L.E., 1986. Fluorescence detection in automated DNA sequence analysis. Nature 321 (6071), 674–679.

Spang, A., Saw, J.H., Jørgensen, S.L., Zaremba-Niedzwiedzka, K., Martijn, J., Lind, A.E., van Eijk, R., Schleper, C., Guy, L., Ettema, T.J.G., 2015. Complex Archaea that bridge the gap between prokaryotes and eukaryotes. Nature 521, 173–179.

Spiegel, F.W., 1990. Phylum plasmodial slime molds: Class Protostelids. In: Margulis, L., Corliss, J.O., Melkonian, M.J., Chapman, D. (Eds.), Handbook of Protoctista. Jones and Bartlett Publishers Inc, Boston, pp. 484–497.

Spiegel, F.W., 1991. A proposed phylogeny of the flagellated protostelids. BioSystems 25, 113–120.

Steitz, T.A., 2008. A structural understanding of the dynamic ribosome machine. Nat. Rev. Mol. Cell Biol. 9, 242–253.

Torrend, C., 1907. Les Myxomycètes. Broteria VI (II), 5–349.

Torres-Machorro, A.L., Hernández, R., Cevallos, A.M., López-Villaseñor, I., 2010. Ribosomal RNA genes in eukaryotic microorganisms: witnesses of phylogeny? FEMS Microbiol. Rev. 34, 59–86.

Vogt, V.R., Braun, R., 1976. Structure of ribosomal DNA in *Physarum polycephalum*. J. Mol. Biol. 106, 567–587.

Walker, L.M., Leontyev, D.V., Stephenson, S.L., 2015. *Perichaena longipes*, a new myxomycete from the Neotropics. Mycologia 107 (5), 1012–1022.

Walsh, P.S., Metzger, D.A., Higuchi, R., 1991. Chelex 100 as a medium for simple extraction of DNA for PCR-based typing from forensic material. BioTechniques 10 (4), 506–513.

Wikmark, O.G., Haugen, P., Lundblad, E.W., Haugli, K., Johansen, S.D., 2007. The molecular evolution and structural organization of group I introns at position 1389 in nuclear small subunit rDNA of myxomycetes. J. Eukaryot. Microbiol. 54, 49–56.

Woese, C., 1990. Towards a natural system of organisms: proposal for the domains Archaea, Bacteria, and Eucarya. Proc. Natl. Acad. Sci. U.S.A. 87 (15), 4576–4579.

Yatchevskiy, A.A., 1907. Mycological flora of European and Asian Russia. The Slime MoldsRichter, Moscow.

Zmitrovich, I.V., 2010. Epimorphology and tectomorphology of higher fungi. Folia Cryptogamica Petropolitana 5, 272.

FURTHER READING

Burki, F., Shalchian-Tabrizi, K., Minge, M., 2007. Phylogenomics reshuffles the eukaryotic supergroups. PLoS ONE 2 (8), 790.

Cavalier-Smith, T., Chao, E.E-Y., Oates, B., 2004. Molecular phylogeny of Amoebozoa and the evolutionary significance of the unikont *Phalansterium*. Eur. J. Protistol. 40, 21–48.

Hoppe, T., Kutschera, U., 2010. In the shadow of Darwin: Anton de Bary's origin of myxomycetology and a molecular phylogeny of the plasmodial slime molds. Theory Biosci. 129, 15–23.

Kitching, I.J., Forey, P.L., Humphries, C.J., Williams, D.M., 2000. Cladistics: The Theory and Practice of Parsimony Analysis, second ed. University Press, Oxford, pp. 228.

Lado, C., Pando, F., 1997. Flora Mycologica Iberica. vol. 2: Myxomycetes, I. Ceratiomyxales, Echinosteliales, Liceales, Trichiales. Real Jardín Botánico, J. Cramer, Madrid.

Lado, C., 2005–2016. An on line nomenclatural information system of Eumycetozoa. Available from: http://www.nomen.eumycetozoa.com

Massee, G., 1892. A Monograph of the Myxogasteres. Methuen & Co, London.

McNiel, J., Barrie, F.R., Buck, W.R., Demoulin, V., Greuter, W., (Eds.), 2011. International Code of Nomenclature for Algae, Fungi, and Plants (Melbourne Code). Regnum Vegetabile, vol. 154. Koeltz Scientific Books, Melbourne.

Ride, W.D.L., Cogger, H.G., Dupuis, C., Kraus, O., Minelli, A. (Eds.), 1999. International Code of Zoological Nomenclature. fourth ed. The Natural History Museum, London.

Schnittler, M., 2001. Ecology of Myxomycetes from a winter-cold desert in western Kazakhstan. Mycologia 93, 653–669.

Stephenson, S.L., 2003. Myxomycetes of New Zealand. Fungi of New Zealand, vol. 3, Fungal Diversity Press, Hong Kong.

Stephenson, S.L., Schnittler, M., Novozhilov, Y.K., 2008. Myxomycete diversity and distribution from the fossil record to the present. Biodivers. Conserv. 17, 285–301.

Chapter 4

Genomics and Gene Expression in Myxomycetes

Dennis Miller, Ramesh Padmanabhan, Subha N. Sarcar
University of Texas at Dallas, Richardson, TX, United States

INTRODUCTION

The *Physarum* genome consists of several types of DNA. Early studies of the DNA content of the haploid nucleus indicated the presence of about 40–45 chromosomes (Mohberg, 1977) with an average length of 6–7 Mb, giving a chromosomal DNA content of the haploid nucleus of 240–315 Mb. In addition to the nuclear chromosomes, in the nucleolus the rDNA genes are located on 60 kb extrachromosomal "minichromosomes" (Ferris, 1985; Ferris and Vogt, 1982; Vogt and Braun, 1976). There are approximately 150 of these minichromosomes per nucleus, thus adding about 9 Mb to the nuclear DNA content. Consistent with these observations, the DNA content of the *Physarum* haploid nucleus was determined to be about 0.3 pg, or about 290 Mb (Melera, 1980; Mohberg, 1977; Mohberg and Rusch, 1971; Mohberg et al., 1973). In addition to the nuclear DNA content, the DNA within mitochondria (mtDNA) contributes an additional 24 Mb of DNA (400 copies of a 60 kb mtDNA per nucleus; Bohnert et al., 1975; Braun and Evans, 1969; Evans, 1966; Evans and Suskind, 1971; Guttes and Guttes, 1969). Of the total DNA content of *Physarum* about 10% is mtDNA, about 5% is minichromosome (rDNA), and about 85% is chromosomal DNA.

NUCLEOLAR DNA: MINICHROMOSOMES rDNA GENES

The nucleolar minichromosome or extrachromosomal rDNA molecule in *Physarum polycephalum* is about 60 kb in length (Ferris, 1985; Ferris and Vogt, 1982; Vogt and Braun, 1976). There are about 150 copies of the rDNA minichromosome, constituting about 9 Mb or about 5% of the nuclear DNA (Hardman, 1985). This DNA is more G + C-rich than chromosomal DNA (or mitochondrial DNA) and forms a satellite band in CsCl–Bisbenzimide density gradients. Each minichromosome contains two rDNA transcription units

Myxomycetes: Biology, Systematics, Biogeography, and Ecology. http://dx.doi.org/10.1016/B978-0-12-805089-7.00004-4

oriented as inverted repeats. The rDNA molecule itself consists of two 29.6 kb inverted repeats, each with one rDNA transcription unit about 13.3 kb in length. The rDNA transcription units are divergently transcribed in a direction away from the central axis of symmetry toward the termini. The transcription units are separated by about 22 kb, consisting of two 11 kb tandem inverted repeats. This region contains the origins of replication where rDNA replication initiates and proceeds bidirectionally from the origin. Each 11 kb region has two origins of replication for a total of four origins per rDNA molecule. On average, only one of these origins is initiated in each mitotic cycle, resulting in a doubling of the number of rDNA molecules in each cycle. Replication of the rDNA molecule occurs in the last two-thirds of S-phase and all of G2-phase. An untranscribed region of about 5 kb is located at each terminus (Ferris, 1985; Ferris and Vogt, 1982; Hardman, 1985; Vogt and Braun, 1976). The termini (telomeres) of the minichromosome are somewhat variable in length and are composed of six-nucleotide repeats identical in sequence to the chromosomal telomeres in *Physarum* and to the human telomere repeats generated by the telomerase (Forney et al., 1987).

The identical rDNA transcription units produce a 45S polycistronic rDNA primary transcript, which is processed into the 19S rRNA, 5.8S rRNA, and 26S rRNA. The transcription unit is arranged: 5′, TIS (transcription initiation site), ETS (external transcribed spacer), 19S rRNA, ITS1 (internal transcribed spacer 1), 5.8S rRNA, ITS2 (internal transcribed spacer 2), 26S rRNA, 3′ (Hardman, 1985). Group I introns have been found in the 19S and 26S rRNAs (Ferris, 1985; Ferris and Vogt, 1982; Gubler et al., 1979; Nandipati et al., 2012). These "self-splicing" group I introns have been shown to be mobile elements that readily "home" to rDNA sequences (Johansen et al., 1992; Muscarella and Vogt, 1989; Muscarella et al., 1990; Nandipati et al., 2012). Due to this mobility, there is variation in the presence and type of group I intron found in rDNAs of *Physarum* and other myxomycetes. This variation has been exploited to develop robust phylogenies of myxomycetes (Chapter 3 and Nandipati et al., 2012).

CHROMOSOMES AND CHROMOSOMAL DNA

The chromosomal DNA consists of about 270 Mb of DNA in the form of about 40 chromosomes. The genetic complexity of the chromosomal DNA is significantly smaller (about 200 Mb) due to the large amount of moderately and highly repetitive DNA sequences present. This has made sequence analysis and assembly unusually difficult. Abundant highly repetitive DNA sequences, which are pyrimidine-rich, contain mononucleotide and dinucleotide repeats, are organized in long tandem clusters, and are located in intergenic regions at numerous sites and within introns. The large amount of these highly repetitive sequences has prevented complete assembly of chromosome sequences. As some of these highly repetitive sequences are located in introns, even the assembly of genes can be problematic. Single-copy sequences constitute about 190 Mb (about 70% of the chromosomal DNA). These single-copy sequences have been sequenced and assembled into 69,687 scaffolds with an N50 scaffold length of 54,474 kb (Schaap et al., 2016; 10.0 draft assembly). Contaminating sequences and contigs shorter than 200 bp have been removed from the sequences submitted to GenBank. The sequences in GenBank constitute 7707 scaffolds with a median length of about 6500 bp. These scaffolds range in length from about 1,500 to 648,085 bp. The fact that the longest scaffold length (about 0.65 Mb) is only about 10% of the chromosome length bears witness to the large number of sites containing highly repetitive sequences.

Chromosome Number and Telomeres

Based on microscopic studies, the haploid chromosome number has been variously reported as 40–43, with a diploid number of about 90 and an average molecular weight per chromosome of 4×10^9 Da (about 6 Mb) (Mohberg, 1977), although there have been reports that the chromosome number varies (Adler and Holt, 1974; McCullough et al., 1973; Mohberg, 1977; Mohberg et al., 1973) indicating nuclear instability. Forney et al. (1987) reported that the telomeres on *Physarum* chromosomes are typical six-nucleotide repeats that happen to be identical to the six-nucleotide repeats on human chromosome telomeres (5′-TTAGGG-3′). A search of the "*Physarum* V10 NCBI supercontigs wo bacteria" (http://www.Physarum-blast.ovgu.de) with 7 tandem repeats of 5′-TTAGGG-3′ identifies 78 scaffolds with these sequences, consistent with a haploid chromosome number of about 40. As expected for telomere sequences, 33 of the 78 scaffolds had the tandem repeats at the end of the scaffold sequence. 5′-TTAGGG-3′ repeats were found at the 3′ end of the scaffold sequence in 17 scaffolds; while 5′-CCCTAA-3′ repeats were found at the 5′ end of the scaffold sequence in 13 scaffolds, indicating that the structure of *Physarum* chromosomes is:

$$(5'\text{-CCCTAA-}3')n\text{-----------------}(5'\text{-TTAGGG-}3')n$$

$$(3'\text{-GGGATT-}5')n\text{-----------------}(3'\text{-AATCCC-}5')n$$

One small scaffold contained only repeats of these sequences (32 repeats) and appears to be a telomere fragment. Surprisingly, 45 (58%) scaffolds had repeats of these sequences located internally within the scaffold. It is not clear if these internal sequences are naturally occurring or an artifact of sequence assembly. It is possible that these sequences have been internalized through some mechanism and have become internal repetitive elements. It is also possible that the similarity of these sequences has resulted in incorrect assembly of these sequences. If the internal sequences in these scaffolds are fused telomeres, then the number of unique telomeres would increase to 123 ($45 \times 2 = 90 + 33$), which is consistent with about 62 chromosomes. In 13 scaffolds there are 2 separate tandem repeat sequences, and in 1 scaffold there are 3 separate tandem repeat sequences. One or both of these sequence repeats can be internal, and for one scaffold, the two tandem repeats are located at each terminus. These repeats are separated by as little as three nucleotides and as much as 4484 bp.

The number of repeating units range from 5 (the lowest number of repeats that could be detected by a search) to 39 (although these 39 repeats were internal and may have been a fusion of 2 smaller terminal repeats). The largest number of repeats located at the terminus of the scaffold is 37. The majority of the internal repeats range from 5 to 7 repeat sequences, a number smaller than predicted from the fusion of terminal sequences that average about 20 repeating units.

Gene Content and Chromosomal DNA

Schaap et al. (2016) used gene prediction algorithms and transcriptome data to identify the number of protein coding genes in the 188.75 Mb of unique chromosomal DNA. These authors defined 34,438 genes. These genes were assembled using transcriptome data that identified numerous small spliceosomal introns. The number of introns was predicted to be 676,387 with an average size of 231 bp giving about 20 introns per gene (Schaap et al., 2016).

Analysis of the gene content of *Physarum* chromosomes reveals a remarkable sophistication (Schaap et al., 2016). A number of genes coding for sensory and regulatory proteins characterized in animals and yeast are also found in *Physarum*. Genes coding for proteins, such as tyrosine kinases, and cAMP and cGMP proteins that are involved in regulatory and developmental processes in animals have obvious analogs in *Physarum.*

The Mitochondrial RNA Polymerase Gene, an Example of Protein Coding Genes on Nuclear DNA

The mitochondrial RNA polymerase gene (rpoM) produces a single subunit protein of 998 amino acids (Miller et al., 2006). The gene is located in NCBI scaffold 1,548 (20,895 nucleotides). Initiation of transcription occurs 21 nucleotides downstream of a putative TATA box (AAATAAAA) and 138 nucleotides upstream of the AUG translational initiation codon (after removal of a 1520 nucleotide intron). The theoretical primary transcript would have a length of over 11,500 nucleotides. The mature, processed transcript from cDNA analysis is 3162 nucleotides (27% of the theoretical primary transcript) with a 138 nucleotide 5′ untranslated sequence and a 30 nucleotide 3′ untranslated sequence flanking a protein coding region of 2994 nucleotides (998 codons). A consensus sequence cleavage site for polyA addition (AATAAA) is located 14 nucleotides downstream of the TAA (UAA in RNA) termination codon and 10–14 nucleotides upstream of the site of polyA addition. (The actual site of polyA addition is ambiguous, since it occurs before or within a string of four As in the coding sequence.) A comparison of the amino acid sequence inferred from the codons of the open reading frame using the classical genetic code produces 12 consensus sequences highly conserved in all mitochondrial RNA polymerases (mtRNAP 1a, 1b, 2, 3a, 3b, 4, 5, 6, 7, 8, 9, 10; Miller et al., 2006). This indicates that *Physarum* has a standard genetic code. Interrupting the protein coding sequence are 15 introns that appear to be spliceosomal introns. These 15 introns define 16 exons. The size of the exons ranges from 90 nucleotides (E10) to 693 nucleotides (E4). Thirteen of the 16 exons are between 90 and 180 nucleotides. The remaining 3 exons (E4, E12, E13) are significantly larger with 693, 422, and 306 nucleotides, respectively.

The 15 introns range in size from 51 nucleotides (I14) to over 5600 nucleotides (I8). The introns can be put into three size groups. These are (1) 50–200 nucleotides (8 introns), (2) 300–600 nucleotides (4 introns), and (3) 800–5600 nucleotides (3 introns, I1, I8, and I12). The introns appear to be spliceosomal, since all of the introns have spliceosomal-like consensus sequences located at the intron boundaries. These consensus sequences are not identical to the classic GT-AG spliceosomal intron consensus sequences, but are quite similar. The 5′ GT and the 3′ AG seem to be absolutely conserved. The consensus sequence for the 15 introns is:

$$5'\text{-}\mathrm{R\,g|\underline{GT}A\,t}\text{----------}\mathrm{a\,C\,\underline{AG}|R}\text{-}3'$$

where | is the intron/exon boundary and R is a purine (A/G). The lowercase g, a, and t indicate G, A, and T nucleotides that are less conserved in the consensus sequence. This consensus sequence is similar to the consensus determined by Glockner et al. (2008) for 30 splicing donor/acceptor sites in *Physarum.*

Beyond the boundary consensus sequences, the interior of the introns have highly repetitive simple sequences. These sequences vary among the introns present. They include very CT-rich sequences (I1, I2, I3, I6, I7), very AT-rich sequences (I8, I9), very GA-rich sequences (I3), GT-rich sequences (I12), long stretches of C nucleotides (I5, I8, I13), or long stretches of T nucleotides (I4, I9, I12). These sequences are arranged as homopolymers, such as C_{20} (I13) or T_{15} (I4), alternating homopolymers, such as $(GT)_{112}$ (I12) or more complex sequences, for example, $[T_{2\text{–}6}\,(CT)_{2\text{–}4}\,C_{2\text{–}9}]_6$ followed by $[A(UC)_{1\text{–}7}]_{16}$ in intron 1.

As might be expected, these highly repetitive simple sequences are not stable. While the exons are relatively conserved and the sites of introns are conserved, the content of the introns beyond the consensus sequences can vary dramatically in sequence and length between different genera of myxomycetes.

The sequence of the mtRNAP gene of *Fuligo septica* in the region corresponding to Introns I13, I14, and I15 in *P. polycephalum* has been determined. Comparison of these sequences show that the exon sequences are quite conserved, but that the intron sequences vary dramatically. Intron I13

(121 nucleotides in *Physarum*) is absent in *Fuligo*, and introns I14 (51 nucleotides in *Physarum*) and I15 (163 nucleotides in *Physarum*), while at the same relative positions in the gene, have different length and completely different sequences. I14 in *Physarum* is 51 nucleotides long and contains a long stretch (20 nucleotides) of C homopolymer, while the *Fuligo* intron at the analogous position is 106 nucleotides in length and is CT-rich with multiple tandem repeats of CT and CTT. I15 in *Physarum* is at least 163 nucleotides long and contains a long stretch (greater than 38 nucleotides) of an A homopolymer, while the *Fuligo* intron at the analogous position is 223 nucleotides in length and contains several long T homopolymer stretches.

SMALL, NONCODING RNAs

In addition to protein coding genes, numerous small noncoding RNAs have been identified in the *Physarum* nuclear genome. These include tRNAs, and 5S rRNAs, spliceosomal RNAs (U1, U2, U4, U5, U6), sno RNAs, and lnc RNAs. Hall and Braun (1977) predicted that about 270 5S rRNA genes were present on *Physarum* chromosomes, and thus far, 15 full-length genes for 5S rRNA have been identified on 15 separate scaffolds (Schaap et al., 2016).

tRNA genes were identified at 319 locations on 308 scaffolds. Eleven pairs of the putative tRNA genes were located on the same scaffold and so are probably linked. Spacing between the linked tRNA genes varies between 17 and 5375 nucleotides. Ten of the pairs are separated by less than 500 bp. Two of the linked pairs have the same anticodon (Leu both with IAG and Gly both with GCC) and the third pair is presumably of the same isotype (Tyr) but has different anticodons (IUA and GUA). All of the other linked pairs are different isotypes.

tRNA Gene Structure

Of the 319 tRNA genes identified, 295 have standard structure and have been assigned an isotype based on their anticodon. Twenty-four of the putative tRNA genes lack the potential for standard tRNA structure and have been tentatively designated as tRNA pseudogenes. Six tRNA genes appear to code for tRNAs with an anticodon 5′-UCA-3′, which recognizes the 5′-UGA-3′ standard termination codon and have been tentatively assigned as selenocysteine tRNAs. Twenty-six of the tRNA genes have potential introns in the anticodon loop of the tRNA. Each potential intron is located two nucleotides 3′ of the anticodon. tRNA intron sizes ranged from 10 to 30 nucleotides, except for the potential intron in one tRNA (Arg UCU) which is 205 nucleotides. The 26 potential introns are located in 9 different isoacceptors (Ile, Gly, Tyr, Gln, Arg, Ala, Lys, Leu, and Sec) with 11 different anticodon types. Multiple tRNAs (2–7) from 6 of the anticodon types had introns, but in no case did every tRNA within an anticodon type have introns. tRNAs with introns within an anticodon type might have the same intron or different introns. For example, seven tRNA isoacceptors (Tyr GUA) each had an intron of the same length (10 nucleotides) at the same position, 2 nucleotides 3′ of the anticodon. Four of the introns had the same sequence and two others were close variants of this sequence with one or two nucleotide differences. The seventh 10-nucleotide intron sequence was not related to the other 6, and 2 tRNAs with this anticodon did not have an intron. Of the 9 tRNAs (Ala) with 5′-UGC-3′ as the anticodon, 4 had introns of 26–30 nucleotides and 5 had no intron. The four introns were similar in length and identical near the 5′ and 3′ boundary sequences, but variable in the central portion of the intron.

tRNA Anticodons, Codon/Anticodon Interactions, and the Genetic Code

Of the 64 possible anticodon sequences, 51 have been observed. Analysis of these anticodons can provide information about the genetic code, and the nature of codon–anticodon interactions (wobble rules) in *Physarum.* Assuming that all A nucleotides at the 5′ position of the anticodon are deaminated to Inosine, the 5′ position of the anticodons should be I, G, U, or C. The 13 anticodons that are not observed are: IAA (Phe), GAU (Ile), ICU (Ser), GGC (Ala), IUA (Tyr), ICA (Cys), IUG (His), IUC (Asp), IUU (Asn), ICC (Gly), GCG (Arg), UUA (term), and CUA (term). The observation of a tRNA with a UAU anticodon that could potentially interact with both the AUG (Met) codon and the AUA (Ile) codon through wobble base pairing interactions as it can in *E. coli*, argues that a U at the 5′ position of an anticodon will not wobble base pair with a G at the 3′ position in codons. The absence of anticodons with an I at the 5′ position in two-codon families (IAA, IUA, ICA, IUG, IUU, IUC, ICU) argues that I can base pair with U, C, or A at the 3′ position in codons similar to *E. coli.* The presence of tRNAs in two-codon families with only G at the 5′ position of the anticodon (GAA, GUA, GCA, GUG, GUU, GCU, GUC) indicates that G at the 5′ position in anticodons is able to base pair with U or C at the 3′ position in codons. A C at the 5′ position of anticodons must be specific to a G at the 3′ position of the codon, since AUG (Met) and UGG (Trp) codons are recognized specifically by CAU and CCA. The absence of anticodons (UUA, CUA) that would recognize termination codons (UAA, UAG) is consistent with the observation that *Physarum* has a standard, classical genetic code, since there are not normally tRNAs with anticodons that will recognize termination codons. The one exception to this is the presence of six tRNA genes that will produce tRNAs with UCA in the anticodon. These tRNAs would be expected to compete with termination factors that recognize the UGA termination codon. It is likely that these tRNA are charged with selenocysteine, rather than cysteine or tryptophan, and that they are minor tRNAs that compete weakly with termination factors for UGA. These considerations argue that the codon/anticodon (wobble) rules for *Physarum* are:

5′ position of anticodon	3′ position of codon
I	U, C, A
G	U, C
U	A
C	G

These codon/anticodon (wobble) rules are different from either *E. coli* (Crick, 1966) or *S. cerevisiae* (Guthrie and Abelson, 1982).

MITOCHONDRIAL DNA IN MYXOMYCETES

The mitochondrial DNA (mtDNA) of the myxomycetes is among the most complex found in any eukaryote in terms of its potential gene complement and the expression of the mitochondrial genes through a novel type of RNA editing that is both unique to the myxomycetes and seems to be a common feature of all myxomycetes studied. The prototype mtDNA of the myxomycetes is the mtDNA of *P. polycephalum.* It has a genetic complexity of about 60 kbp (although this varies among *P. polycephalum* strains from 56 kb to over 70 kb) and has a circular restriction enzyme map (Jones et al., 1990). Electron microscopy has been used to detect 60 kbp circular DNA in *Physarum* mtDNA isolates indicating that the mtDNA in *Physarum* is physically circular rather than arranged as long tandem repeats (Bohnert, 1977). Variations of this single circular mtDNA may be present in other myxomycetes (variation in the structure of the mtDNA from different strains of *Physarum*). There are about 400 copies of the mtDNA per haploid genome making the mtDNA a moderately repetitive DNA element in the *Physarum* genome, constituting about 8% of the total DNA content of *Physarum.* This DNA along with nuclear DNA sequences (especially the rDNA sequences that are also moderately repetitive) have been useful in developing myxomycete phylogenies (Fiore-Dunno et al., 2005; Krishnan et al., 2007) and establishing the relationship of the myxomycetes among the other eukaryotic organisms (Miller, 2014; see later).

The Relationship of Myxomycetes Relative to Other Clades in the Amoebozoa Based on Mitochondrial DNA Characteristics

Amoebozoa is a supergroup of eukaryotes that unifies the slime molds (including the myxomycetes), archamoebae, and lobose amoeba (Cavalier-Smith, 1998; Cavalier-Smith et al., 2015). The monophylyl of the Amoebozoa is supported by molecular phylogenies based on small subunit rRNA genes (Bolivar et al., 2001; Fahrni et al., 2003; Milyutina et al., 2001) and combined analyses of protein sequences (Arisue et al., 2002; Baldauf et al., 2000; Bapteste et al., 2002; Forget et al., 2002). The Amoebozoa supergroup branches with the common ancestor of animals and fungi (Opisthokonts), which is thought to constitute a monophyletic group (Unikonts). These molecular phylogenies indicate that the Amoeboezoa is composed of two related, but anciently divergent groups, the Lobosea and Conosea. The Conosea include archamoebae (genera *Entamoeba* and *Mastigamoeba*), the acellular (plasmodial) slime molds (Myxomyecetes), the cellular slime molds (Dictyostellids), Group 1 protostelids, Cavostellids, Filamoebae, and Schizoplasmodiids. The Lobosea include Tubulinea, Thecamoeba, Acanthamoebids, Vannellids, and Dactylopodida (Katz, 2012; Shadwick et al., 2009).

Mitochondrial DNAs (mtDNAs) have been sequenced from a number of representative organisms from the Amoebozoa (Gray et al., 2004; Miller, 2014). In the Conosea, the archamoeba are amitochondriate or contain relict mitochondria (Aguilera et al., 2008; Gill et al., 2007; Roger and Simpson, 2008), but the complete sequence of the mtDNA of the prototype organism of the myxomycetes, *P. polycephalum*, has been reported (Takano et al., 2001) and the mtDNA of the closely related myxomycete, *Didymium iridis*, has been partially sequenced (Hendrickson and Silliker, 2010a,b; Kopp, 2012; Traphagen et al., 2010). mtDNAs from two dictyostelid genera have been completely sequenced. These are *Dictyostelium discoideum* (Ogawa et al., 2000) and *Polysphondylium pallidum* (Burger et al., unpublished data). Portions of the mtDNA from several other species of *Dictyostelium* (*D. citrinum*, *D. muceroides*, and *D. fasciculatum*) have also been sequenced (Bullerwell et al., 2010). Two representative mtDNAs have been sequenced from the Lobosea, *Acanthamoeba castellani*, the prototype of the acanthamoebids (Burger et al., 1995), and *Hartmanella veriformis*, a representative of the Tubulinea (Burger et al., unpublished data). The mtDNAs from *Physarum*, *Dictyostelium*, and *Acanthamoeba* have a number of features in common but also display some fundamental differences. Based on nuclear SSU rRNA gene sequence comparisons, a schematic molecular phylogeny of these three Amoebozoans is shown in Fig. 4.1A. Surprisingly, based on mtDNA features discussed later, a different phylogenetic topology can be inferred (Fig. 4.1B). Using these mtDNA features as evolutionary criteria, myxomycetes appear to be related to, but anciently divergent from, the other members of the Amoebozoa, while the mtDNA of *Dictyostelium* and *Acanthamoeba* have more similarities with each other than either of them have with the mtDNA of *P. polycephalum*.

Comparison of the Mitochondrial DNA Genomes of Three Representative Amoebozoans

The well-characterized mtDNA of three prototype Amoebozoans—*A. castellani*, *P. polycephalum* and *D. discoideum*—are compared later using a number of criteria. Similarities and differences are summarized in Table 4.1.

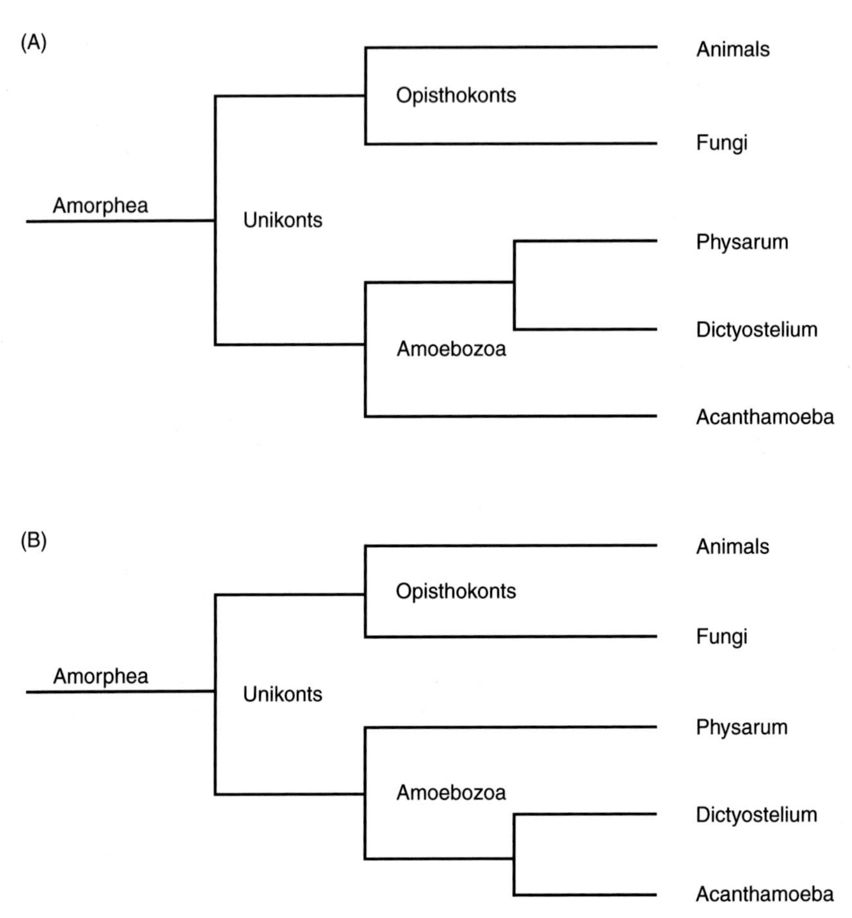

FIGURE 4.1 Schematic phylogenetic trees of Unikonts showing possible relationships of the myxomycetes (*Physarum*) to other Amoebozoa (*Dictyostelium* and *Acanthamoeba*) and Opisthokonts (animals and fungi). Amphorea, the common ancestor of the Unikonts, is shown at left in both panels: (A) tree topology based on nuclear DNA sequences, such as Baldauf et al. (2000), and (B) tree topology based on mtDNA features.

mtDNA size and gene content: All three prototype mtDNAs are double-stranded, circular DNAs between 40 and 65 kbp. *A. castellani* is 41,591 bp with 57 genes, while *D. discodeum* is 55,564 bp with 59 genes. *P. polycephalum* mtDNAs range from 60 to 63 kbp and contain at least 45 identified genes and 27 significant (greater than 100 codons), but unassigned, open reading frames (Fig. 4.2). Thirty-five genes are common to all three mtDNAs. Each mtDNA has genes involved in oxidative phosphorylation and electron transport, and mitochondrial protein synthesis (rRNAs, tRNAs, and ribosomal proteins), and unassigned open reading frames (Tables 4.2 and 4.3). Four tRNA genes, 3

TABLE 4.1 Comparison of Three Prototype Amoebozoan mtDNAs

	Acanthamoeba castellani	*Dictyostelium discoideum*	*Physarum polycephalum*
mtDNA size (bp)	41,591	55,564	61,019
Identified protein coding genes	33	34	37
Unassigned ORFs	5	5	27
tRNA genes	16	18	5
rRNA genes	3	3	3
Fused cox1/2 gene	+	+	–
All genes in same transcriptional orientation	+	+	–
Group 1 introns	+	+	–
Inertional RNA editing	–	–	+
5′-Replacement	+	+	+

rRNA genes, 14 ribosomal protein genes, and 14 genes necessary for oxidative phosphorylation and electron transport are in common among the 3 mtDNAs. Fifty genes are shared in common between *A. castellani* and *D. discoideum*, the unique genes in each mtDNA being primarily tRNA genes and unassigned ORFs.

Genome organization and gene order: While both *A. castellani* and *D. discoideum* have all genes and ORFs in the same transcriptional orientation, the genes of *P. polycephalum* are transcribed from both strands (52 genes and ORFs are oriented clockwise, and 20 genes and ORFs are oriented counter-clockwise). *A. castellani* and *D. discoideum* also have several blocks of genes with a common gene order. The gene sequences rpl16-rpl14-rpl5-rps14-rps8-rpl6-rps13-rps11, rps12-rps7-rpl2-rps19, and nad9-nad7-atp6 are conserved in both mtDNAs. The gene sequences rps12-rps7-rpl2-rps19 and rpl14-rpl5-rps14-rps8-rpl6-rps13 are also conserved in *P. polycephalum* and are therefore conserved in all three amoebozoan mtDNAs. Both *A. castellani* and *D. discoideum* have cox1–cox2 fused in a single open reading. The presence of this unusual feature in both organisms indicates that this feature arose in a common ancestor. This fusion is absent in *P. polycephalum.* Another common feature of *A. castellani* and *D. discoideum* that is absent in *P. polycephalum* is the presence of group 1 introns in the large subunit ribosomal RNA gene. *A. castellani* has three group 1 introns in the LSU rRNA gene, while *D. discoideum* has one group 1 intron in the LSU rRNA gene and four group 1 introns in the cox1/2

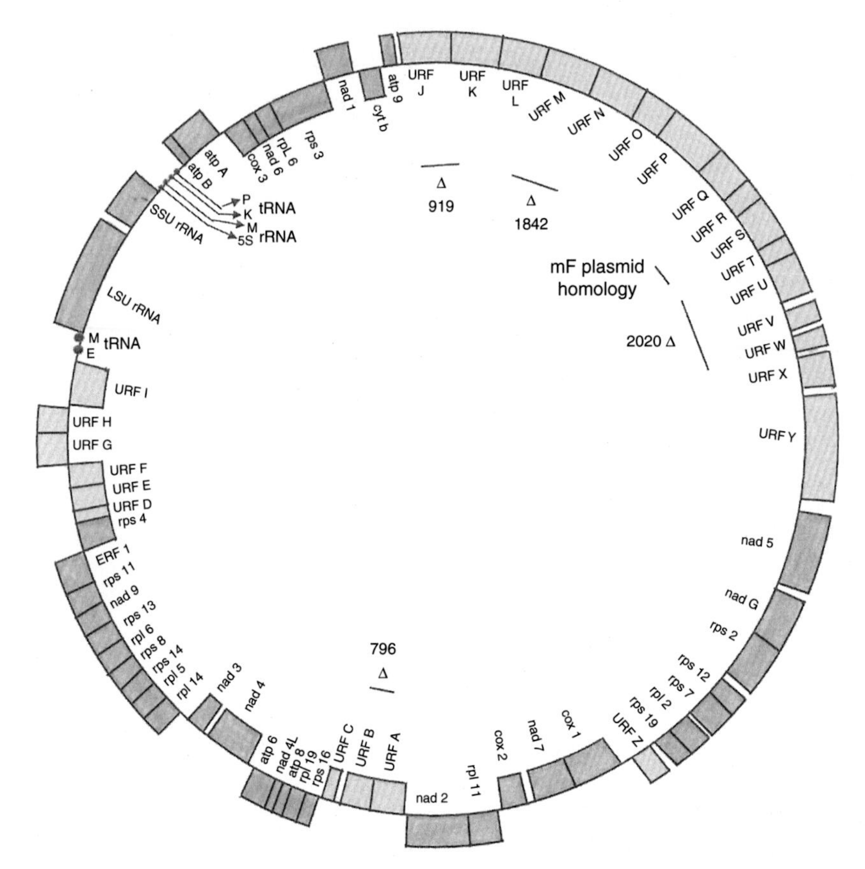

FIGURE 4.2 Composite map of the mtDNA from several strains of *Physarum polycephalum*. The mtDNA has 72 genes indicated within the circle. Genes designated by URF (shown in *yellow*) are unassigned reading frames. Each URF is designated by a letter after the convention of the mtDNA map of the M3 strain of *Physarum*. Classic mitochondrial genes are shown in *pink*. Genes shown as boxes on the outside of the circle are transcribed in a clockwise direction; genes shown as boxes on the inside of the circle are transcribed in a counterclockwise direction. The composite map is 62,862 bp, identical to the Takano et al. (2001) sequence. Four sites of deletion in mtDNA of other strains of *Physarum* are shown within the circle. The sizes in base pairs of the deletions and their extent relative to URFs are indicated. The bar beside URF R indicates the location of mF plasmid homology.

fusion gene. Group I introns have not been detected in the mtDNA of *P. polycephalum* or other myxomycete mtDNA.

Mitochondrial Gene Expression in the Amoebozoa

mtDNA transcription, promoters: Mitochondrial DNAs throughout eukaryotes are primarily transcribed by a highly conserved and dedicated mitochondrial RNA polymerase (Le et al., 2009; Miller et al., 2006). However, promoter

TABLE 4.2 Gene Content Comparison of Three Prototype Amoebozoan mtDNAs

	Acanthamoeba castellani	*Dictyostelium discoideum*	*Physarum polycephalum*
nad1	+	+	+
nad2	+	+	+
nad3	+	+	+
nad4	+	+	+
nad4L	+	+	+
nad5	+	+	+
nad6	+	+	+
nad7	+	+	+
nad9	+	+	+
nad11(G)	+	+	+
cob	+	+	+
cox1	+	+	+
cox2	+	+	+
cox3	+	+	+
atp1	+	+	+
atp6	+	+	+
atp8	–	+	+
atp9	+	+	+
rps2	+	+	+
rps3	+	+	+
rps4	+	+	+
rps7	+	+	+
rps8	+	+	+
rps11	+	+	+
rps12	+	+	+
rps13	+	+	+
rps14	+	+	+
rps16	–	–	+
rps19	+	+	+
rpl2	+	+	+
rpl5	+	+	+
rpl6	+	+	+
rpl11	+	+	+
rpl14	+	+	+
rpl16	+	+	+
rpl19	–	–	+

TABLE 4.3 RNA Gene Content Comparison of Three Prototype Amoebozoan mtDNAs

	Acanthamoeba castellani	*Dictyostelium discoideum*	*Physarum polycephalum*
tRNA-A	+	+	–
tRNA-C	–	+	–
tRNA-D	+	–	–
tRNA-E	+	+	+
tRNA-F	+	+	–
tRNA-G	–	–	–
tRNA-H	+	+	–
tRNA-I1	+	+	–
tRNA-I2	+	+	–
tRNA-I3	–	+	–
tRNA-K	+	+	+
tRNA-L1	+	+	–
tRNA-L2	+	+	–
tRNA-M1	+	+	+
tRNA-M2	–	–	+
tRNA-N	–	+	–
tRNA-P	+	+	+
tRNA-Q	+	+	–
tRNA-R	–	+	–
tRNA-S	–	–	–
tRNA-T	–	–	–
tRNA-V	–	–	–
tRNA-W	+	+	–
tRNA-X	+	–	–
tRNA-Y	+	+	–
LSU rRNA	+	+	+
SSU rRNA	+	+	+
5S rRNA	+	+	+

consensus sequences and transcriptional patterns of mtDNAs vary significantly. The gene arrangement and asymmetric location of the genes on one strand of the mtDNA of *A. castellani* and *D. discoideum* indicate the possibility of polycistronic transcripts from a single promoter or small number of promoters. Barth et al. (2001) detected eight major polycistronic transcripts from the mtDNA of *D. discoideum*, which Le et al. (2009) showed to be derived from rapid cotranscriptional processing of a single primary transcript initiated from a single promoter. The single transcription initiation site, as detected by 5′ capping analysis, is located upstream of the LSU rRNA gene (Le et al., 2009).

The gene arrangement and distribution of genes on both strands of the mtDNA of *P. polycephalum* implies the need for multiple sites of transcription initiation in *Physarum.* Bundschuh et al. (2011) characterized the mitochondrial transcriptome of the plasmodial form of *P. polycephalum.* They detect the potential for at least eight polycistronic transcripts and show that many of the adjacent genes are indeed transcribed as a single polycistronic RNA. Analysis of the regions upstream of the potential initiation sites has not revealed a possible promoter consensus sequence. Interestingly, 25 of the 26 significant unassigned open reading frames do not appear to be transcribed in the plasmodial (diploid) form of *P. polycephalum.* Consistent with this lack of transcription is the observation that the mtDNA of some strains of *P. polycephalum* have deletions in these reading frames without apparent phenotypic effect (Fig. 4.2). Further analysis of the mtDNA of related myxomycetes should reveal whether these ORFs are transcribed under other developmental or environmental conditions (e.g., in the amoeboflagellate form of myxomycetes) or are an evolutionary artifact resulting from the capture of DNA through recombination with a non-mtDNA, such as a mitochondrial plasmid as described by Nakagawa et al. (1998) and Nomura et al. (2005). The mF mitochondrial plasmid has been shown to enhance recombination of mtDNA causing mtDNA gene rearrangement in some strains of *Physarum.*

RNA self-splicing by Group I introns: A necessary step of gene expression in mtDNA of *A. castellani* and *D. discoideum* is the self-splicing of group I introns to produce the mature LSU rRNA in both organisms and the cox1/2 transcript in *D. discoideum.* The three introns of the LSU rRNA gene of *A. castellani* each have an open reading frame (ORF), as do introns 2 (2 ORFs), 3 and 4 of the cox1/2 gene in *D. discoideum.* These ORFs presumably code for maturases or homing endonucleases necessary for efficient splicing and/or intron mobility. These introns probably have the potential for mobility based on the presence of an open reading frame internal to intron 2 of the cox1/2 gene, which codes for a site-specific DNA homing endonuclease (Ogawa et al., 2000).

RNA editing in mitochondria of the Amoebozoa: Several types of RNA editing are necessary to produce functional, mature rRNAs, tRNAs, and mRNAs in mitochondria of the Amoebozoa. One type of RNA editing present in all three prototype organisms is a 5′-replacement tRNA editing activity, which repairs tRNAs that overlap other genes or have acceptor end mismatches. First characterized in mitochondrial tRNAs of *A. castellani* (Lonergan and Gray, 1993) 5′

replacement editing of mitochondrial tRNAs has been inferred in *D. discoideum* based on tRNA gene sequence and demonstration of 5′-tRNA editing activity in vitro (Abad et al., 2011), and 5′-edited mitochondrial tRNAs have been characterized in *P. palladium* (Schindel and Gray, unpublished data) and recently in *P. polycephalum* (Gott et al., 2010). This 5′-replacement tRNA editing is based on activity that can add nucleotides to the 5′ end of RNAs in a template-dependent manner (3′–5′ polymerase). In *D. discoideum* a His-tRNA guanylyl-transferase (Thg-1)-like protein has been implicated in the 5′ editing of tRNAs (Abad et al., 2011). While 13 out of 16 tRNAs encoded in the *A. castellani* genome are edited (Lonergan and Gray, 1993; Price and Gray, 1999), only 2 of the 5 tRNAs encoded by the *P. polycephalum* mtDNA are edited by 5′ replacement editing. In each case, a single nucleotide is replaced by a G to create a G–C base pair at the 1:72 position of the acceptor stem of the tRNA (Gott et al., 2010).

C to U RNA editing has also been identified at a limited number of sites in mitochondrial RNAs of the Amoebozoa [four sites in *P. polycephalum* (Gott et al., 1993) and one site in the SSU rRNA gene of *D. discoideum* mtDNA (Barth et al., 1999)]. Horton and Landweber (2000) have shown that C to U editing occurs in the cox1 mRNA in other myxomycetes.

The most extensive RNA editing is the insertional RNA editing found exclusively in mitochondria of the myxomycetes. This type of RNA editing is a signature characteristic of the myxomycetes. The absence of this type of RNA editing in the other groups of the amoebozoa is further evidence of the ancient divergence of the Myxomycetes from the other organisms in the Amoebozoa. This type of RNA editing has been demonstrated in all five orders in the myxomycetes (Physarales, Stemonitales, Echinosteliales, Tricheales, and Liceales) and in more than 60 species of myxomycetes (Antes et al., 1998; Hendrickson and Silliker, 2010a,b; Horton and Landweber, 2000; Krishnan et al., 2007; Traphagen et al., 2010). This broad, but exclusive, distribution of insertional RNA editing makes it a candidate for a defining characteristic of the myxomycetes. RNAs produced from the mtDNA of *P. polycephalam*, have insertions at 1324 sites relative to the mtDNA sequence producing the RNAs (Bundschuh et al., 2011). These inserted nucleotides constitute about 4% of mRNAs and 2% of tRNAs and rRNAs. Insertions are primarily single cytidines (1255 sites) or single uridines (43 sites) or a subset of the possible dinucleotides (AA 4 sites, UU 2 sites, UG/GU 4 sites, UC/CU 9 sites, UA 2 sites, GC/CG 2 sites; 23 sites total). At three sites there is a single purine insertion (a G at two sites and an A at one site). Of the 47 transcribed genes and ORFs in the mtDNA of the plasmodial form of *P. polycephalum*, 45 require insertional RNA editing to produce functional tRNAs, rRNAs, and mRNAs (Bundschuh et al., 2011). These nucleotide insertions are made cotranscriptionally as the RNA is produced by adding nontemplated nucleotides to the 3′ end of the nascent RNA (Cheng et al., 2001; Miller and Miller, 2008). RNA editing sites are distributed relatively uniformly throughout edited genes. The distribution of sites is consistent with the random creation of sites at any location, constrained only by a nine-nucleotide limitation in the proximity of adjacent sites (Krishnan et al., 2007; Rhee et al., 2009). The observed pattern has an average spacing of about 25 nucleotides between editing sites in mRNA, consistent with this model. This dynamic model of editing site fixation is consistent with the variation of editing site location in analogous genes of related myxomycetes (Antes et al., 1998; Hendrickson and Silliker, 2010a,b; Horton and Landweber, 2000; Krishnan et al., 2007; Traphagen et al., 2010). The mechanism and evolution of this unique type of RNA editing will be discussed in more detail later.

Evolutionary Implications of mtDNA Structure and Features

A strong case can be made for the common ancestry of the organisms in the Amoebozoa based on their mtDNA similarities. They have a similar size and gene content, each with between 16 and 18 genes for oxidative phosphorylation and electron transport (14 genes in common), 10 and 11 genes coding for proteins of the small subunit of the mitoribosome (10 genes in common), and 6 genes coding for proteins of the large subunit of the mitoribosome (5 genes in common), and 3 rRNA genes for the LSU rRNA, the SSU rRNA, and a 5S rRNA (Bullerwell et al., 2010). The surprising observation in light of phylogenies based on the nuclear SSU rRNA genes or other nonmitochondrial criteria, is the number of similarities between the mtDNA features of *A. castellani* and *D. discoideum* and the number of differences in mtDNA features between *D. discoideum*, the prototype of the dictyostelids and *P. polycephalum*, the prototype of the myxomycetes. To determine whether these differences are due to a relatively ancient divergence of the dictyostelids and myxomycetes with the development of some convergent features and morphology, or due to a relatively rapid evolution in the mtDNA of the myxomycetes, will require characterization of the mtDNAs of additional representatives of the Amoebozoa, especially the group 1 protostellids and the Cavostellids, which are currently thought to be monophyletic with the dictyostelids and myxomycetes (Shadwick et al., 2009).

RELATIONSHIPS WITHIN THE MYXOMYCETES BASED ON mtDNA STRUCTURE

Structure, gene content, and organization of the Physarum mtDNA: The mtDNA of *P. polycephalum* has a circular restriction map and a genetic complexity of about 60,000 bp (although the size varies in several strains of *Physarum*; see section further on). The mtDNA of *Physarum* has been completely sequenced in several strains (Miller, unpublished data; Takano et al., 2001) and the transcriptome has been studied (Bundschuh et al., 2011). These studies revealed 72 potential genes that can be grouped in several categories (Table 4.4). There are 8 genes coding for RNAs: a large and a small subunit rRNA, a 5S rRNA, and 5 tRNAs (Table 4.4). The remaining 64 genes are potentially protein coding and can be grouped in 2 general categories. The classical genes of mtDNAs coding for protein (37 genes) and potentially protein-coding genes that are not classical

TABLE 4.4 Gene Content of *Physarum polycephalum* mtDNA

Gene	Insertional RNA editing	Transcribed direction	Genomic location	Gene length	Protein length for ORFs
tRNA-E	+	+R	48,163–48,231	69	
tRNA-M1	+	+R	48,237–48,305	69	
tRNA-M2	–	+C	53,273–53,344	72	
tRNA-K	+	+C	53,364–53,435	72	
tRNA-P	+	+C	53,454–53,524	71	
LSU rRNA	+	+C	48,432–51,149	2,718	
SSU rRNA	+	+C	51,353–53,166	1,814	
5S rRNA	–	+C	53,178–53,273	96	
ERF 1 (php 24)	+	+C	42,721–43,372	652	
ORF A	–	–R	32,169–32,882	714	238
ORF B	–	–R	32,885–34,117	1,233	411
ORF C	–	–R	34,396–34,617	222	74
ORF D	–	–R	44,413–44,679	267	89
ORF E	–	–R	44,684–45,295	612	204
ORF F	–	–R	45,432–45,872	441	147
ORF G	–	–C	45,774–46,411	738	246
ORF H	–	–C	46,572–47,168	597	199
ORF I	–	–R	47,693–48,083	390	130
ORF J	–	–C	61,650–196	1,412	471
ORF K	–	–C	196–1,260	1,065	355
ORF L	–	–C	1,269–2,354	1,086	362
ORF M	–	–C	2,354–4,570	1,217	739
ORF N	–	–C	4,557–5,756	1,200	400
ORF O	–	–C	5,771–6,349	579	193
ORF P	–	–C	6,400–7,908	1,509	503
ORF Q	–	–C	7,908–9,074	1,167	389
ORF R	–	–C	9,082–9,744	663	221
ORF S	–	–C	9,740–11,367	1,629	543
ORF T	–	–C	11,517–11,738	222	74
ORF U	–	–C	11,731–12,585	1,128	376
ORF V	–	–C	13,096–13,431	336	112
ORF W	–	–C	13,543–13,935	393	131
ORF X	–	–C	13,938–14,605	669	223
ORF Y	–	–C	14,639–16,810	2,172	724
ORF Z (php 15)	–	+C	24,416–25,471	1,056	352

TABLE 4.4 Gene Content of *Physarum polycephalum* mtDNA (*cont.*)

Gene	Insertional RNA editing	Transcribed direction	Genomic location	Gene length	Protein length for ORFs
nad1	+	+C	58,893–59,821	929	
nad2	+	+C	30,699–32,105	1,407	
nad3	+	+R	38,435–38,808	374	
nad4	+	+R	36,933–38,315	1,383	
nad4L	+	+C	35,788–36,062	275	
nad5	+	+C	17,259–19,152	1,894	
nad6	+	+R	56,280–56,758	479	
nad7	+	+R	27,535–28,600	1,065	
nad9	+	+C	41,520–41,994	475	
nad11(G)	+	+C	19,300–20,316	1,017	
cob (cyt b)	+	+R	59,903–61,038	1,136	
cox1	+	+R	25,816–27,534	1,719	
cox2	+	+R	28,983–29,666	864	
cox3	+	+R	55,517–56,278	762	
atp1(A)	+	+C	53,845–55,380	536	
atp B ?	–	+C	53,565–53,858	293	
atp6	+	+C	36,067–36,774	708	
atp8	+	+C	35,567–35,788	222	
atp9	+	+C	61,225–61,467	243	
rps2	+	+C	20,278–21,633	1,356	
rps3	+	+R	57,431–58,800	1,370	
rps4	+	+R	43,440–44,229	790	
rps7	+	+C	22,246–23,009	764	
rps8	+	+C	40,088–40,509	422	
rps11	+	+C	41,997–42,717	721	
rps12	+	+C	21,746–22,241	496	
rps13	+	+C	40,978–41,517	539	
rps14	+	+C	39,823–40,087	265	
rps16	+	+C	34,704–34,988	284	
rps19	+	+C	23,774–24,139	366	
rpl2	+	+C	23,009–23,777	769	
rpl5	+	+C	39,338–39,822	485	
rpl6	+	+R	40,506–40,972	467	
rpl11	+	+C	29,776–30,652	877	
rpl14	+	+C	38,985–39,338	354	
rpl16	+	+C	56,910–57,434	525	
rpl19	+	+C	34,988–35,540	553	

mtDNA genes (27 genes) (Table 4.4). Of the 37 classical protein-coding genes, all but one requires RNA editing to produce the protein. They produce proteins that are part of the 5 complexes of the electron transport chain and ATP synthesis or they code for protein components of the mitoribosome (11 subunits of the small subunit designated rps and 7 units of the large subunit designate rpl). The mitochondrially encoded subunits of the electron transport chain and ATP synthase (Table 4.4) are: 10 subunits of the NADH dehydrogenase complex (complex I) produced from genes designated *nad*, 1 gene (*cob* or *cyt b*) coding for cytochrome b of the cytochrome bc1 complex (complex III), 3 genes designated with *cox* coding for subunits of cytochrome oxidase (complex IV), and 5 genes designated with *atp* coding for subunits of the ATP Synthase complex (complex V).

The function of the 27 nonclassical genes (Table 4.4) is unclear. While these potential protein-coding genes have significant open reading frames (ORFs), 25 of the 27 ORFs are not transcribed (Bundschuh et al., 2011; Jones et al., 1990). Of the two that are transcribed (ORF Z and ERF1), only ERF1 requires RNA editing.

Both strands of the *Physarum* mtDNA are transcribed. Twenty potential genes (13 classical and 7 nonclassical) are located on one strand, while 52 potential genes (34 classical and 18 nonclassical) are located on the other strand. The genes are clustered in 15 groups that are potentially transcribed from the same strand. These regions could be transcribed from a single promoter and produce polycistronic transcripts. The largest of these groups covers 18,022 bp and is composed of the nonclassical ORFs J-Y. These very significant ORFs are separated by only a few nucleotides and some even overlap with overlapping initiation and termination codons. Although the potential for a polycistronic transcript exists, this region does not appear to be transcribed (Bundschuh et al., 2011; Jones et al., 1990). Several of the potential polycistronic regions display a gene sequence conserved in other members of the Amoebozoa (rps12-rps7-rpl2-rps19 and rpl14-rpl5-rps14-rps8-rpl6-rps13). Bundschuh et al. (2011) provide evidence consistent with a polycistronic transcript including the 23S LSU rRNA, 17S SSU rRNA, 5S rRNA, tRNA Met2, tRNA Lys, tRNA pro, atp1.

Linear mF Mitochondrial Plasmid

In addition to the mtDNA, the mitochondria of some strains of *Physarum* contain an additional genetic element, the linear mF plasmid. Although mitochondrial plasmids have been identified and characterized in fungi and plants, the mF linear plasmid is one of the largest mitochondrial plasmids at 14.5 kb. It has terminal inverted repeats (TIRs) at the ends with proteins bound to its 5′ ends, resembling the replication mechanism of linear DNA viruses. These TIRs flank nine significant open reading frames, all potentially transcribed in the same direction to potentially produce a single polycistronic transcript encoding nine proteins. Transcription is initiated just inside the end of the left TIR, producing three major transcripts, corresponding to ORFs 1 and 2, 1–3, and 1–5, respectively. Two minor (low abundance) transcripts are also seen, one corresponding to ORF 6 and the other to the full-length polycistronic transcript containing all nine ORFs (Takano et al., 1996). The amino acid sequences of the proteins do not match other proteins in GenBank except for the protein produced from ORF 9, which resembles a DNA polymerase (Takano et al., 1994). Significantly, ORF 7 has partial identity to ORF R in mtDNA (Fig. 4.2). These identical regions provide a site for recombination between the mtDNA and the mF plasmid. Takano et al. (1992) have shown that this recombination occurs in some *P. polycephalum* strains producing 77 kb linear mtDNAs with mF plasmid TIRs at the termini. These mtDNAs presumably replicate using the linear plasmid mechanism. This alternate mechanism of replication may have implications for plasmodial senescence (Nakagawa et al., 1998), similar to the fungal senescence caused by mitochondrial plasmids in *Neurospora*.

Whether the mF plasmid conveys a selective advantage to strains of *Physarum* that contain it or is a true molecular parasite (selfish DNA) adapted to autonomous replication and transmission in mitochondria is not known. It apparently does not have any obligate function, since strains lacking the plasmid have no apparent growth or respiration defects. Kawano et al. (1991) have shown that different strains of *Physarum* have one of two phenotypes, mif^+ (mitochondrial fusion) and mif^- (no mitochondrial fusion) and that the mF plasmid correlates with the mif^+ (mitochondrial fusion) phenotype. Mitochondrial fusion allows mtDNA recombination between heteroplasmic zygotes. This can alter the transmission of mtDNA alleles from its typical isogametic, uniparental inheritance.

Variation in the Structure of the mtDNAs From Different Strains of *Physarum*

The mtDNA of myxomycetes evolves rapidly. Different genera have very different structures and gene orders, presumably due to inter- and intramolecular recombination. Even different strains of *Physarum* have different mtDNA sizes with different gene content. These variations have shed light on the necessity of the nonclassical, but significant ORFs in *Physarum* mtDNA. The observation that these ORFs do not code for classical mitochondrial proteins and that they do not seem to be transcribed in plasmodia under the conditions studied, argues that they may be nonfunctional pseudogenes. However, the presence of highly significant open reading frames implies a mechanism of selection to maintain the open reading frame. It is possible that these sequences were recently acquired and have not had time to accumulate mutations that might interrupt the reading frame. To determine if these ORFs are present in other strains of *P. polycephalum* the mtDNA from these strains have been analyzed and partially sequenced.

The mitochondrial DNAs of several strains of *P. polycephalum* have been mapped and partially sequenced (Miller et al., personal communication) to identify regions with significant insertions or deletions relative to the mtDNA sequence of Takano et al. (2001). Currently, five regions have been identified with altered mtDNA structure in five different strains (Fig. 4.2). Interestingly, all of these insertions or deletions fall within nonclassical open reading frames. The M3 strain of *Physarum* has an 1842 bp deletion in ORF M relative to the Takano mtDNA sequence. This deletion appears to have occurred through an intragenic recombination of small direct repeat sequences. A second deletion of 919 bp in ORF J is present in some strains derived from the M3 strain. A *Physarum* strain purchased from Carolina Biological Supply Company contains a 2020 bp deletion of the ORF T, ORF U, ORF V region and also has altered sequences in the ORF P, ORF Q, ORF R region at the site of the mF mitochondrial plasmid homology. Alterations in this same region were also characterized by Nakagawa et al. (1998) for strains auxS and auxI. These strains also had an insertion in ORF B of 796 bp. All of the strains with these deletions were checked to determine if the excised DNA was autonomously replicated or transferred to nuclear DNA. None of the excised sequences could be detected, indicating that these sequences were lost in these strains. Taken together, these observations indicate that the ORFs are optional sequences in these strains, since these strains showed no associated growth phenotypes under the conditions grown.

mtDNA Structure in *Didymium Iridis*

All or most of the mtDNA in *Didymium iridis* has been sequenced by the Silliker laboratory (Kopp, 2012). The sequence complexity is at least 69,050 bp, 7 kb larger than the Takano et al. (2001) mtDNA sequence for *Physarum*. The classical mitochondrial gene content is nearly identical to *Physarum* mtDNA (Hendrickson and Silliker, 2010a,b; Silliker, personal communication; Traphagen et al., 2010; Fig. 4.2). While several contigs have been produced; assembly of these contigs into a single circular genome has been problematic. It is likely that intragenic recombination has created autonomously replicating subgenomic circular mtDNAs, since some of the contigs appear to be circular (Kopp, 2012). This may indicate that origins of replication are positioned at multiple locations allowing recombinant circles to autonomously replicate.

Gene order on the *Didymium* mtDNA contigs seems to be generally different than on the mtDNA of *Physarum* with a few notable exceptions, that is, rps12/rps7/rpl2/rps19; atp8/nad4l/atp6; cox3/nad6; and rpl16/rps3, which are conserved as potential polycistronic transcript regions in both mtDNAs. The categories of genes in *Didymium* also seem similar to those in *Physarum*. There are unassigned ORFs, which do not correspond to classical mitochondrial genes, and the full complement of classical mitochondrial genes, all of which (except nad 3) require RNA editing for their expression.

Mitochondrial Bar Coding and mtDNA Phylogenies of Myxomycetes

Analysis of variation in mtDNA sequences has been useful in generating phylogenies. Because mtDNA is a moderately repetitive DNA relative to single-copy genes in the chromosomes, it is easy to isolate and is particularly good for ancient DNA analysis since some of the multiple copies of the mtDNA are able to persist intact relative to single-copy genes. mtDNA-generated phylogenies and phylogenies generated from nuclear sources, such as SSU rRNA sequences, are generally similar in topology and branch length, indicating that these DNAs acquire change in essentially the same way and act as separate molecular clocks, albeit clocks often running at different rates.

Krishnan et al. (2007) developed a bar-coding technique useful for field work, based on the polymerase chain reaction, which can specifically amplify a region of the mitochondrial small subunit ribosomal DNA. This region contains flanking sequences that are universally conserved among myxomycetes and many other organisms, such as fungi, making them ideal for PCR primers. The primers used to amplify the mitochondrial DNA SSU rRNA were chosen so as to specifically amplify the mitochondrial core SSU rRNA in preference to the nuclear SSU rRNA. Internal to these highly conserved flanking sequences are sequences that are less highly conserved, moderately conserved, and relatively variable, and so these sequences can be used to compare both closely related and distantly related species. The amplification products produced from these primers correspond to coordinates 676–1127 (452 nucleotides) of the mitochondrial SSU rRNA of *P. polycephalum* (Mahendran et al., 1994). Amplification products from myxomycetes generally vary from about 435 to 520 nucleotides depending on the size of the variable region within the amplification product. These sequences contain characterized RNA editing sites and so can be used to both study the evolution of RNA editing (see section further on) and to generate phylogenies.

Miller et al. (personal communication) has used these bar-coding amplification products to generate a phylogeny of 70 different myxomycetes corresponding to the six myxomycete orders (Physarales, Stemonitales, Echinosteliales, Trichiales, Liceales, and Ceratiomyxales) and outgroup organisms (Fig. 4.3). This tree generally recapitulates phylogenies produced using nuclear sequences from myxomycetes (Fiore-Dunno et al., 2005). It is reassuring to see that these phylogenies generally match the phylogenetic trees generated using myxomycete morphological characters, although some variations exist. In general, the genera within each of the six orders group together. The tree is rooted through sequences from *Acanthamoeba*, *Dictyostelium*, *Protostelium*, and *E. coli*. The Ceratiomyxales form a single clade separate from the sister clade that contains the other myxomycetes. This sister clade is formed by two separate clades, one containing the organisms of the orders Liceales and Trichiales, the other clade contains the organisms of the orders Physarales, Stemonitales, Echinosteliales.

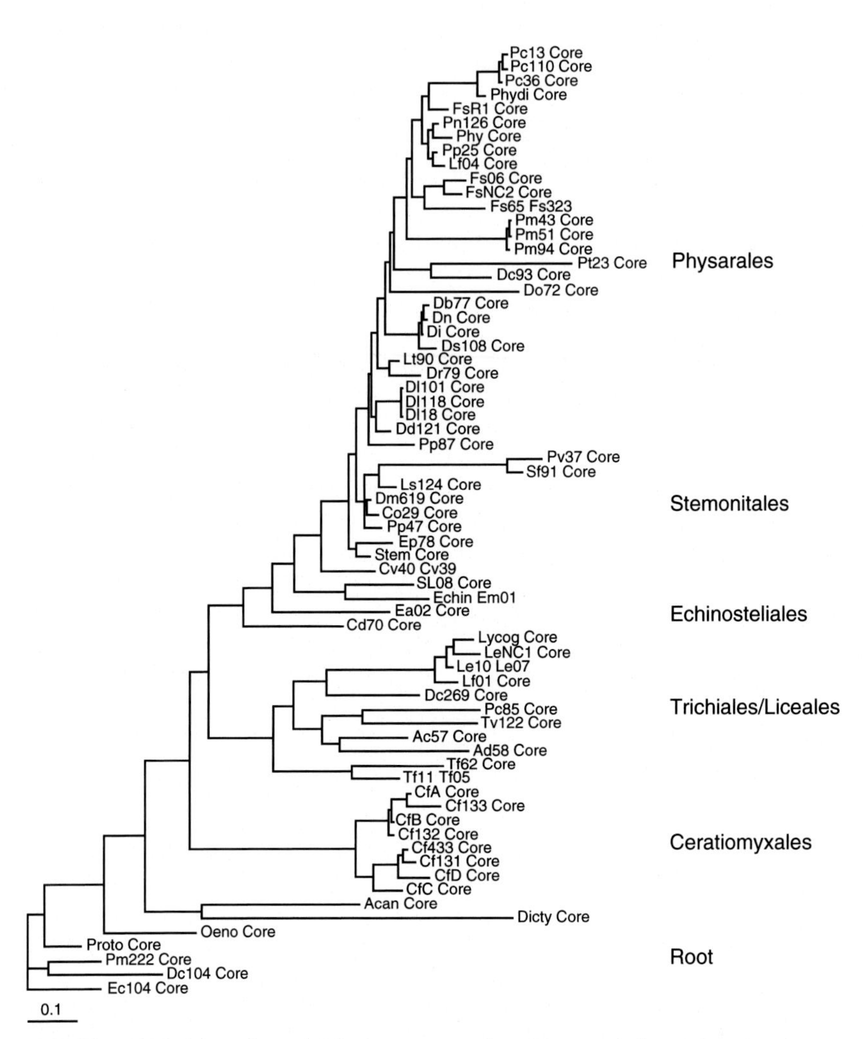

FIGURE 4.3 Phylogenetic tree of the myxomycetes based on comparisons of core sequences of the small subunit rRNA gene on mtDNA from 61 species of myxomycetes and 7 outgroup species. Species within the six orders of myxomycetes are abbreviated to the right. Pc13, Pc110, Pc36: *Physarum compressum*; Phydi: *Physarum didermoides*; Pn126: *Physarum necaraguense*; Pp: *Physarum polycephalum*; Pp25: *Physarum pusillum*; Lf04: *Leocarpus fragilis*; Fs06, FsNC2, Fs65, Fs323, FsR1: *Fuligo septica*; Pm43, Pm43, Pm94: *Physarum mellum*; Dc93: *Dianema corticatum*; Db77: *Didymium bahiense*; Dn: *Didymium nigripes*; Di: *Didymium iridis*; Ds: *Didymium squamulosum*; Lt90: *Lepidoderma tigrinum*; Dr79: *Diderma radiatum*; Dl101, Dl118, Dl18: *Diachea leucopodia*; Ls124: *Lamproderma sauteri*; Co29: *Colloderma*; Ep78: *Enerthenema papillatum*; Stem: *Stemonitis flavogenita*; Cv40: *Cribraria violacea*; Echino, Em01: *Echinostelium minutum*; Cd70: *Clastoderma debaryanum*; Lycog, LeNC1, Le10: *Lycogala epidendrum*; Lf01: *Lycogala flavofuscum*; Dc269: *Dictydium cancellatum*; Pc85: *Perichanena corticalis*; Tv122: *Trichia varia*; Ac57: *Arcyria cinerea*; Ad58: *Arcyria denudata*; Tf62, Tf11, Tf05: *Tubifera ferruginosa*; CfA, Cf133, CfB, Cf132, Cf433, Cf131, CfD: *Ceratiomyxa fruticulosa*; Acan: *Acanthamoeba*; Dicty: *Dictyostelium discoideum*; Proto: *Protostelium*; Oeno: *Oenothera* (plant); Ec: *Escherichia coli*.

The Liceales/Trichiales clade is not well resolved into the separate orders, the Trichiales and Liceales, which do not form separate clades. Within the Liceales/Trichiales (LT) clade, *Arcyria*, *Trichia*, and *Perichaena* form a single clade, and species of *Lycogala* and *Tubifera* group together but do not form a single clade that would separate the members of the Liceales from the Trichiales.

Within the Physarales/Stemonitales/Echinosteliales (PSE) clade the organisms of the Echinosteleales (Echinosteliaceae and Clastodermaceae) do not resolve into a single clade but do branch closely together. The organisms of the order Stemonitales group as a single clade with two branches, one branch with the two genera *Stemonitis* and *Enerthenema* as a clade and the other branch with *Lamproderma*, *Colloderma* and *Diachea*. The members of the Physarales are paraphyletic within the PSE clade and do not form a single sister clade with the members of the Stemonitales. Within the PSE clade, members of the Physarales cluster together in separate smaller clades which separately group five species of *Physarum* [*P. polycephalum*, *P. didermoides*, *P. compressum*, *P. pusillum*, and *P. nutans* (=*P. album*)] and *Leocarpus*, four species of *Didymium* species (*D. iridis*, *D. nigripes*, *D. bahiense*, and *D. squamulosum*), along with *Fuligo septica* and *Physarum melleum*. Nandipati et al. (2012) have also reported a polyphyletic origin of the genus *Physarum* (Physarales), based on data from nuclear rDNA minichromsome analysis and group I intron comparisons.

INSERTIONAL RNA EDITING IN THE MYXOMYCETES

Insertional RNA editing in the myxomycetes is a novel form of RNA editing discovered and characterized in *P. polycephalum* (Mahendran et al., 1991; Miller et al., 1993a,b,c) and present in the mitochondria of every myxomycete studied to date (Krishnan et al., 2007). This unique type of RNA editing has not been observed in any organism outside the myxomycetes and thus is a molecular candidate for a defining characteristic of the myxomycetes.

This type of RNA editing is characterized by the specific insertion of a nucleotide or nucleotides in nascent mitochondrial RNAs relative to the mitochondrial DNA template to create open reading frames in mRNAs and classical secondary and tertiary structures in tRNAs and rRNAs. Two examples of deletional RNA editing have been observed in the myxomycetes (Gott et al., 2005; Krishnan et al., 2007), but it is not clear if the mechanism of these deletions is similar to insertional RNA editing or if this constitutes an additional type of RNA editing in myxomycetes. Although insertional RNA editing has been identified in mitochondria of other organisms (notably, the trypanosomes), this insertional editing in these organisms is posttranscriptional and is based on a gRNA template and editosomes which insert (and delete) uridines in primary transcripts produced from cryptogenes on maxicircle mtDNA (Stuart et al., 2005). This posttranscriptional mechanism is in contrast to the cotranscriptional mechanism used in myxomycete RNA editing. It is this cotranscriptional mechanism that makes the RNA editing in the myxomycetes unique. This cotranscriptional editing is based on the mitochondrial RNA polymerase which can add nontemplated nucleotides to the 3′ end of RNAs (Miller and Miller, 2008). The current model predicts that *Physarum* mitochondrial RNA polymerase is able to pause in transcription at precise sites, add nontemplated nucleotides to the 3′ end of nascent RNA, and then resume templated transcription (Cheng et al., 2001; Miller and Miller, 2008). The inserted nucleotide is usually a single cytidine, but at some sites the nucleotide can be a uridine, and at rare sites adenosine or guanosine nucleotides are inserted (Bundschuh et al., 2011). Less frequently a dinucleotide can be inserted relative to the mtDNA template. The sequence of the dinucleotide is a subset of the possible dinucleotide sequences. Most common are AA, CU (or UC), and GU (or UG), but GC (or CG), UU, and UA have been observed, 6 of the possible 10 dinucleotide combinations (Bundschuh et al., 2011).

Bundschuh et al. (2011) have used deep transcriptome analysis to identify the complete complement of insertional RNA editing sites in mitochondria of *P. polycephalum*. They detected 1324 sites, 1301 single nucleotide insertions, and 23 dinucleotide insertions. Of the single insertion sites, 1255 are cytidine, 43 are uridine, 2 are guanosine, and 1 is adenosine. This shows that any nucleotide can be specified at a site but that pyrimidines and particularly cytidines are favored. All four nucleotides can also be inserted in the form of dinucleotides, although not all possible dinucleotide combinations are observed and there is a bias in the frequencies with which the dinucleotides are observed. The most common dinucleotide is CU (or UC), inserted at nine sites with AA and UG (or GU), each inserted at four sites, and UU, UA, and GC (or CG), each inserted at two sites.

These 1314 RNA editing sites are distributed among 43 genes on the *Physarum* mtDNA (10 sites are extragenic; Bundschuh et al., 2011). All of these genes, except one (ERF I), have been identified as classical mitochondrial genes and include 37 protein-coding genes, 2 rRNA genes, and 4 tRNA genes (Table 4.4). This means that there are multiple RNA editing sites in each gene. In mRNAs the editing sites are distributed rather evenly throughout the reading frame and there is on average about 25 nucleotides between editing sites, so that about 4% of mRNAs are nontemplated nucleotides. In ribosomal rRNAs and tRNAs, the spacing distribution also averages about 25 nucleotides, but only in some regions. The overall editing site density in rRNAs is about 1 for every 50 nucleotides, resulting in about 2% of rRNAs being nontemplated nucleotides.

RNA editing consists of the RNA polymerase adding specific nontemplated nucleotides to the 3′ end of nascent RNA transcripts. Mitochondrial RNA polymerases are highly conserved among all eukaryotic organisms and are a member of a super family of single subunit polymerases including bacteriophage RNA polymerases, DNA-directed DNA polymerases, and RNA-directed RNA polymerases (reverse transcriptases) (Cermakian et al., 1977; Lazcano et al., 1988; Miller et al., 2006). Many of these polymerases have been shown to add nontemplated nucleotides to the 3′ end of nucleic acids but only the *Physarum* mitochondrial polymerase has been shown to add a specific non-DNA templated nucleotide at a specific location during transcription resulting in non-DNA templated internal insertions (RNA editing). How this specificity of location and nucleotide identity is achieved and how transcription can be resumed after a nontemplated insertion are major questions of the mechanism of RNA editing in myxomycetes that remain to be answered.

Specificity of RNA Editing Site Location

Editing sites in mRNAs must have the specific nucleotide added, at precisely the correct site, with 100% efficiency to produce mRNAs that will specify the correct amino acid, without reading frame shifts. How the RNA polymerase is caused to pause precisely at the proper editing site and insert the specific non-templated nucleotide is not yet understood. Early attempts to identify editing site consensus sequences were not fruitful, although a dinucleotide bias was observed upstream of RNA editing sites [i.e., many, but not all, editing sites are preceded by a purine-pyrimidine dinucleotide (especially a purine, U dinucleotide) but this sequence is present at many sites that are not editing sites]. The role, if any, of this bias is not known, but it is clearly not sufficient to explain RNA editing site specificity. It is unlikely that any consensus sequence will be identified since editing site location can change in different myxomycetes even within regions highly conserved in sequence (see Section "Evolution of RNA Editing, Variation in Editing Site Location").

Another possible method of specifying editing site location would be the marking of editing sites through chemical modifications of the mtDNA at editing sites. No evidence of such modifications which correlate with RNA editing sites has been reported.

An alternate method of marking RNA editing sites would be the use of an antisense sequence complementary to the mRNA sequence at the editing site. This RNA would have the additional advantage of being able to specify the identity of the inserted nucleotide through base pairing. The insertional RNA editing in mitochondria of trypanosomes provides a precedent for such a mechanism. Here small guide RNAs (gRNAs) antisense to the mRNA mark editing sites and template uridine insertion and deletion (Stuart et al., 2005). These gRNAs are produced from templates on maxicircle (mtDNA) and specialized minicircle DNAs within trypanosome mitochondria. Sensitive searches for such antisense RNAs or DNA sequences on mtDNA that could code for gRNAs in *Physarum* have not revealed any candidates for a gRNA-like mechanism. Transcriptome analysis using deep sequencing has also not detected any candidate gRNAs (Bundschuh et al., 2011).

The complexity and specificity of RNA editing in *Physarum* and the maintenance of editing site location during replication of mtDNA argues that the genetic information for functional mitochondrial proteins, rRNAs, and tRNAs must be maintained. This information is not present in the mtDNA and no antisense template able to transfer the genetic information for editing site location and nucleotide specificity through base pairing has been detected. Only fully edited, mature RNAs have this genetic information and it is not clear how this information could be transmitted and maintained without an antisense RNA sequence, since this type of genetic information is classically transmitted in replication and transcription through base pairing interactions with the template strand of DNA. One way in which editing sites on the mtDNA could be marked is through base pairing interactions between the template strand of the mtDNA and the fully edited, mature RNA. The extra nucleotides in the RNA resulting from nontemplated nucleotide insertions during transcription would not base pair with mtDNA and could be flipped out of the duplex, marking editing sites. These extra bases in the mRNA-mtDNA duplex could cause the mitochondrial RNA polymerase to pause and initiate nontemplated nucleotide addition. This model would require the nascent RNA to remain associated with the template strand of the DNA after synthesis, forming a type of DNA/RNA/DNA triplex. This triplex could consist of a DNA/RNA duplex associated through Watson-Crick base pairing with the displaced DNA strand plectonemically associated with the duplex in the major groove and stabilized by Hoogsteen base pairing of pyrimidines with the template strand. While this model can explain how editing sites are identified and maintained, it does not explain how the identity of the inserted nucleotide is determined.

Specificity of Inserted Nucleotide Identity

Although there is variability in the nontemplated nucleotide that can be inserted at RNA editing sites, the type of nucleotide inserted at a particular site is constant. How this nucleotide specificity is achieved is not known. Since 1255 of the 1324 (95%) of the insertional RNA editing sites in *Physarum* are single cytidines, it is tempting to propose that once the RNA polymerase is paused in transcription at a potential editing site, a cytidine-specific terminal transferase activity adds a single cytidine to the 3′ end of the nascent transcript and then DNA-templated RNA synthesis continues by adding a templated nucleotide to the 3′ end of the RNA. However, this then begs the question of how the remaining 5% of the insertions are specified as different nucleotides or dinucleotides. Whether these noncytidine insertions are exceptions to the rule and must be specified in some way or whether the identity of all inserted nucleotides are specified in the same way, the need for some type of template is clear. The nature of this template is not clear. Preliminary evidence (Miller et al., personal communication) from another member of the polymerase super family of which mitochondrial RNA polymerases are a part (T7 RNA polymerase) may provide some clues to the way that this specificity is achieved. T7 RNAP is known to add a random nontemplated nucleotide to the 3′ end of RNAs (or DNAs) in the presence of complex DNA or RNA sequences. This addition can be controlled in vitro by the addition of simple, defined nucleic acid sequences to the T7 RNA polymerase. In this in vitro system the identity of the templated nucleotides can be specified by creating the potential for intramolecular (or intermolecular) base pairing interactions that create recessed 3′ ends that can be extended by one or a few nucleotides on the template provided by the extended 5′ end (limited primer extension). Indeed, the addition of a nucleotide to the 3′ end is dependent on this potential. This activity can occur in the absence of transcription when only a single rNTP is provided. Whether this type of intra- or intermolecular templating can occur in RNA during RNA editing to specify nucleotide identity is under study.

Evolution of RNA Editing, Variation in Editing Site Location

Dependable phylogenies based on nuclear and mitochondrial DNA sequences have allowed analysis of the evolution of RNA editing patterns in different strains of myxomycetes and has provided insight into the constraints on RNA editing site location in mitochondria of these organisms. Studies comparing RNA editing patterns in different genera of myxomycetes with the RNA editing pattern in *Physarum* (Antes et al., 1998; Hendrickson and Silliker, 2010a,b; Horton and Landweber, 2000; Krishnan et al., 2007; Traphagen et al., 2010) have revealed that the same type of insertional RNA editing is present in all of the myxomycetes examined, but that the location of the editing sites varied within the analogous, and often very conserved sequence, was studied. This variation indicates an unanticipated dynamic in the location of RNA editing sites over evolutionary time periods and provides insights to the constraints on editing site location, as well as the mechanism of editing site establishment and removal.

Krishnan et al. (2007) compared editing site location among six myxomycetes by comparison of cDNA and mtDNA sequences in the core region of the mitochondrial SSU rRNA (Fig. 4.4). This core region included 452 nucleotides and 10 characterized editing sites in *Physarum* [eight sites with single cytidine insertions and two sites with dinucleotides (AA and CU) insertions]. Alignment of the mtDNA and cDNA sequences from these six myxomycetes revealed a total of 29 different editing sites within this region, although the number of editing sites in this region in any one organism varied from 8 to 10. While some of these 29 potential RNA editing sites were located closer to each other than 9 nucleotides, the distance between actual editing sites was greater than 9 nucleotides in each individual organism. In this conserved region, all potential editing sites that did not have a nontemplated insertion, had a templated nucleotide at that position so the cDNA sequences of these organisms aligned without gaps, and gaps in the mtDNA sequence alignments were strongly predictive of RNA editing sites. In most, but not all of the sites, the nontemplated nucleotide was identical to the templated nucleotide. All of these editing sites had similar constraints to RNA editing sites in *Physarum*, that is, (1) they were never located closer to each other than 9 nucleotides, (2) they were predominately single cytidine insertions with occasional single uridine and dinucleotide insertions, and (3) many had purine-pyrimidine nucleotides immediately upstream of the site.

```
Pp mtDNA   T AGGTACTAAAGTATGGGGAT AAATAGGATTAGAGA CCTAGTAGTCCATACCTTA
Pp  cDNA   TCAGGTACTAAAGTATGGGGATCAAATAGGATTAGAGACCCTAGTAGTCCATACCTTA
Dn mtDNA   T AGGTACTAAAGCATGGGTAT GAAAAGGATTAGAGA CCTTGTAGTCCATGCTGTA
Dn  cDNA   TCAGGTACTAAAGCATGGGTATCGAAAAGGATTAGAGACCCTTGTAGTCCATGCTGTA
Pd mtDNA   TAAGGTACGAAAGTATGGGGAT AAATGGGATTAGAGACCCCAGTAGTCCATA CTTA
Pd  cDNA   TAAGGTACGAAAGTATGGGGATCAAATGGGATTAGAGACCCCAGTAGTCCATACCTTA
Sf mtDNA   TAAGGTACTAAAGCATGGGTAT GAAAAGGATTAGAGA CCTTGTAGTCCATGCCTTA
Sf  cDNA   TAAGGTACTAAAGCATGGGTATCGAAAAGGATTAGAGACCCTTGTAGTCCATGCCTTA
Le mtDNA   TGAGGTATGAAGGTATGGGTATCGATCGGGATTAGAGACCCCAGTAGTCCATA AGTA
Le  cDNA   TGAGGTATGAAGGTATGGGTATCGATCGGGATTAGAGACCCCAGTAGTCCATACAGTA
Em mtDNA   TACGGTACGAAAG GTGGGGAGCAAATGGGATTAGAGACCCCTGTAGTCCATG CTTA
Em  cDNA   TACGGTACGAAAGCGTGGGGAGCAAATGGGATTAGAGACCCCTGTAGTCCATGCCTTA
            C           C        C                C             C

Pp mtDNA   AACAATGAG..AGCTAACGCGTGAAACA  CCGCCTGGGGAATGTGGCCGCAAGGT T
Pp  cDNA   AACAATGAG..AGCTAACGCGTGAAACACTCCGCCTGGGGAATGTGGCCGCAAGGTCT
Dn mtDNA   AACCATGAG..AGCTAACGCGTGAAACA  CCGCCTGGGGAATGTGGCCGCAAGGT T
Dn  cDNA   AACCATGAG..AGCTAACGCGTGAAACACTCCGCCTGGGGAATGTGGCCGCAAGGTCT
Pd mtDNA   AACAATGAG..AGCTAACGCGTAGAACA  CCGCCTGGGGAATGTGGCCGCAAGGT T
Pd  cDNA   AACAATGAG..AGCTAACGCGTAGAACACTCCGCCTGGGGAATGTGGCCGCAAGGTCT
Sf mtDNA   AACAATGAG..ACGTAACGCGTGAAACA  CCGCCTGGGGAATGTAGCCGCAAGGT T
Sf  cDNA   AACAATGAG..ACGTAACGCGTGAAACACTCCGCCTGGGGAATGTAGCCGCAAGGTCT
Le mtDNA   AACGCTGCA..AGCTAACGCGTTAAATATGCCACCTGGGCAGTA GGCCGCAAGGTCA
Le  cDNA   AACGCTGCA..AGCTAACGCGTTAAATATGCCACCTGGGCAGTACGGCCGCAAGGTCA
Em mtDNA   AACGATGAG..AGCTAACGCGAGAAATA  CCGCCTGGGAACTATGG CGCAAGGTCG
Em  cDNA   AACGATGAG..AGCTAACGCGAGAAATACTCCGCCTGGGAACTATGGCCGCAAGGTCG
           Variable region omitted     CU              C  C        C

Pp mtDNA   AAACTCAAAGGAATTGACGGAGACTTATACAAGGGGTGGAG ATGTGGTTTAATTCAA
Pp  cDNA   AAACTCAAAGGAATTGACGGAGACTTATACAAGGGGTGGAGCATGTGGTTTAATTCAA
Dn mtDNA   AAACTCAAAGGAATAGACGGAGACTTATACAAGGGGTGGAG ATGTGGTTTAATTCAA
Dn  cDNA   AAACTCAAAGGAATAGACGGAGACTTATACAAGGGGTGGAGCATGTGGTTTAATTCAA
Pd mtDNA   AAACTCAAAGGAATTGACGGAGACTTATACAAGGGGTGGAG ATGTGGTTTAATTCAA
Pd  cDNA   AAACTCAAAGGAATTGACGGAGACTTATACAAGGGGTGGAGCATGTGGTTTAATTCAA
Sf mtDNA   AAACTCAAAGGAATAGACGGAGGCTTGTACAAGGGGTGGAGCATGTGGTTTAATTCAA
Sf  cDNA   AAACTCAAAGGAATAGACGGAGGCTTGTACAAGGGGTGGAGCATGTGGTTTAATTCAA
Le mtDNA   AAACTCAAAAGAATAGACGGAGA TTGTCCAAGGGGTGGAGCATGTGGTTTAATTCAA
Le  cDNA   AAACTCAAAAGAATAGACGGAGACTTGTCCAAGGGGTGGAGCATGTGGTTTAATTCAA
Em mtDNA   AAACTCAAAGGAATAGACGGAGACTTGTA AAGGGGTGGAGCATGTGGTTTAATTCAA
Em  cDNA   AAACTCAAAGGAATAGACGGAGACTTGTACAAGGGGTGGAGCATGTGGTTTAATTCAA
                                  C     C           C
```

FIGURE 4.4 **RNA editing sites—cDNA/mtDNA alignment pairs for six myxomycetes.** Pp, Dn, Pd, Sf, Le, and Em indicate *Physarum polycephalum, Didymium nigripes, Physarum didermoides, Stemonitis flavogenetia, Lycogola epidendrum,* and *Echinostelium minutum*, respectively. Gaps in the cDNA/mtDNA indicate RNA editing sites. RNA editing sites present in some strains are indicated with C or CU beneath the six alignment pairs. A variable region in which alignment was difficult has been omitted.

Despite these similarities, the locations of editing sites in these organisms varies dramatically. Although *P. polycephalum* and *Lycogala epidendrum* have almost the same number of editing sites in this region (10 and 9, respectively), they do not have any editing sites in common. In general, more closely related organisms have a greater number of common sites in this region. These differences in editing site location implies that editing sites may be created and/or eliminated over time resulting in the divergence of editing patterns. To determine the sequence in which these changes occur, the pattern of editing sites in these six organisms was superimposed on the topology of the mitochondrial phylogenetic tree. Several different patterns of editing site change consistent with the tree topology are equally parsimonious, but all of them predict both the creation and deletion of sites to produce the editing pattern in the contemporary organism.

How insertional editing sites are gained or lost in myxomycetes is not known. The fact that most of the inserted non-DNA templated nucleotides in RNAs correspond to templated nucleotides already present at the analogous site in other organisms that are not edited at that site, implies that a mechanism must preexist to add nucleotides to RNAs to compensate for deletions in the mtDNA. Conversely, this preexisting activity must not function if the deletion is eliminated through some mechanism. Krishnan et al. (2007) have proposed a model of the evolution of RNA editing in the myxomycetes based on the general model of RNA editing proposed by Gray et al. (Covello and Gray, 1993; Gray, 2001; Price and Gray, 1998), which explains the linkage between these two processes and the observed variation in editing site patterns among myxomycetes. Analysis of editing site location in the same sequence of different myxomycetes shows that the editing can be essentially at any location. Krishnan et al. (2007) propose that editing sites accumulate through corrections in the RNA of random deletions in the DNA template and that the location of these corrected deletions is only constrained by three criteria. These are (1) the purine-pyrimidine bias, (2) the cytidine preference, and (3) the proximity constraint of at least nine nucleotides between editing sites. The first two criteria are not particularly limiting, but the proximity constraint limits the density of editing sites and could ultimately slow or stop editing site accumulation in an organism. This constraint would not allow editing sites to be created between existing editing sites separated by 9–17 nucleotides, and sites separated by 18–26 nucleotides would have small target sites (1–10 nucleotides) for an additional editing site to be created. The probability of random deletions occurring within these permissible zones would become smaller with each addition until the rate of accumulation approached zero. This saturation density would be expected at an average spacing of about 22 nucleotides, very close to the observed density of RNA editing sites (about 25 nucleotides) in *Physarum*. The fact that many of the myxomycetes studied have this density may indicate that most of them are at or near saturation. There is also evidence that editing patterns could be altered through the deletion of editing sites. Landweber (1992) and Simpson et al. (Maslov et al., 1994; Simpson and Maslov, 1994) have proposed retrotransposition as a mechanism of eliminating editing sites in Trypanosomes. Integration of cDNAs produced from reverse transcription of edited RNAs would remove the deletions and restore the genetic information to the mtDNA. This process would tend to eliminate the editing site saturation and allow rapid fixation of new editing sites at random locations creating new patterns of editing these organisms.

An additional constraint on the location of RNA editing sites seems to be the functional importance of the nucleotide added in RNA editing. This constraint affects thc distribution of RNA editing sites in rRNAs. Highly conserved and presumably functionally important sequences accumulate editing sites at near saturation density, while editing sites in poorly conserved sequences generally can be absent or present at a low density. This implies that editing site fixation is affected by natural selection as might be expected if editing restores the function of a critical RNA. Consistent with this idea is the observation that editing sites are fixed at virtually all locations in mRNAs, since any deletions within this region will cause a frameshift that would generally prevent the production of a functional protein.

CONCLUDING REMARKS AND AREAS OF FUTURE RESEARCH

This chapter has summarized the recent advances in the genomics and gene expression of myxomycetes with special emphasis on the novel mechanism of gene expression in mitochondria of the myxomycetes (RNA editing). To date, with a few exceptions, most of the information about the molecular genetics of myxomycetes has come from research on the prototype organism of the myxomycetes, *P. polycephalum*. The recent completion of the genomic sequence of *Physarum* has helped to define the relationship of the myxomycetes to other members of the Amoebozoa and the place of myxomycetes in the evolution of eukaryotic organisms. This first genomic sequence in the myxomycetes should spur further genomic research in the myxomycetes, provide a reference for genomic comparisons, and aid in myxomycete genome assemblies.

The mitochondrial DNA of myxomycetes also promises to be a fertile area of research. The *Physarum* mtDNA has the potential to express one of the largest number of genes of any mtDNA. Many of these potential genes are novel among all mtDNAs and their function, if any, is unknown. Some do not appear to be transcribed under the conditions studied. Whether these novel genes are preserved in the mtDNA of other myxomycetes is still to be determined. If they are preserved, the function of these apparently myxomycete-specific genes will be of particular interest. Also, many of the classic features of mtDNAs, such as promoters, origins of replication, and so forth have not yet been characterized in *Physarum*. Comparison of the mtDNA sequences from additional myxomycetes should help to identify these features and to provide additional information about evolution within the myxomycetes and the relationship of the myxomycetes to other organisms within the Amoebozoa.

Finally, studies to elucidate the mechanism and evolution of the novel type of RNA editing present in mitochondria of myxomycetes are crucial to our understanding of the evolution and uniqueness of these organisms, as well as to the evolution of gene expression, in general. These studies should also provide insights into the origin of mitochondria and could be relevant to the mechanism of gene expression in mitochondria of all eukaryotes, including humans.

REFERENCES

Abad, M.G., Long, Y., Willcox, A., Gott, J.M., Gray, M.W., Jackman, J.E., 2011. A role for tRNA His guanylyltransferase (Thg1)-like proteins from *Dictyostelium discoideum* in mitochondrial 5′-tRNA editing. RNA 17, 613–623.

Adler, P.N., Holt, C.E., 1974. Genetic analysis in the Colonia strain of *Physarum polycephalum*: heterothallic strains that mate with and are partially isogenic to the Colonia strain. Genetics 78, 1051–1062.

Antes, T., Costandy, H., Mahendran, R., Spottswood, M., Miller, D., 1998. Insertional editing of tRNAs of *Physarum polycephalum* and *Didymium nigripes*. Mol. Cell. Biol. 18, 7521–7527.

Aguilera, P., Barry, T., Tovar, J., 2008. Entamoeba histolytica mitosomes: organelles in search of a function. Exp. Parasitol. 118, 10–16.

Arisue, N., Hashimoto, T., Lee, J.A., Moore, D.V., Gordon, P., Sensen, C.W., Gaasterland, T., Hasegawa, M., Muller, M., 2002. The phylogenetic position of the pelobiont *Mastigamoeba balamuthi* based on sequences of rDNA and translation elongation factors EF-1 alpha and EF-2. J. Eukaryot. Microbiol. 49, 1–10.

Baldauf, S.L., Roger, A.J., Wenk-Siefert, I., Doolittle, W.F., 2000. A kingdom level phylogeny of eukaryotes based on combined protein data. Science 290, 972–977.

Bapteste, E., Brinkmann, H., Lee, J., Moore, D., Sensen, C., Gordon, P., Durufle, L., Gaasterland, T., Lopez, P., Muller, M., Phillippe, H., 2002. The analysis of 100 genes support the grouping of three highly divergent amoebae: *Dictyostelium*, *Entamoeba*, and *Mastigamoeba*. Proc. Natl. Acad. Sci. USA 99, 1414–1419.

Barth, C., Greferath, U., Kotsifas, M., Fisher, P.R., 1999. Polycistronic transcription and editing of the mitochondrial small subunit (SSU) ribosomal RNA in *Dictyostelium discoideum*. Curr. Genet. 36, 55–61.

Barth, C., Greferath, U., Kotsifas, M., Tanaka, Y., Alexander, S., Alexander, H., Fisher, P.R., 2001. Transcript mapping and processing of mitochondrial RNA in *Dictyostelium discoideum*. Curr. Genet. 39, 355–364.

Bohnert, H.J., 1977. Size and structure of the mitochondrial DNA of the slime mold, *Physarum polycephalum*. Exp. Cell Res. 106, 426–430.

Bohnert, H.J., Schiller, B., Bohme, R., Sauer, H.W., 1975. Circular DNA and rolling circles in nuclear r-DNA from mitotic nuclei of *Physarum polycephalum*. Eur. J. Biochem. 57, 361–369.

Bolivar, I., Fahrni, J.F., Smirnov, A., Pawlowski, J., 2001. SSU rRNA-based phylogenetic position of the genera amoeba and chaos (Lobosea, Gymnamoebia): the origin of gymnamoebae revisited. Mol. Biol. Evol. 18, 2306–2314.

Braun, R., Evans, T.E., 1969. Replication of nuclear satellite and mitochondrial DNA in the mitotic cycle of *Physarum*. Biochim. Biophys. Acta 182, 511–522.

Bullerwell, C.E., Burger, G., Gott, J.M., Kourennaia, O., Schnare, M.N., Gray, M.W., 2010. Abundant 5S rRNA-like transcripts encoded by the mitochondrial genome in amoebozoa. Eukaryot. Cell 9, 762–773.

Bundschuh, R., Antmuller, J., Becker, C., Nurnburg, P., Gott, J.M., 2011. Complete characterization of the edited transcriptome of the mitochondrion of *Physarum polycephalum* using deep sequencing of RNA. Nucleic Acids Res. 39, 6044–6055.

Burger, G., Plante, I., Lonergan, K.M., Gray, M.W., 1995. The mitochondrial DNA of the amoebid protozoa, *Acanthamoeba castellani*: complete sequence, gene content and genome organization. J. Mol. Biol. 245, 522–537.

Cavalier-Smith, T., 1998. A revised six-kingdom system of life. Biol. Rev. Camb. Philos. Soc. 73, 203–266.

Cavalier-Smith, T., Fiore-Dunno, A.M., Chao, E., Kudryavtsev, A., Berney, C., Snell, E.A., Lewis, R., 2015. Multigene phylogeny resolves deep branching of Amoebozoa. Mol. Phylogenet. Evol. 83, 293–304.

Cermakian, N., Ikeda, T.M., Miramontes, P., Lang, B.F., Gray, M.W., 1977. On the evolution of the single-subunit RNA polymerases. J. Mol. Evol. 45, 671–681.

Cheng, Y.W., Visomirski-Robic, L.M., Gott, J.M., 2001. Non-templated addition of nucleotides to the 3′ end of nascent RNA during RNA editing in *Physarum*. EMBO J. 20, 1405–1414.

Covello, P.S., Gray, M.W., 1993. On the evolution of RNA editing. Trends Genet. 9, 265–268.

Crick, F.H.C., 1966. Codon–anticodon pairing: the wobble hypothesis. J. Mol. Biol. 19, 548–555.

Evans, T.E., 1966. Synthesis of a cytoplasmic DNA during the G2 interphase of *Physarum polycephalum*. Biochem. Biophys. Res. Commun. 22, 678–683.

Evans, T.E., Suskind, D., 1971. Characterization of mitochondrial DNA of the slime mold, *Physarum polycephalum*. Biochim. Biophys. Acta 228, 350–364.

Fahrni, J.F., Bolivar, I., Berney, C., Nassonova, E., Smirnov, A., Pawlowski, J., 2003. Phylogeny of lobose amoebae based on action and small-subunit ribosomal RNA genes. Mol. Biol. Evol. 18, 418–426.

Ferris, P.J., 1985. Nucleotide sequence of the central non-transcribed spacer region of *Physarum polycephalum* rDNA. Gene 39, 203–211.

Ferris, P.J., Vogt, V.M., 1982. Structure of the central spacer region of extrachromosomal ribosomal DNA in *Physarum polycephalum*. J. Mol. Biol. 159, 359–381.

Fiore-Dunno, A., Berney, C., Pawlowski, J., Baldauf, S.L., 2005. High-order phylogeny of plasmodial slime molds (Myxogastria) based on elongation factor 1-A and small subunit rRNA gene sequences. J. Eukaryot. Microbiol. 52, 201–210.

Forget, L., Ustinova, J., Wang, Z., Huss, V.A.R., Lang, B.F., 2002. *Hyaloraphidium curvatum*: a linear mitochondrial genome, tRNA editing, and an evolutionary link to lower fungi. Mol. Biol. Evol. 19, 310–319.

Forney, J., Henderson, E.R., Blackburn, E.H., 1987. Identification of the telomeric sequence of the acellular slime molds *Didymium iridis* and *Physarum polycephalum*. Nucleic Acids Res. 15, 9143–9152.

Gill, E.E., Diaz-Trivino, S., Barbara, M.J., Silberman, J.D., Stechmann, A., Gaston, D., Tamas, I., Roger, A.J., 2007. Novel mitochondria-related organelles in the anaerobic amoeba *Mastigamoeba balamuthi*. Mol. Microbiol. 66, 1306–1320.

Glockner, G., Golderer, G., Werner-Felmayer, G., Meyer, S., Marwan, W., 2008. A first glimpse at the transcriptome of *Physarum polycephlum*. BMC Genomics 9, 6–16.

Gott, J.M., Parimi, N., Bundschuh, R., 2005. Discovery of new genes and deletion editing in *Physarum* mitochondria enabled by a novel algorithm for finding edited RNAs. Nucleic Acids Res. 33, 5063–5072.

Gott, J.M., Somerlot, B.H., Gray, M.W., 2010. Two forms of RNA editing are required for tRNA maturation in *Physarum* mitochondria. RNA 16, 482–488.

Gott, J.M., Visomirski, L.M., Hunter, J.L., 1993. Substitutional and insertional RNA editing of the cytochrome c oxidase subunit I mRNA of *Physarum polycephalum*. J. Biol. Chem. 268, 25483–25486.

Gray, M.W., 2001. Speculations of the origin and evolution of editing. In: Bass, B.L. (Ed.), RNA Editing. Oxford University Press, Oxford, pp. 160–184.

Gray, M.W., Lang, B.F., Burger, G., 2004. Mitochondria of protists. Annu. Rev. Genetet. 38, 477–524.

Gubler, U., Tyler, T., Braun, R., 1979. The gene for the 26S rRNA in *Physarum* contains two insertions. FEBS Lett. 100, 347–350.

Guthrie, C., Abelson, J., 1982. Organization and expression of tRNA genes in *Saccharomyces cerevisiae*. In: Strathern, J.N., Jones, E.W., Broach, J.R. (Eds.), Molecular Biology of the Yeast, *Saccharomyces cerevisiae*. Cold Spring Harbor Laboratory Press, New York, NY, pp. 487–528.

Guttes, E., Guttes, S., 1969. Thymidine incorporation by mitochondria in *Physarum polycephalum*. Science 145, 1057–1058.

Hall, L., Braun, R., 1977. The organization of genes for transfer RNA and ribosomal RNA in amoebae and plasmodia of *Physarum polycephalum*. Eur. J. Biochem. 76, 165–174.

Hardman, N., 1985. Molecular organization of the *Physarum* genome. In: Dove, W.F., Dee, J., Hatano, S., Haugli, F.B., Wohlfarth-Botterman, K. (Eds.), The Molecular Biology of *Physarum polycephalum*. Plenum Press, New York, NY/London, pp. 39–66.

Hendrickson, P.G., Silliker, M.E., 2010a. RNA editing in six mitochondrial ribosomal protein genes of *Didymium iridis*. Curr. Genet. 56, 203–213.

Hendrickson, P.G., Silliker, M.E., 2010b. RNA editing is absent in a single mitochondrial gene of *Didymium iridis*. Mycologia 102, 1288–1294.

Horton, T., Landweber, L.F., 2000. Evolution of four types of RNA editing in myxomycetes. RNA 6, 1339–1346.

Johansen, S., Johansen, T., Haugli, F., 1992. Structure and evolution of myxomycete nuclear group I introns: a model for horizontal transfer by intron homing. Curr. Genet. 22, 297–304.

Jones, E.P., Mahendran, R., Spottswood, M.R., Yang, Y.-C., Miller, D.L., 1990. Mitochondrial DNA of *Physarum*: physical mapping, cloning, and transcription mapping. Curr. Genet. 17, 331–337.

Katz, L.A., 2012. Origin and diversification of eukaryotes. Annu. Rev. Microbiol. 66, 411–427.

Kawano, S., Takano, H., Mori, K., Kuriowa, T., 1991. A mitochondrial plasmid that promotes mitochondrial fusion in *Physarum polycephalum*. Protoplasma 160, 167–169.

Kopp, D., 2012. Assembly of the *Didymium iridis* mitochondrial genome by genome walking. College of Science and Health Theses and Dissertations, De Paul University Library, Chicago, IL.

Krishnan, U., Barsamian, A., Miller, D.L., 2007. Evolution of RNA editing sites in the mitochondrial small subunit rRNA of the myxomycetes. Methodos Enzymol. 424, 197–220.

Landweber, L.F., 1992. The evolution of RNA editing in kinetoplastid protozoa. Biosystems 28, 41–45.

Lazcano, A., Fastag, J., Gariglio, P., Ramirez, C., Oro, J., 1988. On the early evolution of RNA polymerase. J. Mol. Evol. 27, 365–376.

Le, P., Fisher, P.R., Barth, C., 2009. Transcription of the *Dictyostelium discoideum* mitochondrial genome occurs from a single initiation site. RNA 15, 2321–2330.

Lonergan, K.M., Gray, M.W., 1993. Editing of transfer RNAs in *Acanthamoeba castellani* mitochondria. Science 259, 812–816.

Mahendran, R., Spottswood, M.S., Ghate, A., Ling, M.L., Jeng, K., Miller, D.L., 1994. Editing of the mitochondrial small subunit rRNA in *Physarum polycehalum*. EMBO J. 13, 232–240.

Mahendran, R., Spottswood, M.S., Miller, D.L., 1991. RNA editing by cytidine insertion in mitochondria of *Physarum polycephalum*. Nature 349, 434–438.

Maslov, D.A., Avila, H.A., Lake, J.A., Simpson, L., 1994. Evolution of RNA editing in kinetoplastid protozoa. Nature 368, 345–348.

McCullough, C.H.R., Cooke, D.J., Foxon, S.L., Subbury, P.E., Grant, W.D., 1973. Nuclear DNA content and senescence in *Physarum polycephalum*. Nat. New Biol. 245, 263–265.

Melera, P.W., 1980. Transcription in the myxomycete, *Physarum polycephalum*. In: Dove, W.F., Rusch, H.P. (Eds.), Growth and Differentiation in *Physarum polycephalum*. Princeton University Press, Princeton, NJ, pp. 64–97.

Miller, D.L., 2014. Mitochondrial genomes in Amoebozoa. In: Gray, M.W. (Ed.), ASBMB Encyclopedia of Molecular Life Science, Mitochondrial Genomes. Springer Science + Business Media, New York, NY.

Miller, M.L., Antes, T.J., Qian, F., Miller, D.L., 2006. Identification of a putative mitochondrial RNA polymerase from *Physarum polycephalum*: characterization, expression, purification and transcription *in vitro*. Curr. Genet. 49, 259–271.

Miller, M.L., Miller, D.L., 2008. Non-DNA-templated addition of nucleotides to the 3′ end of RNAs by the mitochondrial RNA Polymerase of *Physarum polycephalum*. Mol. Cell Biol. 28, 5795–5802.

Miller, D.L., Mahendran, R., Spottswood, M.S., Costandy, H., Ling, M.L., Yang, N., 1993a. Insertional editing in mitochondria of *Physarum*. Semin. Cell Biol 4, 261–266.

Miller, D.L., Mahendran, R., Spottswood, M.S., Ling, M.L., Wang, S., Yang, N., Costandy, H., 1993b. RNA editing in mitochondria of *Physarum polycephalum*. In: Benne, R. (Ed.), RNA Editing: The Alteration of Protein Coding Sequences of RNA. Ellis Horwood, Chichester, pp. 87–103.

Miller, D.L., Mahendran, R., Spottswood, M.S., Ling, M.L., Wang, S., Yang, N., Costandy, H., 1993c. RNA editing in mitochondria of *Physarum polycephalum*. In: Brennicke, A., Kuck, U. (Eds.), Plant Mitochondria. VHC, Weinheim, pp. 53–62.

Milyutina, I.A., Aleshin, V.V., Mikrjukov, K.A., Kedrova, O.S., Petrov, N.B., 2001. The unusually long small subunit ribosomal RNA gene found in amitochondriate amoeboflagellate *Pelomyxa palustris:* its rRNA predicted structure and phylogenetic implication. Gene 272, 131–139.

Mohberg, J., 1977. Nuclear DNA content and chromosome numbers throughout the life cycle of the Colonia strain of the myxomycete, *Physarum polycephalum*. J. Cell Sci. 24, 95–108.

Mohberg, J., Babcock, K.L., Haugli, F.B., Rusch, H.P., 1973. Nuclear DNA content and chromosome numbers in the myxomycete *Physarum polycephalum*. Dev. Biol. 34, 228–245.

Mohberg, J., Rusch, H.P., 1971. Isolation and DNA content of nuclei of *Physarum*. Exp. Cell Res. 66, 305–316.

Muscarella, D.E., Ellison, E.L., Rouff, B.M., Vogt, V.M., 1990. Characterization of I-Ppo, an intron-encoded endonuclease that mediates homing on a group I intron in the ribosomal DNA of *Physarum polycephalum*. Mol. Cell Biol. 10, 3386–3396.

Muscarella, D.E., Vogt, V.M., 1989. A mobile group I intron in the nuclear DNA of *Physarum*. Cell 56, 443–454.

Nakagawa, C.C., Jones, E.P., Miller, D.L., 1998. Mitochonrial DNA rearrangements associated with mF plasmid integration and plasmodial longevity in *Physarum polycephalum*. Curr. Genet. 33, 178–187.

Nandipati, S.C.R., Haugli, K., Coucheron, D.H., Haskins, E.F., Johansen, S.D., 2012. Polyphyletic origin of the genus *Physarum* (Physarales, Myxomycetes) revealed by nuclear rDNA mini-chromosome analysis and group I intron synapomorphy. BMC Evol. Biol. 12, 166–175.

Nomura, H., Moriyama, Y., Kawano, S., 2005. Rearrangements in the *Physarum polycephalum* mitochondrial genome associated with a transition from linear mF-mtDNA recombinants to circular molecules. Curr. Genet. 47, 100–110.

Ogawa, S., Yoshino, R., Angata, K., Iwamoto, M., Pi, M., Kuroe, K., Matsuo, K., Morio, T., Urushihara, H., Yanagisawa, K., Tanaka, Y., 2000. The mitochondrial DNA of *Dictyostelium discoideum*: complete sequence, gene content and genome organization. Mol. Gen. Genet. 263, 514–519.

Price, D.H., Gray, M.W., 1998. Editing of tRNA. In: Grosjean, H., Benne, R. (Eds.), Modification and Editing of RNA. ASM Press, Washington, DC, pp. 289–305.

Price, D.H., Gray, M.W., 1999. Confirmation of predicted edits and demonstration of unpredicted edits in *Acanthamoeba castellani* mitochondrial tRNAs. Curr. Genet. 35, 23–29.

Rhee, A.C., Somerlot, B.H., Parmi, N., Gott, J.M., 2009. Distinct roles for sequences upstream of and downstream from *Physarum* editing sites. RNA 15, 1753–1765.

Roger, A.J., Simpson, A.G.B., 2008. Evolution: revisiting the root of the eukaryotic tree. Curr. Biol. 19, R165–R167.

Schaap, P., Barrantes, I., Minx, P., Sasaki, N., Anderson, R.W., Benard, M., Biggar, K.K., Buchler, N.E., Bundschuh, R., Chen, X., Fronick, C., Fulton, L., Golderer, G., Jahn, N., Knoop, V., Landweber, L.F., Maric, C., Miller, D.L., Noegel, A., Peace, R., Pierron, G., Sasaki, T., Schallenberg-Rudinger, M., Schleicher, M., Singh, R., Spaller, T., Storey, K.B., Suzuki, T., Tomlinson, C., Tysom, J.J., Warren, W.C., Werner, E.R., Werner-Felmayer, G., Wilson, R.K., Winkler, T., Gott, J.M., Glockner, G., Marwan, W., 2016. The *Physarum polycephalum* genome reveals extensive use of prokaryotic two-component and metazoan-type tyrosine kinase signaling. Genome Biol. Evol. 8, 109–125.

Shadwick, L.L., Spiegel, F.W., Shadwick, J.D.L., Brown, M.W., Silberman, J.D., 2009. *Eumycetozoa=Amoebozoa*?: SSU rDNA phylogeny of prostelid slime molds and its significance for the amoebozoan supergroup. PLoS One 4 (8), e6754.

Simpson, L., Maslov, D.A., 1994. Ancient origin of RNA editing in kinetoplastid protozoa. Curr. Opin. Genet. Dev. 4, 887–894.

Stuart, K.D., Schnaufer, A., Ernst, N.L., Panigrahi, A.K., 2005. Complex management: RNA editing in *Trypanosomes*. Trends Biochem. Sci. 30, 97–105.

Takano, H., Abe, T., Sakurai, R., Moriyama, Y., Miyazawa, Y., Nozaki, H., Kawano, S., Sasaki, N., Kuroiwa, T., 2001. The complete DNA sequence of the mitochondrial DNA of *Physarum polycephalum*. Mol. Gen. Genet. 264, 539–545.

Takano, H., Kawano, S., Kuriowa, T., 1992. Constitutive homologous recombination between mitochondrial DNA and a linear mitochondrial plasmid in *Physarum polycephalum*. Curr. Genet. 22, 221–227.

Takano, H., Kawano, S., Kuriowa, T., 1994. Complex terminal structure of a linear mitochondrial plasmid from *Physarum polycephalum*: three terminal inverted repeats and an ORF encoding DNA polymerase. Curr. Genet. 25, 252–257.

Takano, H., Mori, K., Kawano, S., Kuroiwa, T., 1996. Rearrangements of mitochondrial DNA and the mitochondrial fusion-promoting plasmid (mF) are associated with defective mitochondrial fusion in *Physarum polycephalum*. Curr. Genet. 29, 257–264.

Traphagen, S.J., Dimarco, M.J., Silliker, M.E., 2010. RNA editing of 10 *Didymium iridis* mitochondrial genes and comparison with the homologous genes in *Physarum polycephalum*. RNA 16, 828–838.

Vogt, V.M., Braun, R., 1976. Structure of ribosomal DNA in *Physarum polycephalum*. J. Mol. Biol. 106, 567–587.

Chapter 5

Molecular Techniques and Current Research Approaches

Laura M. Walker*, Thomas Hoppe, Margaret E. Silliker[†]**
**Smith College, Clark Science Center, Northampton, MA, United States;*
***Institute of Botany and Landscape Ecology, Ernst Moritz Arndt University, Greifswald, Germany; [†]DePaul University, Chicago, IL, United States*

PART A: A SYNTHESIS OF THE CURRENT KNOWLEDGE ON MOLECULAR TECHNIQUES USED FOR THE RECORDING OF BIODIVERSITY AND ECOLOGICAL ANALYSES

INTRODUCTION

Myxomycetes have been studied by scientists and enthusiasts since the 18th century (Chapter 2); throughout this time, the most commonly employed method for studying myxomycete ecology has been the use of field observations and moist chamber cultures, both of which were followed by morphological identification of fruiting bodies. Although simple and cost effective, these traditional methods do not account for the array of complex life histories found throughout the group nor do they reveal reproductive isolation within a morphologically defined species (morphospecies), thereby limiting our understanding of myxomycete biodiversity and ecology (Walker and Stephenson, 2016). For example, some ubiquitous species of myxomycetes may have lost the ability to mate and to form fruiting bodies altogether (Fiore-Donno et al., 2010a), thereby leaving them undetected by traditional methods (Chapter 1, Fig. 1.8 for a diagram of a typical myxomycete life cycle and a thorough discussion of life cycle variations). Although the proportion of myxomycetes with incomplete life cycles is unknown, a recent environmental sampling study of grassland soils suggests that the presence of all myxomycetes has been grossly underestimated by older, more traditional methods (Fiore-Donno et al., 2016). Molecular approaches offer alternative means of study, and further reveal the biological complexity of these organisms.

Myxomycetes: Biology, Systematics, Biogeography, and Ecology. http://dx.doi.org/10.1016/B978-0-12-805089-7.00005-6

The first major movement toward molecular study of myxomycetes took place in the late 1960s with the use of isozyme patterns to investigate relationships between closely related species (Betterley and Collins, 1983; Franke, 1967; Franke and Berry, 1972; Franke et al., 1968). Attempts were made to use isozyme profiles to distinguish between similar morphospecies and even between different strains of a single morphospecies. For example, Berry and Franke (1973) observed that within a single morphospecies, *Fuligo septica,* distinct isozyme patterns could be detected between white and yellow strains. They also outlined the limited information available for other species and strains. Isozyme patterns of 44 isolates of *Didymium iridis* correlated with three distinct heterothallic breeding groups and many reproductively isolated nonheterothallic (probably apomictic) isolates (Betterley and Collins, 1983). Although this technique did make some early contributions, it soon became clear that the inter- versus intraspecific variation repeatedly found within morphospecies limited its usefulness in species delimitation outright (El Hage et al., 2000; Franke, 1973; Franke and Berry, 1972), and the technique was eventually abandoned.

The most influential development toward the use of molecular methods came in the 1980s with DNA sequencing for the purpose of phylogenetic investigations. These early works were taxonomically limited and typically involved the sequencing of a limited number of gene sequences (Baldauf, 1999; Baldauf and Doolittle, 1997; Cavalier-Smith, 1993; Otsuka et al., 1983; Rusk et al., 1995). Nonetheless, as more and more gene sequences emerged and the evolutionary history of myxomycetes became more apparent, molecular biology tools and applications were also rapidly advancing. In 2003 DNA was first successfully recovered from field-collected myxomycete fruiting bodies (Martín et al., 2003), substantially increasing research possibilities, especially since myxomycetes are notoriously difficult to culture. Less than 10% are easily cultivated in the laboratory (Haskins and Wrigley de Basanta, 2008). The use of molecular methods to study myxomycetes is especially valuable because it provides an alternative to the morphological identification of species of myxomycetes (the traditional and standard method of identification). Morphological identification of species is time consuming, requires a substantial amount of training, and, furthermore, is complicated by morphological plasticity and the presence of cryptic species (genetically distinct species with shared morphology) (Walker and Stephenson, 2016). Together these complications limit the value of morphological data as a stand-alone method for identification of taxa. Therefore, it is no surprise that molecular research methods have continued to be pursued and enhanced over the years.

The ability to use field-collected myxomycetes as a source of DNA substantially increased the opportunities for molecular work due to a large quantity of available herbarium specimens and the opportunities available to target myxomycetes in environmental samples and from any stage of the life cycle. The availability of molecular data and the use of various molecular techniques to study myxomycetes have been on a steep and steady incline.

NUCLEIC ACID ISOLATION

DNA Isolation

Nucleic acids can be isolated from any stage of the myxomycete life cycle (e.g., amoeboflagellates, plasmodia, or spores). The choice of which life stage and nucleic acid depends upon the research question and the source of starting material. Herein, we discuss the isolation of nucleic acids from the various myxomycete life stages and sample types.

Plasmodia represents one of two trophic stages in the myxomycete life cycle. A plasmodium contains up to 800,000 identical nuclei per mm^3 (Gray and Alexopoulos, 1968), and sometimes reaches considerable dimensions (e.g., a meter or more in total extent in a few species). Isolation of DNA from this stage is easily performed with an alkaline lysis protocol (Birnboim and Doly, 1979; Winter, 2016) and therefore is an especially good source for the isolation of large quantities of genetic material. Amoeboflagellate cells represent the other of the two trophic stages in the myxomycete life cycle. Similar to plasmodia, amoeboflagellate cells can be used to easily isolate genetic material. Amoeboflagellate cells are smaller (on an average approximately 10 μM), and typically contain a single nucleus. The isolation of nucleic acids from amoeboflagellate cells can be especially valuable in the context of single-cell isolations and environmental sampling. An alternative method can be used for both amoeboflagellates and plasmodia (Silliker et al., 2002); a longer lysis period for the plasmodial stage reduces carbohydrate contamination.

Spore-containing fruiting bodies are the most commonly used source of nucleic acids, particularly, DNA. Isolation of DNA from fruiting bodies is especially valuable because the molecular data can be directly interpreted in light of morphological identification of myxomycetes (species), which can only be carried out with mature fruiting bodies. However, unless the fruiting bodies have been generated through sterile, axenic culture, it is important to consider the possibility of contaminating DNA (e.g., from fungi or other myxomycetes) that may be present inside or on the surface of the fruiting body. Microcysts represent the remaining life stage from which nucleic acids can be isolated. Both spores and microcysts (a resting stage formed by amoeboflagellate cells) are two resistant stages in the myxomycete life cycle that can remain viable for long periods of time (Stephenson and Feest, 2012). Because spores and microcysts cannot be easily broken with enzymatic treatments, they must instead be physically disrupted to release genetic material for isolation. The most common method for breaking spore and microcyst walls is manual disruption using a sterile pestle or homogenization in the presence of glass beads, sand, or another similar material (Hoppe et al., 2010). Disruption of the cell wall in this way can be used to isolate nucleic acids from a very small number of spores or microcysts to many thousands of spores from a mature, spore-containing fruiting body.

Although there are no commercially available nucleic acid isolation kits specifically designed for myxomycetes at this time, a variety of kits are available that can be used to isolate and purify myxomycete nucleic acids. Isolation kit protocols typically begin with a cell disruption/homogenization stage in a protective lysis buffer. Often times, the kits can be modified (and often with advice from the manufacturer) to best suit the researchers needs based on the sample type. For example, many researchers have found success with the use of DNA isolation kits that were initially developed for the isolation of plant tissues or stool samples, simply by modifying the lysis step to include a bead-shaking step to break open spore cell walls.

RNA Isolation

Unlike genomic DNA, which can be isolated from both active and inactive cells (as well as cell-free DNA) in a sample, the isolation of RNA allows one to view the genes that are being actively transcribed. So where DNA analyses can be used to identify the presence of a certain taxon, RNA analyses measure gene expression and may suggest the functional role of a gene or taxon. Transcriptome analyses of individual cells and metatranscriptomes of communities are extremely powerful tools with great potential for studying myxomycetes. To date, however, far fewer groups have undertaken RNA analyses as compared to DNA analysis.

The isolation of RNA requires greater care than the isolation of DNA because it is far more easily degraded. For example, samples need to be prepared in a sterile and RNase-free environment. These precautions are not specific to myxomycetes; therefore, a great deal of literature is available for one to learn best practices. Otherwise, in principle, the isolation of RNA from myxomycetes is very similar to the isolation of DNA. Like DNA, RNA can potentially be isolated from any stage of the myxomycete life cycle, although obviously transcription levels will be drastically different between the different stages (e.g., active versus resting), so again, the choice of the appropriate life stage is dependent upon the research question. As described earlier for DNA isolation kits, a variety of RNA isolation kits are commercially available and can be modified for use with myoxmycetes (Kamono et al., 2009).

NUCLEIC ACID AMPLIFICATION AND APPLICATIONS

Nucleic acid amplification with the polymerase chain reaction (PCR) allows a quick and easy generation of an exponential number of copies of a targeted gene sequence. Amplified target sequences can then be used in a number of downstream applications (e.g., DNA cloning and sequencing to generate gene trees for phylogenetics and ecological surveys). In the following section some of the gene sequences most commonly targeted for amplification will be described.

Ribosomal Small Subunit (SSU)

The most widely used gene for molecular analysis in the myxomycetes is the ribosomal small subunit (SSU) RNA gene. Across many major taxa the SSU is an extremely slowly evolving gene, making it very important in phylogenetic studies, among other applications. In contrast, the myxomycete SSU is a highly variable marker and contains a large number of group I introns of various lengths (Fiore-Donno et al., 2010b, 2012; Nandipati et al., 2012; Pawlowski et al., 2012; Shadwick et al., 2009).

Due to a large number of introns, full-length sequences of the myxomycete SSU can be difficult to obtain, especially with a single PCR. However, many researchers target an intron-free region in the 5′ SSU of approximately 750 bp in length that can be amplified with a single pair of PCR primers (Fiore-Donno et al., 2012). This region of the myxomycete SSU has proven powerful in the context of phylogenetics and particularly in ecological studies (Fiore-Donno et al., 2016; Novozhilov et al., 2013a). Intraspecific SSU sequence variation can be significant between clades and even between multiple isolates of a single myxomycete morphospecies from a single population. When SSU sequence data are viewed parallel to morphological data, one may find identical SSU sequences for multiple collections of a single morphospecies (Fiore-Donno et al., 2011) or alternatively, SSU sequence diversity can be twice as great as morphological diversity (Novozhilov et al., 2013b).

Elongation Factor-1α

Elongation factor-1α (EF-1α) is another commonly sequenced gene in myxomycetes. Due to its role in protein synthesis, EF-1α is highly conserved (Baldauf, 1999), and is therefore especially valuable for constructing myxomycete phylogenetic trees (Baldauf and Doolittle, 1997; Feng and Schnittler, 2015; Fiore-Donno et al., 2005; Fiore-Donno et al., 2010b). The nearly complete sequence of EF-1α of *Physarum polycephalum* can be amplified with a single pair of universal primers (Baldauf and Doolittle 1997), although multiple primer pairs are often required for other taxa (Fiore-Donno et al., 2005; Hoppe and Kutschera, 2010). Furthermore, primers can be designed across an insertion that is found solely in animals and fungi (Baldauf and Palmer, 1993), thereby eliminating any contamination by fungal or animal DNA (Fiore-Donno et al., 2005).

Internal Transcribed Spacer (ITS)

The internal transcribed spacer (ITS) lies between the small and large subunits of the ribosomal RNA (rRNA) gene sequences and contains the fastest evolving region of the rRNA in many organisms (White et al., 1990). Therefore, where the slowly evolving EF-1α sequence has been most useful for determining deeper evolutionary relationships, ITS have proven to be valuable for analyzing intraspecific variation in some taxa. ITS was the first molecular marker to be examined for the purpose of identifying intraspecific variation in myxomycetes (Martín et al., 2003; Winsett, 2010). However, unlike the successful use of ITS sequencing in fungi (Schoch et al., 2012; White et al., 1990) and some other groups, the use of ITS sequence analysis in myxomycetes has been rather limited. Myxomycete ITS sequences can be extremely variable in both sequence and length (Winsett and Stephenson, 2008) and in many cases cannot even be aligned among closely related taxa (Fiore-Donno et al., 2011; Martín et al., 2003). The high level of sequence variation limits its usefulness as a stand-alone marker in the myxomycetes, although as a supplementary marker it can be valuable for differentiating between some species (Baba et al., 2015; Winsett and Stephenson 2008).

Cytochrome *c* Oxidase (COI)

Due to its high rate of evolution, and its universality among eukaryotes (mitochondriate), the Cytochrome *c* Oxidase (COI) gene was proposed as a barcoding marker that could aid in the identification of all already known species and as a way to reveal previously unknown biodiversity (Brown et al., 1979; Hebert et al., 2003; Moore, 1995). However, in myxomycetes success with use of the COI gene sequence has been limited due to the high levels of sporadic variation among lineages.

The universal primers proposed for COI barcoding (Folmer et al., 1994) do not reliably amplify myxomycete DNA, and even with primers specifically designed to amplify the first half of the myxomycete COI (Walker et al., 2011), sequences can be difficult to obtain and provide varying results (Feng and Schnittler, 2015; Liu et al., 2015). Similar to ITS, when COI is used as a supplemental tool, it has proven successful for some amoebae (Nassonova et al., 2010) and other protists (Barth et al., 2006; Chantangsi et al., 2007; Evans et al., 2007). It is possible that COI sequencing in the myxomycetes will prove to have some applications; however, the application does not seem to be in the broad identification of species as has been suggested, nor can it be used for reconstructing deeper evolutionary relationships among myxomycetes.

Sequencing and Applications of Gene Amplification

DNA sequencing has undergone substantial changes since the earliest technologies were developed in the 1960s (Heather and Chain, 2016). No matter the technology utilized, the goal remains to determine the order and composition of nucleotides in a piece of DNA. The ability to decode DNA sequences is now an essential tool in biological research as it can be used to infer a variety of information, such as biochemical properties and phylogenetic relationships. The first sequences of myxomycete origin were generated in the context of phylogenetic studies (Baldauf and Doolittle, 1997). These early studies were limited by the sequencing technology of the time and the availability of myxomycete DNA. Now it is possible to obtain myxomycete DNA from any life stage, either from individual isolates or from communities represented in environmental samples. And with drastically improved sequencing methods, it is now possible to obtain

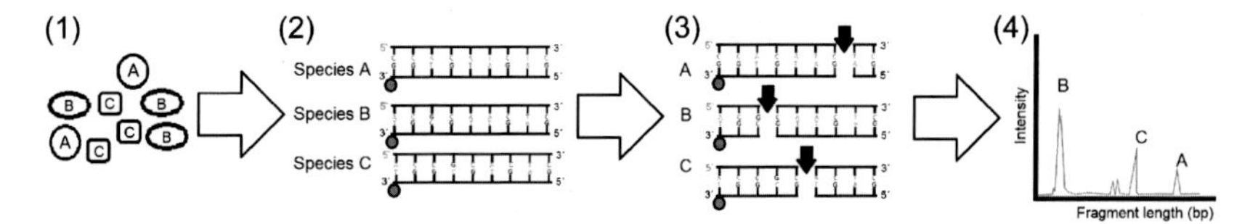

FIGURE 5.1 Workflow of a terminal restriction fragment length polymorphism analysis. (1) Soil community with three different ribotypes (A, B, and C), (2) amplification of a homologous gene and labeling of the ends, (3) digestion with restriction enzyme to reveal length polymorphism by, and (4) visualization of the different sized fragments.

millions of reads from a sample with the use of high-throughput sequencing (HTS). In this section, several applications are described that utilize one or more DNA sequencing technologies (Sanger sequencing or HTS technology) to study myxomycete biodiversity or ecology.

Terminal Restriction Fragment Length Polymorphism (TRFLP)

Terminal restriction fragment length polymorphism (TRFLP) is based upon the presence of specific PCR-fragment lengths for individual genotypes after digestion with a restriction enzyme (Fig. 5.1). This technique allows a quick and economical way to compare different communities; it has been used to investigate prokaryotic and fungal communities (Hoppe and Schnittler, 2015).

In TRFLP fluorescently labeled PCR primers are designed to target and amplify a gene of interest. The resulting amplicons are subjected to digestion with a restriction enzyme, which identifies sequence polymorphisms and results in fragments of various lengths. The fluorescently labeled fragments are then fractionated using a DNA sequencer (Liu et al., 1997). TRFLP with myxomycete SSU was used to identify substrate specific variation in myxomycete communities (Hoppe and Schnittler, 2015).

Denaturing Gradient Gel Electrophoresis (DGGE)

Denaturing gradient gel electrophoresis (DGGE) is a technique that has been used to separate a mixture of DNA fragments according to their melting point, to analyze microbial communities without cultivation. This technique was first used for investigations of prokaryotic communities (Muyzer et al., 1993) but was later adopted to detect the presence of airborne myxomycetes (Kamono and Fukui, 2006) and to investigate myxomycete communities inhabiting the soil (Kamono and Fukui, 2006). Similar to TRFLP, primers are first designed to amplify target gene sequences from the samples of interest (typically SSU for myxomycetes). Next, PCR products are run along a urea gradient in a polyacrylamide gel to separate fragments. Partially melted products migrate more slowly than compact molecules through the gel, allowing the detection of sequence variation.

Not only is this method valuable for making a visual comparison of communities (similar to a fingerprint), but also, well-separated bands can be extracted from the gel to be sequenced and identified. By isolating and sequencing bands from a DGGE analysis, Ko et al. (2009) generated 13 new SSU sequences from decaying wood and leaves on the forest floor. DGGE is still a valuable tool for obtaining environmental sequences, allowing for detection and even identification of myxomycetes without the requirement of a fruiting body. However, many find this technique to be clumsy and rather time consuming. Additionally, as sequencing technologies improve and prices decline, DGGE analysis is declining in popularity as more and more researchers turn to HTS.

High-Throughput Sequencing

Unlike Sanger sequencing, the latest generation of truly HTS technologies allows one to simultaneously generate millions of sequences (Kircher and Kelson, 2010). Several HTS sequencers are currently available (e.g., Illumina's MiSeq and Life Technologies' Ion Torrent); the various technologies utilized by each continue to increase the ease and accuracy of generating large amounts of sequence data with reduced time and cost (Heather and Chain, 2016). Examples of HTS technologies being utilized to study myxomycetes are only now emerging but show great promise.

The first published example of HTS of myxomycetes was by Fiore-Donno et al. (2016), wherein the authors investigated natural populations of myxomycetes and *Acanthamoeba* in soils from three sites across Germany. The authors obtained more than 900,000 reads. Stringent filtering of the myxomycete reads was applied to remove poor quality reads, duplicates, and chimeras; the authors identified 338 operational taxonomic units (OTUs) which represented 35 unique BLAST hits. These hits represent 10 myxomycete genera and 2 unidentified taxa. Their study confirms that myxomycetes are abundant soil predators in grassland soils and also identified a biogeographic pattern of the communities between the three sites investigated.

As this example highlights, the use of HTS technology is likely to have a great impact on the study of myxomycetes. Until recently, myxomycetes could be identified only through morphology of the fruiting body. Therefore, specimens in the soil (or any microhabitat) that have lost the ability to form fruiting bodies or could not be cultured to fruiting body formation for any reason, could not be identified or enumerated. However, when total DNA is isolated from an environmental sample, all life stages of the individuals present can be detected, eliminating the requirement for a mature fruiting body. Targeted environmental sequencing in this way allows a glimpse into the whole, natural myxomycete community. There are still limitations with the use of HTS and environmental sequencing that need to be considered and improved upon, such as PCR bias and a limited number of reference sequences. Nonetheless, implementation of HTS particularly with environmental samples will allow us to gain information about hitherto unknown communities, genetic diversity, and will likely reveal undescribed species (especially uncultivable and rare or non-fruiting species).

CONCLUDING REMARKS

Although the use of molecular techniques for studying myxomycetes occurred rather late in comparison to most other groups of organisms, already significant advances have been made. DNA sequencing of conserved genes for phylogenetic reconstruction has revealed the monophyly of the myxomycetes and their placement on the eukaryotic tree of life in the Amoebozoan clade (Adl et al., 2005; Baldauf and Doolittle, 1997; Cavalier-Smith, 1998; Fiore-Donno et al., 2005, 2010b; Shadwick et al., 2009). Molecular gene trees have confirmed the presence of two major groups of myxomycetes commonly referred to as the dark and bright-spored clades (Cavalier-Smith, 2013; Fiore-Donno et al., 2005). Molecular data are also revealing unexpected branching patterns which in many cases contrast with previous understanding of myxomycete phylogeny based largely on morphological data within both the dark-spored (Fiore-Donno et al., 2008, 2012) and bright-spored clades (Fiore-Donno et al., 2013; Leontyev et al., 2014; Walker et al., 2015).

The sequence of the very first complete myxomycete genome (of *Physarum polycephalum*) was recently reported (Schaap et al., 2016). This is a significant accomplishment and provides a wealth of valuable data for various areas of study, from myxomycete physiology to our understanding of the last common ancestor of all eukaryotes. DNA sequencing has also revealed the existence of at least one non-fruiting group of myxomycetes, the previously named "hyper-amoebae" (Fiore-Donno et al., 2010a; Walochnik et al., 2004). The high level of genetic diversity identified in natural populations of myxomycetes hints toward the existence of cryptic species (Aguilar et al., 2014; Novozhilov et al., 2013b). From the early environmental studies that utilized DGGE and TRFLP, we learned about the diversity of myxomycetes inhabiting soils (Kamono and Fukui, 2006) and the importance of wind-dispersal of spores (Kamono and Fukui, 2006). More recently, environmental soil samples have been subject to metatranscriptomics (Geisen et al., 2015; Urich et al., 2008) and targeted sequencing approaches (Fiore-Donno et al., 2016) utilizing HTS to reveal the great abundance and diversity of myxomycetes in those microhabitats. As HTS technologies continue to increase, the availability and amount of data that can be generated, the processing and analyses of the resulting data has become more and more complex and computer intensive. However, due to a significant value of such techniques, efforts should focus on improving the use of this technology for the study of myxomycetes. Annotated reference sequence databases are particularly in need of expansion. Also, the use of single-cell genomics and transcriptomics should be investigated. The use and improvement upon the techniques described herein and other newly emerging, powerful methods will allow us to gain a deeper understanding of myxomycete habitat preference and biogeography.

PART B: COMPARATIVE MOLECULAR BIOLOGY AND USE OF MYXOMYCETES AS MODEL ORGANISMS

MYXOMYCETES AS MODEL ORGANISMS: PROMISES, DISAPPOINTMENTS, AND DISCOVERIES

The peculiar biology of *P. polycephalum* and its myxomycete relatives has often promised opportunities for making progress in understanding basic eukaryotic cell and molecular biology; however, various experimental challenges have frequently defeated the fulfillment of these promises. Yet, the continued study of this organism and its relatives has yielded unexpected biological wonders, such as complex mobile introns, unique RNA editing patterns in the mitochondria (reviewed in Chapter 4), and "brainless" coordinated activity (Chapter 11).

In an excellent review of *P. polycephalum* biology, Burland et al. (1993b) conjectured: "Were Jane Austen a late 20th-century molecular biologist, she might well write a novel, as she did nearly two centuries ago, entitled *Pride and Prejudice*, modifying the first sentence only slightly, to read… It is a truth universally acknowledged that a single-celled organism in possession of good cell biology must be in want of DNA transformation."

Were Jane Austen a 21st-century molecular biologist, her character would surely be in want of a genome. This section reviews molecular work done mostly with *P. polycephalum* in the absence of a seamless transformation system and reference genome. The recent publication of the *P. polycephalum* genome (Schaap et al., 2016), transcriptome data (Barrantes et al., 2010, 2012; Bundschuh et al., 2011; Glöckner et al., 2008, 2016; Watkins and Gray, 2008), and the ability to disrupt gene function by RNAi (Haindl and Holler, 2005; Materna and Marwan, 2005; Pinchai et al., 2006) bring renewed motivation and tools for pursuing the study of *P. polycephalum* and its relatives.

The Cell Cycle

Rusch (1980) described seizing upon *P. polycephalum* in the 1950s as the ideal research organism for studying cancer since the synchronous mitotic divisions in the one-celled, massive, multinucleated plasmodium should allow biochemical dissection of the cell cycle and then factors which disrupt regulated cell division. Early on, a phosphorylating activity was found in *P. polycephalum* plasmodia that fluctuated during the cell cycle and could be a mitotic trigger (Bradbury et al., 1974). Detailed descriptions were made of the mitotic cycle; experiments manipulated the timing of the cycle by somatically fusing plasmodia at different stages (Holt, 1980; Rusch et al., 1966; Wolf and Sauer, 1982). However, by 1985 researchers were frustrated at the lack of established cell cycle mutants in *P. polycephalum* (Laffler and Tyson, 1986). An understanding of the cell cycle was instead obtained by working with other organisms (Nurse, 2000), but subsequent work confirms that cell cycle control in *P. polycephalum* is similar to most eukaryotes (Ducommun et al., 1990; Li et al., 2004; Schaap et al., 2016).

With new tools and genomic data, cell cycle studies with *P. polycelphalum* may be worth revisiting, particularly with respect to regulation of closed versus open mitosis that occurs at different stages of the life cycle (Solnica-Krezel et al., 1991).

Somatic Incompatibility Systems

Observations of *Didymium iridis* (Collins, 1966) and *P. polycephalum* (Carlile and Dee, 1967) revealed that diploid plasmodia could fuse. Since plasmodia engulf food, they must be able to meet and fuse with themselves around a food source. When two compatible haploid myxamoebal clones are mixed, the resulting plasmodial population is genetically identical and can also fuse. Extensive studies with both *D. iridis* and *P. polycephalum* have defined the incompatibility systems that govern self/nonself recognition between plasmodia (for reviews, see Carlile and Gooday, 1978; Collins and Betterley, 1982; Schrauwen, 1984). In both *D. iridis* and *P. polycephalum* multiple compatibility loci were identified. Differences at strong loci result in fast avoidance reactions while differences at weak loci cause a slow response that allows some cytoplasmic mixing and cytotoxicity in the fusion region. In some cases, the plasmodia appear to fuse compatibly but ultimately one of the nuclear types is eliminated (Carlile and Gooday, 1978). These genetic studies have not been pursued at the molecular level, though a study by Schrauwen (1979) presents evidence that incompatibility reactions trigger new gene expression.

While these observations have not greatly contributed to our understanding of cell–cell recognition, they have provided a useful tool for studying certain biological processes. For example, fusions between plasmodia at different stages in the cell cycle resulted in synchronous mitosis at a time intermediate to what would be predicted for the individual plasmodia (Rusch et al., 1966). Plasmodial fusions have also been used to perform complementation experiments between sporulation mutants, or between plasmodia that have been differently induced to sporulate. These studies are described later in the section on development (Marwan, 2003a,b; Marwan et al., 2005; Sujatha et al., 2005). Plasmodial fusions between plasmodia of different ages, age heterokaryons, have been used to study senescence in *D. iridis* (Clark and Hakim, 1980; Clark and Lott, 1989; Collins and Betterley, 1982). Fragmented microplasmodia fuse to form networks in a pattern consistent with general network theory (Fessel et al., 2012). The plasmodium, and its ability to readily undergo somatic fusions, provides a unique tool for studying cell biology.

Plasmodial Structure, Motility, Behavior, and Learning

The spectacular multinucleate plasmodium is a defining feature of the myxomycetes. A review by Kessler (1982) focusing on *P. polycephalum*, summarized what was known about the structure and mechanism of plasmodial motility up to that time. The unicellular, multinucleate plasmodium is organized into an ectoplasmic gel layer that defines the tubular channels surrounding the fluid sol endoplasm. The plasma membrane is highly invaginated and closely associated with internal fibrils that are longitudinally and radially arrayed along the vein-like tubules. Actin- and myosin-powered contractions create hydraulic pressure within the tubules, which move the endoplasm at velocities as high as 1.3 mm/sec. The mass flow of cytoplasm is then reversed, resulting in shuttle-streaming. The timing of flow reversals occurs about every 1.5 min, though net flow is in the direction of movement. Macroscopically, a migrating plasmodium appears as an advancing intricate fan of protoplasm with trailing veins. The probable role of calcium in controlling the contractions was proposed (Kessler, 1982). Yoshiyama et al. (2010) experimentally showed the relationship between internal calcium concentrations and contraction–relaxation. Kessler (1982) proposed a stretch entrainment model to explain the rhythm of contractions. More recent work refined the reciprocal relationship between cell shape and contraction (Nakagaki et al., 2000b) and models of plasmodial morphology in relation to force (Alim et al., 2013; Lewis et al., 2015).

While early studies concentrated on plasmodial chemotaxis in relation to various attractants and repellents (Ueda and Kobatake, 1982), the focus has shifted to viewing the plasmodium as a decision making adaptive network, with primitive learning. Nakagaki et al. have demonstrated how *P. polycephalum* plasmodia reorganize their structure to maximize foraging efficiency. In one experiment, a plasmodium was grown to fill a maze; when a food source was positioned at the entrance and exit of the maze, the plasmodium reorganized itself to mass upon the food sources but remain connected through the maze by the shortest distance (Nakagaki et al., 2000a). Regions of the plasmodium inhabiting dead ends in the maze were retracted to decrease the path length between the food sources. Similar experiments were conducted with two, three, and multiple food sources (Nakagaki et al., 2001, 2004; Tero et al., 2010); the behavior determining computational calculations of plasmodia have been described by mathematical models (Tero et al., 2006, 2010). In further studies, plasmodia were confronted with multiple kinds of inputs. Latty and Beekman (2010) challenged plasmodia to discriminate between high- and low-quality food sources in a shaded environment. When the higher quality food was in a lighted environment, which plasmodia typically avoid, they showed that plasmodia would "risk" foraging in the lighted area, but only when the reward was significant (5 times greater food concentration). They called this "behavioral titration," indicating that plasmodia were making multiobjective foraging decisions. Remarkably, Dussutour et al. (2010) showed that *P. polycephalum* plasmodia would select different nutrient food patches that precisely optimized its nutritional requirements. More complicated decision making scenarios (Beekman and Latty, 2015; Reid et al., 2016) highlight the multiple sensory inputs that need to be integrated. It is perhaps no surprise that the genome of this adaptive foraging organism has been found to be highly enriched in genes for sensory receptors and signaling pathways (Schaap et al., 2016).

Group I Introns in the Myxomycetes

The myxomycetes are a rich source of nuclear group I introns; their study has provided insights into their modes of dispersal and evolution, as well as their biological mechanisms and functions, (Haugen et al., 2005; Hedberg and Johansen, 2013; Nielsen and Johansen, 2009). Group I introns are identified by their characteristic secondary and tertiary structures; however, within this major class of introns distinctive subgroups exist (Michel and Westhof, 1990; Zhou et al., 2008). Nuclear group I introns (as opposed to those that occur in mitochondria, chloroplasts, or viruses) occur only in the ribosomal DNA genes of eukaryotic microbes (Hedberg and Johansen, 2013).

Introns were first discovered in the myxomycetes in the large subunit (LSU) ribosomal DNA (rDNA) genes of *P. polycelphalum* by DNA/RNA hybridization and electron microscopy (Gubler et al., 1979). Johansen et al. (1992) determined the secondary structure of these group I introns in *P. polycephalum* along with two homologous introns in *D. iridis* that had similar secondary structures and identical insertion locations within the LSU rDNA gene. As more sequence data accumulated (Vader et al., 1994; Wikmark et al., 2007), it became clear that the introns first discovered in *P. polycephalum* occur throughout the Physarales in the LSU rDNA genes at base pair positions 1949 and 2449 (this is based on the numbering system in the *E. coli* LSU rDNA gene; Raue et al., 1988). Phylogenies of the Physarales based on introns at LSU positions L1949 and L2449 are in agreement with the LSU rDNA phylogeny of host strains, suggesting that the introns were present in the ancestor of the Physarales and that the introns are vertically inherited within this group (Wikmark et al., 2007). A phylogenetic analysis of the Physarales based on the nuclear large and small (LSU and SSU) rDNA sequences indicated that the Physaraceae is actually polyphyletic, consisting of three distinct clades (Nandipati et al., 2012). Interestingly, variation in a conserved region of the LSU L2449 intron in members of the Physaraceae also supported the three clades. Within the Didymiaceae (another family within the Physarales) the core conserved regions of the L1949 and L2449 introns are nearly identical; however, the peripheral loop regions show significant variation of possible taxonomic value (Wikmark et al., 2007). Another group I intron was found exclusively in the Didymiaceae at position 1389 of the SSU (S1389); it appears to be vertically inherited within this group (Wikmark et al., 2007). Other recent papers have demonstrated the utility of introns in revealing cryptic species within morphological species (Feng et al., 2016; Feng and Schnittler, 2015). Known insertion sites for nuclear group I introns in the myxomycetes have been summarized by Nandipati et al. (2012). Some of these are obligatory and vertically inherited, whereas others are optional.

Sporadically occurring, optional group I introns, located at certain fixed positions in the rDNA, are present in very diverse organismal groups. For example, the S516 group I intron found in the SSU of *Fuligo septica, Badhamia gracilis,* and *Physarum flavicomum* (Fse.S516, Bgr.S516, and Pfl.S516, respectively) appears to be related to S516 introns found in amoebae and brown algae (Haugen et al., 2003). An emerging understanding of the biology of introns is beginning to explain their distribution and evolution. The vertically inherited introns discussed earlier are not self-splicing in vitro, and instead are likely dependent on host proteins for splicing. The persistence of these introns may even argue that they have evolved to benefit the host (Hedberg and Johansen, 2013; Nielsen and Johansen, 2009).

Vertically inherited introns may represent an endpoint in a cycle proposed by Goddard and Burt (1999), see also Haugen et al. (2005) and Nielsen and Johansen (2009). The cycle begins with an initial intron acquisition by horizontal transmission catalyzed by an intron encoded homing endonuclease gene (HEG). Once present in the genome, intron-lacking rDNA genes are converted into intron-containing genes by the activity of the homing endonuclease. The homing endonuclease makes a double-stranded DNA break at the target site that is then repaired by gene conversion to include the intron. Once all of the intron-lacking genes are converted to intron-containing genes, the activity of the endonuclease may be lost by mutation over time. Subsequent loss of the intron may occur or it may be stably/vertically inherited. Hedberg and Johansen (2013) reviewed different types of myxomycete group I introns that could represent these various stages proposed by Goddard and Burt (1999).

One other mode of horizontal transmission has been identified involving an RNA intermediate. Haugen et al. (2003), working with *F. septica,* showed the presence in vitro of a circular RNA molecule containing the full-length intron which could be a substrate for reverse-splicing or retro-homing, whereby the RNA is reverse transcribed into DNA and inserted into the genome. The *F. septica* Fse.S516 intron does not contain an HEG, but it is related to other HEG containing introns, so it may represent a degenerate homing intron (Haugen et al., 2003). Another *D. iridis* intron Di.S956-1 (discussed in detail further) appears to use both DNA and RNA mediated modes of transmission (Birgisdottir and Johansen, 2005). *F. septica* has been found to contain twelve group I introns in its rDNA genes, all of which are non-HEG containing, and, all but the vertically inherited L1949 and L2449 introns, are self-splicing (Lundblad et al., 2004). Twenty group I introns have been found in the rDNA genes of *Diderma niveum*, representing the various types of introns discussed so far (Nielsen and Johansen, 2009).

Two myxomycete introns in particular have contributed to the larger discussion of intron biology. The first of these (Ppl.L1925) is a strain-specific group I intron found in *P. polycephalum,* which exactly matches the insertion site of a self-splicing group I intron in the well-studied ciliate *Tetrahymena thermophila* (Cech and Bass, 1986; Muscarella and Vogt, 1989). The core structure of the introns in both groups were similar; however, only the *P. polycephalum* intron was shown to encode a site-specific endonuclease capable of making a double-stranded cut at the intron insertion site (Muscarella et al., 1990). Further, they showed evidence for intron transposition in matings between intron-containing and intron-lacking strains, resulting in the conversion of intron-lacking sites (Muscarella and Vogt, 1989). The Ppl.L1925 intron (see Johansen and Haugen (2001) for group I intron nomenclature) has been intensively studied as the first nuclear transposable/mobile group I intron. The intron-encoded endonuclease I-*Ppo* has been characterized (Muscarella et al., 1990); when the intron was expressed in the yeast *Saccharomyces cerevisiae*, it led to transposition into the yeast rDNA gene where it retained its mobility (Muscarella and Vogt, 1993).

The second group I intron of note is Dir.S956-1 found in *D. iridis*. When first described (Johansen and Vogt, 1994), three components were identified in Dir.S956-1, two ribozymes and one open reading frame (ORF). It was thought that this compound intron was formed by an intron inserting into another intron to create a twintron. Dir.S956-1 was the first example of a group I twintron (Decatur et al., 1995). Further studies (Decatur et al., 1995) characterized the activities of the two ribozymes, GIR1 and GIR2 (the ORF is between the two ribozymes). The GIR2 ribozyme showed the typical group I activity involved in intron self-splicing. The GIR1 ribozyme appeared to be involved in processing or expression of the I-*Dir*I endonuclease encoded by the ORF. The authors noted that protein coding mRNAs are typically transcribed by RNA polymerase II and further processed in preparation for translation, however, rDNA genes, and their introns, are transcribed by RNA polymerase I where the transcript is the final end product. This raised questions about the mode of processing of the I-*Dir*I transcript.

In vivo, the Dir.S956-1 intron was shown to be mobile and able to convert intron-lacking rDNA genes into intron-containing genes in matings between intron$^+$ and intron$^-$ haploid strains (Johansen et al., 1997). When I-*Dir*I was expressed in *E. coli*, a site-specific endonuclease could be isolated that cleaved intron-lacking DNA at the intron insertion site. The maturation of the I-*Dir*I mRNA was further elucidated by Vader et al. (1999). The GIR1 ribozyme appeared to be responsible for cleaving itself from the I-*Dir*I transcript. Then the I-*Dir*I transcript is processed by truncation at the 3′ end and polyadenylation. In addition, a small intron was spliced from the I-*Dir*I transcript. This suggested that the GIR1 ribozyme's role was a first step in maturation of the I-D*ir*I transcript, resulting in a substrate for the RNA polymerase II processing pathway. In fact, the GIR1 ribozyme cleaves by a transesterification branching reaction, which forms a three-nucleotide lariat on the 5′ end of the released endonuclease transcript (Nielsen et al., 2005). The authors proposed that the small lariat (or lariat cap) would stabilize the mRNA in a way similar to the me^7-G-cap added to the 5′ end of RNAs transcribed by RNA polymerase II.

Remarkably, the sequence of the processes described earlier is regulated by a conformational switch, or riboswitch, in the Di.S956-1 intron (Nielsen and

Johansen, 2009). When the intron is transcribed a hairpin structure called HEG P1 (homing endonuclease gene P1) is formed at the juncture between the GIR1 ribozyme and the I-*Dir*I coding region. This conformation interferes with the intramolecular folding that favors the transesterification branching cleavage, thereby insuring that transcription of the entire twintron occurs before GIR1 acts. Complete transcription of the entire twintron produces the GIR2 ribozyme responsible for splicing out of the complete intron. Once the complete intron is spliced out, a new folding pattern is favored which disrupts the inhibitory HEG P1 hairpin and favors GIR1 cleavage activity which cuts and caps the I-*Dir*I transcript. At this stage, formation of the HEG P1 hairpin at the 5′ end of the I-*Dir*I is again favored and may facilitate the release of the transcript from the GIR1 ribozyme.

While group I introns may help to sort out some of the evolutionary history and taxonomy of the myxomycetes, their introns have also contributed greatly to our understanding of the biology of introns.

Molecular Approaches to Studying Development in Myxomycetes

The complex life cycle of these relatively simple organisms provides distinct stages where developmental transitions can be studied. Descriptions of these stages have been reviewed (Rusch, 1980; Sauer, 1982; Wick and Sauer, 1982) and include the following transitions: myxamoebae to cysts, myxamoebae to flagellated cells, myxamoebae to plasmodia, plasmodia to spherules (sclerotia), and plasmodia to fruiting bodies.

Gorman and Wilkins (1980) provided a detailed picture of the characteristics that define each of these stages. For example, plasmodia have closed mitosis, produce mucopolysaccharide slime, are pigmented, can phagocytize amoebae, readily undergo somatic fusion, move by oscillatory cytoplasmic streaming, and can form spherules (also known as sclerotia), or fruiting bodies. Myxamoebae, on the other hand, have open mitosis, move by pseudopodia, can become flagellated in the presence of a liquid, can encyst, and can become plasmodia by mating or selfing (apogamy). Finer distinctions can also be made. For example, plasmodia in a sporulation-competent state can have three different fates—growth, sporulation, or spherulation—depending on environmental cues. A plasmodium that has been starved for a period of time and then receives a brief period of illumination becomes *committed* to sporulation within a few hours. Myxamoebae, on the other hand, divides vegetatively at low cell densities but are induced to become *competent* to mate by substances that accumulate at high cell densities. A variety of developmental mutants have been identified to further understand these transitions (Gorman and Wilkins, 1980; Haugli et al., 1980).

Pallotta et al. (1986) was the first to describe cloning of stage specific cDNAs from *P. polycephalum* myxamoebae, plasmodia, spherules, and spores. During the 1980s and 1990s several working groups created cDNA libraries from various stages of *P. polycephalum* (summarized in Table 5.1). The cDNAs,

TABLE 5.1 Stage Specific cDNA Libraries

RNA source(s)	References
Myxamoebae, plasmodia, spherules, spores	Pallota et al. (1986a)
Myxamoebae, plasmodia	Pallota et al. (1986b)
Sporulation, spherulation	Schreckenbach and Werenskiold (1986)
Myxamoebae, plasmodia	Sweeney et al. (1987)
Sporulation	Martel et al. (1988)
Myxamoebal–plasmodial transition	Bailey et al. (1999)
Sporulation	Kroneder et al. (1999)

Bailey, J., Cook, L.J., Kilmer-Barber, R., Swanson, E., Solnica-Krezel, L., Lohman, K., Dove, W.F., Dee, J., Anderson, R.W., 1999. Identification of three genes expressed primarily during development in *Physarum polycephalum*. Arch. Microbiol. 172, 364–376.
Kroneder, R., Cashmore, A.R., Marwan, W., 1999. Phytochrome-induced expression of *lig1*, a homologue of the fission yeast cell-cycle checkpoint gene *hus1*, is associated with the developmental switch in *Physarum polycephalum* plasmodia. Curr. Genet. 36, 86–93.
Martel, R., Tessier, A., Pallota, D., Lemieuz, G. 1988 Selective gene expression during sporulation of *Physarum polycephalum*. J. Bacteriol. 170, 4784–4790.
Pallotta, D., Bernier, F., Hamelin, M., Martel, R., Lemieux, G., 1986a. cDNA Cloning of *Physarum polycephalum* stage specific mRNAs. In: Dove, W.F., Dee, J., Hatano, S., Haugli, F., Wohlfarth-Bottermann, K.E. (Eds.), The Molecular Biology of *Physarum polycephalum*. Plenum Press, New York, pp. 315–327.
Pallotta, D., Laroche, A., Tessier, A., Shinnick, T., Lemieux, G., 1986b. Molecular cloning of stage specific mRNAs from amoebae and Plasmodia of *Physarum polycephalum*. Biochem. Cell Biol. 64, 1294–1302.
Schreckenbach, T., Werenskiod, A.-K., 1986. Gene expression during plasmodial differentiation. In: Dove, W.F., Dee, J., Hatano, S., Haugli, F., Wohlfarth-Bottermann, K.E. (Eds.), The Molecular Biology of *Physarum polycephalum*. Plenum Press, New York, pp. 131–150.
Sweeney, G.E., Watts, D.I., Turnock, G., 1987. Differential gene expression during the amoebal-plasmoidal transition. Nucl. Acids Res. 15, 933–945.

reversed transcribed from RNA at different time points, provided some evidence of differential gene expression.

Developmental transitions that were once studied a few genes at a time can now be studied by comparing changes to transcriptomes, the collection of genes expressed at a certain time point. *P. polycephalum* plasmodia are considered competent to sporulate after a period of starvation; exposure to light commits plasmodia to sporulation after an incubation period. The transcriptomes of competent plasmodia and light-induced committed plasmodia have been compared (Barrantes et al., 2010). The transcripts were classified into three categories (biological processes, molecular function, and cellular components) using BLAST2GO (Gotz et al., 2008). Down- and upregulated genes were identified; cell death–associated transcripts, in particular, were upregulated. A related study carried out the same analysis obtaining RNA from a single macroplasmodium at the two developmental stages; the study was in good agreement with the previous study using batches of cells (Barrantes et al., 2012).

The transcriptome studies with competent and committed plasmodia build upon earlier studies where various sporulation mutants were characterized by demonstrating time-resolved somatic complementation (Marwan, 2003a,b; Marwan et al., 2005; Sujatha et al., 2005). These experiments exploit the ability of plasmodia to fuse somatically. For example, a *pho-1* mutant plasmodium that cannot be induced by light to sporulate, can be fused with a fruiting body morphogenesis mutant *vac-1* that has been induced by light, in combination, complementation occurs and normal sporulation results (Marwan, 2003b).

The expression of sporulation important genes identified in these transcriptome studies was also later found to be a reliable predictor of sporulation among individual plasmodia (Hoffmann et al., 2012). In addition, the expression of these genes was found to be altered in sporulation mutants (Rätzel et al., 2013). When plasmodia of sporulation mutants were fused and followed over time, the expression levels of the sporulation marker genes synchronized and determined the developmental fate in an all-or-nothing response (Walter et al., 2013). A detailed study of correlated gene expression patterns suggests alternative programming of differentiation (Rätzel and Marwan, 2015). These studies take full advantage of the biology of *P. polycephalum* as a model developmental system; nearly genetically identical somatic cells can be somatically fused, the multinucleate plasmodial cells are large and synchronize rapidly, a plasmodium can be subdivided for different treatments, or subcultured to follow a time course of development. Another advantage of working with *P. polycephalum* is that it is holocarpic, meaning the entire cell mass is converted into spores. This is in contrast to most organisms where germ cells are produced within somatic tissue.

Transfection of *Physarum polycephalum* and RNAi

First attempts to introduce foreign DNA into *P. polycephalum* were reported by Haugli and Johansen (1985). They chose yeast and mammalian vectors with the kanamycin resistance gene, which confers resistance to G418 (an antibiotic to which *P. polycephalum* is sensitive). They added the origin of replication from the *P. polycephalum* rDNA minichromosomes to the vectors (Ferris and Vogt, 1982). Myxamoebal cells were treated with either $CaCl_2$ or polyethylene glycol (PEG) prior to treatment with plasmid DNA. Although the frequency of transformants was low, four out of six strains tested retained resistance to G418 in the absence of selection.

McCurrach et al. (1988) achieved transient chloramphenicol transferase (*cat*) expression by either electroporation or $CaCl_2$ treatment of *P. polycephalum* myxamoebae. The *cat* gene was placed downstream of a putative promoter from the long terminal repeat (LTR) element in the *P. polycephalum* genome. This plasmid also included the origin of replication from the *P. polycephalum* rDNA minichromosome.

To identify other *P. polycephalum* genomic sequences with promoter function, yeast/bacterial shuttle vectors were constructed with either a 5′ truncated or intact hygromycin resistance (*hph*) gene inserted into the vector multiple cloning site (Burland et al., 1991). *P. polycephalum* genomic DNA fragments (1–5 kb in size) were then inserted upstream of the hygromycin insert to capture promoter sequences that were tested by transfection of the constructs into yeast cells followed by hygromycin selection. In addition, a 1.1 kb fragment was fused upstream of the truncated hygromycin gene, which contained the region upstream of the *P. polycephalum ardC* actin gene along with the first eight codons of the *ardC* gene. This sequence was named *PardC*, promoter of the *ardC* gene. One of the constructs with a randomly selected *P. polycephalum* fragment conferred resistance to hygromycin in fission yeast, but it functioned poorly as a promoter in budding yeast. In contrast, the *PardC-hph* construct conferred resistance in both yeast types, suggesting that this was a strong promoter in a variety of hosts. Pierron et al. (1999) later showed that the *PardC* sequence functions as an origin of replication, as well as a promoter.

A series of experiments by Burland et al. used electroporation and the *PardC* promoter to improve transfection methods for *P. polycephalum* myxamoebae. The chloramphenicol transferase (*cat*) gene could be transiently expressed in *P. polycephalum* when a construct containing the *PardC* promoter, the *cat* gene, and *TardC* (the terminator of the *ardC* gene) was electroporated into myxamoebae (Burland et al., 1992). Stably resistant *P. polycephalum* transformants could be isolated when myxamoebae were transfected with a *PardC-hph-TardC* construct (Burland et al., 1993a). A *PardC*-luciferase construct was used to test the effect of adding a putative *P. polycephalum* origin of replication (AC1-A, Accession number X74751) to the vector (Bailey et al., 1994). Although the 1.2 kb AC1-A fragment did not increase the persistence of luciferase activity over time, it greatly increased, nine-fold, the peak luciferase activity. This enhancer-like activity could be attributed to a 760-bp sequence within the larger fragment. Characteristic of enhancers, the effect of the AC1-A fragment was seen regardless of its orientation on the vector.

Homologous gene replacement in *P. polycephalum* has been demonstrated using a construct with a deleted *ardD* actin gene, *ardDΔ1*, and the *PardC* driven *hph* gene (Burland and Pallotta, 1995). Only constructs with the *PardC-hph* gene downstream of the *ardDΔ1* gene produced transformants. Constructs with the *PardC-hph* gene inserted into the *ardDΔ1* gene did not produce transformants. Since the *ardD* gene is not expressed in myxamoebae, expression of *PardC-hph* might have been reduced when embedded in *ardDΔ1* sequence. The optimized parameters for myxamoebal cell pre-treatment and electroporation established by these studies have been reviewed (Burland and Bailey, 1995). Liu et al. (2009) electroporated *P. polycephalum* plasmodia to achieve transient expression from genes in a *PardC*-gene-*TardC* cassette.

Despite these advances, transfection of genes into myxomycetes is not routine. Another approach to studying gene function has been to use RNA interference (RNAi) to knockdown gene expression. Materna and Marwan (2005) were the first to demonstrate the effect of antisense RNA on gene expression in *P. polycephalum*. They prepared vectors with enhanced yellow fluorescent protein (EYFP) in sense and antisense orientation driven by the *PardC* promoter. When flagellate cells were transfected with both sense and antisense vectors, the EYFP expression levels were similar to untreated cells (no DNA added). Thus the antisense construct completely canceled the expression seen when only the sense EYFP vector was present.

A 90% decrease in gene expression was seen when dsRNA or siRNA, based on the *P. polycephalum* polymalatase gene, were injected into plasmodia (Haindl and Holler, 2005). The suppression of the polymalatase mRNA lasted for 144 h, after which time expression began to recover. However, repeated injections maintained the same level of suppression with no sign of resistance to suppression. Expression of the mRNA for the spherulin 3b gene of *P. polycephalum* was decreased to 1% when dsRNA, based on the spherulin 3b gene, was injected into plasmodia (Pinchai et al., 2006). Suppression of the spherulin 3b gene also resulted in a decrease of polymalic acid (PMLA) in plasmodia to 12% of untreated cell levels, demonstrating a regulatory role for the spherulin 3b gene. Attempts to use siRNA in *P. polycephalum* plasmodia to suppress mitochondrial DNA packaging proteins Glom and Glom2 failed, however, injected morpholino antisense oligomers effectively and reversibly suppressed these genes in plasmodia (Itoh et al., 2011).

The First Complete Myxomycete Genome

Early attempts to clone genomic DNA were frustrated by the instability of highly repetitive DNA sequences (Nader, 1986). The high number of repeats and intron insertion sites again presented a challenge recently during the assembly of the *P. polycephalum* genome (Schaap et al., 2016), the first complete myxomycete genome. Lacking a reference genome, the assembly was assisted by reference to transcriptome data, which verified the transcription of ORFs and supplied information on splice junctures and splicing variants.

Comparative analysis of genome data and transcriptomes in the Amoebozoa (Adl et al., 2005) and beyond, have identified genes ancestral to the Amoebozoa and Metazoa, genes specific to the Amoebozoa, genes specific to certain clades within the Amoebozoa, as well as the clade-specific loss of certain pathways, and evidence for lateral gene transfers (Glöckner et al., 2008, 2016; Watkins and Gray, 2008).

The advent of the first complete myxomycete genome (Schaap et al., 2016), available transcriptomes from various developmental stages, and the ability to disrupt gene function in vivo by RNAi should renew interest in the myxomycetes as a model system. Preliminary analysis of the *P. polycephalum* genome has already revealed the molecular richness and evolutionary distinctiveness myxomycetes offer (Schaap et al., 2016).

REFERENCES

Adl, S.M., Simpson, A.G.B., Farmer, M.A., Andersen, R.A., Anderson, O.R., Barta, J.R., Bowser, S.S., Brugerolle, G.U.Y., Fensome, R.A., Fredericq, S., James, T.Y., Karpov, S., Kugrens, P., Krug, J., Lane, C.E., Lewis, L.A., Lodge, J., Lynn, D.H., Mann, D.G., McCourt, R.M., Mendoza, L., Moestrup, O., Mozley-Standridge, S.E., Nerad, T.A., Shearer, C.A., Smirnov, A.V., Spiegel, F.W., Taylor, M.F.J.R., 2005. The new higher level cassification of eukaryotes with emphasis on the taxonomy of protists. J. Eukaryot. Microbiol. 52 (5), 399–451.

Aguilar, M., Fiore-Donno, A.M., Lado, C., Cavalier-Smith, T., 2014. Using environmental niche models to test the "everything is everywhere" hypothesis for *Badhamia*. ISME J. 8, 737–745.

Alim, K., Amselem, G., Peaudecerf, F., Brenner, M.P., Pringle, A., 2013. Random network peristalsis in *Physarum polycephalum* organizes fluid flows across an individual. Proc. Natl. Acad. Sci. 110, 13306–13311.

Baba, H., Kolukirik, M., Zümre, M., 2015. Differentiation of some Myxomycetes species by ITS sequences. Turk. J. Botany 39, 377–382.

Bailey, J., Benard, M., Burland, T.G., 1994. A luciferase expression system for *Physarum* that facilitates analysis of regulatory elements. Curr. Genet. 26, 126–131.

Baldauf, S.L., 1999. A search for the origins of animals and fungi: comparing and combining molecular data. Am. Nat. 154, 178–188.

Baldauf, S.L., Doolittle, W.F., 1997. Origin and evolution of the slime molds (Mycetozoa). Proc. Natl. Acad. Sci. 94, 12007–12012.

Baldauf, S.L., Palmer, J.D., 1993. Animals and fungi are each other's closest relatives: congruent evidence from multiple proteins. Proc. Natl. Acad. Sci. USA 90, 11558–11562.

Barrantes, I., Glockner, G., Meyer, S., Marwan, W., 2010. Transcriptomic changes arising during light-induced sporulation in *Physarum polycephalum*. BMC Genom. 11, 115.

Barrantes, I., Leipzig, J., Marwan, W., 2012. A next-generation sequencing approach to study the transcriptomic changes during the differentiation of *Physarum* at the single-cell level. Gene Reg. Syst. Biol. 6, 127–137.

Barth, D., Krenek, S., Fokin, S.I., Berendonk, T.U., 2006. Intraspecific genetic variation in *Paramecium* revealed by mitochondrial cytochrome *c* oxidase I sequences. J. Eukaryot. Microbiol. 53, 20–25.

Beekman, M., Latty, T., 2015. Brainless but multi-headed: decision making by the acellular slime mould *Physarum polycephalum*. J. Mol. Biol. 427, 3734–3743.

Berry, J.A., Franke, R.G., 1973. Taxonomic significance of intraspecific isozyme patterns of the slime mold *Fuligo septica* produced by disc electrophoresis. Am. J. Botany 60, 976–986.

Betterley, D.A., Collins, O.N.R., 1983. Reproductive systems, morphology, and genetical diversity in *Didymium iridis* (Myxomycetes). Mycologia 75, 1044–1063.

Birgisdottir, A.B., Johansen, S., 2005. Site-specific reverse splicing of a HEG-containing group I intron in ribosomal RNA. Nucl. Acids Res. 33, 2042–2051.

Birnboim, H.C., Doly, J., 1979. A rapid alkaline extraction procedure for screening recombinant plasmid DNA. Nucl. Acids Res. 7, 1513–1523.

Bradbury, E.M., Inglis, R.J., Matthews, H.R., 1974. Control of cell division by very lysine rich histone (F1) phosphorylation. Nature 247, 257–261.

Brown, W.M., George, M., Wilson, A.C., 1979. Rapid evolution of animal mitochondrial DNA. Proc. Natl. Acad. Sci. USA 76, 1967–1971.

Bundschuh, R., Altmüller, J., Becker, C., Nürnberg, P., Gott, J.M., 2011. Complete characterization of the edited transcriptome of the mitochondrion of *Physarum polycephalum* using deep sequencing of RNA. Nucl. Acids Res. 39, 6044–6055.

Burland, T.G., Bailey, J., 1995. Electroporation of *Physarum polycephalum*. In: Nickoloff, J.A. (Ed.), Electroporation Protocols for Microorganisms. Humana Press, Totowa, pp. 303–320.

Burland, T.G., Bailey, J., Adam, L., Mukhopadhyay, M.J., Dove, W.F., Pallotta, D., 1992. Transient expression in *Physarum* of chloramphenicol acetyltransferase gene under the control of actin gene promoters. Curr. Genet. 21, 393–398.

Burland, T.G., Bailey, J., Pallotta, D., Dove, W.F., 1993a. Stable, selectable, integrative DNA transformation in *Physarum*. Gene 132, 207–212.

Burland, T.G., Pallotta, D., 1995. Homologous gene replacement in *Physarum*. Genetics 139, 147–158.

Burland, T.G., Pallotta, D., Tardif, M.C., Lemieux, G., Dove, W.F., 1991. Fission yeast promoter-probe vectors based on hygromycin resistance. Gene 100, 241–245.

Burland, T., Solnica-Krezel, L., Bailey, J., Cunningham, D., Dove, W., 1993b. Patterns of inheritance development and the mitotic cycle in the protist *Physarum polycephalum*. Adv. Microb. Physiol. 35, 1–69.

Carlile, M.J., Dee, J., 1967. Plasmodial fusion and lethal interaction between strains in a myxomycete. Nature 215, 832–834.

Carlile, M.J., Gooday, G.W., 1978. Cell fusion in myxomycetes and fungi. In: Poste, G., Nicolson, G.l. (Eds.), Membrane Fusion. Elsevier/Nort-Holland Biomedical Press, Amsterdam, pp. 219–265.

Cavalier-Smith, T., 1993. Kingdom protozoa and its 18 phyla. Microbiol. Rev. 57, 953–994.

Cavalier-Smith, T., 1998. A revised six-kingdom system of life. Biol. Rev. Cambridge Philos. Soc. 73, 203–266.

Cavalier-Smith, T., 2013. Early evolution of eukaryote feeding modes, cell structural diversity, and classification of the protozoan phyla Loukozoa, Sulcozoa, and Choanozoa. Eur. J. Protist. 49, 115–178.

Cech, T.R., Bass, B.L., 1986. Biological catalysis by RNA. Ann. Rev. Biochem. 55, 599–629.

Chantangsi, C., Lynn, D.H., Brandl, M.T., Cole, J.C., Hetrick, N., Ikonomi, P., 2007. Barcoding ciliates: a comprehensive study of 75 isolates of the genus *Tetrahymena*. Int. J. Syst. Evol. Microbiol. 57, 2412–2425.

Clark, J., Hakim, R., 1980. Aging of plasmodial heterokaryons in *Didymium iridis*. Mol. Gen. Genet. 178, 419–422.

Clark, J., Lott, T., 1989. Age heterokaryon studies in *Didymium iridis*. Mycologia 81, 636–638.

Collins, O.N.R., 1966. Plasmodial compatibility in heterothallic and homothallic isolates of *Didymium iridis*. Mycologia 58, 362–372.

Collins, O.N.R., Betterley, D.A., 1982. *Didymium iridis* in past and future research. In: Aldrich, H., Daniel, J.W. (Eds.), Cell Biology of *Physarum* and *Didymium*. Academic Press, New York, pp. 25–57.

Decatur, W.A., Einvik, C., Johansen, S., Vogt, V.M., 1995. Two group I ribozymes with different functions in a nuclear rDNA intron. EMBO J. 14, 4558–4568.

Ducommun, B., Tollon, Y., Gares, M., Beach, D., Wright, M., 1990. Cell cycle regulation of p34cdc2 kinase activity in *Physarum polycephalum*. J. Cell Sci. 96 (Pt 4), 683–689.

Dussutour, A., Latty, T., Beekman, M., Simpson, S.J., 2010. Amoeboid organism solves complex nutritional challenges. Proc. Natl. Acad. Sci. USA 107, 4607–4611.

El Hage, N., Little, C., Clark, J.D., Stephenson, S.L., 2000. Biosystematics of the *Didymium squamulosum* complex. Mycologia 92, 54–64.

Evans, K.M., Wortley, A.H., Mann, D.G., 2007. An assessment of potential diatom "Barcode" genes (cox1, rbcl, 18S and ITS rDNA) and their effectiveness in determining relationships in *Sellaphora bacillariophyta*. Protist 158, 349–364.

Feng, Y., Klahr, A., Janik, P., Ronikier, A., Hoppe, T., Novozhilov, Y.K., Schnittler, M., 2016. What an intron may tell: several sexual biospecies coexist in *Meriderma* spp. (Myxomycetes). Protist 167, 234–253.

Feng, Y., Schnittler, M., 2015. Sex or no sex? Group I introns and independent marker genes reveal the existence of three sexual but reproductively isolated biospecies in *Trichia varia* (Myxomycetes). Organ. Div. Evol. 15, 631–650.

Ferris, P.J., Vogt, V.M., 1982. Structure of the central spacer region of extrachromosomal ribosomal DNA in *Physarum polycephalum*. J. Mol. Biol. 159, 359–381.

Fessel, A., Oettmeier, C., Bernitt, E., Gauthier, N.C., Dobereiner, H.G., 2012. *Physarum polycephalum* percolation as a paradigm for topological phase transitions in transportation networks. Phys. Rev. Lett. 109, 078103.

Fiore-Donno, A.M., Berney, C., Pawlowski, J., Baldauf, S.L., 2005. Higher-order phylogeny of plasmodial slime molds (Myxogastria) based on elongation factor 1-A and small subunit rRNA gene sequences. J. Eukaryot. Microbiol. 52, 201–210.

Fiore-Donno, A.M., Clissmann, F., Meyer, M., Schnittler, M., Cavalier-Smith, T., 2013. Two-gene phylogeny of bright-spored myxomycetes (slime moulds, superorder Lucisporidia). PLoS One 8, e62586.

Fiore-Donno, A.M., Kamono, A., Chao, E.E., Fukui, M., Cavalier-Smith, T., 2010a. Invalidation of *Hyperamoeba* by transferring its species to other genera of Myxogastria. J. Eukaryot. Microbiol. 57, 189–196.

Fiore-Donno, A.M., Kamono, A., Meyer, M., Schnittler, M., Fukui, M., Cavalier-Smith, T., 2012. 18S rDNA phylogeny of *Lamproderma* and allied genera (Stemonitales, Myxomycetes, Amoebozoa). PLoS One 7, e35359, 9.

Fiore-Donno, A.M., Meyer, M., Baldauf, S.L., Pawlowski, J., 2008. Evolution of dark-spored Myxomycetes (slime-molds): molecules versus morphology. Mol. Phylogenet. Evol. 46, 878–889.

Fiore-Donno, A.M., Nikolaev, S.I., Nelson, M., Nikolaev, S.I., Nelson, M., Pawlowski, J., Cavalier-Smith, T., Baldauf, S.L., 2010b. Deep phylogeny and evolution of slime moulds (mycetozoa). Protist 161, 55–70.

Fiore-Donno, A.M., Novozhilov, Y.K., Meyer, M., Schnittler, M., 2011. Genetic structure of two protist species (Myxogastria, Amoebozoa) suggests asexual reproduction in sexual amoebae. PLoS One 6, e22872.

Fiore-Donno, A.M., Weinert, J., Wubet, T., Bonkowski, M., 2016. Metacommunity analysis of amoeboid protists in grassland soils. Sci. Report 6, 19068.

Folmer, O., Black, M., Hoeh, W., Lutz, R., Vrijenhoek, R., 1994. DNA primers for amplification of mitochondrial cytochrome *c* oxidase subunit I from diverse metazoan invertebrates. Mol. Marine Biol. Biotechnol. 3, 294–299.

Franke, R.G., 1967. Preliminary investigation of the double-diffusion technique as a tool in determining relationships among some myxomycetes, order Physarales. Am. J. Botany 54, 1189–1197.

Franke, R.G., 1973. Symposium on the use of electrophoresis in the taxonomy of algae and fungi. V. Electrophoresis and the taxonomy of saprophytic fungi. Bull. Torrey Botan. Club 100, 287–296.

Franke, R.G., Balek, R.W., Visentin, L.P., 1968. Taxonomic significance of isozyme patterns of some Myxomycetes, order Physarales, produced with starch gel electrophoresis. Mycologia 60, 31–339.

Franke, R.G., Berry, J.A., 1972. Taxonomic application of isozyme patterns produced with disc electrophoresis of some ss order Physarales. Mycologia 64, 830–840.

Geisen, S., Tveit, A.T., Clark, I.M., Richter, A., Svenning, M.M., Bonkowski, M., Urich, T., 2015. Metatranscriptomic census of active protists in soils. ISME J. 9, 2178–2190.

Glöckner, G., Golderer, G., Werner-Felmayer, G., Meyer, S., Marwan, W., 2008. A first glimpse at the transcriptome of *Physarum polycephalum*. BMC Genom. 9, 6.

Glöckner, G., Lawal, H.M., Felder, M., Singh, R., Singer, G., Weijer, C.J., Schaap, P., 2016. The multicellularity genes of dictyostelid social amoebas. Nat. Comm. 7, 12085.

Goddard, M.R., Burt, A., 1999. Recurrent invasion and extinction of a selfish gene. Proc. Natl. Acad. Sci. USA 96, 13880–13885.

Gorman, J.A., Wilkins, A.S., 1980. Developmental phases in the life cycle of *Physarum* and related Myxomycetes. In: Dove, W.F., Rusch, H.P. (Eds.), Growth and Differentiation in *Physarum polycephalum*. Princeton University Press, Princeton, pp. 157–202.

Gotz, S., Garcia-Gomez, J.M., Terol, J., Williams, T.D., Nagaraj, S.H., Nueda, M.J., Robles, M., Talon, M., Dopazo, J., Conesa, A., 2008. High-throughput functional annotation and data mining with the Blast2GO suite. Nucl. Acids Res. 36, 3420–3435.

Gray, W.D., Alexopoulos, C.J., 1968. Biology of the Myxomycetes. The Ronald Press Company, New York.

Gubler, U., Wyler, T., Braun, R., 1979. The gene for the 26 S rRNA in *Physarum* contains two insertions. FEBS Lett. 100, 347–350.

Haindl, M., Holler, E., 2005. Use of the giant multinucleate plasmodium of *Physarum polycephalum* to study RNA interference in the myxomycete. Anal. Biochem. 342, 194–199.

Haskins, E.F., Wrigley de Basanta, D., 2008. Methods of agar culture of myxomycetes—an overview. Revista Mexicana de Micología 27, 1–7.

Haugen, P., Coucheron, D.H., Ronning, S.B., Haugli, K., Johansen, S., 2003. The molecular evolution and structural organization of self-splicing group I introns at position 516 in nuclear SSU rDNA of myxomycetes. J. Eukaryot. Microbiol. 50, 283–292.

Haugen, P., Simon, D.M., Bhattacharya, D., 2005. The natural history of group I introns. Trends Genet. 21, 111–119.

Haugli, F.B., Cooke, D., Sudbery, P., 1980. The genetic approach in the analysis of the biology of *Physarum polycephalum*. In: Dove, W.F., Rusch, H.P. (Eds.), Growth and Differntiation in *Physarum polycephalum*. Princeton University Press, Princeton, pp. 129–156.

Haugli, F., Johansen, T., 1985. Toward a DNA transformation system for *Physarum polycephalum*. In: Dove, W.F., Dee, J., Hatano, S., Haugli, F., Wohlfarth-Bottermann, K.E. (Eds.), The Molecular Biology of *Physarum polycephalum*. Plenum Press, New York, pp. 337–340.

Heather, J.M., Chain, B., 2016. The sequence of sequencers: the history of sequencing DNA. Genomics 107, 1–8.

Hebert, P.D.N., Cywinska, A., Ball, S.L., deWaard, J.R., 2003. Biological identifications through DNA barcodes. Proc. Royal Soc. London B 270, 313–321.

Hedberg, A., Johansen, S.D., 2013. Nuclear group I introns in self-splicing and beyond. Mobile DNA 4, 17.

Hoffmann, X.K., Tesmer, J., Souquet, M., Marwan, W., 2012. Futile attempts to differentiate provide molecular evidence for individual differences within a population of cells during cellular reprogramming. FEMS Microbiol. Lett. 329, 78–86.

Holt, C.E., 1980. The Nuclear Replication Cycle in *Physarum polycephalum*. In: Dove, W.F., Rusch, H.P. (Eds.), Growth and Differentiation in *Physarum polycephalum*. Princeton University Press, Princeton, pp. 9–63.

Hoppe, T., Kutschera, U., 2010. In the shadow of Darwin: Anton de Bary's origin of myxomycetology and a molecular phylogeny of the plasmodial slime molds. Theor. Biosci. 129, 15–23.

Hoppe, T., Müller, H., Kutschera, U., 2010. A new species of *Physarum* (Myxomycetes) from a boreal pine forest in Thuringia (Germany). Mycotaxon 114, 7–14.

Hoppe, T., Schnittler, M., 2015. Characterization of myxomycetes in two different soils by TRFLP analysis of partial 18S rRNA gene sequences. Mycosphere 6, 216–227.

Itoh, K., Izumi, A., Mori, T., Dohmae, N., Yui, R., Maeda-Sano, K., Shirai, Y., Kanaoka, M.M., Kuroiwa, T., Higashiyama, T., Sugita, M., Murakami-Murofushi, K., Kawano, S., Sasaki, N., 2011. DNA packaging proteins Glom and Glom2 coordinately organize the mitochondrial nucleoid of *Physarum polycephalum*. Mitochondrion 11, 575–586.

Johansen, S., Elde, M., Vader, A., Haugen, P., Haugli, K., Haugli, F., 1997. In vivo mobility of a group I twintron in nuclear ribosomal DNA of the myxomycete *Didymium iridis*. Mol. Microbiol. 24, 737–745.

Johansen, S., Haugen, P., 2001. A new nomenclature of group I introns in ribosomal DNA. RNA 7, 935–936.

Johansen, S., Johansen, T., Haugli, F., 1992. Structure and evolution of myxomycete nuclear group I introns: a model for horizontal transfer by intron homing. Curr. Genet. 22, 297–304.

Johansen, S., Vogt, V.M., 1994. An intron in the nuclear ribosomal DNA of *Didymium iridis* codes for a group I ribozyme and a novel ribozyme that cooperate in self-splicing. Cell 76, 725–734.

Kamono, A., Fukui, M., 2006. Rapid PCR-based method for detection and differentiation of Didymiaceae and Physaraceae (myxomycetes) in environmental samples. J. Microbiol. Methods 67, 496–506.

Kamono, A., Matsumoto, J., Kojima, H., Fukui, M., 2009. Characterization of myxomycetes communities in soil by reverse transcription polymerase chain reaction (RT-PCR)-based method. Soil Biol Biochem 41, 1324–1330.

Kessler, D., 1982. Plasmodial structure and motility. In: Aldrich, H., Daniel, J.W. (Eds.), Cell Biology of *Physarum* and *Didymium*. Academic Press, New York, pp. 145–208.

Kircher, M., Kelson, J., 2010. High-throughput DNA sequencing concepts and limitations. Bioessays 32, 524–536.

Ko, T.W.K., Stephenson, S.L., Jeewon, R., Lumyong, S., Hyde, K.D., 2009. Molecular diversity of myxomycetes associated with decaying wood and forest floor leaf litter. Mycologia 100, 592–598.

Laffler, T.G., Tyson, J.J., 1986. The *Physarum* cell cycle. In: Dove, W.F., Dee, J., Hatano, S., Haugli, F., Wohlfarth-Bottermann, K.E. (Eds.), The Molecular Biology of *Physarum polycephalum*. Plenum Press, New York, pp. 79–109.

Latty, T., Beekman, M., 2010. Food quality and the risk of light exposure affect patch-choice decisions in the slime mold *Physarum polycephalum*. Ecology 91, 22–27.

Leontyev, D.V., Schnittler, M., Moreno, G., Stephenson, S.L., Mitchell, D.W., Rojas, C., 2014. The genus Alwisia (Myxomycetes) revalidated, with two species new to science. Mycologia 106, 936–948.

Lewis, O.L., Zhang, S., Guy, R.D., del Álamo, J.C., 2015. Coordination of contractility, adhesion and flow in migrating *Physarum* amoebae. J. Royal Soc. Interf. 12, 20141359.

Li, G.Y., Xing, M., Hu, B., 2004. A PSTAIRE CDK-like protein localizes in nuclei and cytoplasm of *Physarum polycephalum* and functions in the mitosis. Cell Res. 14, 169–175.

Liu, S., Cheng, C., Lin, Z., Zhang, J., Li, M., Zhou, Z., Tian, S., Xing, M., 2009. Transient expression in microplasmodia of *Physarum polycephalum*. Sheng Wu Gong Cheng Xue Bao 25, 854–862.

Liu, W.T., Marsh, T.L., Cheng, H., Forney, L.J., 1997. Characterization of microbial diversity by determining terminal restriction fragment length polymorphisms of genes encoding 16S rRNA. Appl. Environ. Microbiol. 63, 4516–4522.

Liu, Q.S., Yan, S.H., Chen, S.L., 2015. Further resolving the phylogeny of Myxogastria (slime molds) based on COI and SSU rRNA genes. Russ. J. Genet. 51, 39–45.

Lundblad, E.W., Einvik, C., Ronning, S., Haugli, K., Johansen, S., 2004. Twelve Group I introns in the same pre-rRNA transcript of the myxomycete *Fuligo septica*: RNA processing and evolution. Mol. Biol. Evol. 21, 1283–1293.

Martín, M.P., Lado, C., Johansen, S., 2003. Primers are designed for amplification and direct sequencing of ITS region of rDNA from Myxomycetes. Mycologia 95, 474–479.

Marwan, W., 2003a. Detecting functional interactions in a gene and signaling network by time-resolved somatic complementation analysis. Bioessays 25, 950–960.

Marwan, W., 2003b. Theory of time-resolved somatic complementation and its use to explore the sporulation control network in *Physarum polycephalum*. Genetics 164, 105–115.

Marwan, W., Sujatha, A., Starostzik, C., 2005. Reconstructing the regulatory network controlling commitment and sporulation in *Physarum polycephalum* based on hierarchical Petri Net modelling and simulation. J. Theor. Biol. 236, 349–365.

Materna, S.C., Marwan, W., 2005. Estimating the number of plasmids taken up by a eukaryotic cell during transfection and evidence that antisense RNA abolishes gene expression in *Physarum polycephalum*. FEMS Microb. Lett. 243, 29–35.

McCurrach, K., Glover, L.A., Hardman, N., 1988. Transient expression of a chloramphenicol acetyltransferase gene following transfection of *Physarum polycephalum* myxamoebae. Curr. Genet. 13, 71–74.

Michel, F., Westhof, E., 1990. Modelling of the three-dimensional architecture of group I catalytic introns based on comparative sequence analysis. J. Mol. Biol. 216, 585–610.

Moore, W.S., 1995. Inferring phylogenies from mtDNA variation: mitochondrial-gene trees versus nuclear-gene trees. Evolution 49, 718–726.

Muscarella, D.E., Ellison, E.L., Ruoff, B.M., Vogt, V.M., 1990. Characterization of I-Ppo, an intron-encoded endonuclease that mediates homing of a group I intron in the ribosomal DNA of *Physarum polycephalum*. Mol. Cell. Biol. 10, 3386–3396.

Muscarella, D.E., Vogt, V.M., 1989. A mobile group I intron in the nuclear rDNA of *Physarum polycephalum*. Cell 56, 443–454.

Muscarella, D.E., Vogt, V.M., 1993. A mobile group I intron from *Physarum polycephalum* can insert itself and induce point mutations in the nuclear ribosomal DNA of *Saccharomyces cerevisiae*. Mol. Cell. Biol. 13, 1023–1033.

Muyzer, G., De Waal, E.C., Uitterlinden, A.G., 1993. Profiling of complex microbial populations by denaturing gradient gel electrophoresis analysis of polymerase chain reaction-amplified genes coding for 16S rRNA. Appl. Environ. Microbiol. 59, 695–700.

Nader, W.F., 1986. Gene cloning and construction of genomic libraries in *Physarum*. In: Dove, W.F., Dee, J., Hatano, S., Haugli, F.B., Wohlfarth-Bottermann, K.E. (Eds.), The Molecular Biology of *Physarum polycephalum*. Springer, New York, pp. 291–300.

Nakagaki, T., Kobayashi, R., Nishiura, Y., Ueda, T., 2004. Obtaining multiple separate food sources: behavioural intelligence in the *Physarum plasmodium*. Proc. Royal Soc. 271, 2305–2310.

Nakagaki, T., Yamada, H., Toth, A., 2000a. Intelligence: maze-solving by an amoeboid organism. Nature 407, 470.

Nakagaki, T., Yamada, H., Tóth, Á., 2001. Path finding by tube morphogenesis in an amoeboid organism. Biophys. Chem. 92, 47–52.

Nakagaki, T., Yamada, H., Ueda, T., 2000b. Interaction between cell shape and contraction pattern in the *Physarum plasmodium*. Biophys. Chem. 84, 195–204.

Nandipati, S.C., Haugli, K., Coucheron, D.H., Haskins, E.F., Johansen, S.D., 2012. Polyphyletic origin of the genus *Physarum* (Physarales, Myxomycetes) revealed by nuclear rDNA mini-chromosome analysis and group I intron synapomorphy. BMC Evol. Biol. 12, 1–10.

Nassonova, E., Smirnov, A., Fahrni, J., Pawlowski, J., 2010. Barcoding amoebae: comparison of SSU, ITS and COI genes as tools for molecular identification of naked lobose amoebae. Protist 161, 102–115.

Nielsen, H., Johansen, S.D., 2009. Group I introns: moving in new directions. RNA Biol. 6, 375–383.

Nielsen, H., Westhof, E., Johansen, S., 2005. An mRNA is capped by a 2', 5' lariat catalyzed by a group I-like ribozyme. Science 309, 1584–1587.

Novozhilov, Y.K., Okun, M.V., Erastova, D., Shchepin, O.N., Zemlyanskaya, I.V., García-Carvajal, E., Schnittler, M., 2013a. Description, culture and phylogenetic position of a new xerotolerant species of *Physarum*. Mycologia 105, 1535–1546.

Novozhilov, Y.K., Schnittler, M., Erastova, D., Okun, M.V., Schepin, O.N., Heinrich, E., 2013b. Diversity of nivicolous myxomycetes of the Teberda State Biosphere Reserve (Northwestern Caucasus, Russia). Fungal Divers. 59, 109–130.

Nurse, P., 2000. A long twentieth century of the cell cycle and beyond. Cell 100, 71–78.

Otsuka, T., Nomiyama, H., Yoshida, H., Kukita, T., Kuhara, S., Sakaki, Y., 1983. Complete nucleotide sequence of the 5.8S rRNA gene of *Physarum polycephalum*: its significance in the gene evolution. Proc. Natl. Acad. Sci. USA 80, 3163–3167.

Pallotta, D., Bernier, F., Hamelin, M., Martel, R., Lemieux, G., 1986. cDNA cloning of *Physarum polycephalum* stage specific mRNAs. In: Dove, W.F., Dee, J., Hatano, S., Haugli, F., Wohlfarth-Bottermann, K.E. (Eds.), The Molecular Biology of *Physarum polycephalum*. Plenum Press, New York, pp. 315–327.

Pawlowski, J., Audic, S., Adl, S., Bass, D., Belbahri, L., Berney, C., Bowser, S.S., Cepicka, I., Decelle, J., Dunthorn, M., Fiore-Donno, A.M., Gile, G.H., Holzmann, M., Jahn, R., Jirků, M., Keeling, P.J., Kostka, M., Kudryavtsev, A., Lara, E., Lukeš, J., Mann, D.G., Mitchell, E.A.D., Nitsche, F., Romeralo, M., Saunders, G.W., Simpson, A.G.B., Smirnov, A.V., Spouge, J.L., Stern, R.F., Stoeck, T., Zimmermann, J., Schindel, D., de Vargas, C., 2012. CBOL Protist Working Group: barcoding eukaryotic richness beyond the animal, plant, and fungal kingdoms. PLoS Biol. 10, e1001419.

Pierron, G., Pallotta, D., Benard, M., 1999. The one-kilobase DNA fragment upstream of the ardC actin gene of *Physarum polycephalum* is both a replicator and a promoter. Mol. Cell. Biol. 19, 3506–3514.

Pinchai, N., Lee, B.S., Holler, E., 2006. Stage specific expression of poly(malic acid)-affiliated genes in the life cycle of *Physarum polycephalum*. Spherulin 3b and polymalatase. FEBS J. 273, 1046–1055.

Rätzel, V., Ebeling, B., Hoffmann, X.-K., Tesmer, J., Marwan, W., 2013. *Physarum polycephalum* mutants in the photocontrol of sporulation display altered patterns in the correlated expression of developmentally regulated genes. Dev. Growth Differ. 55, 247–259.

Rätzel, V., Marwan, W., 2015. Gene expression kinetics in individual plasmodial cells reveal alternative programs of differential regulation during commitment and differentiation. Dev. Growth Differ. 57, 408–420.

Raue, H.A., Klootwijk, J., Musters, W., 1988. Evolutionary conservation of structure and function of high molecular weight ribosomal RNA. Prog. Biophys. Mol. Biol. 51, 77–129.

Reid, C.R., MacDonald, H., Mann, R.P., Marshall, J.A., Latty, T., Garnier, S., 2016. Decision-making without a brain: how an amoeboid organism solves the two-armed bandit. J. Royal Soc. Interface 13, 20160030.

Rusch, H.P., 1980. Introduction. In: Dove, W.F., Rusch, H.P. (Eds.), Growth and Differentiation in *Physarum polycephalum*. Princeton University Press, Princeton, pp. 1–8.

Rusch, H.P., Sachsenmaier, W., Behrens, K., Gruter, V., 1966. Synchronization of mitosis by the fusion of the plasmodium of *Physarum polycephalum*. J. Cell Biol. 31, 204–209.

Rusk, S.A., Spiegel, F.W., Lee, S.B., 1995. Design of polymerase chain reaction primers for amplifying nuclear ribosomal DNA from slime molds. Mycologia 87, 140–143.

Sauer, H.W., 1982. Developmental Biology of Physarum. Cambridge University Press, Cambridge.

Schaap, P., Barrantes, I., Minx, P., Sasaki, N., Anderson, R.W., Benard, M., Biggar, K.K., Buchler, N.E., Bundschuh, R., Chen, X., Fronick, C., Fulton, L., Golderer, G., Jahn, N., Knoop, V., Landweber, L.F., Maric, C., Miller, D., Noegel, A.A., Peace, R., Pierron, G., Sasaki, T., Schallenberg-Rudinger, M., Schleicher, M., Singh, R., Spaller, T., Storey, K.B., Suzuki, T., Tomlinson, C., Tyson, J.J., Warren, W.C., Werner, E.R., Werner-Felmayer, G., Wilson, R.K., Winckler, T., Gott, J.M., Glockner, G., Marwan, W., 2016. The *Physarum polycephalum* genome reveals extensive use of prokaryotic two-component and metazoan-type tyrosine kinase signaling. Genome Biol. Evol. 8, 109–125.

Schoch, C.L., Seifert, K.A., Huhndorf, S., Robert, V., Spouge, J.L., Levesque, C.A., Chen, W., 2012. Nuclear ribosomal internal transcribed spacer (ITS) region as a universal DNA barcode marker for Fungi. Proc. Natl. Acad. Sci. USA 109, 6241–6246.

Schrauwen, J.A.M., 1979. Post-fusion incompatibility in *Physarum polycephalum*. Arch. Microbiol. 122, 1–7.

Schrauwen, J.A.M., 1984. Cellular interaction in plasmodial slime moulds. In: Linskens, H.F., Heslop-Harrision, J. (Eds.), Cellular Interactions. Springer-Verlag, Berlin, pp. 291–308.

Shadwick, L.L., Spiegel, F.W., Shadwick, J.D.L., Brown, M.W., Silberman, J.D., 2009. Eumycetozoa= Amoebozoa?: SSUrDNA phylogeny of protosteloid slime molds and its significance for the Amoebozoan supergroup. PLoS One 4, e6754.

Silliker, M.E., Liles, J.L., Monroe, J.A., 2002. Patterns of mitochondrial inheritance in the myxogastrid *Didymium iridis*. Mycologia 94, 939–946.

Solnica-Krezel, L., Burland, T.G., Dove, W.F., 1991. Variable pathways for developmental changes of mitosis and cytokinesis in *Physarum polycephalum*. J. Cell Biol. 113, 591–604.

Stephenson, S.L., Feest, A., 2012. Ecology of soil eumycetozoans. Acta Protozool. 51, 201–208.

Sujatha, A., Balaji, S., Devi, R., Marwan, W., 2005. Isolation of *Physarum polycephalum* plasmodial mutants altered in sporulation by chemical mutagenesis of flagellates. Eur. J. Protisol. 41, 19–27.

Tero, A., Kobayashi, R., Nakagaki, T., 2006. Physarum solver: a biologically inspired method of road-network navigation. Physica A 363, 115–119.

Tero, A., Takagi, S., Saigusa, T., Ito, K., Bebber, D.P., Fricker, M.D., Yumiki, K., Kobayashi, R., Nakagaki, T., 2010. Rules for biologically inspired adaptive network design. Science 327, 439–442.

Ueda, T., Kobatake, Y., 1982. Chemotaxis in plasmodia of *Physarum polycephalum*. In: Aldrich, H., Daniel, J.W. (Eds.), Cell Biology of *Physarum* and *Didymium*. Academic Press, New York, pp. 111–143.

Urich, T., Lanzén, A., Qi, J., Huson, D.H., Schleper, C., Schuster, S.C., Ward, N., 2008. Simultaneous assessment of soil microbial community structure and function through analysis of the meta-transcriptome. PLoS One 3, e2527.

Vader, A., Naess, J., Haugli, K., Haugli, F., Johansen, S., 1994. Nucleolar introns from *Physarum flavicomum* contain insertion elements that may explain how mobile group I introns gained their open reading frames. Nucl. Acids Res. 22, 4553–4559.

Vader, A., Nielsen, H., Johansen, S., 1999. In vivo expression of the nucleolar group I intron-encoded I-dirI homing endonuclease involves the removal of a spliceosomal intron. EMBO J. 18, 1003–1013.

Walker, L.M., Dewsbury, D.R., Parks, S.S., Winsett, K.E., Stephenson, S.L., 2011. The potential use of mitochondrial cytochrome *c* oxidase I for barcoding myxomycetes. In: VII International Congress on Systematics and Ecology of Myxomycetes, Recife, Brazil, September 11–16, 138.

Walker, L., Leontyev, D., Stephenson, S., 2015. *Perichaena longipes*, a new myxomycete from the Neotropics. Mycologia 107, 1012–1022.

Walker, L.M., Stephenson, S.L., 2016. The species problem in myxomycetes revisited. Protist 167, 319–338.

Walochnik, J., Michel, R., Aspöck, H., 2004. A molecular biological approach to the phylogenetic position of the genus *Hyperamoeba*. J. Eukaryot. Microbiol. 51, 433–440.

Walter, P., Hoffmann, X.K., Ebeling, B., Haas, M., Marwan, W., 2013. Switch-like reprogramming of gene expression after fusion of multinucleate plasmodial cells of two *Physarum polycephalum* sporulation mutants. Biochem. Biophys. Res. Comm. 435, 88–93.

Watkins, R.F., Gray, M.W., 2008. Sampling gene diversity across the supergroup Amoebozoa: large EST data sets from *Acanthamoeba castellanii*, *Hartmannella vermiformis*, *Physarum polycephalum*, *Hyperamoeba dachnaya* and *Hyperamoeba sp.* Protist 159, 269–281.

White, T.J., Bruns, T., Lee, S., Taylor, J., 1990. Amplification and direct sequencing of fungal ribosomal RNA genes for phylogenetics. PCR Protocol, 315–322.

Wick, R.J., Sauer, H.W., 1982. Developmental biology of slime molds: an overview. In: Aldrich, H.C., Daniel, J.W. (Eds.), Cell Biology of *Physarum* and *Didymium*. Academic Press, New York, pp. 3–19.

Wikmark, O.-G., Haugen, P., Haugli, K., Johansen, S.D., 2007. Obligatory group I introns with unusual features at positions 1949 and 2449 in nuclear LSU rDNA of Didymiaceae myxomycetes. Mol. Phylogenet. Evol. 43, 596–604.

Winsett, K.E., 2010. Intraspecific variation in two cosmopolitan myxomycetes, *Didymium squamulosum* and *Didymium difforme* (physarales: Didymiaceae). Doctoral Thesis, University of Arkansas, Fayetteville. Theses and Dissertations, Paper 202. Availabe from:http://scholarworks.uark.edu/etd/202

Winsett, K.E., Stephenson, S.L., 2008. Using ITS sequences to assess intraspecific genetic relationships among geographically separated collections of the myxomycete *Didymium squamulosum*. Revista Mexicana de Micología 27, 59–65.

Winter, M., 2016. Comparing the diversity of myxomycetes along a forest disturbance gradient in Costa Rica using moist chamber cultures and molecular methods. Master Thesis, Ernst-Moritz-Arndt-University of Greifswald, Germany.

Wolf, R., Sauer, H.W., 1982. Time-lapse analysis of mitosis in vivo in macroplasmodia of *Physarum polycephalum*. In: Aldrich, H., Daniel, J.W. (Eds.), Cell Biology of *Physarum* and *Didymium*. Academic Press, New York, pp. 261–264.

Yoshiyama, S., Ishigami, M., Nakamura, A., Kohama, K., 2010. Calcium wave for cytoplasmic streaming of *Physarum polycephalum*. Cell Biol. Int. 34, 35–40.

Zhou, Y., Lu, C., Wu, Q.-J., Wang, Y., Sun, Z.-T., Deng, J.-C., Zhang, Y., 2008. GISSD: group I intron sequence and structure database. Nucl. Acids Res. 36, D31–D37.

Chapter 6

Physiology and Biochemistry of Myxomycetes

Qi Wang, Yu Li, Pu Liu
Engineering Research Center of Chinese Ministry of Education for Edible and Medicinal Fungi, Jilin Agricultural University, Changchun, Jilin, PR China

INTRODUCTION

As outlined in some detail in Chapter 1 and referred to throughout the other chapters in this volume, the life cycle of myxomycetes involves three morphologically distinct stages (Chen et al., 2013; Clark, 1995; Clark and Collins, 1976; Collins, 1979; Gray and Alexopoulos, 1968; Liu et al., 2010). These are the amoeboflagellate stage (which consists of naked amoebae or flagellated cells), the plasmodium, and the fruiting body containing spores. In addition, there are two resistant stages—the microcyst and the sclerotium—that allow myxomycetes to survive under unfavorable environmental conditions. The former is derived from the amoeboflagellate stage and the latter from the plasmodium. In the first portion of this chapter, the physiological aspects of each of these stages will be considered. Since the life cycle of a particular myxomycete begins with the germination of a spore, the discussion presented herein will begin with this structure.

Spore Germination

In myxomycetes, the time required for spore germination and the percentage of spores actually germinating varies with the conditions present, the age of the spores, the species involved, the particular strain, and even with the particular fruiting body from which the spores were derived. For most species of myxomycetes, the optimal temperature for spore germination is 22–30°C and the optimal pH is 4.5–7.0 (Smart, 1937). However, the spores of snowbank (nivicolous) species can germinate at lower temperatures (Shchepin et al., 2014). Oxygen is a requirement for the germination of spores of *Fuligo septica* (Nelson and Orlowski, 1981), with those kept under anaerobic conditions failing to germinate until they had been exposed to air. Smart (1937) studied

Myxomycetes: Biology, Systematics, Biogeography, and Ecology. http://dx.doi.org/10.1016/B978-0-12-805089-7.00006-8

seventy species and varieties of myxomycetes representing all of the types of fruiting bodies known for this group of organisms, in an effort to correlate the relationship of external factors to spore germination. His research showed that there are some factors that affect spore germination in cultures. These include such things as the nutritional condition of the medium, the hydrogen-ion concentration of the medium, temperature, and the mutual effect of spores sown in mass as indicated by single spore and multispored cultures. This supposed mass effect has been hypothesized to be due to a soluble autocatalytic factor, but some recent reports indicate that it might not be a real effect (Haskins and Hinchee, 1992).

Spore germination is accomplished by one of two methods. Either the spore cracks (or splits) open or a minute pore dissolves in the spore wall and the protoplast emerges. The method of germination evidently is constant for each species. Lipid bodies appear to provide the primary energy source during germination (Alexopoulos and Mims, 1979). Gilbert (1928) found that osmosis was most important in the split type of germination while enzymes were involved in both types of germination. The change in split germination in *Physarum gyrosum* to spore germination by the addition of cellulose lends some support to his hypothesis (Koevenig, 1964).

To accelerate spore germination, various wetting agents have been tried. For example, Elliott (1948) indicated that germination was induced or improved by using a solution of bile salts as a wetting agent. Erbisch (1964) reported "abundant" germination after 72 h in 75-year-old spores of *Hemitrichia clavata* treated with bile salts. Nutrient extracts from natural substrates, including humus, corn, and bean seeds, decaying wood, fallen pine needles, and leaves have been shown to improve the percentage of spore germination (Shi and Li, 2003). Elliott (1949), who worked with 59 species, using (presumably herbarium) specimens of various ages, some of them as old as 61 years, induced the spores of all species but one to germinate by employing sodium taurocholate as a wetting agent. However, since the cultures were not used to produce a complete life cycle, it is highly likely that contamination of herbarium specimens with newer spores occurred, since most recent studies indicate that germination rapidly decreases with age (Winsett, 2011).

The percentage of spores germinating and the time required for germination vary among different families of myxomycetes. In a comparative study of the families of myxomycetes with respect to the ability of the spores to germinate in distilled water, it was found that nearly all members of the Reticulariaceae showed an excellent percentage of spore germination. However, in the Physaraceae, Amaurochaetaceae, Heterodermaceae, Trichiaceae, and Arcyriaceae, the ability to germinate appears to differ among genera and species, but species in the Tubulinaceae did not germinate. The average time of spore germination in the Reticulariaceaes, 1–6 h, was shorter than for most myxomycetes, and this family as a whole stands far above in terms of the germination ability of spores.

Amoeboflagellate Cells

In the normal course of events, both flagellated cells (traditionally referred to as swarm cells) and naked amoeboid cells (myxamoebae) are formed during the life cycle of a myxomycete. It is usually convenient to use the term "amoeboflagellate cell" to encompass both forms. The general characteristics of amoeboflagellate cells were observed in many species of myxomycetes by Gilbert (1929) and Smith (1929). Normally, water plays an important role in the formation of flagellated cells or myxamoebae. Flagellated cells tend to be associated with microenvironments in which water is present. In contrast, myxamoebae are usually the form associated with dryer conditions. In artificial cultures, the flagellated stage may be completely suppressed, at least in some species, by germinating the spores on a moist agar surface in the absence of free water (Alexopoulos, 1960a). By adding water to a culture or permitting a wet culture to dry out, a shift from flagellated cells to myxamoebae or the reverse may sometimes be induced over a long period of time before zygote formation begins.

Flagellated cells withdraw their flagella before dividing, since cell division apparently occurs only in the myxamoebae stage. The myxamoebae stage of *Didymium nigripes* may be prolonged and plasmodium formation prevented by the addition of 2% glucose or 0.2% brucine to the medium (Kerr and Sussman, 1958), but zygote formation is not inhibited, at least not by glucose (Therrien, 1966). After a population of myxamoebae reaches a certain cell density, they become competent to form zygotes (Shipley and Holt, 1982). Zygote formation, either by sexual fusion of apogamic development (Clark and Haskins, 2013), involves changes in the cell surface of the outer membranes, since zygotes coalesce with other zygotes but engulf genetically identical yet haploid myxamoebae (Ross, 1967).

Myxamoebae motility involves lobose pseudopodia, which apparently use an actinomyosin contractile system similar to that found in the plasmodium (Taniguchi et al., 1978). A study of the characteristics of myxamoebae in *Hemitrichia calyculata*, *H. clavata*, *Physarum melleum*, *Stemonitis flavogenita*, and *Physarum globuliferum* showed that the structure of myxamoebae affected their patterns of movement. The morphology of myxamoebae varies slightly within these five species, with the primary difference being the presence or absence of a transparent ectoplasm and the regularity or irregularity of the overall shape of the cell itself. Transparent ectoplasm may be responsible for the pattern of movement of myxomoebae. It is difficult to form a typical pseudopodium for myxamoebae with a thick ectoplasm, and such myxamoebae exhibit a gliding motility with slow speed. The other movement patterns include flowing motility and creeping motility (Li et al., 2013).

Amoeboflagellate nutrition occurs via the engulfment of bacteria, other microorganisms and organic matter into food vacuoles. While a few species have been grown in axenic culture, on a relatively simple medium of glucose, amino acids, vitamins, and hematin (Clark et al., 1990) or a more complex medium

(Haskins, 1970), most species have resisted all attempts to culture them without a living food organism present.

Plasmodium

The single most distinctive stage of a myxomycete is the assimilative structure, the plasmodium. The latter is essentially a naked, free-living, multinucleate, motile mass of protoplasm which varies in size and in certain morphological details with age, the particular type of plasmodium involved, the species involved, and to a certain extent with the nature of the substrate upon which it is growing (Gray and Alexopoulos, 1968). Three different types of plasmodia—protoplasmodia, aphanoplasmodia, and phaneroplasmodia—are recognized based on gross morphology (Alexopoulos, 1960b).

When nuclei and nucleoli were extracted and purified and sclerotia were induced to form from plasmodia of *F. septica*, observations under TEM indicate that the nucleus possesses a central nucleolus with fibrillar centers, dense fibrillar component, and a granular component. A large number of grume granules exist in plasmodia. The sclerotium, with a double-membrane structure, contains organelles and lipid drops (Wang et al., 2007).

Plasmodia can be attracted by food sources, such as oats (Nakagaki et al., 2000), bacteria (Konijn and Koevenig, 1971), and the fruiting bodies of macrofungi (Emoto, 1932; Madelin et al., 1975), which they engulf and then use to synthetize carbohydrate and protein (Chet et al., 1977; Kincaid and Mansour, 1978; Knowles and Carlile, 1978; McClory and Coote, 1985). However, the plasmodia of most species cannot be grown without a living food source, and only *Physarum polycephalum* and a few other species have been grown in axenic culture (Clark et al., 1990; Daniel and Rusch, 1961). Surprisingly, these species can be grown on a relatively simple medium consisting of glucose, amino acids, vitamins, and hematin. The plasmodium of *P. polycephalum* can make complex nutritional decisions, despite lacking a coordination center and consisting of only a single vast multinucleate cell (Chapter 11). It is able to grow to contact patches of different nutrient quality in the precise proportions necessary to compose an optimal diet. That such organisms have the capacity to maintain the balance of carbon- and nitrogen-based nutrients by selective foraging has considerable implications not only for our understanding of nutrient balancing in distributed systems but for the functional ecology of soils, nutrient cycling, and carbon sequestration (Dussutour et al., 2010). This aspect of myxomycetes is covered in detail in another chapter in this volume.

The rhythmic, reversible streaming of the protoplasm is characteristic of the plasmodia of most myxomycetes and is a well-known phenomenon. An early theory seeking to explain protoplasmic streaming implicated the changes in viscosity of myxomyosin when it interacts with ATP (Kamiya, 1959).

Ultrastructural studies have refined this premise by clarifying the structures and mechanisms involved. Myxomyosin, actin, and ATP were demonstrated in the plasmodium of *P. polycephalum* (Ts'o et al., 1956a,b, 1957a,b), and the reaction of these substances appears to be similar to that of the actomyosin-ATP system in muscle. More recent experiments with the phaneroplasmodia of *P. polycephalum* have shown that protoplasmic streaming is caused by a hydraulic pressure flow mechanism generated by the contractions of the protoplasm (Komnick et al., 1973). This is a function of the assembly and disassembly of actin-containing cytoplasmic fibrils in the ectoplasm (Hinssen, 1981). The differentiation into gel-like ectoplasm and fluid endoplasm depends on the degree of polymerization of the actin, which in turn has been found to be regulated by a polymerization-inhibiting protein (actin modulating protein). Just what mechanism directs streaming in aphanoplasmodia and protoplasmodia apparently has not yet been determined.

Streaming of the protoplasm in the plasmodium is directly related to locomotion. When the plasmodium is moving over the substrate, the total volume of protoplasm transported during a given period of time will be somewhat greater in the general direction of movement than in the reverse (Kamiya, 1950, 1959). This is obviously a simple but effective method of circulation. Polarity of the plasmodium appears to be closely associated with potassium concentration, with a greater concentration prevailing in the anterior as opposed to the posterior region of a migrating plasmodium (Anderson, 1962, 1964). During streaming and migration of the plasmodium, it continually divides and reunites to form a reticulate structure. However, unless two different plasmodia are genetically identical, they do not fuse to coalesce into a single plasmodium. This ability to recognize self from nonself is controlled by a multigene complex that determines membrane fusion and the survival of mixed cytoplasm (Clark and Haskins, 2012).

Turnock et al. (1981) identified some 300 types of proteins contained in plasmodia and the amoeboflagellates of *P. polycephalum*, of which approximately three-fourths were present in both stages, but synthesized in different proportions and at different rates. The nature of the pigments in myxomycete plasmodia has attracted the attention of many researchers, but beyond the fact that many of these pigments act as indicators, changing color with changes in pH, little definite knowledge has been obtained. In nature, pigmentation is affected by a number of environmental conditions, but under controlled laboratory conditions it appears to be a stable factor (Collins, 1979). However, Collins et al. have isolated several color mutants in the laboratory. It has been both suggested and denied that the yellow pigments have the properties of anthracenes, flavones, pteridines, polypeptides, or polyenes. Czeczuga (1980) found carotenoids in the plasmodia of various species and xanthophylls in *F. septica* plasmodia, as well as in developing fruiting bodies of other species. Pigments also have been postulated to be photoreceptors playing an important role in the process of sporulation (Daniel and Rusch, 1962a; Lieth and Meyer, 1957; Wolf, 1959; Wormington and Weaver, 1976).

The presence of various enzymes, vitamins, sterols, and other organic substances has been detected in the plasmodium of *P. polycephalum*, and the production of antibiotics by several other species of myxomycetes has been reported (Chassain, 1980; Considine and Mallette, 1965; Locquin, 1948; Sobels, 1950; Taylor and Mallette, 1978). The responses of plasmodia to various external factors, such as anaesthetics, low and high temperatures, gravity, light, and irradiation have been studied to some extent and considerable knowledge has accumulated on this subject. Rakoczy (1973), working with *Physarum nudum*, found that light affected direction of migration, pigment composition, and sporulation, and inhibited plasmodial growth. Furthermore, plasmodial age can influence the direction of migration (with respect to light) and the length of exposure to light required to induce sporulation. Chemotaxis (Carlile, 1970; Madelin et al., 1975), the nutritional state of the plasmodium, and humidity also affect migration (Hüttermann, 1973a).

The velocity of vesicles during cytoplasmic streaming has been measured by counting all of the vesicles passing through a designated window. During streaming, the vesicles are distributed in their moving velocities, and the distribution itself varies with time. The mean velocity of vesicles and its standard deviation were found to exhibit a linear relationship, suggesting a possibility that vesicles in the cytoplasm would also be involved in the generation of force (Honda et al., 1990).

The extracellular matrix of the *Physarum* plasmodium is secreted by the exocytosis of vesicles that contain a slime precursor. The nature of this matrix slime in *Physarum* is somewhat in dispute, with Simon and Henney (1970) concluding that it is a glycoprotein whose carbohydrate is galactose, while McCormick et al. (1970) identify a polysaccharide consisting of galactose, sulfate, and traces of rhamose. Using an antibody produced in response to biochemically purified slime, it is possible to detect the intracellular localization of the slime vesicles. Slime vesicles are abundant in the advancing front of the plasmodium, as confirmed by electron microscopic observation of two different cross-sectional angles. By screening various reagents, it was found that rhodamine-phosphatidylethanolamine (Rh-PE) binds specifically to slime in both its intravesicular and extracellular forms, as confirmed by immunoelectron microscopy using an antibody against fluorochrome rhodamine. Plasmodia vitally stained with Rh-PE exhibit dynamic fluorescent patterns during the course of locomotion. The fluorescence is conspicuous at the periphery of the leading pseudopods and oscillated according to the shuttle streaming that accompanied the relaxation and contraction of the periphery. It was most intense in the relaxation phase when pseudopods extended, and becomes weak in the contraction phase when pseudopods contract. These results suggest that the slime vesicles carried by cytoplasmic streaming accumulate prior to secretion at the advancing margin of the plasmodium (Sesaki and Ogihara, 1997).

Sporulation

In most cases, the sporulation of myxomycetes in nature seems to occur largely at night (Gray and Alexopoulos, 1968). Sporulation is certainly affected by a number of factors, including temperature, moisture, availability of food, light, plasmodial size, and pH, but the initial stimulus that induces this process is still unknown. The plasmodia continue to grow as long as the supply of food is sufficient, but when the nourishment is exhausted, they quickly pass into the fruiting stage (Camp, 1937). Gray (1939) showed that temperature and pH were interrelated factors in his study of sporulation in *P. polycephalum*. Within certain limits, the higher the temperature, the lower the pH required for sporulation. Gray (1938, 1941, 1953), using *P. polycephalum* as an example, also showed that light is necessary for sporulation. An optimal growth age also appears to be a factor (Daniel and Rusch, 1962a,b; Olive, 1975).

Starvation is essential for sporulation in some species of myxomycetes but apparently not in all. For example, ultrastructural studies of certain species in the family Trichiaceae revealed viable bacteria from the culture medium contained within food vacuoles in the protoplasm of young fruiting bodies (Mims, 1969), the slime coat surrounding the developing fruiting body (Charvat et al., 1973), and within the capillitium (Charvat et al., 1974). Daniel and Rusch (1962a,b), working with bacterium-free cultures of *P. polycephalum*, found that the conditions necessary for sporulation included a medium containing niacin and niacinamide or certain substitutes, such as tryptophane.

A dark incubation period of 4 days and a subsequent exposure to light of wavelengths between 350 and 500 μm are essential for sporulation in *P. polycephalum* (Daniel and Rusch, 1962a,b). Sporulation-competent plasmodia may also be obtained without illumination, by injection of cytoplasm from another plasmodium that was activated by light, or by small amounts of salt solutions (Gorman and Wilkins, 1980). Starvation in darkness leads to spherule formation (Sauer et al., 1969). In this study, these authors enumerated the synthesis of DNA before illumination, a continued synthesis of proteins, and RNA synthesis until 3 h after the end of illumination as essential processes in sporulation. At that time the plasmodium (of *P. polycephalum*) became irreversibly committed to sporulation. Mitosis also appears to be a prerequisite for this transformation.

METABOLITES OF MYXOMYCETES

Myxomycetes are known to produce various metabolites, and the bioactivities of these metabolites have been subjected to study, especially over the past couple of decades. Some of those metabolites have been demonstrated to possess antibacterial, cytotoxic, and antioxidant activities. A preliminary understanding of the chemical composition and structure of myxomycete metabolites will be included in this part of the chapter, and the relationships that exist between biological activity, physiological function, and their specific structure (if any) will also be described.

Since large amounts of fruiting bodies and plasmodia of myxomycetes or at least a sufficient quality of sample material required for chemical analysis and bioactivity tests is often difficult to obtain, reports of metabolites and bioactivities from myxomycetes are somewhat limited. However, at present almost 100 different compounds, primarily alkaloids, terpenoids, fatty acids, aromatic compounds, amino acids, esters, and naphthoquinone have been extracted from 33 species of myxomycetes (Dembitsky et al., 2005; Göttler and Holler, 2006; Ishibashi, 2007; Jiang et al., 2014; Shintani et al., 2009, 2010; Steglich, 1989; Zhu, 2012).

Myxomycetes have been thought to represent a potential source of natural active products that have unique chemical structures, and studies carried out to date have proved this to be the case. However, additional investigations are still required because the number of species examined thus far is still relatively small. Nevertheless, myxomycetes have a unique life cycle and are expected to produce many useful metabolites.

Metabolites From Members of the Ceratiomyxales

Ceratiomyxa fruticulosa is the major species that has been subjected to study in the Ceratiomyxales. This is not surprising, since it is by far the most commonly encountered member of this order. So far, a pyrone pigment (ceratiopyrone) with aliphatic side chains has been isolated from the plasmodium of *C. fruticulosa* (Fig. 6.1). On the pyrone ring structure of this compound, there are alicyclic carbon chains with two double bonds, which form a conjugate structure with the ring, thus making the compound ultraviolet absorbent in the ultraviolet region of the spectrum. The light sensitivity of plasmodia was considered to be related to the presence of this compound (Steglich, 1989).

Metabolites From Members of the Liceales

More than 40 compounds have been isolated from 10 species representing 6 genera of the Liceales. These are mainly fatty acids, alkaloids, naphthoquinone pigment derivatives that have related structures, as well as small amounts of esters and glycoside (Table 6.1). In *Lycogala*, many compounds, including

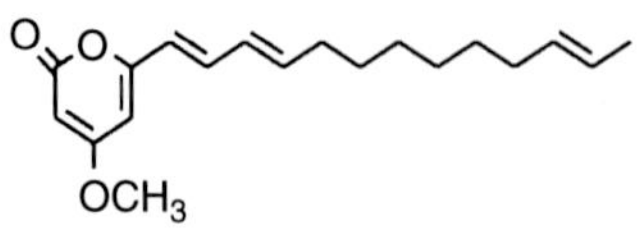

FIGURE 6.1 Ceratiopyrone isolated from *Ceratiomyxa fruticulosa*.

TABLE 6.1 Chemical Constituents From Members of the Liceales (Buchanan et al., 1996; Fröde et al., 1994; Hashimoto et al., 1994; Hosoya et al., 2005; Ippongi et al., 2011; Ishikawa et al., 2002; Iwata et al., 2003; Kamata et al., 2004, 2005; Misono et al., 2003a; Nakatani et al., 2003; Naoe et al., 2003; Řezanka and Dvořáková, 2003; Řezanka et al., 2004)

Myxomycete	Chemical compounds
Enteridium lycoperdon	Enteridinines A, B
Lycogala epidendrum	Polyacetylene triglycerideslycogarides A–C (Fig. 6.2) Acyl glyceride D–G Ploypropioniclactone glycoside lycogalinosides A–B (Fig. 6.3) 3,4-(Indole-3-2)pyrrole-2,5-dicarboxylic acid derivatives Bisindole alkaloid lycogarubin B–C Bisindole alkaloid arcyriaflavin A–B Bisindole alkaloid staurosporinone Bisindole alkaloid lycogaric acid A Bisindole alkaloid lycogarubin C derivant Bisindole alkaloid lycogarubin B derivant Bisindole alkaloid arcyriarubin A Bisindole alkaloid 6-hydroxyl staurosporinone Bisindole alkaloid 5,6-dyhydroxyl arcyriaflavin A
Tubulifera arachnoidea	Tubiferic acid
T. casparyi	Bisindole alkaloid arcyriaflavin C Bisindole alkaloid arcyriaflavin B
T. dimorphotheca	TerPene lactone tubiferal A–B
Lindbladia tubulina	Naphthoquinone pigment lindbladione Naphthoquinone pigment lindbladiapyrone 7-Methoxyl lindbladione 6,7-Dimethoxy lindbladione 6,7-Dimethoxy dihydro lindbladione Dihydro lindbladione 6-Methoxy dihydro–lindbladione
Cribraria intricata	Naphthoquinone pigment lindbladione
C. purpurea	Dihydrofuran naphthoquinone Cribrarione A (Fig. 6.4)
C. cancellata	Naphthoquinone pigment cribrarione B (Fig. 6.5)

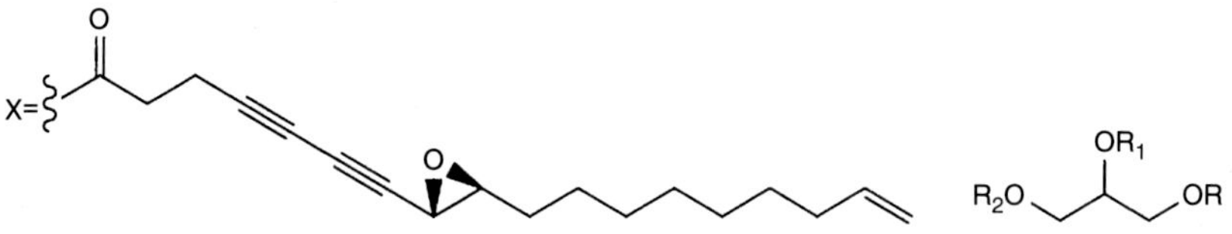

FIGURE 6.2 Polyacetylene triglycerides.

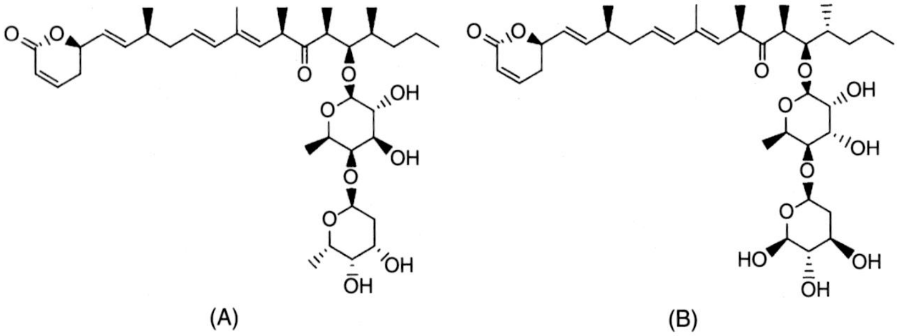

(A) (B)

FIGURE 6.3 Polypropionate lactone glycoside (A) and (B).

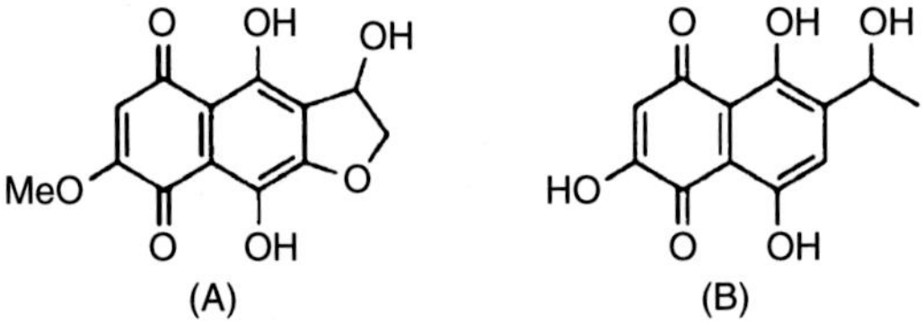

(A) (B)

FIGURE 6.4 Cribrarione (A) and cribrarione (B) (Ishibashi, 2007).

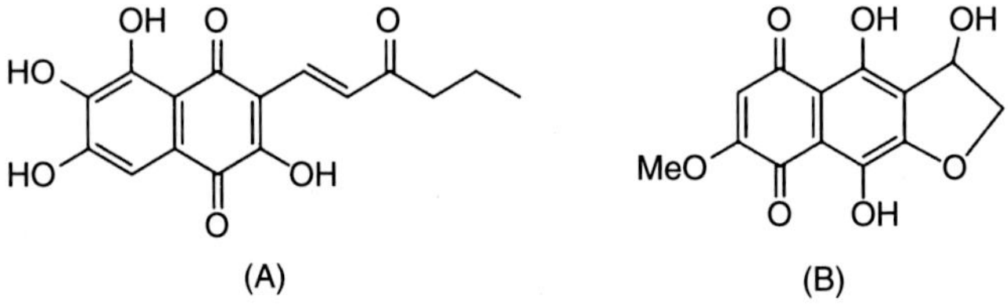

(A) (B)

FIGURE 6.5 Naphthoquinone pigments (A) and (B).

cyclohexene, octadien, decenal, hexadecatrienal, benzene, benzofuran, asarone, and ethul citrate, have been extracted from the fruiting bodies using the GCMS method (Fig. 6.6) (Wang, 2014). Indole alkaloids (Fig. 6.7) were the major compounds isolated from *Lycogala epidendrum*. Two new compounds (6-hydroxystaurosporinone and 5, and 6-dihydroxyarcyriaflavin A) isolated from *L. epidendrum* showed cytotoxicity against HeLa, Jurkat, and vincristine-resistant KB/VJ300 cells, and 6-hydroxystaurosporinone particularly inhibited protein tyrosine kinase activity (Hosoya et al., 2005).

Tubiferic acid, which is a triterpenoid acid with a side chain containing 2,6-dimethyl-4,5-dihydroxy-2-hexenoic acid moiety, was isolated from fruiting bodies of *Tubulifera* (*Tubifera*) *arachnoidea* (Ippongi et al., 2011). Tubiferal A and B along with triterpenoid lactone were also isolated from *T. dimorphotheca*. Tubiferal A exhibited a reversal effect of vincristine (VCR) resistance, but tubiferal B did not exhibit such an effect (Kamata et al., 2004).

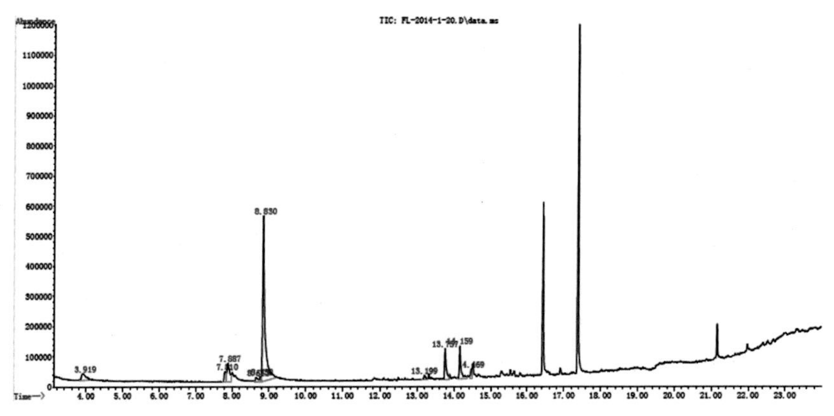

FIGURE 6.6 GC–MS spectra of petroleum ether extraction from the fruiting bodies of *Lycogala epidendrum*.

FIGURE 6.7 Indole alkaloids (A) and (B).

FIGURE 6.8 Lindbladione (Ishibashi, 2007).

Enteridinines A and B, two deoxysugar esters with the absolute configurations of the hydroxyl and methyl groups, have been isolated from *Enteridium lycoperdon*. The unique structures, 1,7-dioxaspiro[5.5]undecanes with an O-b-d-mycarosyl-(1→4)-b-d-olivosyl and an O-b-l-olivomycosyl-(1→4)-b-d-amicetosyl-(1→4)-b-l-digitoxosyl unit, were associated with enteridinines A and B, respectively (Řezanka et al., 2004).

Naphthoquinone pigments (Fig. 6.8) were the primary compounds isolated from *Lindbladia* and *Cribraria*. Lindbladione, the common naphthoquinone pigment, is one of the familiar red pigments found in the Cribrariaceae (Ishikawa et al., 2002; Misono et al., 2003a).

Metabolites From Members of the Trichiales

At present, nearly 40 compounds have been isolated from 9 species in 3 genera of the Trichiales. These compounds are mainly alkaloids and fatty acids but also include a small amount of naphthoquinone pigments (Table 6.2). Analysis of fatty acid composition has been done for *A. cinerea*, *A. denudata*, *A. obovlata,*

TABLE 6.2 Chemical Constituents From Members of the Trichiales (Kamata et al., 2005, 2006; Nakatani et al., 2003; Řezanka, 1993; Steglich, 1989; Steglich et al., 1980; Kopanski et al., 1982, 1987)

Myxomycete	Chemical constituents
Arcyria ferruginea	Bisindole alkaloid dihydro arcyriarubin C Bisindole alkaloid arcyriarubin C Bisindole alkaloid arcyriaflavin C
A. cinerea	Fatty acid Bisindole alkaloid cinereapyrrole A Bisindole alkaloid cinereapyrrole B
A. denudata	Fatty acids Bisindole alkaloid arcyriarubin A–C (Fig. 6.9) Bisindole alkaloid arcyriaflavins A–D (Fig. 6.9) Pigment arcyroxepins A–B Indole alkaloid arcyriacyanin A Indole alkaloid dihydroarcyriacyanin A Indole alkaloid dihydroarcyrioxocin A Indole alkaloid arcyroxocins A–B Indole alkaloid arcyroxindole A
A. nutans	Fatty acid
A. obvelata	Dihydroarcyriacianin A Arcyriaflavin B Arcyroxocin B Hydroarcyriacianin A
Trichia favogiena	z,z-5,9-Hexadecadienoicacid 7,13-Docosadienoic acid 7,15-Docosadienoic acid 5,11,14-Epoxyeicosatrienoic acids 5,11,14,17-Arachidonic acid (Fig. 6.10)
T. favoginea var. persimilis	Kehokorin A–C
T. varia	z,z-5,9-Hexadecadienoicacid 7,13-Docosadienoic acid 7,15-Docosadienoic acid 5,11,14-Epoxyeicosatrienoic acids 5,11,14,17-Arachidonic acid (Fig. 6.10)
T. floriformis	2,3,5-Trimethynaphthoquinone
Metatrichia vesparium	Arcyriaflavin C Trichione Vesparione

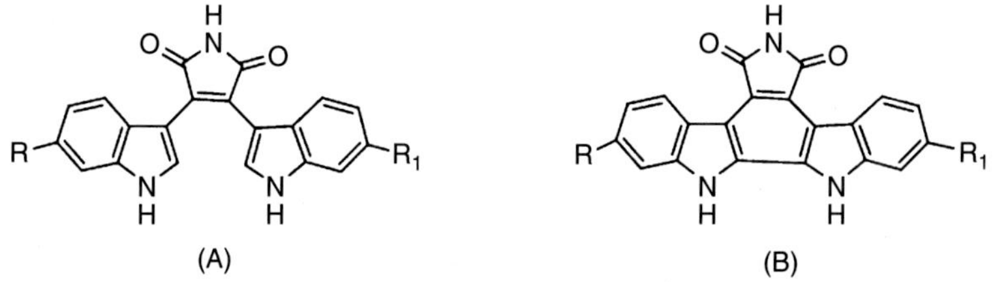

FIGURE 6.9 Indole alkaloids (A) and (B).

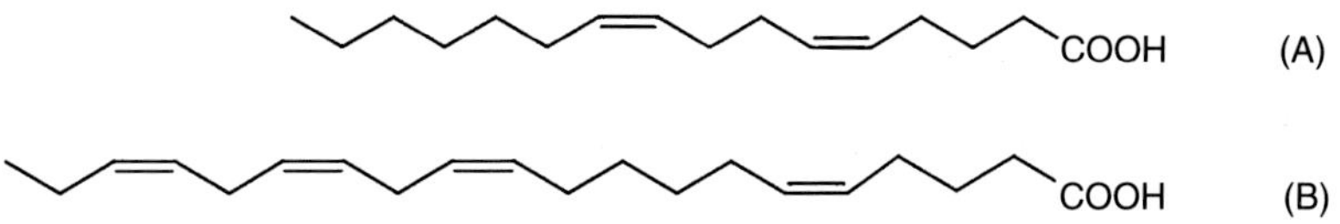

FIGURE 6.10 Fatty acid (A) and (B).

T. favoginea, and *T. varia*, and the results obtained are listed in Table 6.2. Most of these species have common fatty acids except for *T. favogiena* and *T. varia*, for which the fatty acids are not found in other species of the same family. This suggests a closer genetic relationship between these two species than with the others. The fatty acids listed in Table 6.2 also represent their characteristic components and could be used as the basis for chemical classification in Trichiales (Řezanka, 1993). Indole skeleton structures of the alkaloids isolated in *Arcyria* can be considered as characteristic for this genus.

The lipid soluble compounds obtained from members of the Trichiales have been separated and analyzed by GC/MS. The compounds are diverse but specific for each species, which reflects the relationships between species and the presence of their unique compounds and thus would seem to have potential use in studies of myxomycete taxonomy (Zhu and Wang, 2005).

Some bioactive compounds have been isolated from species in the Trichiales because of the structure of the bisindole skeleton in these compounds, such as arcyriaflavin C from *A. ferruginea* (Nakatani et al., 2003), arcyroxocin B, and dihydroarcyriacyanin A from *A. denudata* and *A. obvelata*, respectively (Kamata et al., 2006).

Metabolites From Members of the Physarales

More than 50 compounds have been isolated from 12 species in 4 genera of the Physarales. These are fatty acids, sterols, amino acids, esters, quinone pigments, and a small amount of glucosides (Table 6.3). Most compounds were isolated only from the Physarales, but about one-fifth of the compounds are included among the whole assemblage of compounds isolated from myxomycetes in general. Chemical analysis by using GCMS found that esters and steroids were the major chemical components isolated from seven species in

TABLE 6.3 Chemical Constituents From Members of the Physarales (Bullock and Dawson, 1976; Casser et al., 1987; Comes and Kleinig, 1973; Eisenbarth and Steffan, 2000; Göttler and Holler, 2006; Hamana and Matsuzaki, 1984; Ishibashi et al., 1999, 2001; Murakami-Murofushi et al., 2002; Korn et al., 1965; Lenfant et al., 1970; Misono et al., 2003b,c; Nakatani et al., 2004; Nakatani et al., 2005a,b; Nowak and Steffan, 1998; Řezanka et al., 2005; Steglich, 1989; Shintani et al., 2009; Simon and Henney, 1970; Steffan et al., 1987; Zhu, 2012)

Myxomycete	Chemical constituent
Didymium bahiense	Bahiensol Makaluvamine A–B (Fig. 6.11)
D. squamulosum	Clionasterol
D. iridis	Makaluvamine I Damirone C
Diderma chondrioderma	β-Sitosterol Phthalate-(2-)ethyl hexyl ester
Fuligo septica	Fuligorubin A
F. candida	Cycloanthranilylproline 4-Aminobenzoyltryptophan Fuligocandin A Fuligocandin B Fuligoic acid
F. cinerea	Fulicineroside
Leocarpus fragilis	Acyltetramic acids L-Tyrosine
Physarum flavicomun	Lanosterol Poriferasterol 22-Dihydroporiferasterol Amino acid
P. polycephalum	Oleate Linoleate 11-Eicosenoate 11,14-Eicosadienoate 8,11,14-Eicosatricnoate Arachidonate (Fig. 6.12) Stigmasterol β-Sitosterol Stigmastanol Campestanol Campesterol Cholesterol Lanosterol (Fig. 6.13) 24-Methylene dihydrolanosterol Poriferasterol Ergostanol Δ^5–Ergostenol

TABLE 6.3 Chemical Constituents From Members of the Physarales (*cont.*)

Myxomycete	Chemical constituent
	Phospholipase D Cyclic phosphatidic acid (cPA) (Fig. 6.14) Chrysophysarin A Physarochrome A Physarorubinic acid A–B Polycephalin B–C Amino acid 1,3-Diaminopropane Poly(β-L-malic acid)
P. rigidum	Physarigins A–C Amino acid
P. melleum	Melleumin A Melleumin B (Figs. 6.15 and 6.16)

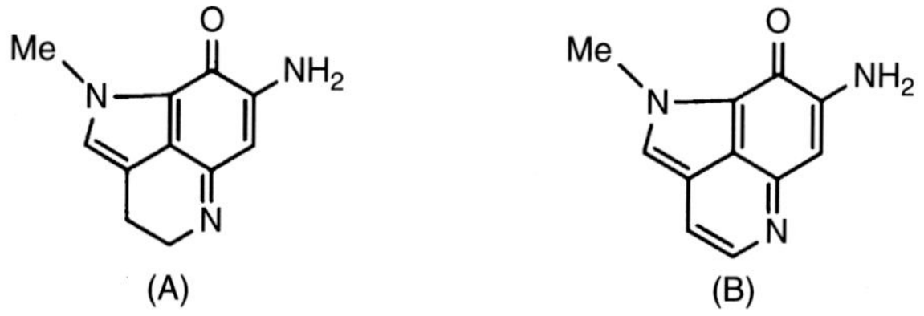

FIGURE 6.11 Pyrroloiminoquinones (A) and (B).

FIGURE 6.12 Fatty acid from *Physarum polycephalum* (A), (B), and (C).

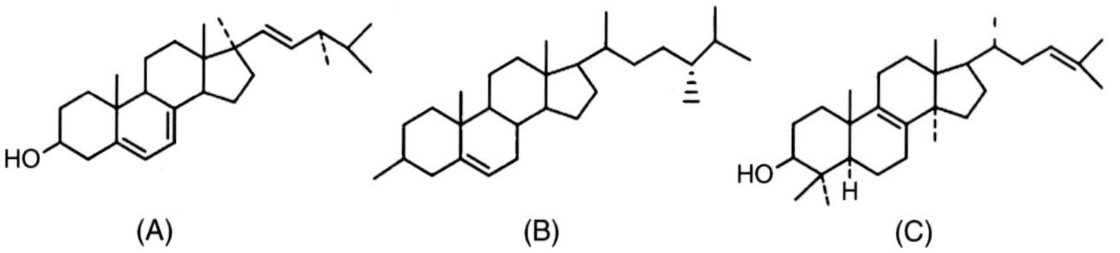

FIGURE 6.13 Stigmasterol campesterol lanosterol (A), (B), and (C).

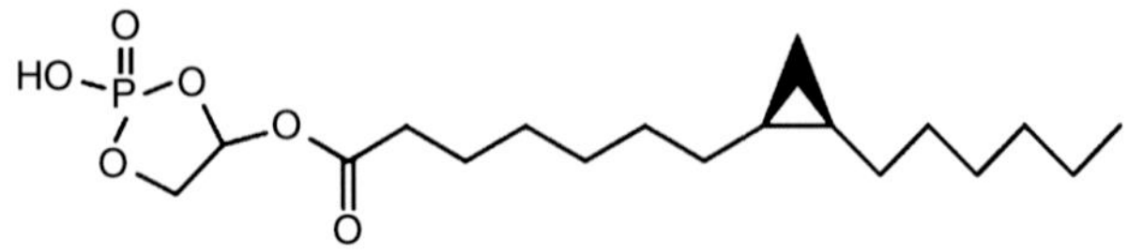

FIGURE 6.14 Cyclic phosphatidic acid.

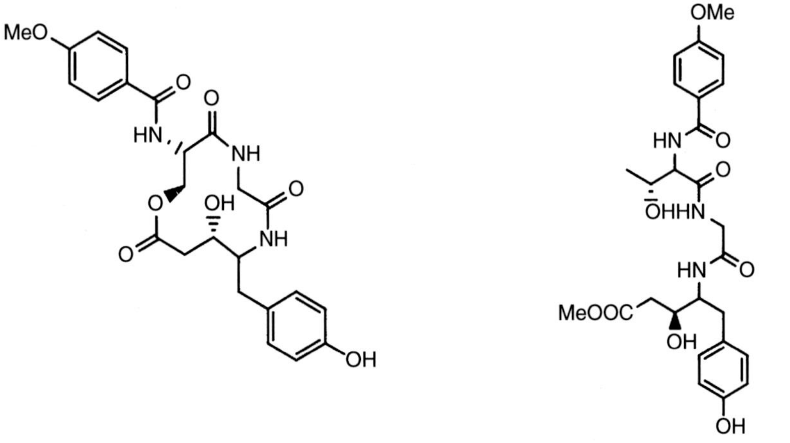

FIGURE 6.15 Peptides lactone (A) and (B).

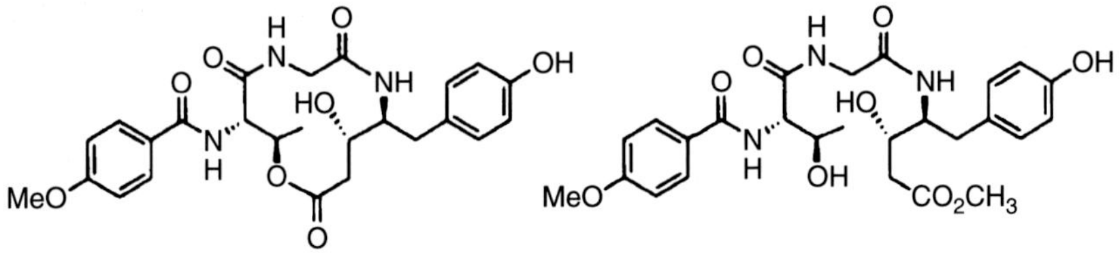

FIGURE 6.16 Melleumin (A) and melleumin (B) (Ishibashi, 2007).

the Physaraceae. Furan nucleus is a unique compound in *F. septica* (Fig. 6.17) (Jiang, 2013).

The main fatty acids are hexadecanoic acid, 9-octadecenoic acid, and 9,12-octadecadienoic acid. The molecular structure of fatty acids from the Physarales is closer to that of the true amoebae, which supports the conclusion that myxomycetes are more closely related to protozoans than other organisms (Korn et al., 1965). The pigments makaluvamine A–B (Fig. 6.11), makaluvamine I, and damirone C were isolated from *Didymium bahiense* and *D. iridis*, respectively. All of these pigments have a pyrrole imino quinone skeleton structure. Similar compounds have also been extracted from marine sponges.

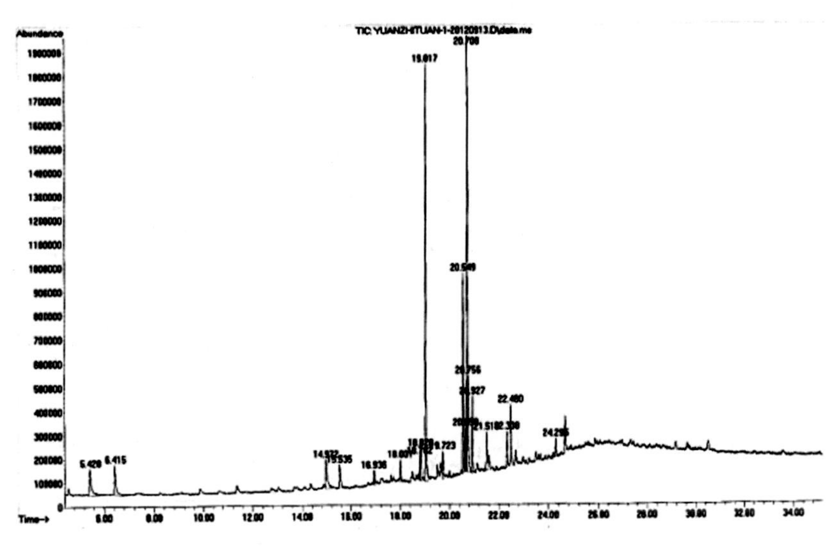

FIGURE 6.17 GC–MS spectra of plasmodia petroleum ether extraction of *Fuligo septica.*

The β-sitosterol with cyclopentane hydrogen phenanthrene skeleton structure and phthalate-(2-ethyl) hexyl ester of peptide acid esters were isolated from the fruiting bodies of *Diderma chondrioderma* (Zhu, 2012). A variety of sterols were isolated from *P. polycephalum* and *P. flavicomun*, but 22 dihydro-porous sterol was isolated only from the latter species. As such, this compound can be used as a chemical characteristic to distinguish between these two species of myxomycetes (Bullock and Dawson, 1976), although it should be noted that these two species are also morphologically rather different. An ester compound (3-octene acid propyl ester) has been isolated from the plasmodia of *F. septica.* The yellow pigment chrysophysarin A (Eisenbarth and Steffan, 2000) and physarochrome A (Steffan et al., 1987) have been isolated from the plasmodia of *P. polycephalum*, as well as fuligorubin A (Casser et al., 1987) from *F. septica.* These pigments were related to photoreception and energy conversion in the life cycle of these myxomycetes. While physarigins A–C, three yellow pigments, were isolated from a cultured myxomycete *P. rigidum*, the chemical structure of which is similar to the yellow pigments mentioned earlier (Misono et al., 2003b).

Metabolites From Members of the Stemonitales

Hexadecanoic acid, octadecadienoic acid, oleic acid, stigmasterol, and ergosta, which are mainly lipophilic compounds have been identified from *S. flavogenita* by the GC–MS method (Fig. 6.18) (Wang, 2014). Stigmasterol, a sterol compound, has been isolated from the fruiting body of *S. splendens* (Fig. 6.19) (Zhu, 2012).

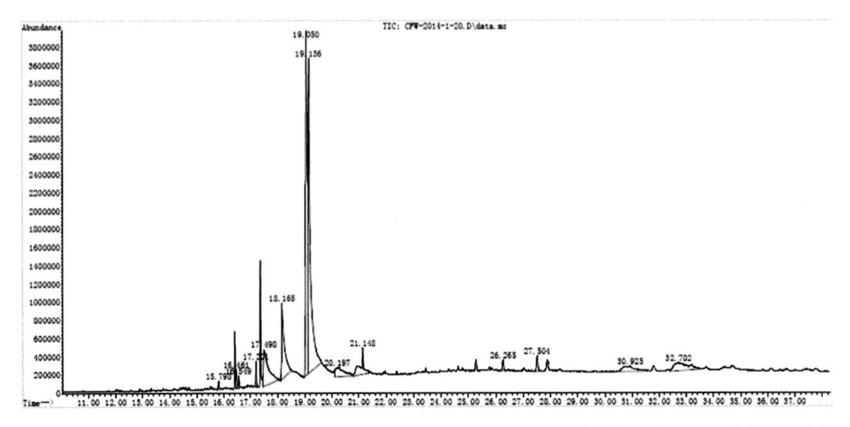

FIGURE 6.18 GC–MS spectra petroleum ether extraction of fruiting bodies of *Stemonitis flavogenita*.

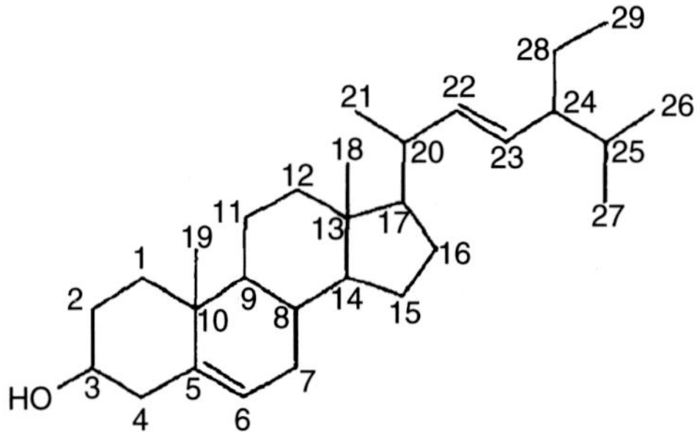

FIGURE 6.19 Sigmasterol.

RESEARCH METHODS USED IN THE STUDY OF METABOLITES

Both the solvent extraction method and the chromatographic separation method are used in investigations of the chemical constituents of the secondary metabolites of myxomycetes. The majority of GC–MS analyses have been used to analyze the lipid soluble components isolated from fruiting bodies or plasmodia by ligroin (Jiang, 2013; Zhu, 2012). Although the separation of other secondary metabolites is relatively complex, chromatographic separation has proved to be an efficient method. Based on the solubility of various components, targeted products can be obtained using the chromatographic separation method with gradient elution (Řezanka et al., 2005). The structures of these compounds then can be determined by elemental analysis, such as IR spectra, 1 HNMR, 13 CNMR, and two-dimensional NMR spectroscopy. Each monomeric compound structure can be identified by its unique spectrum (Casser et al., 1987).

Antibacterial Activity

A new bisindole alkaloid (6-hydroxy-9'-methoxystaurosporinone) has been isolated from field-collected fruiting bodies of *Perichaena chrysosperma*, and this alkaloid was later found to have hedgehog signal inhibitory activity (Shintani et al., 2010). Crude extracts from *F. septica* had a good inhibition effect on *Escherichia coli* and *Bacillus subtilis*, with inhibition rates of 68.00 and 59.45%, respectively. The same extracts had only a minor effect on *Salmonella typhimurium* and *Staphylococcus aureus* and no effect on *Pesudomonas pyocyaneum*. Water extracts did not show antibacterial activity (Jiang, 2013).

Two compounds (lycogalinosides A and B) that were isolated from *L. epidendrum* showed growth inhibitory activities against Gram-positive bacteria. The compound z,z-5,9 16 carbon diene acid was isolated from *Trichia favogina* and *T. varia*, but also can be found in marine sponges and sea anemones. Some research has shown that compounds with similar structures have a negative effect on Gram-positive bacteria but this is not the case for Gram-negative bacteria (Dembitsky et al., 2005).

A crude extract derived from *Cribraria purpurea* exhibited biological activity against *B. subtilis*. This antibacterial activity appears to be related to the presence of the naphthoquinone pigment cribrarione A in myxomycetes (Naoe et al., 2003). Bahiensol isolated from a cultured plasmodium of the myxomycete *D. bahiense* var. *bahiense* had an inhibitory effect on *B. subtilis*. Fulicineroside, a glycosidic dibenzofuran metabolite isolated from *Fuligo* has demonstrated a high level of activity against Gram-positive bacteria (Řezanka et al., 2005).

Antitumor Activity

Polymailc acid (PMLA), a natural product extracted from *P. polycephalum* (Braud and Vert, 1992; Gasslmaier et al., 2000; Lee et al., 2002), is a promising drug with antitumor activity. PMLA as a carrier matrix lacks toxicity in vitro and in vivo, is characterized by nonimmunogenicity, biodegradability, and versatility for drug loading (Ljubimova et al., 2008a). With these properties, nanocomposites based on PMLA demonstrated inhibition activities relating to tumor angiogenesis (Ljubimova et al., 2008a) and tumor targeting of human breast cancer cells (Inoue et al., 2011, 2012) and human glioma cells (Ding et al., 2010; Ding et al., 2013; Lee et al., 2006; Ljubimova et al., 2008b, 2013; Portilla-Arias et al., 2010). In addition, methanol extracts from the plasmodia and sclerotia of *F. septica* have been shown to display an inhibitory effect on the B16 F1 murine melanoma cells (Jiang, 2013).

Cyclic phosphatidic acid (cPA), which was isolated originally from *Physarum polycephalum*, has a cyclic phosphate at the *sn*-2 and *sn*-3 positions of the glycerol carbons, and this is a key structure for its activities. This substance has been reported to have antimitogenic activities on cell cycle regulation, actin stress fiber formation and rearrangement, cell invasion and metastasis, differentiation and viability of neuronal cells, and mobilization of intracellular calcium (Murakami-Murofushi et al., 2002).

Two compounds (makaluvamine A and B) were isolated from *D. bahiense* (Ishibashi et al., 2001). They have considerable cytotoxic activity on the human colon carcinoma cell line HCT 116 and displayed inhibition activity on topoisomerase in vitro (Barrows et al., 1993). The compounds makaluvamine A and C have been reported to display inhibition activity on human ovarian cancer cells (Matsumoto et al., 1999).

Arcyriarubins A–C and arcyriaflavins A–D have been extracted from *A. denudata*. There are many compounds that have an indolo-2,3-carbazole ring structure. Its derivatives display biological activity against *Bacillus cereus* and antileukemia activity, inhibition of protein kinase A and C (Pereira et al., 1996), tyrosine kinases, and serine kinase activity (Sancelme et al., 1994). The analogues of arcyriaflavin have been tested as anticancer drugs to carry out clinical evaluation tests (Sancelme et al., 1994). Arcyriarubin C was isolated from *Arcyria ferruginea*, and the product of its CIS configuration has an inhibitory activity on Wnt signal transfer, which can provide a new way for antitumor therapy (Nakatani et al., 2003).

Fulicineroside, a glycosidic dibenzofuran metabolite from *Fuligo*, is highly active against crown gall tumors (Řezanka et al., 2005). Two compounds (makaluvamine A and C) isolated from *D. bahiense* have been demonstrated to display activity against colon cancer and ovarian cancer in vitro (Barrows et al., 1993; Matsumoto et al., 1999).

Cytotoxic Activity

Two compounds (6-hydroxystaurosporinone and 5,6-dihydroxyarcyriaflavin A) isolated from fruiting bodies of *L. epidendrum* have been shown to display cytotoxicity against Hela cells, vincristine resistant KB/VJ300 cells and 6-hydroxystaurosporinone. In particular, it can inhibit the activity of protein tyrosine kinase (Hosoya et al., 2005). The naphthoquinone pigment derived from *Lindbladia tubulina* and its derivatives have toxic effects on mice leukemia cells P388, and it is considered as an active substance. The compound 6-methoxy-2-hydrogenation-lindbladione also showed a great resistance reversal effect (Misono et al., 2003a). The comound arcyriaflavin C isolated from *Tubifera casparyi* was toxic to Hela cells (Ishibashi et al., 2001).

Arcyroxocin B has been isolated from fruiting bodies of *Arcyria denudata*, and dihydroarcyriacyanin A was derived from fruiting bodies of *Arcyria obvelata*. The information provided herein represents the first report of the characterization of arcyroxocin B and dihydroarcyriacyanin A. These two compounds have been shown to exhibit cytotoxicity against Jurkat cells (Kamata et al., 2006).

Antioxidant Activity

Three methods have been used to measure antioxidant activities of plasmodia of *F. septica*. These are the total reducing capacity and the scavenging abilities of DPPH and O_2^-. The results showed that methanol extracts from plasmodia and sclerotia of *F. septica* all demonstrated good antioxidant activities (Jiang et al., 2014).

RESEARCH METHODS IN THE STUDY OF BIOACTIVITY

Research on the bioactivity of myxomycetes is primarily related to antibacterial activity, antitumor activity, antioxidant activity, and cytotoxicity. The main method used in the study of antimicrobial activity is the filter paper diffusion method. The antibacterial activity of methanol and petroleum ether extracts of myxomycete can be determined by the filter paper diffusion method, the results indicate whether or not the extracts being evaluated have inhibitory effects (Jiang et al., 2014).

The MTT cell proliferation assay is the main method used to study the anticancer activity of compounds derived from myxomycetes. The crude extract from the plasmodia and sclerotia of *F. septica*, when assayed on mouse melanoma B16F1 value by MTT, showed that it has an inhibitory effect on tumor cell proliferation (Nakatani et al., 2004).

Chemical Characteristics

The alkaloid with the structure of the skeleton can be obtained in *Arcyria* and can be regarded as the characteristic compound of the genus. The sterol content of two myxomycetes, *P. polycephalum* and *P. flavicomum*, has been examined. The compound 22-dihydrolanosterol was only present in *P. flavicomum*, which means that it can be used to differentiate these two species (Bullock and Dawson, 1976).

The yellow optically active pigments physarochrome A and chrysophysarin A have been isolated from microplasmodia of *P. polycephalum* (Eisenbarth and Steffan, 2000). Fuligorubin A is responsible for the yellow color of the plasmodia of *F. septica*. The new pigments physarigins A–C (1–3) have been isolated from a cultured plasmodium of *P. rigidum*, which has a similar chemical structure (Hashimoto et al., 1994).

Fatty acids of nine different myxomycetes have been analyzed. In addition to the common fatty acids, polyunsaturated and methylene noninterrupted polyunsaturated fatty acids, for example those with 5,9-and/or 5,11-double bonds, were identified by GC–MS of their corresponding oxazolines. However, these compounds have been identified in only a single species of myxomycete. Therefore, their presence offers a new concept in the biosynthesis of these compounds and chemotaxonomy of myxomycetes (Řezanka, 1993).

Dienoic acids formed a large portion of the total fatty acids and varied from 31.43% in *Physarum* sp. to 54.09% in *Trichia favoginea* being represented by four major isomers of 18:2 and trace isomers of 20:2. The total amount of polyenoic acids varied widely, from 8.22% in *T. favogiena* to 18.4% in *F. septica*, and were represented only by tri- and tetraenoic acids, with 9, 12, 15–18:3 (varying from 1.95% to 6.66%) and 5, 8, 11, 14–20:4 (varying from 0.28% to 5.36%) acids most abundant.

GC–MS has been used in analyzing the liposoluble constituents of fruiting bodies and plasmodia of *D. chondrioderma* along with the fruiting bodies of *Didymium crustaceum*, *C. leucocephalum*, and *S. splendens*. The same compounds were found in both plasmodia and fruiting bodies of the same species, which demonstrates that the characteristic chemical elements actually existed. However, the components from different species were diverse, which reflects the relationships among species and, as such, presumably can be used in studies of the taxonomy of myxomycetes (Li et al., 2007; Zhu and Wang, 2005).

Trace chemical elements in the fruiting bodies of seven representative species of myxomycetes belonging to the order Stemonitales have been studied with the technique of energy dispersive X-ray microanalysis. The results indicated that there are differences in the trace chemical element composition of the seven species. Calcareous granules in the fruiting bodies of members of the Stemonitales could not be observed under the microscope. However, the concentration of the element calcium was consistently high in all seven species (Chen et al., 2008).

Plasmodia of six species of myxomycetes (*P. melleum*, *F. septica*, *Physarella oblonga*, *D. megalosporum*, *D. squamulosum*, and *D. melanospermum*) have been examined to study the type and relative content of the various chemical elements, using energy dispersive X-ray analysis (EDX) and a consideration of their morphological characteristics. The concentrations of chemical elements in the plasmodium of a particular species appear to be closely related to its morphological characteristics. Eight elements (including carbon, oxygen, sodium, magnesium, phosphorus, sulfur, potassium, and calcium) were present in the plasmodia of all six species, but three elements (aluminium, silicon, and chlorine) were detected in only a portion of a given plasmodium. Moreover, the relative content of chemical elements was found to be different in each plasmodium. The relative contents of carbon, sodium, phosphorus, and calcium in each plasmodium are close for members of a particular family, and those of oxygen, sulfur, potassium, and magnesium are similar in plasmodia that share the same morphological characteristics and have been cultured under the same condition (Song et al., 2014).

When analyzed using EDX technology, both Ca and P were detected in 11 species belonging to the Physaraceae. From these, 10 species in the Didymiaceae had Ca, but only 1 species had P. As noted earlier in this discussion, exploring the differences in chemical elements could have potential significance in the classification of myxomycetes (Schoknecht, 1975). Moreover, as a general observation, based on chemical composition analysis, it is apparent that similarities in the physiological and chemical characteristics of myxomycetes correspond to their genetic relationships.

ACKNOWLEDGMENTS

Appreciation is extended to Jim Clark for providing some of the information used to develop the first part of this chapter.

REFERENCES

Alexopoulos, C.J., 1960a. Morphology and laboratory cultivation of *Echinostelium minutum*. Am. J. Bot. 47, 37–43.

Alexopoulos, C.J., 1960b. Gross morphology of the plasmodium and its possible significance in the relationships among the myxomycetes. Mycologia 52 (1), 1–20.

Alexopoulos, C.J., Mims, C.W., 1979. Introductory Mycology, third ed. Wiley, New York, NY.

Anderson, J.D., 1962. Potassium loss during galvanotaxis of slime mold. J. Gen. Physiol. 45, 567–574.

Anderson, J.D., 1964. Regional differences in ion concentration in migrating plasmodia. In: Allen, P.J., Kamiya, N. (Eds.), Primitive Motile Systems in Cell Biology. Academic Press, New York, NY, pp. 125–136.

Barrows, L.R., Radisky, D.C., Copp, B.R., Swaffar, D.S., Kramer, R.A., Wafters, R.L., Ireland, C.M., 1993. Makaluvamines, marine natural products, are active anti-cancer agents and DNA topo II inhibitors. Anticancer Drug Des. 8, 333–347.

Braud, C., Vert, M., 1992. Degradation of poly(p-malic acid)-monitoring of oligomers formation by aqueous SEC and HPCE. Polym. Bull. 29, 177–183.

Buchanan, M.S., Hashimoto, T., Asakawa, Y., 1996. Acylglycerols from the slime mold, *Lycogala epidendrum*. Phytochemistry 41, 791–794.

Bullock, E., Dawson, C.J., 1976. Sterol content of the myxomycetes *Physarum polycephalum* and *P. flavicomum*. J. Lipid Res. 17, 565–571.

Camp, W.G., 1937. The fruiting of *Physarum polycephalum* in relation to nutrition. Am. J. Bot. 24 (5), 300–303.

Carlile, M.J., 1970. Nutrition and chemotaxis in the myxomycete *Physarum polycephalum*: the effect of carbohydrates on the plasmodium. J. Gen. Microbiol. 63, 221–226.

Casser, I., Steffan, B., Steglich, W., 1987. The chemistry of the plasmodial pigments of the slime mold *Fuligo septica* (myxomycetes). Angew. Chem. Int. Ed. Engl. 26, 586–587.

Charvat, I., Cronshaw, J., Ross, I.K., 1974. Development of the capillitiumin *Perichaena vermicularis*, a plasmodial slime mold. Protoplasma 80, 207–221.

Charvat, I., Ross, I.K., Cronshaw, J., 1973. Ultrastructure of the plasmodial slime mold *Perichaena vermicularis*. II. Peridium formation. Protoplasma 78, 1–20.

Chassain, M., 1980. Essai sur la place ecologique des myxomycetes. Doc. Mycol. 11, 47–57.

Chen, X., Gu, S., Zhu, H., Li, Z., Wang, Q., Li, Y., 2013. Life cycle and morphology of *Physarum pusillum* (myxomycetes) on agar culture. Mycoscience 54 (2), 95–99.

Chen, S.L., Zhong, C.G., Wu, M.Q., Li, Yu, 2008. A preliminary analysis of trace chemical elements in the fruiting bodies of Stemonitales. J. Fungal Res. 6 (4), 220–225.

Chet, I., Naveh, A., Henis, Y., 1977. Chemotaxis of *Physarum polycephalum* towards carbohydrates, amino acids and nucleotides. J. Gen. Microbiol. 102, 145–148.

Clark, J., 1995. Myxomycete reproductive systems: additional information. Mycologia 87, 779–786.

Clark, J., Brown, D., Hu, F.-S., 1990. Growth of the myxomycete *Stemonitis flavogenita* on a defined minimum medium. Mycologia 82, 385–386.

Clark, J., Collins, R.O., 1976. Studies on the mating systems of eleven species of myxomycetes. Am. J. Bot. 63, 783–789.

Clark, J., Haskins, E.F., 2012. Plasmodial incompatibility in the myxomycetes: a review. Mycosphere 3, 143–155.

Clark, J., Haskins, E.F., 2013. The nuclear reproductive cycle in the myxomycetes: a review. Mycosphere 4, 233–248.

Collins, O.R., 1979. Myxomycete biosystematics: some recent developments and future research opportunities. Bot. Rev. 45 (2), 145–201.

Comes, P., Kleinig, H., 1973. Phospholipids and phospholipase D in the true slime mold *Physarum polycephalum*. Biochim. Biophys. Acta 316, 13–18.

Considine, J.M., Mallette, M.F., 1965. Production and partial purification of antibiotic materials formed by *Physarum gyrosum*. Appl. Microbiol. 13, 464–468.

Czeczuga, B., 1980. Investigations on carotenoids in fungi VII. Representatives of the myxomycetes genus. Nova Hedwigia 32, 347–352.

Daniel, J.W., Rusch, H.P., 1961. The pure culture of *Physarum polycephalum* on a partial defined soluble medium. J. Gen. Microbiol. 25, 47–59.

Daniel, J.W., Rusch, H.P., 1962a. Method for inducing sporulation of pure cultures of the myxomycete *Physarum polycephalum*. J. Bacteriol. 83, 234–240.

Daniel, J.W., Rusch, H.P., 1962b. Niacin requirement for sporulation of *Physarum polycephalum*. J. Bacteriol. 83, 1244–1250.

Dembitsky, V.M., Tomas, R., Jaroslav, S., Hanus, L.O., 2005. Secondary metabolites of slime molds (myxomycetes). Phytochemistry 66 (7), 747–769.

Ding, H., Helguera, G., Rodríguez, J.A., Markman, J., Luria-Pérez, R., Gangalum, P., Portilla-Arias, J., Inoue, S., Daniels-Wells, T.R., Black, K., Holler, E., Penichet, M.L., Ljubimova, J.Y., 2013. Polymalic acid nanobioconjugate for simultaneous immunostimulation and inhibition of tumor growth in HER2/neu-positive breast cancer. J. Control. Release 171 (3), 322–329.

Ding, H., Inoue, S., Ljubimov, A.V., Patil, R., Portillaarias, J., Hu, J., Konda, B., Wawrowsky, K.A., Fujita, M., Karabalin, N., Sasaki, T., Black, K.L., Holler, E., Ljubimova, J.Y., 2010. Inhibition of brain tumor growth by intravenous poly(β-l-malic acid) nanobioconjugate with ph-dependent drug release. Proc. Natl. Acad. Sci. 107 (42), 18143–18148.

Dussutour, A., Latty, T., Beekman, M., Simpson, S.J., 2010. Amoeboid organism solves complex nutritional challenges. Proc. Natl. Acad. Sci. USA 107 (10), 4607–4611.

Eisenbarth, S., Steffan, B., 2000. Structure and biosynthesis of chrysophysarin A, a plasmodial pigment from the slime mould *Physarum polycephalum* (myxomycetes). Tetrahedron 56, 363–365.

Elliott, E.W., 1948. The swarm-cells of myxomycetes. J. Wash. Acad. Sci. 38 (4), 133–137.

Elliott, E.W., 1949. The swarm cells of myxomycetes. Mycologia 41, 141–170.

Emoto, Y., 1932. Über die Chemotaxis der Myxomyceten-Plasmodien. Proc. Imp. Acad. Japan 8, 460–463.

Erbisch, F.H., 1964. Myxomycete spore longevity. Mich. Bot. 3, 120–121.

Fröde, R., Hinze, C., Josten, I., Schmidt, B., Steffan, B., Steglich, W., 1994. Isolation and synthesis of 3,4-bis(indol-3-yl)pyrrole-2,5-dicarboxylic acid derivatives from the slime mould *Lycogala epidendrum*. Tetrahedron Lett. 35, 1689–1690.

Gasslmaier, B., Krell, C.M., Seebach, D., Holler, E., 2000. Synthetic substrates and inhibitors of beta-poly(L-malate)-hydrolase (polymalatase). Eur. J. Biochem. 267 (16), 5101–5105.

Gilbert, F.A., 1928. A study of the method of spore germination in myxomycetes. Am. J. Bot. 15 (6), 345–352.

Gilbert, F.A., 1929. Spore germination in the myxomycetes: a comparative study of spore germination by families. Am. J. Bot. 16, 421–432.

Gorman, J.A., Wilkins, A.S., 1980. Developmental phases in the life cycle of Physarum and related myxomycetes. In: Dove, W.F., Rusch, H.P. (Eds.), Growth and Differentiation in *Physarum polycephalum*. Princeton University Press, Princeton, NJ, pp. 157–202.

Göttler, T., Holler, E., 2006. Screening for beta-poly(L-malate) binding proteins by affinity chromatography. Biochem. Biophys. Res. Commun. 341 (4), 1119–1127.

Gray, W.D., 1938. The effect of light on the fruiting of myxomycetes. Am. J. Bot. 25, 511–522.

Gray, W.D., 1939. The relation of pH and temperature to the fruiting of *Physarum polycephalum*. Am. J. Bot. 26, 709–714.

Gray, W.D., 1941. Some effects of the heterochromatic ultra-violet radiation on myxomycete plasmodia. Am. J. Bot. 28, 212–216.

Gray, W.D., 1953. Further studies on the fruiting of *Physarum polycephalum*. Mycologia 45, 817–824.

Gray, W.D., Alexopoulos, C.J., 1968. Biology of the Myxomycetes. Ronald Press, New York, NY.

Hamana, K., Matsuzaki, S., 1984. Unusual polyamines in slime molds *Physarum polycephalum* and *Dictyostelium discoideum*. J. Biochem. 95 (4), 1105–1110.

Hashimoto, T., Akazawa, A., Tori, M., Kan, Y., Kusumi, T., Takahashi, H., Asakawa, Y., 1994. Three novel polyacetylene triglycerides, lycogarides A-C, from the myxomycetes *Lycogala epidendrum*. Chem. Pharm. Bull. 42, 1531–1533.

Haskins, E.F., 1970. Axenic culture of myxamoebae of the myxomycete *Echinostelium minutum*. Can. J. Bot. 48, 663–664.

Haskins, E.F., Hinchee, A.A., 1992. Spore germination in *Echinostelium minutum*. Mycologia 84, 916–920.

Hinssen, H., 1981. An actin-modulating protein from *Physarum polycephalum*.2. Calcium-dependence and other properties. Eur. J. Cell Biol. 23, 234–240.

Honda, H., Sakuma, T., Saeki, N., Takamatsu, K., Matsuno, K., 1990. Movement of vesicles in cytoplasmic streaming in plasmodium. Biochem. Biophys. Res. Commun. 172 (3), 1236–1238.

Hosoya, T., Yamamoto, Y., Uehara, Y., Hayashi, M., Komiyama, K., Ishibashi, M., 2005. New cytotoxic bisindole alkaloids with protein tyrosine kianse inhibitory activity from a myxomycete *Lycogala epidendrum*. Bioorg. Med. Chem. Lett. 15 (11), 2776–2780.

Hüttermann, A., 1973a. *Physarum polycephalum* object of research in cell biology. Ber. Dtsch. Bot. Ges. 86, 1–4.

Inoue, S., Ding, H., Portilla-Arias, J., Hu, J., Konda, B., Fujita, M., Espinoza, A., Suhane, S., Riley, M., Gates, M., Patil, R., Penichet, M.L., Ljubimov, A.V., Black, K.L., Holler, E., Ljubimova, J.Y., 2011. Polymalic acid-based nanobiopolymer provides efficient systemic breast cancer treatment by inhibiting both HER2/neu receptor synthesis and activity. Cancer Res. 71 (4), 1454–1464.

Inoue, S., Patil, R., Portilla-Arias, J., Ding, H., Konda, B., Espinoza, A., Mongayt, D., Markman, J.L., Elramsisy, A., Phillips, H.W., Black, K.L., Holler, E., Ljubimova, J.Y., 2012. Nanobiopolymer for direct targeting and inhibition of EGFR expression in triple negative breast cancer. PLoS One 7 (2), 1211–1227.

Ippongi, Y., Ohtsuki, T., Toume, K., Arai, M.A., Yamamoto, Y., Ishibashi, M., 2011. Tubiferic acid, a new 9,10-secocycloartane triterpenoid acid isolated from the myxomycete *Tubulifera arachnoidea*. Chem. Pharm. Bull. 59 (2), 279–281.

Ishibashi, M., 2007. Study on myxomycetes as a new source of bioactive natural products. Yakugaku Zasshi 127 (9), 1369–1381.

Ishibashi, M., Iwasaki, T., Imai, S., Sakamoto, S., Yamaguchi, K., Ito, A., 2001. Laboratory culture of the myxomycetes: formation of fruiting bodies of *Didymium bahiense* and its plasmodial production of makaluvamine A. J. Nat. Prod. 64, 108–110.

Ishibashi, M., Mitamura, M., Ito, A., 1999. Laboratory culture of the myxomycete: *Didymium squamulosum* and its production of lionasterol. Nat. Med. 53 (6), 316–318.

Ishikawa, Y., Ishibashi, M., Yamamoto, Y., Hayashi, M., Komiyama, K., 2002. Lindbladione and related naphthoquinone pigments from a myxomycete *Lindbladia tubulina*. Chem. Pharm. Bull. 50, 1126–1127.

Iwata, D., Ishibashi, M., Yamamoto, Y., 2003. Cribrarione B, a new napthoquinone pigment from the Myxomycete *Cribraria cancellata*. J. Nat. Prod. 66, 1611–1612.

Jiang, N., 2013. Study on chemical ingredients and activities of major representative species of physarale myxomycetes. Jilin Agricultural University. (In Chinese).

Jiang, N., Liu, Y., Zhu, Y.Y., Wang, Q., 2014. Study on antibacterial activity of plasmodium and sclerotium from *Fuligo septica*. J. Fungal Res. 12 (3), 160–163, (In Chinese).

Kamata, K., Kiyota, M., Naoe, A., Nakatani, S., Yamamoto, Y., Hayashi, M., Komiyama, K., Yamori, T., Ishibashi, M., 2005. New bisindole alkaloids isolated from myxomycete *Arcyria cinerea* and *Lycogala epidendrum*. Chem. Pharm. Bull. 53 (5), 594–597.

Kamata, K., Onuki, H., Hirota, H., Yamamoto, Y., Hayashi, M., Komiyama, K., Sato, M., Ishibashi, M., 2004. Tubiferal A, a backbone-rearranged triterpenoid lactone isolated from the myxomycete *Tubifera dimorphotheca*, possessing reversal of drug resistance activity. Tetrahedron 60, 9835–9839.

Kamata, K., Suetsugu, T., Yamamoto, Y., Hayashi, M., Komiyama, K., Ishibashi, M., 2006. Bisindole alkaloids from myxomycetes *Arcyria denudata* and *Arcyria obvelata*. J. Nat. Prod. 69 (8), 1252–1254.

Kamiya, N., 1950. The protoplasmic flow in the myxomycete plasmodiumas revealed by a volumetric analysis. Protoplasma 39, 344–357.

Kamiya, N., 1959. Protoplasmic streaming. Protoplasmatologia 8 (3a), 199.

Kerr, N.S., Sussman, M., 1958. Clonal development of the true slime mould *Didymium nigripes*. J. Gen. Microbiol. 19, 173–177.

Kincaid, R.L., Mansour, E., 1978. Chemotaxis toward carbohydrates and amino acids in *Physarum polycephalum*. Exp. Cell Res. 116, 377–385.

Knowles, D.J.C., Carlile, M.J., 1978. The chemotactic response of plasmodia of the myxomycete *Physarum polycephalum* to sugars and related compounds. J. Gen. Microbiol. 108, 17–25.

Koevenig, J.L., 1964. Studies on life cycle of *Physarum gyrosum* and other myxomycetes. Mycologia 56 (2), 170–184.

Komnick, H., Stockem, W., Wohlfarth-Botterman, K.E., 1973. Cell motility: mechanisms in protoplasmic streaming and amoeboid movement. Int. Rev. Cytol. 34, 169–249.

Konijn, T.M., Koevenig, J.L., 1971. Chemotaxis in myxomycetes or true slime molds. Mycologia 63, 901–906.

Kopanski, L., Karbach, G., Selbitschka, G., Steglich, W., 1987. Pilzfarbstoffe, 53. Vesparion, ein Naphtho(2,3-b)pyrandion-Derivat aus dem Schleimpilz *Metatrichia vesparium* (myxomycetes). Liebigs Ann. Chem. 9, 793–796.

Kopanski, L., Li, G.R., Besl, H., Steglich, W., 1982. Pilzpigmente, 41. Naphthochinon-Farbstoffe aus den Schleimpilzen *Trichia floriformis* and *Metatrichia vesparium* (myxomycetes). Liebigs Ann. Chem. 9, 1722–1729.

Korn, E.D., Greenbla, C.L., Lees, A.M., 1965. Synthesis of unsaturated fatty acid in the slime molds *Physarum polycephalum* and the zoo flagellates *Leishmania tarentolae, Trypanosoma lewisi*, and *Crithidia* sp.: a comparative study. J. Lipid Res. 6, 43–50.

Lee, B.S., Fujita, M., Khazenzon, N.M., Wawrowsky, K.A., Wachsmann-Hogiu, S., Farkas, D.L., Black, K.L., Ljubimova, J.Y., Holler, E., 2006. Polycefin, a new prototype of multifunctional nanoconjugate based on poly(beta-l-malic acid) for drug delivery. Bioconjug. Chem. 17 (2), 317–326.

Lee, B.S., Vert, M., Holler, E., 2002. Water-soluble aliphatic polyesters: poly(malic acid)s. Doi, Y., Steinbüchel, A. (Eds.), Biopolymers: Polyesters I, Vol. 3a, Wiley VCH, New York, NY, pp. 75–103.

Lenfant, M., Lecompte, M.F., Farrugia, G., 1970. Identification des sterols de *Physarum polycephalum*. Phytochemistry 9, 2529–2535.

Li, Y., Li, H.Z., Wang, Q., Chen, S.L., 2007. Flora Fungorum Sinicorum Myxomycetes I. Science Press, Beijing, (In Chinese).

Li, C., Wang, X.L., Wang, X.L., Li, Y., 2013. Morphology and behavior of myxamoebae of several myxomycete species. Mycosystema 32 (5), 913–918, (In Chinese).

Lieth, H., Meyer, G.F., 1957. Über den Bau der Pigmentgranula bei den Myxomyceten. Naturwissenschaften 44, 449.

Liu, P., Wang, Q., Li, Y., 2010. Spore-to-spore agar culture of the myxomycete *Physarum globuliferum*. Arch. Microbiol. 192 (2), 97–101.

Ljubimova, J.Y., Fujita, M., Khazenzon, N.M., Lee, B.S., Wachsmann-Hogiu, S., Farkas, D.L., Black, K.L., Holler, E., 2008a. Nanoconjugate based on polymalic acid for tumor targeting. Chem. Biol. Interact. 171 (2), 195–203.

Ljubimova, J.Y., Fujita, M., Ljubimov, A.V., Torchilin, V.P., Black, K.L., Holler, E., 2008b. Poly (malic acid) nanoconjugates containing various antibodies and oligonucleotides for multitargeting drug delivery. Nanomedicine 3 (2), 247–265.

Ljubimova, J.Y., Portilla-Arias, J., Patil, R., Ding, H., Inoue, S., Markman, J.L., Rekechenetskiy, A., Konda, B., Gangalum, P.R., Chesnokova, A., Ljubimov, A.V., Black, K.L., Holler, E., 2013. Toxicity and efficacy evaluation of multiple targeted polymalic acid conjugates for triple-negative breast cancer treatment. J. Drug Target. 21 (10), 956–967.

Locquin, M., 1948. Culture des Myxomycètes et production de substances antibiotiques par ces champignons. C. R. Acad. Sci. 227, 149–150.

Madelin, M.F., Audus, F., Knowles, D., 1975. Attraction of plasmodia of the myxomycete, *Badhamia utricularis*, by extracts of the basidiomycete, *Stereum hirsutum*. J. Gen. Microbiol. 89, 229–234.

Matsumoto, S.S., Haughey, H.M., Schmehl, D.M., Venables, D.A., Ireland, C.M., Holden, J.A., Barrows, L.R., 1999. Makaluvamines vary in ability to induce dose-dependent DNA cleavage via topoisomerase II interaction. Anticancer Drugs 10, 39–45.

McClory, A., Coote, J.G., 1985. The chemotaxtic response of the myxomycete *Physarum polycephalum* to amino acids, cyclic nucleotides and folic acid. FEMS Microbiol. Lett. 26, 195–200.

McCormick, J.J., Blomquist, J.C., Rusch, H.P., 1970. Isolation and characterization of an extracellular polysaccharide from *Physarum polycephalum*. J. Bacteriol. 104, 1110–1118.

Mims, C.W., 1969. Capillitial Formation in *Arcyria cinerea*. Mycologia 61 (4), 784.

Misono, Y., Ito, A., Matsumoto, J., Sakamoto, S., Yamaguchi, K., Ishibashi, M., 2003a. Physarigins A-C, three new yellow pigments from a cultured myxomycete Physarum rigidum. Tetrahedron Lett. 44, 4479–4481.

Misono, Y., Ishibashi, M., Ito, A., 2003b. Bahiensol, a new glycerolipid from a cultured myxomycete Didymium bahiense var. bahiense. Chem. Pharm. Bull. 51, 612–613.

Misono, Y., Ishikawa, Y., Yamamoto, Y., Hayashi, M., Komiyama, K., Ishibashi, M., 2003c. Dihydrolindbladiones, three new naphthoquinone pigments from a myxomycete Lindbladia tubulina. J. Nat. Prod. 66, 999–1001.

Murakami-Murofushi, K., Uchiyama, A., Fujiwara, Y., Kobayashi, T., Kobayashi, S., Mukai, M., Murofushi, H., Tigyi, G., 2002. Biological functions of a novel lipid mediator, cyclic phosphatidic acid. Ann. N. Y. Acad. Sci. 905 (1–3), 319–321.

Nakagaki, T., Yamada, H., Tóth, A., 2000. Maze-solving by an amoeboid organism. Nature 407, 470.

Nakatani, S., Kamata, K., Sato, M., Onuki, H., Hirota, H., Matsumoto, J., Ishibashi, M., 2005a. Melleumin A, a novel peptide lactone isolated from the cultured myxomycete Physarum melleum. Tetrahedron Lett. 46, 267–271.

Nakatani, S., Kiyota, M., Matsumoto, J., Ishibashi, M., 2005b. Pyrroloiminoquinone pigments from Didymium iridis. Biochem. Syst. Ecol. 33, 323–325.

Nakatani, S., Naoe, A., Yamamoto, Y., Yamauchi, T., Yamaguchi, N., Ishibashi, M., 2003. Isolation of bisindole alkaloids that inhibit the cell cycle from myxomycetes *Arcyria ferruginea* and *Tubifera casparyi*. Bioorg. Med. Chem. Lett. 13, 2879–2881.

Nakatani, S., Yamamoto, Y., Hayashi, M., Komiyama, K., Ishibashi, M., 2004. Cycloanthranilylproline-derived constituents from a myxomycete *Fuligo candida*. Chem. Pharm. Bull. 52 (3), 368–370.

Naoe, A., Ishibashi, M., Yamamoto, Y., 2003. Cribrarione A, a new antimicrobial Naphthoquinone pigment from a myxomycete *Cribraria purpurea*. Tetrahedron 59, 3433–3435.

Nelson, R.K., Orlowski, M., 1981. Spore germination and swarm cell morphogenesis in the acellular slime mold *Fuligo septica*. Arch. Microbiol. 130, 189–194.

Nowak, A., Steffan, B., 1998. Polycephalin B and C: unusual tetramic acids from plasmodia of the slime mold *Physarum polycephalum* (myxomycetes). Angew. Chem. Int. Ed. 37, 3139–3141.

Olive, L.S., 1975. The Mycetozoans. Academic Press, New York.

Pereira, E.R., Belin, L., Sancelme, M., Prudhomme, M., Ollier, M., 1996. Structure-activity relationships in a series of substituted indolocarbazoles: topoisomerase I and protein kinase C inhibition and antitumoral and antimicrobial properties. J. Med. Chem. 39, 4471–4477.

Portilla-Arias, J., Patil, R., Hu, J., Ding, H., Black, K.L., García-Alvarez, M., Muñoz-Guerra, S., Ljubimova, J.Y., Holler, E., 2010. Nanoconjugate platforms development based in poly(β,L-malic acid) methyl esters for tumor drug delivery. J. Nanomater. 2010, 8.

Rakoczy, L., 1973. The myxomycete *Physarum nudum* as a model organism for photobiological studies. Ber. Deut. Bot. Ges. 86, 141–164.

Řezanka, T., 1993. Polyunsaturated and unusual fatty acids from slime moulds. Phytochemistry 33, 1441–1444.

Řezanka, T., Dvořáková, R., 2003. Polypropionate lactones of deoxysugars glycosides from slime mold *Lycogala epidendrum*. Phytochemistry 63, 945–952.

Řezanka, T., Dvořáková, R., Hanus, L.O., Dembitsky, V.M., 2004. Enteridinines A and B from slime mold *Enteridium lycoperdon*. Phytochemistry 65 (4), 455–462.

Řezanka, T., Hanus, L.O., Dembitsky, V.M., 2005. The fulicineroside, a new unusual glycosidic dibenzofuran metabolite from the slime mold *Fuligo cinerea* (L.) Wiggers. Eur. J. Org. Chem. 13, 2708–2714.

Ross, I.K., 1967. Syngamy and plasmodiumformation in the myxomycete *Didymium iridis*. Protoplasma 64, 104–119.

Sancelme, M., Fabre, S., Prudhomme, M., 1994. Antimicrobial activities of indolocarbazole and bis-indole protein-kinase-C inhibitors. J. Antibiot. 47, 792–798.

Sauer, H.W., Babcock, K.L., Rusch, H.P., 1969. Sporulation in *Physarum polycephalum*: a model system for studies on differentiation. Exp. Cell Res. 57 (2), 319–327.

Schoknecht, J.D., 1975. SEM and X-ray microanalysis of calcareous deposits in myxomycete fructifications. Trans. Am. Microsc. Soc. 94, 216–223.

Sesaki, H., Ogihara, S., 1997. Secretion of slime, the extracellular matrix of the plasmodium, as visualized with a fluorescent probe and its correlation with locomotion on the substratum. Cell Struct. Funct. 22 (2), 279–289.

Shchepin, O., Novozhilov, Y.K., Schnittler, M., 2014. Nivicolous myxomycetes in agar culture: some results and open problems. Protistology 8, 53–61.

Shi, L.P., Li, Y., 2003. Studied on influence of nutritional decoctions and pH on spore germination in the myxomycetes. J. Jilin Agric. Univ. 25 (3), 275–277, 281. (In Chinese).

Shintani, A., Ohtsuki, T., Yamamoto, Y., Hakamatsuka, T., Kawahara, N., 2009. Fuligoic acid, a new pigment with a chiornated polyene-pyrone acid structure isolated from the myxomycete *Fuligo septica* var. *flava*. Tetrahedron Lett. 10, 1016–1017.

Shintani, A., Toume, K., Rifai, Y., Arai, M.A., Ishibashi, M., 2010. A bisindole alkaloid with hedgehog signal inhibitory activity from the myxomycete *Perichaena chrysosperma*. J. Nat. Prod. 73 (10), 1711–1713.

Shipley, G.L., Holt, C.E., 1982. Cell fusion competence and its induction in *Physarum polycephalum* and *Didymium iridis*. Dev. Biol. 90, 110–117.

Simon, H.L., Henney, H.R., 1970. Chemical composition of slime from three species of myxomycetes. FEBS Lett. 7, 80–82.

Smart, R.F., 1937. Influence of certain external factors on spore germination in the myxomycetes. Am. J. Bot. 24, 145–159.

Smith, E.C., 1929. Some phases of spore germination of myxomycetes. Am. J. Bot. 16 (9), 645–650.

Sobels, J.C., 1950. Nutrition de quelques Myxomycètes en cultures pures et associées et leurs propiétés antibiotiques. Koch & Knuttel, Gouda.

Song, X.X., Wang, Q., Li, Y., 2014. Chemical element characteristics of six plasmodia in myxomycetes. J. Northeast Forest. Univ. 42 (1), 112–115, 121.

Steffan, B., Praemassing, M., Steglich, W., 1987. Physarochrome A, a plasmodial pigment from the slime mould *Physarum polycephalum* (myxomycetes). Tetrahedron Lett. 28, 3667–3670.

Steglich, W., 1989. Slime molds (myxomycetes) as a source of new biological active metabolites. Pure Appl. Chem. 61, 281–288.

Steglich, W., Steffan, B., Kopanski, L., Eckhardt, G., 1980. Pilzpigmente, 36. Indolfarbstoffe aus Fruchtkörpern des Schleimpilzes *Arcyria denudata*. Angew. Chem. Int. Ed. Engl. 19, 459–460.

Taniguchi, M., Yamazaki, K., Ohta, J., 1978. Extraction of contractile proteins from myxamoebae of *Physarum polycephalum*. Cell Struct. Funct. 3, 181–190.

Taylor, R.L., Mallette, M.F., 1978. Purification of antibiotics from *Physarum gyrosum* by high-pressure liquid chromatography. Prep. Biochem. 8, 241–257.

Therrien, C.D., 1966. Microspectrophotometric measurement of nuclear deoxyribonucleic acid content in two myxomycetes. Can. J. Bot. 44, 1667–1675.

Ts'o, P.O.P., Bonner, J., Eggman, L., Vinograd, J., 1956a. Observations on an ATP-sensitive protein systemfromthe plasmodia of a myxomycete. J. Gen. Physiol. 39, 325–347.

Ts'o, P.O.P., Eggman, L., Vinograd, J., 1956b. The isolation of myxomyosin, an ATP-sensitive protein from the plasmodium of a myxomycete. J. Gen. Physiol. 39, 801–812.

Ts'o, P.O.P., Eggman, L., Vinograd, J., 1957a. Physical and chemical studies of myxomyosin, an ATP-sensitive protein in cytoplasm. Biochim. Biophys. Acta 25, 532–542.

Ts'o, P.O.P., Eggman, L., Vinograd, J., 1957b. The interaction of myxomyosin with ATP. Arch. Biochem. Biophys. 66, 64–70.

Turnock, G., Morris, S.R., Dee, J., 1981. A comparison of the proteins of the amoebal and plasmodial phases of the slime mould, *Physarum polycephalum*. Eur. J. Biochem. 115, 533–538.

Wang, W., 2014. Research on biological activity and phylogeny from certain myxomycetes of Trichialeas and Physarales. Jilin Agricultural University, Changchun. (In Chinese).

Wang, X.L., Li, Y.S., Li, Y., 2007. Ultrastructure of nucleus and sclerotium of *Fuligo septica* phanero plasmodium. Mycosystema 26 (1), 135–138, (In Chinese).

Winsett, K.E., 2011. Intraspecific variation in response to spore-to-spore cultivation in the myxomycete, *Didymium squamulosum*. Mycosphere 2, 555–564.

Wolf, F.T., 1959. Chemical nature of the photoreceptor pigment inducing fruiting of plasmodia of *Physarum polycephalum*. In: Withrow, R. (Ed.), Photoperiodism and Related Phenomena in Plants and Animals. American Association for the Advancement of Science, Washington, DC, pp. 321–326, (Publ. no. 55).

Wormington, W.M., Weaver, R.F., 1976. Photo receptor pigment that induces differentiation in the slime mold *Physarum polycephalum*. Proc. Natl. Acad. Sci. USA 73, 3896–3899.

Zhu, H., 2012. Studies on ontogeny and chemical constituents of myxomycetes from representative regions in northern China. Phd dissertation, Jilin Agricultural University, Changchun. (In Chinese).

Zhu, H., Wang, Q., 2005. Study on gel electrophoresis of protein from the species of Trichiales. J. Jilin Agric. Univ. 27 (6), 614–616, (In Chinese).

Chapter 7

Taxonomy and Systematics: Current Knowledge and Approaches on the Taxonomic Treatment of Myxomycetes

Carlos Lado*, Uno Eliasson**
**Royal Botanic Garden (CSIC), Madrid, Spain; **University of Gothenburg, Gothenburg, Sweden*

INTRODUCTION

The word taxonomy is derived from the Greek τάξις *taxis*, "arrangement," and νομία *nomia*, "method," and, in general, it can be considered as the science that defines and gives names to the groups of organisms according to their shared features. These features may be of different origin—morphological, cellular, molecular, structural, or genetic—but morphological features have traditionally been most frequently used, due to their easy observation and comparison.

Taxonomy, as we understand it today, begins in the middle of the 18th century when Linnaeus (1753), in his *Systema Naturae*, proposed the first modern taxonomic classification of all living organisms based on a binomial nomenclature system with a generic name followed by a species name identifying the species. Since then, taxonomy has become one of the oldest branches of biology and an essential pillar of biological sciences.

The taxonomy of the myxomycetes is derived from features of their life cycle, especially from the sexual phase, when they generate the fruiting bodies that produce the spores. The different forms, structures, dimensions and ornamentation of the structural elements of the fruiting bodies, such as the hypothallus, the stipe, the peridium, the columella, the capillitium and the spores, serve as a basis for the taxonomy of the group. In the life cycle of myxomycetes, spore-producing fruiting bodies alternate with an assimilative motile stage. Normally haploid spores grow into amoeboflagellate cells. After coalescence of two cells into a cell with a diploid nucleus, repeated mitotic divisions without wall formation result in a multinucleate cell, the plasmodium, surrounded by a slime sheath that eventually gives rise to fruiting bodies. The living protoplast is

Myxomycetes: Biology, Systematics, Biogeography, and Ecology. http://dx.doi.org/10.1016/B978-0-12-805089-7.00007-X

motile inside the sheath, which is continuously produced at the anterior end and collapses as an empty sheath at the posterior end. The characters derived from these assimilative stages, due to their variability and difficult observation, have not been taken into account in the taxonomy of myxomycetes, but can be a new source of information.

Myxomycetes have been considered a natural group of organisms and their taxonomy has remained relatively stable over the years, but their relationships with other organisms have been controversial since they have been treated as special groups of fungi and as protozoa (Fig. 7.1).

Due to their small size, most myxomycetes are normally overlooked by nonspecialists or people not especially searching for them, but a few common species have large and conspicuous plasmodia and fruiting bodies, and are well recognized by mycologists or people interested in nature in general (Fig. 7.2).

DEFINITION OF TAXONOMY, SYSTEMATICS, AND NOMENCLATURE

Both taxonomy and systematics aim at finding and identifying the living units in the web of biodiversity and tracing their phylogeny and evolutionary pathways. *Taxonomy* involves adoption of a scientific name of the biological unit, classification, and study of relationships, and (nowadays) finding the taxon's place in a phylogenetic system. The term *systematics* is often used in a broader sense, an overview of the biodiversity and the placing of species and groups along the branches of an evolutionary tree, but there is no sharp distinction between the terms *taxonomy* and *systematics*.

The *species* is the fundamental unit and the Holy Grail of systematic biology. The species name always consists of two parts, the first name, the *generic name*, indicating to which genus the organism belongs, the second name, the *specific epithet*, defining the species. Subspecific taxa below the species level may be used where appropriate and needed. Genera are grouped into categories of higher hierarchical levels, such as *families*, *orders*, *classes* and, finally, *kingdoms*, *divisions* or *phyla* (singular: *phylum*). The system with two words necessary to define a species is called the *binary nomenclatural system* and is internationally well established in biological sciences. The nomenclature is concerned with the principles, rules, and recommendations that are used for the formation and use of these names. Its rules may occasionally be subjected to minor adjustments decided at international congresses. The current *International Code of Nomenclature for algae, fungi, and plants* (ICN) (McNeill et al., 2012), the so-called Melbourne Code, covers, specifically, myxomycetes (slime molds) as established in the "Preamble 8." The myxomycetes have traditionally been studied by the mycologists since their fruiting bodies resemble those of true fungi and are reproduced by spores, but their closer affinities to the protozoa have become increasingly evident. Due to these affinities, some researchers have applied the rules of the *International Code of Zoological Nomenclature*

FIGURE 7.1 Various examples of fruiting bodies. (A) *Ceratiomyxa fruticulosa*, columnar, and dendroid fruiting bodies (×24) (23267 Lado). (B) *Ceratiomyxa morchella*, morchelloid fruiting body (×48) (9745 Lado). (C) *Echinostelium arboreum* sporocarp (×144) (2937 dwb). (D) *Clastoderma debaryanum* sporocarp (×72) (23250 Lado). (E) *Cribraria piriformis* sporocarp (×24) (MA-Fungi 35666). (F) *Lycogala conicum* aethalium (×24) (11651 Lado). (G) Tubifera *microsperma pseudoaethalium* (×12) (11865 Lado). (H) *Calomyxa metallica* sessile sporocarp (×48) (20398 Lado). (I) *Arcyria affinis* group of sporocarps (×18) (MA-Fungi 65288).

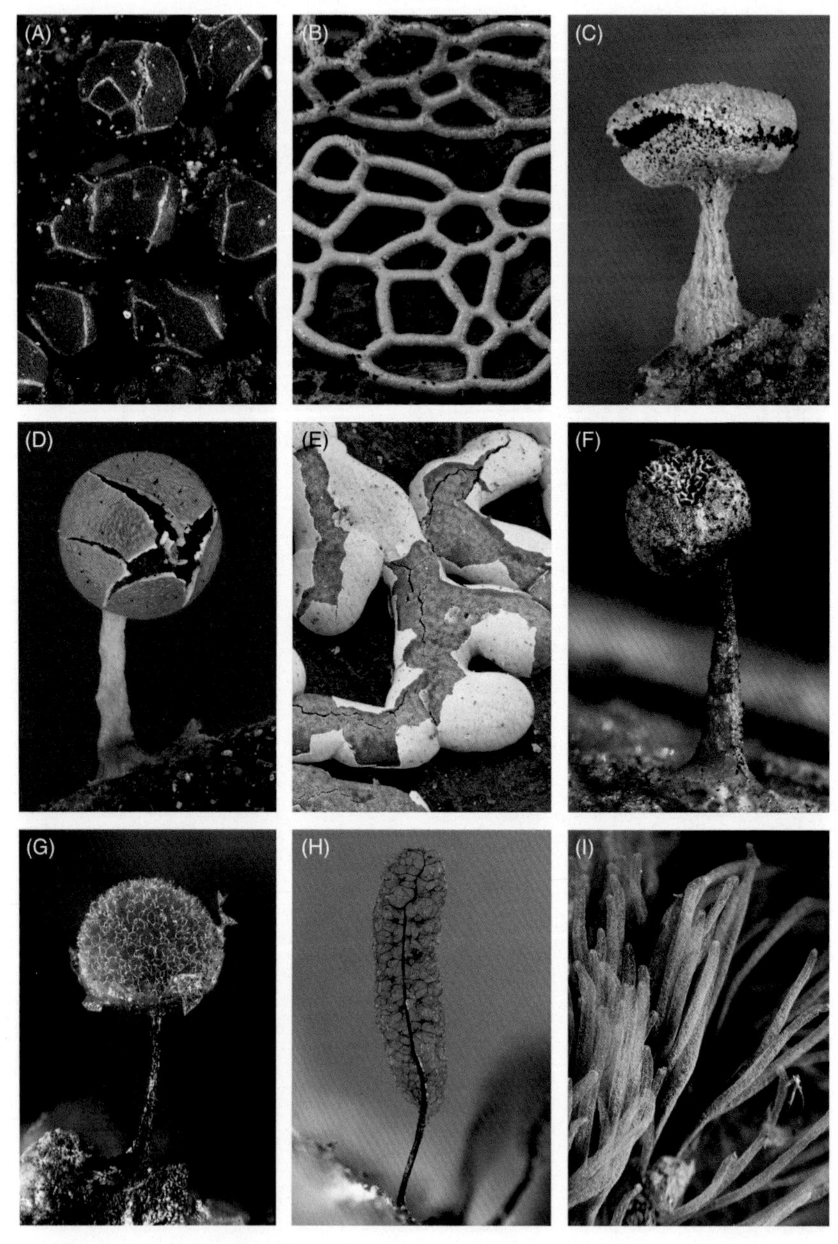

FIGURE 7.2 Various other fruiting bodies. (A) *Trichia agaves* group of sessile sporocarp (×36) (22637 Lado). (B) *Hemitrichia serpula* plasmodicoarps (×6) (11608 Lado). (C) *Didymium operculatum* sporocarp (×48) (3142 dwb). (D) *Diderma miniatum* sporocarp (×48) (24483 Lado). (E) *Diderma effusum* plasmodicoarps (×24) (11866 Lado). (F) *Craterium dictyosporum* sporocarp (×36) (1984 Lado). (G) *Lamproderma splendens* sporocarp (×36) (20736 Lado). (H) *Comatricha suksdorfii* sporocarp (×36) (19949 Lado). (I) *Stemonitis lignicola* group of sporocarps (×6) (20554 Lado).

(ICZN) (Ride, 1999) to naming the hierarchical levels above genera, affecting the end particles of the names. The myxomycetes are an example of an ambiregnal group of organisms, but the ICN is the one that governs its nomenclature until another agreement is reached and decision taken.

Although the word "species" is commonly used and well understood, a theoretical definition is not very easy and the species concept has been defined in different ways. A popular definition is that "a species is what a good taxonomist considers to be a species." Another definition is "a group of morphologically and genetically closely related individuals." But different organisms have different life strategies and it may prove difficult or impossible to find a definition that will cover all groups of organisms. It may well be that *species* is a term that simply cannot be used for some biological systems. Hence, terms like *sibling species* and *morphospecies* have been created, and sometimes we find it appropriate to talk about *species complexes.*

The name of a new species is always formally attached to a *nomenclatural type material*, a carefully selected specimen called the *type specimen.* The type specimen should to the maximum extent possible possess the characteristic features of the species. Nevertheless, it represents merely a sample within the variation range of a species and it is not always the most typical representative in the expression of every morphological character of a species. As stated in Art. 7.2 in Chapter II, Section 2 in the ICN, *the nomenclatural type is not necessarily the most typical or representative element of a taxon.* Therefore, study of a large sample of material is instrumental in establishing limits between species. In other words, field studies and collections taken in the field remain important. Studies of type material are of course important for the correct application of names.

Although the classic species concept has been based on morphological criteria in myxomycetes, as well as in other organisms, other features are also taken into account in modern taxonomic research (Walker and Stephenson, 2016). During the past 2 decades extensive biosystematic studies have been carried out on several species of myxomycetes, among them *Didymium iridis*, *D. squamulosum*, *Arcyria cinerea*, *Physarum compressum* and *Badhamia melanospora* [=*B. gracilis*] (Clark, 1996; Clark et al., 1999, 2002a,b). In summary, it could be said that what is treated in basic monographic works as a species may in certain cases comprise a mixture of nonheterothallic, as well as heterothallic and apomictic strains, which in turn may comprise sexually incompatible biological species. The different isolates may differ to a varying degree in morphology. Although all or most of the isolates generally conform to the standard species description there are often variations from the norm that differ in, for example, the shape of the sporotheca. This has been demonstrated, for example, in *B. melanospora* (Clark et al., 2002a). Morphological variations may also be due to environmental factors. In *Comatricha laxa*, a considerable variation in sporotheca shape has been demonstrated within a single biological species and appears to be due mainly to environmental effects (Clark and Haskins, 2002).

Nevertheless, the species concept is crucial to biodiversity and to large parts of biology. We need morphologically defined species to conduct field studies, inventories, ecological investigations, and to discuss distribution and dispersal. Given the reproductive patterns of myxomycetes, new species should be described only when morphological characters are pronounced and appear constant. Whenever possible, a potential new species should be cultivated under laboratory conditions to prove constancy of characters before being described. Unfortunately, numerous species have proven difficult to grow under laboratory conditions (Haskins and Wrigley de Basanta, 2008; Shchepin et al., 2014; Wrigley de Basanta et al., 2015), and in these cases, molecular investigations have become increasingly instrumental in the recognition of new species of myxomycetes.

BRIEF HISTORY OF THE TAXONOMY OF MYXOMYCETES

Fruiting bodies of several species of myxomycetes are funguslike in appearance and the first species noted in the literature were classified within the fungi (Figs. 7.1, 7.2). A good example is the species now called *Lycogala epidendrum*. Linnaeus (1753) referred it to the genus *Lycoperdon*, apparently based on its gasteromycete-like habit. It was known already in pre-Linnaean times thanks to its common occurrence and large size, mostly around 10 mm in diameter. A figure and description by Panckow (1654) are believed to refer to *L. epidendrum* (Lister, 1925) and may be the oldest myxomycete description possible to identify as to species. Despite the fact that Micheli (1729) demonstrated important distinctions from the gasteromycetes and erected the genus *Lycogala*, the species remained in *Lycoperdon* and was associated with fungi all through the second half of the 18th century.

The Botanical Congress in Vienna in 1905 set the starting point for the nomenclature of the myxomycetes to 1753, the year of the publication of Linnaeus's *Species Plantarum* and the nomenclatural starting point for most groups of organisms. Linnaeus did not focus much attention on organisms regarded as fungi. A scanning of myxomycete names in current use turns up Linnaeus's name in only four combinations, namely *L. epidendrum*, *Diderma radiatum*, *Arcyria denudata*, and *Fuligo septica*. All four species form large fructifications or conspicuous aggregations of fruiting bodies. The true nature of these species was of course unknown at this time and all four were regarded as fungi. The first two species were referred to the genus *Lycoperdon* and were described in the first (Linnaeus, 1753) and second (Linnaeus, 1763) editions, respectively, of his *Species Plantarum*. *Arcyria denudata* was referred to the genus *Clathrus* and described in the first edition, whereas *F. septica* was referred to the genus *Mucor* in the second edition.

Fries (1829) regarded the myxomycetes as related to gasteromycetes but because of their possession of motile cells and a plasmodium in the life cycle they were accommodated in a distinct suborder, the Myxogastres, within the

Order Gasteromycetes in his monumental work *Systema mycologicum*. The name Myxogastres was retained in the monograph by Massee (1892). During the second half of the 18th century and early decades of the 19th century a substantial number of new species were described by various authors whose names persist in author combinations of species accepted today. Among these authors are, Batsch (1783–89), Leers (1789), Persoon (1794), Schrader (1797), and de Schweinitz (1822, 1832). For a more extensive list of authors see Martin and Alexopoulos (1969).

When the life cycle of Myxomycetes was studied in more detail after the middle of the 19th century, thanks to the availability and improved quality of microscopes, it became evident that they were closer to amoeboid protozoa than to fungi in structure and life cycle, and the name Mycetozoa was proposed (de Bary, 1858, 1859, 1864; Rostafiński, 1873). This name reflected the funguslike appearance of the fruiting bodies and the animal-like presence of motile cells. An important advance in the study of the Mycetozoa was the monograph by Rostafiński (1874–75) with supplement (Rostafiński, 1876), considered to be the first published treatise on the taxonomy of these organisms. The name Mycetozoa prevailed for several decades and was used by various authors although sometimes (Macbride, 1899, 1922) with inclusion of organisms not closely associated with myxomycetes in phylogenetic systems of today. The same name was used in the narrower sense by Lister (1894) in the important *A Monograph of the Mycetozoa* and by Hagelstein (1944). Lister's monograph was essentially based on the collections in the British Museum (now called the Natural History Museum, NHM). Apart from keys and descriptions, it contained excellent illustrations including microscopic details. It was a book that set the standard for several forthcoming monographs. Two later editions followed in 1911 and 1925, both revised by his daughter Gulielma Lister and with most of the illustrations in color. Despite the linguistic meaning of the name Mycetozoa, the group remained associated with and included within the fungi s. lat. although at different taxonomic levels. From the middle of the 20th century "myxomycetes" has been the name used for this group of organisms in almost all literature. The name was first employed by Link (1833) and Wallroth (1831–33), but as a suborder and "ordo", and by Winter (1880) as a class, but did not come into common use until the monographs by Macbride and Martin (1934) and Martin (1949). These authors regarded the myxomycetes as a class of the fungi.

Numerous papers with descriptions of new species of myxomycetes have been published during the past few decades. During the past 30 years, more than 380 new species have been described. Field investigations into poorly explored geographical regions, studies of new ecological niches and the application of the moist chamber culture technique have resulted in a growing number of known species. Currently, about 1000 species in 64 genera are recognized (Lado, 2005–16).

North America is an excellent example of how the knowledge of the myxomycete biota of a continent has grown and developed. The first extensive collections and publications were those by de Schweinitz (1822, 1832), which were followed by a large number of other collectors and contributors (for names, see Martin and Alexopoulos, 1969). The first general accounts of the continent's myxomycetes were those by Cooke (1877) and Morgan (1893–1900). In 1899, Macbride published the first edition of his *North-American Slime-Moulds*. This publication, in his own words, had grown out from his previous reports of the myxomycetes of Iowa. A second revised and much extended edition was published in 1922, comprising "all species of myxomycetes hitherto reported from the continent of North America." This was an important book that became the basis for the *Myxomycetes* by Macbride and Martin (1934). In 1944, Hagelstein published *The Mycetozoa of North America*. It was based on the extensive collections in the herbarium of the New York Botanical Garden brought together by earlier collectors, but it also included most of his own large collections. The myxomycete volume (vol. I, part 1) in the *North American Flora* series, was published by Martin in 1949. There is no later monograph specifically dealing with all species found in North America. There are, however, contributions dealing with the myxomycete biota of a particular state, such as California (Kowalski, 1966, 1967, 1973; Kowalski and Curtis, 1968, 1970) and Colorado (Novozhilov et al., 2003). One of the most extensively collected states is Ohio; Keller and Braun (1999) published a summary of all species known from that state and also presented the collectors behind the fieldwork.

Studies of minute organisms that were previously included in the myxomycetes revealed significant differences and diversity in structure and life cycles. The so-called acrasid slime molds had been discovered (Brefeld, 1869) and studied during the second half of the 19th century (van Tieghem, 1880). An increasing number of minute organisms with a naked amoeba-like stage in their life cycle have been found and studied during the 20th and 21st centuries. They differ from myxomycetes in extremely minute fruiting bodies and, often, simpler life cycles. The simplest group is referred to as the protostelids. There are, however, considerable differences among species, and protostelids may not necessarily be a monophyletic group (Shadwick et al., 2009). Spores may be one- or two-celled and flagellated cells present or absent, and the assimilative stadium and fruiting bodies may be different from one genus to another. Based on this, several new orders have been described at the same taxonomic level as the myxomycetes, all within the kingdom Protozoa, distinctly set apart from kingdom Fungi. The evolutionary branch that leads to the myxomycetes and other groups of slime molds, such as protostelids and dictyostelids, is called the Amoebozoa (Cavalier-Smith, 1998, 2013) and is clearly separate from the branch from which fungi and animals have evolved.

GROSS SYSTEMATICS WITHIN MYXOMYCETES

The vast majority of myxomycetes are endosporous, which means that spores are produced inside fruiting bodies, the spore mass being enclosed by a wall, a peridium, of varied structure. In some genera the peridium may be ephemeral, disappearing early during the development of the fruiting body. The endosporous species are currently grouped into five orders. The sole genus *Ceratiomyxa* is exosporous, meaning that the structures traditionally interpreted as spores are borne externally on minute stalks from a columnar or differently shaped hypothallus.

Exosporous Myxomycetes

Order Ceratiomyxales

The different nature of the genus *Ceratiomyxa* in the possession of exospores had been recognized already by earlier workers and the genus had been referred by Rostafiński (1873) to a separate "Cohors" called *Exosporeae*, separate from the endosporous species. The external spores are borne at tips of stalks protruding from erect pillar-shaped structures. Macbride (1899) retained the subclass *Exosporeae* whereas the endosporous species were accommodated in the subclass *Myxogastres*. This version of classification was followed by all monographers prior to the world monograph by Martin and Alexopoulos (1969), where for nomenclatural reasons the subclasses were renamed *Ceratiomyxomycetidae* and *Myxogastromycetidae*, respectively.

Gilbert (1935) compared the exospore on its stalk with a stalked fruitbody of an endosporous species, homologizing the pillar-shaped matrix on which the stalked spores are borne with the hypothallus of an endosporous species. In the same paper, the unique stages in the spore germination process were demonstrated. Although several observations on *Ceratiomyxa* were published earlier by other authors, including a thorough study by Lister (1916), Gilbert was the first to describe the unique "thread phase" stage in the life cycle. Later studies involving SEM and TEM investigations (Nelson and Scheetz, 1975, 1976; Scheetz et al., 1980) have further demonstrated the unique features of the germination process in *Ceratiomyxa.* These clearly set *Ceratiomyxa* apart from all endosporous myxomycetes. A very thin and completely transparent film under the matrix is nowadays interpreted as a true hypothallus.

As further minute organisms with an amoeba-like stage in their life cycle have been discovered and studied, *Ceratiomyxa* seems to be phylogenetically more closely related to species now called protostelids among the so-called cellular slime molds. New species of this group are continuously being discovered and described. The cellular slime molds, as understood today, comprise minute organisms of low structural organization, not necessarily closely interrelated but representing different evolutionary lines. *Ceratiomyxa* has fruiting bodies (the structures originally called exospores) the same size as most of the protostelids. It still deviates in details of the life cycle, but there are significant differences in the life cycles also among protostelids (Olive, 1975; Scheetz, 1972; Scheetz et al., 1980). At least four species of *Ceratiomyxa* are recognized today, but all detailed studies of the life cycle have been based on *C. fruticulosa.* Although not regarded as a true myxomycete in the strictest sense, *Ceratiomyxa* is normally included in monographs and inventories of myxomycetes as a separate subclass, Ceratiomyxomycetidae.

Endosporous Myxomycetes

Fundamental features in the subdivision of the endosporous myxomycetes into orders have been the color of the spores in mass along with the presence or absence of capillitium and calcium in the fruiting bodies. Macbride (1922) recognized five orders that were later reduced to four (Macbride and Martin, 1934), a classification that was followed by Martin (1949) in the North American Flora. The orders were *Liceales* (= *Cribrariales*), *Trichiales*, *Stemonitidales* and *Physarales*. These four orders were still, with minor changes, accepted in the world monograph by Martin and Alexopoulos (1969), but a fifth order, the *Echinosteliales*, was added. The weakest point in this classification is the *Liceales*, where the most important distinguishing feature is the absence of a capillitium. It is a basic principle in modern systematics never to base a taxon on the absence of a character.

Order Liceales (=Cribrariales)

The order Liceales was first recognized by Jahn (1928) and Macbride and Martin (1934) and was retained in more or less the same circumscription throughout the 20th century in monographs. As already mentioned the basic feature of the order is the absence of a capillitium, although a so-called pseudocapillitium is present in several genera. In the classic circumscription of the order, the fruiting bodies exhibit an extremely wide range of variation in size and shape, from sporangia 0.1 mm (sometimes even less) wide in some species of *Licea* to cushion-shaped aethalia 10 cm or more across in *Reticularia*. There is also a great variation in spore color and plasmodium type. The order has long been supposed to be an unnatural taxon (Eliasson, 1977), comprising a mixture of species that simply do not easily fit into the other orders, and there is growing evidence of its polyphyletic origin (Eliasson, 2017).

The genus *Licea*, in the classic circumscription, incorporates forms that lack both a capillitium and a pseudocapillitium. The number of known species has grown during the past 4 decades, thanks to the exploration of new habitats, an increasingly frequent use of the moist chamber culture technique, better technical equipment and more interest in the genus. Martin and Alexopoulos (1969) recognized 19 species. Many new species have been described during the past 30 years and now (Jun. 2016) about 70 species are accepted (Lado, 2005–16). A critical review of the stipitate species has relegated several earlier names to synonymy (Wrigley de Basanta and Lado, 2005). The genus probably incorporates a wide assemblage of forms that are not necessarily closely related. The vast majority of species possess very small fruiting bodies and the minute size of several species makes them technically difficult to handle. The relatively simple morphology of many minute fruiting bodies adds to the difficulties in tracing relationships based only on morphological features.

A few of the minute species have fruiting bodies with outstanding morphological features. For example, *L. kleistobolus* has a specialized dehiscence mechanism (Eliasson, 2017) with a morphologically different lid. Protuberances from the lid into the cavity of the sporotheca may be fingerlike and reminiscent of a capillitium. Several species of the subgenus *Licea* possess processes along the margins of the peridial plates that appear to function in the separation of peridial plates at dehiscence. These processes could be interpreted as a rudimentary capillitium (Gilert, 1985).

Licea fimicola seems to be the only species that has been removed from the genus during the past few decades. It was transferred to a genus of its own (*Kelleromyxa*) based on several important differences (Eliasson et al., 1991) and with the sole species *K. fimicola*. This taxon has a phaneroplasmodium with distinct veins and typical protoplasmic streaming and there are thread-like structures in the sporotheca reminiscent of a true capillitium. The generally fusiform to ovoid fruiting bodies are often clustered in small groups and do not resemble the fruiting bodies of any other species in the genus *Licea*. The spores are also unique in morphological detail, being black in mass and marked with up to 3 μm long thin truncate processes covering more than two-thirds of the spore surface, with the rest of the spore nearly smooth. Molecular data have confirmed the anomaly of *L. fimicola* as a member of the genus *Licea*. Based on molecular data *Kelleromyxa* has recently been referred to a family (Kelleromyxaceae) of its own within the order Physarales (Erastova et al., 2013).

The genus *Licea* may not be a homogeneous genus even after the removal of *L. fimicola*. The two species *Licea variabilis* and *L. retiformis* differ in several respects from the rest of the species in being plasmodiocarpous and possessing a type of phaneroplasmodium in contrast to the protoplasmodium characteristic of other members of the subgenus *Licea* (Gilert, 1987). In structure and morphology, they bear some resemblance to species of *Perichaena* and *Dianema* in the Trichiales.

Recent molecular data (Fiore-Donno et al., 2013) supported the removal of *L. variabilis* from *Licea* and pointed to a closer relationship with the genus *Dianema*. Even apart from *Licea fimicola* and the plasmodiocarpous species *L. variabilis* (and probably *L. retiformis*), the genus *Licea* demonstrates a large variation in the structure of the fruiting bodies. Further molecular studies are encouraged on a wider range of species of *Licea*, in particular those with a bright spore mass and the smallest fruiting bodies. Several specimens filed under *Licea* sp. in herbaria, in particular those with bright yellow spores, may actually be closer to *Perichaena* (order Trichiales).

There is molecular evidence for a close relationship between certain species of the order Liceales and species of the order Trichiales (Fiore-Donno et al., 2005). The order Liceales in its classical circumscription is falling apart.

Fruiting bodies of *Listerella* are very similar in size and in general aspects to species of *Licea* and hardly distinguishable from fruiting bodies of *Licea* subg. *Licea* as viewed under a stereomicroscope. The characteristic moniliform capillitium of *Listerella* has been the main character distinguishing the genus, and some authors have referred it to a separate family (Listerellaceae; Neubert et al., 1993–2000). As Martin and Alexopoulos (1969) remarked, *Listerella* is "essentially a *Licea* with a capillitium," and the sole species *L. paradoxa* "has the peridium and spores of a *Licea*." It was later (Martin et al., 1983) referred to the same family, the Liceaceae. The taxonomic significance of the capillitium in *Listerella* as a distinguishing generic character is probably overemphasized, since internal and sometimes threadlike processes protruding from the peridium into the sporotheca in some species of *Licea* may be interpreted as a form of reduced capillitium, and (as noted earlier) the fingerlike processes protruding from the inner side of the operculum of *L. kleistobolus* may resemble a reduced capillitium (Gilert, 1985). The capillitial threads of *Listerella* are attached near the margins of the peridial lobes and aberrant capillitial outgrowths may be reminiscent of the processes bordering the peridial lobes in some species of *Licea* (Eliasson and Gilert, 1982).

Unfortunately, *Listerella* is difficult to find in the field and the number of specimens in herbaria is limited. Published reports of the structure of the plasmodium in *Listerella* seem to be lacking, although there is good reason to believe that it is of the protoplasmodium type; nor are there any molecular data on the genus. Based on morphology, molecular studies will probably support an inclusion of *Listerella* in *Licea*.

Licea versus Perichaena

Within the genus *Perichaena* (order Trichiales) several species may have a reduced or scantily developed capillitium, for example, *P. liceoides* (Gilert, 1990, 1997). The extent of development of the capillitium may vary among fruiting bodies within the same fruiting. Several species of *Perichaena* described during the past 10 years have minute fruiting bodies that totally lack a capillitium, such as *P. taimyriensis* (Novozhilov and Schnittler, 2000) and *P. polygonospora* (Novozhilov et al., 2008). Based on morphology, they would key out to *Licea* in current determination literature. Affinities between the two genera have been suggested (Eliasson, 1977), but thus far there are no molecular data supporting a close relationship (Fiore-Donno et al., 2013).

The reduction of the capillitium in smaller fruiting bodies can be explained as a strategy in using limited resources as efficiently as possible. This may explain why the capillitium is lacking or scantily developed in several of the minute species found on bark. The corticolous environment is often extreme, with rapid desiccation and limited resources. If the plasmodium is small and the resources thus limited, it is more important for the organism to use available resources for the production of spores, since spores are more important for the reproduction and dispersal of the species than is the capillitium.

Order Echinosteliales

The order Echinosteliales was erected by Martin (1961) for the sole genus *Echinostelium*. The order has later been expanded to include the genera *Clastoderma* and *Barbeyella*. It is a rather vaguely defined order, identified by having minute stalked globose fruiting bodies, a white to brownish spore mass, variously shaped columella and a mostly poorly developed capillitium that may even be totally lacking. The plasmodium, as is generally the case in minute species, is a protoplasmodium. The development of the fruiting body in *Echinostelium* appears to be of a different type than those of the two main types subhypothallic and epihypothallic (Clark and Haskins, 2014), which have been described for myxomycetes. As to the extent of capillitium development in *Echinostelium*, there is a continuous series from being relatively well developed in *E. minutum* to being totally lacking in species like *E. brooksii, E. lunatum*, and several other species. There is an interesting presumed link between the genus *Echinostelium* and the protostelids. The smallest species in the genus (*E. bisporum*) was described originally as a protostelid (Olive and Stoianovitch, 1966). The fruiting bodies are only about 20 μm tall and have only two spores.

Barbeyella has been placed in the Echinosteliales based on general morphology, possession of a protoplasmodium and molecular data (Fiore-Donno et al., 2012), but the epihypothallic sporophore development (Schnittler et al., 2000) and the dark spores raise doubt about its phylogenetic relations. The divergence in the color of the spores, also noted in the genus *Clastoderma*, has been hypothesized by Kretzschmar et al. (2016) as a loss or lack of expression of the genes involved in spore color.

The sole species of the genus *Semimorula* (*S. liquescens*) is a minute organism that for a long time remained enigmatic, with affinities to both myxomycetes and protostelids (Haskins et al., 1983). It was incorporated in *Echinostelium* by Kretzschmar et al. (2016) but differs in unstalked fruiting bodies and the way the spores germinate. However, molecular data place *Semimorula* in the midst of *Echinostelium* (Fiore-Donno et al., 2009).

Order Trichiales

In the order Trichiales, the spore mass is mostly bright-colored, ranging from almost white to yellow or reddish brown; the capillitium is typically well developed and consists of free or attached, simple or branched threads, which are often ornamented. A columella is never present. In the rather few species where the fruiting body development has been studied, it has been found to be subhypothallic. As compared to the treatment in Martin and Alexopoulos (1969), it is now widely accepted that *Listerella* should be removed and placed with *Licea*. A new genus (*Arcyriatella*) has been recognized (Hochgesand and Gottsberger, 1989) which is morphologically intermediate between *Arcyria* and *Minakatella*. As currently accepted, several genera are vaguely defined. Thus, the borderline between *Trichia* and *Hemitrichia* is blurred by plasmodiocarpous forms of *Trichia contorta*. Several species have been referred alternatively to *Arcyria* and *Hemitrichia*. For example, this is the case for *Arcyria stipata* and *A. leiocarpa*.

A capillitium is typically always present in species of this order but highly variable as to extent and structure. It may consist of simple or branched threads or form a net. The threads are solid or tubular, free or attached, almost colorless to brightly colored. Threads may be almost smooth but are mostly distinctly ornamented with spirals, spines, cogs, or a combination of various processes. The structure of the capillitium is diagnostic and generally of the same type within a genus.

Nannenga-Bremekamp (1982) used polarized light to compare selected species of myxomycetes as to their birefringence capacity. Although a birefringent capillitium was found only in species of the Trichiales, several species of the order lacked this capacity. Birefringence was demonstrated in several genera not at all related to the order and in other parts of the fruiting body. It is probably a feature to which not too much taxonomic weight should be given.

The recently established genus *Trichioides* has a unique combination of characters, some of which point to an affinity with the genus *Licea* in the Liceales (Novozhilov et al., 2015).

Order Physarales

Distinguishing features of the order Physarales have traditionally been the dark spore mass combined with the presence of calcium in the peridium, stalk, and/or capillitium. The plasmodium is a phaneroplasmodium and the fruiting body development is subhypothallic (for further references, see Clark and Haskins, 2014). With the inclusion of *Elaeomyxa* and *Diachea* in the order, Martin et al. (1983) recognized 16 genera in the Physarales. *Protophysarum* has been interpreted as a link between the Echinosteliales and the Physarales, but the phaneroplasmodium and the subhypothallic fruiting body development place it closer to the latter order (Blackwell and Alexopoulos, 1975).

Although the presence of calcium, often wrongly referred to as lime in the myxomycete literature, is one of the characteristics of the order Physarales, it may occur in fruiting bodies of species elsewhere in the system, for example, in *Perichaena* (Trichiales), *Cribraria* (Cribrariales), or *Lamproderma* (Stemonitidales). In the Physarales, it is normally noncrystalline and forming granules in the family Physaraceae but mostly crystalline and often "star-shaped" in the family Didymiaceae, one of the classical distinctions between the two families in reference works. Environmental conditions may in certain cases influence the abundance of calcium and the shape of crystals. Calcium may occur as a carbonate in globular or crystalline form but also occurs in other forms—phosphate, oxalate, and silicate (Schoknecht and Keller, 1989).

Several species bridge the gap between *Physarum* and *Badhamia*, for example, *P. decipiens* and to some extent the whole *P. decipiens–P. serpula–P. auriscalpium* complex. Today, there is no molecular support for maintaining the two physaraceous genera as hitherto circumscribed. The vague differences between the genera *Physarum*, *Badhamia*, *Craterium*, and *Fuligo* were discussed by Martin and Alexopoulos (1969), who kept them separate, mainly for practical reasons, since they are generally recognizable. A bit of conservatism may always be justified to avoid unnecessary name changes and new combinations, but provided the phylogenetic evidence is strong and backed by molecular data, the taxonomy used today should whenever possible reflect evolutionary lines. Among the generic names *Physarum*, *Badhamia*, and *Craterium*, the first name is the oldest, dating back to 1794. *Fuligo*, the only consistently aethalioid genus, dates back to 1768. If that genus were to be part of a merger, the number of necessary new combinations would be almost 200. Conservation of generic names may help to keep the number of new combinations low and should be used when appropriate, as has been done with genera of other orders (Doweld, 2015; Lado, 2001; Lado and Pando, 1998; Lado et al., 2005).

Order Stemonitidales

The order Stemonitidales has mostly been regarded as relatively well defined by morphological and developmental characters, in the classical description encompassing species with a dark spore mass, a true columella, a mostly well developed and branched capillitium and, for the majority of species, a special type of plasmodium called an aphanoplasmodium, and an epihypothallic development of the fruiting body (Clark and Haskins, 2014). Nevertheless, great changes have taken place in the inclusion and perception of genera within the order during recent decades. With the exclusion of *Schenella*, which was finally shown to be a gasteromycete (Estrada-Torres et al., 2005), 15 genera were included in the order by Martin and Alexopoulos (1969). This number was later reduced to 11 when *Barbeyella* and *Clastoderma* were transferred to the Echinosteliales and *Elaeomyxa* and *Diachea* were placed in the Physarales (Martin et al., 1983). The two latter genera both combine characteristics of the Stemonitidales and the Physarales. In *Elaeomyxa* the fruiting body development points to a closer affinity with the Physarales, and the structure and organization of the capillitium in *Diachea* resemble the capillitium in certain species of *Didymium* in the Physarales (for further references, see Martin et al., 1983).

Six new genera have been established by the discovery of new species and the splitting of older genera (Lado, 2005–16). The genera *Collaria*, *Paradiachea*, and *Paradiacheopsis* were segregated from former *Comatricha* s. lat. on characters of the peridium, capillitium, and stalk structure, *Meriderma* was excerpted from *Lamproderma* on characters of the capillitium and peridium, and *Stemonaria* and *Stemonitopsis* were split from *Stemonitis* and *Comatricha* based on the structure of the capillitium and stalk.

The creation of the many new genera during the past recent decades reflects the taxonomic difficulties and uncertainties associated with the order. Difficulties relating to the delimitation of genera stem from the great variability and intergradations between the extent to which a few morphological features are expressed (Eliasson et al., 2010). Few characters are absolute, and genera are normally characterized by a combination of characters. A good example is the structure of the peripheral capillitial net in many species. Although a capillitial surface net is a classic character in *Stemonitis*, there are a few species where the surface net is extremely open and wide-meshed and very different from that of the vast majority of species. The partial surface net of *Stemonitopsis* is structurally similar to that of *Stemonitis*, and a peripheral capillitial net is also found in a few species of *Comatricha* s. str. The extent of branching and anostomosing of the internal capillitial net is also highly variable in all of these three genera. The branching pattern of the capillitium has traditionally been regarded as an important character for distinguishing the two genera *Comatricha* and *Lamproderma*, but in the latter genus at least the pattern appears to be plesiomorphic and related to the shape of the sporotheca (Fiore-Donno et al., 2008). Such a key character as a well-developed capillitial surface net appears to intergrade among a series of taxa into simpler branching types. *Comatricha* s. str. is characterized by normally lacking a capillitial surface net and possessing a solid stalk that, at least at the base, is composed of more or less intertwining fibers (Nannenga-Bremekamp, 1967). However, a capillitial surface net is found in *C. ellae*, and there are species that sometimes have a hollow stalk like that in *Stemonitis*.

The difficulties in generic delimitation, as far as morphology is concerned, is well illustrated by *Comatricha nigricapillitia*, which morphologically combines characters of three genera, resembling *Lamproderma* in the robust and abruptly ending columella, *Comatricha* in having the capillitium arising along a large portion of the columella, and *Collaria* in possessing a peridial collar at the base of the sporotheca (Moreno et al., 2004). The situation is further complicated by data from molecular studies (Fiore-Donno et al., 2008), which suggest that there may actually be two different species involved, both belonging to the same clade as *Enerthenema*.

As already mentioned, several genera of the Stemonitidales are rather vaguely defined morphologically and appear to differ from one another merely to the extent to which a few basic features are expressed (Eliasson et al., 2010). Molecular data obtained thus far have indicated that the order is paraphyletic with the Physarales (Fiore-Donno et al., 2008). *Lamproderma* appears to be paraphyletic and possible to divide into two distinct clades (Fiore-Donno et al., 2008). Molecular data are instrumental for tracing the evolutionary relationships among the genera, and additional molecular investigations are likely to suggest several changes in the circumscription of genera within this order.

As to the future, molecular techniques and approaches will almost certainly change our view of genera and delimitation of genera and perhaps yield some big surprises, just as it has among vascular plants. The monumental book by Martin and Alexopoulos (1969) appeared earlier than a half century ago, when molecular techniques were almost unknown. The gross taxonomic system has remained more or less unchanged since then, and there seems to be molecular support for most of our basic ideas on the systematics of the myxomycetes. One of the big challenges is the genus *Licea* and the former order Liceales. As to the order, we already know through molecular evidence that *Cribraria* is a very basal side branch (Fiore-Donno et al., 2013). In the future, we may even find that it deserves the rank of a higher level of its own.

TAXONOMIC LITERATURE ON THE MYXOMYCETES

Almost 50 years have passed since the publication of the world monograph *The Myxomycetes* by Martin and Alexopoulos (1969). The book represented a major taxonomic platform by stimulating interest in these organisms and encouraging further studies. A considerable number of myxomycete papers now appear every year and among them are several books. Many books published during the past 50 years are of a monographic nature but geographically restricted. Examples include *Sluzowce Polski* (Krzemieniewska, 1960), *De Nederlandse Myxomyceten* (Nannenga-Bremekamp, 1974), with the English translation *A Guide to Temperate Myxomycetes* (Nannenga-Bremekamp, 1991), *Flora Neotropica Monograph No. 16. Myxomycetes* (Farr, 1976), *The Myxomycetes of Japan* (Emoto, 1977), *Indian Myxomycetes* (Lakhanpal and Mukerji, 1981), *Definitorium Fungorum Rossiae. Divisio Myxomyctota* (Novozhilov, 1993), the volume *Myxomycetes I* of the *Flora Mycologica Iberica* (Lado and Pando, 1998), *The Myxomycete Biota of Japan* (Yamamoto, 1998), *The Myxomycetes of Britain and Ireland* (Ing, 1999), *Myxomycetes of New Zealand* (Stephenson, 2003), *Flora Fungorum Sinicorum. Myxomycetes* (Li et al., 2008a,b) and *Limasienet* (Finnish myxomycetes) (Härkönen and Sivonen, 2011). The three-volume work *Die Myxomyceten* (Neubert et al., 1993–2000) treats the myxomycetes of Germany and neighboring regions of the Alps. *Les Myxomycètes* (Poulain et al., 2011) essentially covers the species of central Europe but includes numerous species thus far found only in other parts of the world. The two latter works are richly illustrated with color pictures of the highest quality. All published and beautifully illustrated books facilitate and stimulate further studies. However, many taxonomic problems often lurk behind many of the extensive descriptions and beautiful illustrations. There are many re``search projects that can be directed toward species regarded as well known. Our circumscription of several species may well be changed in the future, new species will be recognized and names now in current use will be relegated to synonymy. The online nomenclatural information system of the Eumycetozoa created by Lado (2005–16) greatly facilitates staying up-to-date with the progress made in the taxonomy of myxomycetes.

CLASSIFICATION OF THE MYXOMYCETES (FROM CLASS TO GENERA, WITH REFERENCES TO THE PROTOLOGUE AND A BRIEF DIAGNOSIS OF THE MOST RELEVANT DISTINCTIVE CHARACTERS)

The classification of living organisms is structured in a hierarchy of inclusions, in which a group includes other minor groups and is, in turn, subordinate to a larger one. Each group is assigned a taxonomic rank or taxonomic category that accompanies the name of the group (orders, families, genera, species, etc.). However, there is no single classification since the differentiating characters of the groups may receive different value or interpretation. In the last two centuries, the classification of the myxomycetes has been changed, based on increased scientific knowledge of the group. Persoon in his *Synopsis Methodica Fungorum,* published in 1801, was the first who paid serious attention to the classification of myxomycetes, and included among the fungi in their Angiocarpi class, order Dermatocarpi. Fries (1829), in his *Systema mycologicum*, considered the beginning of modern systematic mycology, classified the myxomycetes as a suborder (Myxogastres) in the order of fungi "Gasteromycetes genuini." With the studies of de Bary (1864, 1884), about the life cycle of myxomycetes, the relationship of the group with the protozoans was put in evidence and they were referred to as the Mycetozoa. But it was Rostafiński who proposed, in his *Versuch eines Systems der Mycetozoen* (1873), the first modern classification of the group. In Table 7.1 some comparative examples of classification systems from Rostafiński to the present day are presented.

Rostafiński (1873) produced the first classification of the Myxomycetes where microscopic characters were considered. This classification served as the basis for his *Śluzowce Monografia* (Rostafiński, 1874–76), considered the first monograph of the myxomycetes in a modern sense. Rostafiński's classification uses some taxonomic ranks unrecognized today, but it is very detailed and many of the cirteria used remain valid. Arthur Lister, in the classic *Monograph of the Mycetozoa* (Lister, 1894), followed basically the criteria of Rostafiński but restructured the classification of myxomycetes. In the second edition of this monograph, published in 1911 and revised by his daughter Gulielma Lister (Lister, 1911), no significant changes occurred. However, in the third edition of 1925, the taxonomic ranks were updated, and were adjusted to current criteria. Martin and Alexopoulos in their monograph *The Myxomycetes* (1969), which is considered an obligate reference and a milestone in the taxonomy of this group, summarized all the taxonomic knowledge of myxomycetes available at the time the book was written. They considered the myxomycetes to be a group of fungi and adapted their classification accordingly. Later, these authors along with Farr published *The Genera of Myxomycetes* (1983), a condensed version of the previous work in which new data of taxonomic value, such as the type of plasmodia and the morphogenesis of the stalks of the fruiting bodies, were incorporated in the classification of these organisms. Recently, Cavalier-Smith (2013), using Zoological nomenclature, proposed a new classification of the myxomycetes based on evolutionary and phylogenetic

TABLE 7.1 Some Proposals of Classification of the Myxomycetes

Rostafiński (1873)	Lister (1925)	Martin et al. (1983)	Cavalier-Smith (2013)
Class Mycetozoa	Class Mycetozoa	Class Myxomycetes	Class Myxomycetes
Cohors Exosporeae	Subclass Exosporeae	Subclass Ceratiomyxomycetidae	Subclass Exosporea
Tribus Ceratiaceae	Family Ceratiomyxaceae	Order Ceratiomyxales	Order Ceratiomyxida
		Family Ceratiomyxaceae	Family Ceratiomyxidae
Cohors Endosporeae	Subclass Endosporeae		
Ordo Enteridieae	Order Lamprosporales	Subclass Myxogastromycetidae	Subclass Mixogastria
Tribus Lycogalaceae	Suborder Anemineae	Order Liceales	Superorder Lucisporidia
Ordo Anemeae	Family Heterodermaceae	Family Cribrariaceae	Order Liceida
Tribus Dictyosteliaceae[a]	Family Liceaceae	Family Enteridiaceae	Family Cribraridae
Tribus Liceaceae	Family Lycogalaceae	Family Liceaceae	Family Dictydiaethalidae
Tribus Licaethaliaceae	Family Reticulariaceae	Order Echinosteliales	Family Liceidae
Ordo Heterodermeae	Family Tubulinaceae	Family Clatodermataceae	Family Listerillidae
Tribus Cribrariaceae	Suborder Calonemineae	Family Echinosteliaceae	Family Tubiferidae
Tribus Dictydiaethaliaceae	Family Arcyriaceae	Order Trichiales	Order Trichiida
Ordo Reticularieae	Family Margaritaceae	Family Dianemaceae	Family Arcyridae
Tribus Reticulariaceae	Family Trichiaceae	Family Trichiaceae	Family Dianematidae
Ordo Calonemea	Order Amaurosporales	Order Physarales	Family Minakatellidae
Tribus Arcyriaceae	Suborder Calcarineae	Family Didymiaceae	Family Trichidae
Tribus Perichaenaceae	Family Didymiaceae	Family Elaeomyxaceae	
Tribus Trichiaceae	Family Physaraceae	Family Physaraceae	Superorder Collumellidia
Ordo Calcareae	Suborder Amaurochaetineae		Order Echinosteliida
Tribus Cienkowskiaceae	Family Amaurochaetaceae	Subclass Stemonitomycetidae	Family Clastodermatidae
Tribus Didymiaceae	Family Collodermaceae	Order Stemonitidales	Family Echinosteliidae
Tribus Physaraceae	Family Stemonitaceae	Family Schenellaceae[b]	Order Fuscisporida
Tribus Spumariaceae		Family Stemonitidaceae	Suborder Physarina
Ordo Amaurochaeteae			Family Didymiidae
Tribus Amaurochaetaceae			Family Physaridae
Tribus Brefeldiaceae			Suborder Lamprodermina
Tribus Echinosteliaceae			Family Lamprodermidae
Tribus Enerthemaceae			Suborder Stemonitina
Tribus Stemonitidaceae			Family Stemonitidae

[a] *Now considered a different class.*
[b] *Now excluded from the Myxomycetes. It is a gasteromycete.*

TABLE 7.2 Hierarchical Position of the Myxomycetes in the Tree of Life (Cavalier-Smith, 2013)

Domain Eukaryota
Supergroup Podiates (Clade Podiates)
Kingdom Protozoa
Phylum Amoebozoa
Subphylum Conosa
Infraphylum Semiconosia
Parvphylum Mycetozoa
Superclass Macromycetozoa
Class Myxomycetes

evidences. This classification is not yet consolidated but incorporates new and interesting information to be taken into consideration to achieve a classification for the myxomycetes that is as natural as possible. Other authors, such as Adl et al. (2012) and Ruggiero et al. (2015), also have published new classifications on the hierarchical position of the myxomycetes in the higher level of the Tree of Life. In Table 7.2, a summary of the proposal of Cavalier-Smith (2013) is given.

With the idea of summarizing and updating, from class to genera, the current knowledge of the nomenclature and taxonomy of myxomycetes, in this chapter we have presented a classification system where a brief history of the names, the ranks proposed, the synonyms, and a brief diagnosis of the morphological distinctive features are included. Adl et al. (2012) provided descriptions with many cellular and ultrastructural characters, and we recommend consulting this publication to supplement some of the diagnoses elaborated here. Many existing names were used by authors in the literature of past centuries, and these are difficult to find, so the list of names could be incomplete, but should still serve as a starting point for future nomenclatural work. Some names also have been improperly used historically, having been attributed to other authors, or are invalid according to the ICN (McNeill et al., 2012); in these cases, for clarification, the correct name and authorship, as well as a reference to the articles that fail to comply, are included. The classification also includes the authors of the taxa and a reference to the protologue (original publication where the name is proposed and the taxon is described), which have been abbreviated following the recommendations of the ICN. In addition, the principle of priority in the rank, which is stipulated by the ICN as obligate for families, as well as for taxa of lower ranks, and recommended for upper ranks, has been taken into account.

For the classification presented, the contributions of Martin (1961, 1966), Martin and Alexopoulos (1969), Olive (1970, 1975), Alexopoulos (1973, 1978), Martin et al. (1983), Lado (2001), and David (2002), who made previous compilations or have analyzed the classification of the myxomycetes throughout history, have been extremely useful. The web page www.eumycetozoa.com, maintained and updated by Lado (2005–16), where detailed information on the nomenclatural status of genera and ranks below genera is compiled, represents another important source of information. Additional information on the nomenclature of myxomycetes was obtained from the following websites or repositories of information www.mycobank.org and www.indexfungorum.org, which regularly provide updated data.

The nomenclature employed in this classification is according to the ICN, but as some authors employ Zoological Nomenclature to facilitate cross-referencing the information, we have added, in square brackets, the equivalent zoological name. The PhyloCode has not been taken into account since it is a developing draft that, in its current version, is specifically designed to regulate the naming of clades but leaves the governance of species names to the Nomenclature codes. Possible changes that may occur in the near future, in view of the results in current phylogenetic papers, are also noted. As an example, Leontyev (2015) recently proposed a new classification of myxomycetes that involves deep changes, where the creation of new subclasses, superorders, orders, and families would bring about a major relocation of taxa. These proposals establish new relationships among taxa that can open up new and interesting research lines. However, the data are still incomplete and are subject to change and the names are invalid because they do not comply with the rules of ICN; therefore, we refer to the names but do not include them in the classification.

Class **MYXOMYCETES** G. Winter, Rabenh. Krypt.-Fl., ed. 2, 1(1): 32 (1880)

Wallroth in Fl. crypt. Germ. 2: xxv (Feb.–Mar. 1833) introduces the name at the rank of "Ordo" within the Class Fungi. Cooke in Contr. mycol. brit. 1 (1877) uses the name at an unspecified supraordinal rank based on Wallroth's original coining of the name. Link in Handbuch 3: 405 (1833) employed the name Myxomycetes but as suborder. Kreisel in Grundz. Natürl. Syst. Pilze 198 (1969) cites the authority as "Link ex Haeckel" but Haeckel in Gen. Morphol. Organ. 2: xxvi (1866) cites as "stamm" (David, 2002). Definitely stated to be a class by Winter (1880). Martin and Alexopoulos, Myxomycetes (1969) and Martin et al., Gen. Myxomycetes (1983) preferred to base the class on Link (1833), although later than Wallroth (by a month), the concept is closer to the "modern" sense of the myxomycetes.

= Mycetozoa de Bary ex Rostaf., Vers. Syst. Mycetozoen 1 (1873)

Name originated with "Mycetozoen" de Bary, Bot. Zeitung (Berlin). 16: 369 (1858). Also used by Engler, in Engler and Prantl, Nat. Pflanzenfam. 1(1): 1 (1889), as division; by L.S. Olive, Mycetozoans: 4 (1975), as subphylum, and Cavalier-Smith, Biol. Rev. 73: 232 (1998), as infraphylum.

= Myxogastres Schröt., in Engler and Prantl, Nat. Pflanzenfam. 1(1): 8 (1889), "as Myxogasteres"

Fries, Syst. mycol. 3: 67 (1829), is the first to use the name Myxogastres but as suborder. Cooke, Handb. Brit. fung. 1: 377 (1871) gives the Myxogastres as an Ordo in the "family" Gasteromycetes. Also given in the account by Engler, in Engler and Prantl, Nat. Pflanzenfam. 1(1): 1 (1889), and Engler, Syllabus Pflanzenfam., ed. 1: 1 (1892) as "Myxogasteres." See also Massee, Monogr. Myxogastr.: 28 (1892), as "(Fries)," no rank given, Macbride, N. Amer. Slime-moulds, ed. 1. 28 (1899) as "Sub-Class," and Schnittler et al., in Frey (ed.) Syllabus Pl. Fam., ed. 13, 1(1): 57 (2012), as "subclass Myxogastria L.S. Olive"; given in the "correct" form by Locquin, Numéro spécial Bull. Soc. Linn. Lyon 1: 235 (Feb. 1974) and Taxia Fung. 6 (Jun. 1974) as "Myxogastromycetes (Fr.) Schr." but subsequently attributed to Feltgen, Vorstud. Pilz-Fl. Luxemburg 3 (1899) by Locquin in Mycol. Gén. Struct. 76 (1984), a mistake Feltgen said "Myxogasteres." As class "Myxogastrea Fries, 1829, stat. nov." in Cavalier-Smith, Microbiol. Rev. 57: 971 (1993) (ending in accordance with the ICZN).

= Eumycetozoa L.S. Olive, Mycetozoans 4 (1975), nom. inval., Art. 39.1. This name was first introduced by Zopf, in Schenk (ed.), Handb. Bot 3(2): 97 (1884), but no rank was given. Schnittler et al., in Frey (ed.), Syllabus Pfl. Fam., ed. 13, 1(1): 46 (2012), cited as "Eumycetozoa (Zopf) L.S. Olive."

Trophic stage a free-living, multinucleate amoeba, the plasmodium, which generates static fruiting bodies with spores.

Martin et al. (1983), based on the type of development of the fruiting bodies and type of plasmodia, recognized, in this class, three subclasses, Ceratiomyxomycetidae, Myxogastromycetidae, and Stemonitomycetidae. This division has been largely accepted and serves as a basis for the general classification presented here. Cavalier-Smith (2013) recognized only two subclasses within the Myxomycetes, the subclass Exosporeae and Myxogastria, the first comprising the order Ceratiomyxales, a group that some authors exclude from the myxomycetes. In the subclass Myxogastria, recent phylogenetic works show two major lineages, light-spored and dark-spored (Fiore-Donno et al., 2005, 2010; Kretzschmar et al., 2016), so called superorder Lucisporidia and superorder Fuscisporidia, a concept not new, since it was used by Rostafiński (1874–75), Cooke (1877), and Lister (1925) to differentiate the two orders, Lamprosporales and Amaurosporales, that were recognized within the Endosporeae subclass.

In the class Myxomycetes 3 subclasses, 6 orders, 15 families, and 64 genera have been recognized and, according to Lado (2005–16), a total of 980 accepted species. Leontyev (2015) proposes a total of 8 orders and 14 families.

Subclass **Ceratiomyxomycetidae** G.W. Martin ex G.W. Martin and Alexop., Myxomycetes 32 (1969)

Mentioned by G.W. Martin, in Ainsworth and Bisby's Dictionary Fungi, ed. 5, 497 (1961), but nom. inval., Art. 39.1. Farr, in Martin et al., Gen. Myxomycetes: 38 (1983) states that the publication of the name by Martin (1961) is invalid and attribute the place of publication to Martin and Alexopoulos, Myxomycetes: 32 (1969). In neither case is a Latin diagnosis provided (David, 2002). Farr and Alexopoulos (1977, p. 214) mentioned that in Martin and Alexopoulos (1969) a reference is made to the description of "Cohors Exosporeae" by Rostafiński and the subclass is, therefore, to be authored by G.W. Martin ex G.W. Martin and Alexop. Also mentioned as class Ceratiomyxomycetes by Hawksworth, Sutton, and Ainsworth, Ainsworth and Bisby's Dictionary of the Fungi, ed. 7, 257 (1983), but nom. inval., Art. 39.1.

= Exosporeae Rostaf. Vers. Syst. Mycetozoen 2.1873, as "cohors"

Stated to be at the rank of subclass by Lister, Monogr. mycetozoa, ed. 1, 21, 25 (1894), without reference to Rostafiński, and Macbride, N. Amer. Slime-moulds, ed. 1. 17 (1899), also by Cavalier-Smith, Eur. J. Protistol. 49: 146 (2013) as subclass "Exosporeae Rostafiński, 1873 stat. n."

Spores borne singly and externally at the tips of threadlike stalks, on columnar, dendroid, or morchelloid fruiting bodies. Consisting of 1 order, 1 family, 1 genus and 4 species.

Order **Ceratiomyxales** G.W. Martin ex M.L. Farr and Alexop., Mycotaxon 6(2): 213 (1977) [Ceratiomyxida]

Based on the order Ceratiomyxales G.W. Martin, N. Amer. Fl. 1(1): 5 (1949), nom. inval., Art. 39.1.

With the characters of the subclass. Consisting of 1 family, 1 genus and 4 species.

Family Ceratiomyxaceae J. Schröt., in Engler and Prantl, Nat. Pflanzenfam. 1(1): 15 (1889) [Ceratiomyxidae]

With the characters of the order (1 genus, 4 species).

Ceratiomyxa J. Schröt., in Engler and Prantl, Nat. Pflazenfam. 1(1): 16 (1889) (4 species)

Nom. cons., see Lado et al., Taxon 54(2): 544 (2005) and Gams, Taxon 54(3): 829 (2005).

Subclass **Myxogastromycetidae** G.W. Martin, in Ainsworth and Bisby's Dictionary of Fungi, ed. 5, 497 (1961), nom. inval., Art. 39.1

This name was also mentioned by Alexopoulos (1973, 1978); Ross (1973) emended the concept of this subclass.

= Myxogastrea L.S. Olive, Bot. Rev (Lancaster). 36: 77 (1970) "as Myxogastria," nom. inval., Art. 39.1

Refers to Macbride, N. Amer. Slime-moulds 20 (1899), as "subclass" Myxogastres (Fr.) T. Macbr., based on Fries, Syst. mycol. 3: 3, 67 (1829) as suborder. Also proposed by Cavalier-Smith, Eur. J. Protistol. 49: 146 (2013) as subclass "Myxogastria Fries, 1829 stat. n."

= Endosporeae Rostaf., Vers. Syst. Mycetozoen 2 (1873), as "cohors"

Stated to be at the rank of subclass by Lister, Monogr. mycetozoa, ed. 1. [21] (1894). Also mentioned by Cooke, Contr. mycol. brit. 10 (1877) as division.

Spores borne in mass, on inside of fruiting bodies of various types. Development of fruiting bodies subhypothallic, the plasmodial protoplasm rising internally through the developing stalk in stipitate forms. Fruiting bodies sporocarpic, plasmodiocarpic, aethalioid, or pseudoaethalioid, stalked or sessile, with or without calcareous deposits. Peridium continuous with stalk and hypothallus, persistent to evanescent. Stalk, when present, filled with calcium compounds, granules, particles of refuse matter, sporelike cells, or with waxy or oil inclusions. Columella usually absent, sometimes present. Assimilative stage of various types (protoplasmodium or phaneroplasmodium), but never a true aphanoplasmodium. Consisting of 4 orders, 13 families, 45 genera, and 765 species.

Cavalier-Smith (2013), applying zoological nomenclature, recognized within the subclass Myxogastria two superorders (Table 7.1), superorder Lucisporidia Caval.-Sm., 2013 (light-spored) and superorder Columellidia Caval.-Sm., 2013 (columella present). The first superorder comprises the orders Liceales E. Jahn (Liceida) (a synonym of Cribrariales T. Macbr.) and Trichiales T.H. Macbr. (Trichiida), characterized by clear or brightly colored spores. The superorder Columellidia, characterized by a columella present or ancestrally present, includes the order Echinosteliales G.W. Martin (Echinosteliida), with pale or dark spores, and the new order Fuscisporida Caval.-Sm., 2013 (dark spores). In this new order the author recognized three suborders Lamprodermina Caval.-Sm., 2013, Stemonitina Caval.-Sm., 2013, and Physarina Caval.-Sm., 2013. Recently, Kretzschmar et al. (2016, p. 5) have raised the order Fuscisporida to the rank of superorder Fuscisporidia Fiore-Donno, 2013, and including the three orders Echinosteliales (Echinosteliida), Stemonitidales (Stemonitida), and Physarales (Physarida). Leontyev (2015) raised the superorder Columellidia to the rank of subclass. None of these authors recognize the different development (morphogenesis) of the stalk of the members of the order Stemonitidales, a character highlighted by Ross (1973) and that we consider of relevant evolutionary value. Furthermore, these authors do not consider that most species of the order Echinosteliales have brightly colored spores, instead of dark spores, and they justified the loss of pigment in the spores as an evolutionary hypothesis.

Order **Echinosteliales** G.W. Martin, Mycologia 52(1): 127 (1961) ["1960"] [Echinosteliida]

The concept of the order was emended by Alexopoulos, in Alexopoulos and Brooks, Mycologia 63(4): 927 (1971).

Fruiting bodies sporangiate, without calcareous deposits in their structures, stalked, usually less than 1000 μm in total height. Stalk filled with granular material. Columella present. Capillitium usually scanty, smooth. Spores in mass white, brightly colored, sometimes brownish to blackish. Consisting of 2 families, 4 genera, and 19 species.

Family Clastodermataceae Alexop. and T.E. Brooks, Mycologia 63(4): 926 (1971) [Clastodermatidae]

Plasmodium of the protoplasmodium type. Fruiting bodies minute, 200–1300 μm high. Peridium partially evanescent, usually remaining as a collar around the stalk, also attached to the peripheral capillitial threads. Stalk filled with granular material in its lower part. Columella present. Capillitium well developed. Spores in mass brownish to blackish (2 genera, 4 species).

Barbeyella Meyl., Bull. Soc. Bot. Genève 6: 89 (1914) (1 species)

Clastoderma A. Blytt, Bot. Zeitung (Berlin) 38: 343 (1880) (3 species)

Family Echinosteliaceae Rostaf. ex Cooke, Contr. mycol. brit. 53 (1877) [Echinosteliidae]

Published by Rostafiński, Vers. Syst. Mycetozoen 7 (1873) as "tribus," first spelt correctly and as family by Cooke, Contr. mycol. brit. 53 (Jan.–Jul. 1877), also by Luerssen, Handb. syst. Bot. 1: 42 (Dec. 1877).

= Heimerliaceae G. Arnaud, Botaniste 34: 35 (1949), as "Heimerliacées," nom. inval., see Art. 18.4, 32.1, 63.

Plasmodium of the protoplasmodium type. Fruiting bodies minute, 20–550 μm high. Peridium completely evanescent, in some cases remaining as a small and rudimentary collar around the stalk. Stalk filled with granular material in its lower part. Columella present or absent, sometimes with a spore-like body attached to the apex of the stalk. Capillitium scanty, rudimentary, or absent. Spores in mass white or brightly colored (2 genera, 15 species).

Echinostelium de Bary, in Rostafiński, Vers. Syst. Mycetozoen 7 (1873) (14 species)

Semimorula E.F. Haskins, McGuinn. and C.S. Berry, Mycologia 75(1): 153 (1983) (1 species)

This genus has just been synonymized with *Echinostelium*, see Kretzschmar et al. (2016).

Order **Cribrariales** T. Macbr., N. Amer. Slime-Moulds, ed. 2. 199 (1922) [Cribrariida]

The name Liceales has been largely applied to this order, but Cribrariales has priority.

= Lycogalales T. Macbr., N. Amer. Slime-Moulds, ed. 2. 232 (1922)

= Lamprosporales Lister, Monogr. mycetozoa, ed. 3. 2 (1925)

Proposed by Lister, Monogr. mycetozoa, ed. 1. 22 (1894) as "cohort."

= Liceales E. Jahn, in Engler and Prantl, Nat. Pflanzenfam. ed. 2, 2: 319 (1928)

Fruiting bodies with or without calcareous deposits in their structures, sessile or stalked. Columella absent. Capillitium absent, rarely very rudimentary. Pseudocapillitium present or absent, when present of tubular, irregular filaments or of perforated plates that may fray out into threads. Spores in mass clear or brightly colored, sometimes olivaceous or brownish. Consisting of 4 families, 9 genera, and 154 species.

The order is not natural; it is a polyphyletic group delimited on the lack of capillitium.

Family Cribrariaceae Corda, Icon. fung. 2: 22 (1838) [Cribrariidae]

= Lindbladiaceae Locq., Syn. gen. fung. (Paris): [1] (1972), nom. inval., Art. 39.1

Fruiting bodies stalked, sporocarpic or rarely pseudoaethalioid. Peridium remaining as a preformed net, with spherical calcic granules 0.5–4 μm diam. Capillitum absent. Pseudocapillitium absent. Spores in mass clear or brightly colored (2 genera, 47 species).

Cribraria Pers., Neues Mag. Bot. 1: 91 (1794) (46 species)

Nom. cons., see Lado et al., Taxon 54(2): 544 (2005) and Gams, Taxon 54(3): 829 (2005).

Lindbladia Fr., Summa veg. Scand. 449 (1849) (1 species)

Family Dictydiaethaliaceae Luerss. Handb. syst. Bot. 1: 42 (1877) [Dictydiaethaliidae]

Published by Rostafiński, Vers. Syst. Mycetozoen 5 (1873), as "tribus." Also published as family by Nannenga-Bremekamp (1982) but nomen. nud.; see also Nannenga-Bremekamp (1985) but nom. inval., art. 38.1. Neubert et al., Myxomyceten 1: 117 (1993) made a validation and author citation correction of the proposal of Nannenga-Bremekamp.

= Clathroptychiaceae Rostaf. ex Cooke, Contr. mycol. brit. 55 (1877)

Published by Rostafiński, Śluzowce monogr. 38, 224 (1875), but not at family rank, first spelt correctly and as family by Cooke, Contr. mycol. brit. 55 (1877).

Fruiting bodies sessile, pseudoaethalioid. Peridium remaining in the upper part as polyhedral plates. Capillitium absent. Pseudocapillitium present, filiform. Spores in mass clear or brightly colored (1 genus, 2 species).

Dictydiaethalium Rostaf., Vers. Syst. Mycetozoen 5 (1873) (2 species)

Family Liceaceae Chevall. Fl. gén. env. Paris 1: 343. 1826, as "Liceae" [Liceidae]

Also mentioned by Rostafiński, Vers. Syst. Mycetozoen 4 (1873) as "tribus."

= Protodermataceae Rostaf. Ex Cooke, Contr. mycol. brit. 2, 10 (Jan.–Jul. 1877), as "Protodermaceae"

First published by Rostafiński, Sluzowce monogr. 90 (1875) but not at family rank.

= Orcadellaceae T. Macbr., N. Amer. Slime-moulds, ed. 2. 203 (1922)

= Listerellaceae E. Jahn ex Neubert et al., Myxomyceten 1: 158 (1993)

Proposed previously by Jahn, Ber. Deutsch. Bot. Ges. 24:540 (1907) "(1906)," as "Listerellaceen" but nom. inval., art. 32.1, see art. 18.4. Also proposed by Locquin., Syn. gen. fung. [1] (1972), but nom. inval., Art. 39.1.

Fruiting bodies sessile or stalked, sporangiate or plasmodiocarpic, often minute. Peridium remaining in the lower part; without spherical calcic granules. Capillitium absent, sometimes very rudimentary and scanty. Pseudocapillitium absent. Spores in mass clear or brightly colored, sometimes olivaceous, brownish or black (2 genus, 73 species).

Licea Schrad., Nov. gen. pl. 16 (1797) (72 species)

Listerella E. Jahn, Ber. Deutsch. Bot. Ges. 24: 540 (1907) "(1906)" (1 species)

The monospecific genus *Listerella* has been considered by some authors as an independent family.

Family Reticulariaceae Chevall. ex Corda, Icon. fung 5: 22 (1842) [Reticularidae]

Published by Chevallier, Fl. gén. env. Paris 1: 341 (1826) as "ordre Reticularieae," nom. inval., Art. 32.1, first correctly spelt by Corda (1842).

= Lycogalaceae Corda, in Opiz, Naturalientausch (Beiträge Naturgeschichte 12: 637 (1828), nom. inval., Art. 32.1

In Corda, Icon. fung. 5: 22 (1842), the name is used at a subfamilial rank.

= Licaethaliaceae Rostaf. ex Luerssen, Handb. syst. Bot. 1: 41 (Dec. 1877),

Mentioned by Rostafiński, Vers. Syst. Mycetozoen 4 (1873) as "tribus," first spelt correctly and as family by Luerssen, Handb. syst. Bot. 1: 41 (Dec. 1877).

= Heterodermaceae Lister, Monogr. mycetozoa, ed. 1. 136 (1894), nom. inval., Art. 32.1, see Art. 18.1

= Tubulinaceae Lister, Monogr. mycetozoa, ed. 1. 152 (1894)

= Tubiferaceae T. Macbr., N. Amer. Slime-moulds, ed. 2. 203 (1922)

= Alwisiaceae Locq., Syn. gen. fung. [1] (1972), nom. inval., Art. 39.1

= Enteridiaceae M.L. Farr, Mycologia 74(2): 339 (1982)

Previously proposed by Locquin, Syn. gen. fung. [1] (1972) but nom. inval., Art. 32.1.

Fruiting bodies sessile, rarely stalked, aethalioid or pseudoaethalioid. Peridium remaining in the lower part. Capillitium absent. Pseudocapillitium present, columnar, tubular or as perforate or frayed membranes. Spores in mass light brown (4 genera, 32 species).

Leontyev, who is carrying out a review of the family Reticulariaceae (Leontyev, 2015; Leontyev et al., 2014), has proposed a new classification. The genus *Alwisia* is claimed to have a true capillitium (Leontyev et al., 2014), and, based on the evolution of fruiting bodies, two new genera have been segregated. A third genus, *Siphoptychium* Rostaf., until now considered a synonym of *Tubifera* (Leontyev, 2016), is reestablished.

Alwisia Berk. and Broome, J. Linn. Soc., Bot. 14: 86 (1873) (4 species)

Lycogala Adans., Fam. pl. 2: 7 (1763) (6 species)

Reticularia Bull., Herb. France 7(78–84): pl. 326 (1787–88) (10 species)

Nom. cons., Art. 14, see Lado et al., Taxon 47: 109 (1998); Lado and Pando, Taxon 47: 453 (1998) and Gams, Taxon 50: 270 (2001).

Tubifera J.F. Gmel., Syst. nat. 2: 1472 (1792) (12 species)

Order **Trichiales** T. Macbr., N. Amer. Slime-moulds, ed. 2, 237 (1922) [Trichiida]

Fruiting bodies sessile or stalked, calcareous deposits absent in their structures but occasionally present in the peridium. Stalk filled with sporelike cells or granular material. Columella absent. Capillitium present, filiform or tubular, ornamented. Spores in mass clear or brightly colored. Plasmodium of the aphanoplasmodium type. Consisting of 4 families, 14 genera, and 184 species.

Family Arcyriaceae Rostaf. ex Cooke, Contr. mycol. brit. 69 (1877) [Arcyriidae]

Published by Rostafiński, Vers. Syst. Mycetozoen 15 (1873) as "tribus," first spelt correctly by Cooke, Contr. mycol. brit. 69 (Jan.–Jul. 1877), also by Luerssen, Handb. syst. Bot. 1: 43 (Dec. 1877).

= Perichaenaceae Rostaf. ex Cooke, Contr. mycol. brit. 77 (1877)

Mentioned by Rostafiński, Vers. Syst. Mycetozoen 15 (1873) as "tribus," first spelt correctly and as family by Cooke, Contr. mycol. brit. 77 (Jan.–Jul. 1877), also by Luerssen, Handb. syst. Bot. 1: 43 (Dec. 1877).

Fruiting bodies sessile or stalked, sporangiate or plasmodiocarpic. Capillitium tubular, hollow, ornamented with warts, spines, cogs, ridges, half-rings, rings or reticulations (5 genera, 87 species).

Arcyodes O.F. Cook, Science 15: 651 (1902) (1 species)

Arcyria F.H. Wigg., Prim. fl. holsat. 109 (1780) (50 species)

A species of the genus (*Arcyria sulcata* Dörfelt and A.R. Schmidt) is a fossil found in Baltic amber from the Eocene (Dörfelt et al., 2003)

Arcyriatella Hochg. and Gottsb., Nova Hedwigia 48(3–4): 485 (1989) (1 species)

Cornuvia Rostaf., Vers. Syst. Mycetozoen 15 (1873) (1 species)

Perichaena Fr., in Fries and Lindgren, Symb. gasteromyc. 2: 11 (1817) (34 species)

Family Dianemataceae T. Macbr., N. Amer. Slime-moulds, ed. 2, 237 (1922), as "Dianemaceae" [Dianemidae]

Previously proposed by Macbride, N. Amer. Slime-moulds, ed. 1, 180 (1899) as "Dianemeae."

= Margaritaceae Lister, Monogr. mycetozoa, ed. 1. 202 (1894)

Based on *Margarita* Lister, a nom. illeg., Art. 53.1, non *Margarita* Gaudin, 1829.

Fruiting bodies sessile or stalked, sporocarpic or plasmodiocarpic. Capillitium filiform, solid, smooth or ornamented with tenuous and irregular spirals or reticulations (2 genera, 15 species).

Calomyxa Nieuwl., Amer. Midl. Naturalist 4: 335 (1916) (2 species)

Dianema Rex, Proc. Acad. Nat. Sci. Philadelphia 43: 397 (1891) (13 species)

Family Minakatellaceae Nann.-Bremek. ex Neubert et al., Myxomyceten 1: 160 (1993) [Minakatellidae]

First proposed by Nannenga-Bremekamp (1982) but nom. inval., Art. 36.1, also proposed by Nannenga-Bremekamp (1985) but nom. inval., Art. 39.1.

Fruiting bodies sessile, pseudoaethalioid. Capillitium tubular, hollow, almost smooth (1 genus, 1 species).

Minakatella G. Lister, J. Bot. 59: 92 (1921) (1 species)

Family Trichiaceae Chevall., Fl. gén. env. Paris 1: 322 (1826) [Trichiidae]

= Prototrichiaceae T. Macbr., N. Amer. Slime-moulds, ed. 2, 258 (1922)

Previously proposed by Macbride, N. Amer. Slime-moulds, ed. 1, 199 (1899) as "Prototrichieae"

Fruiting bodies sessile or stalked, sporocarpic or plasmodiocarpic. Capillitium tubular, hollow, ornamented with spiral bands, sometimes combined with spines, frequently elateriform, ornamentation sometimes faint or irregular to rings or almost smooth (6 genera, 81 species).

Calonema Morgan, J. Cincinnati Soc. Nat. Hist. 16(1): 27 (1893) (6 species)
Hemitrichia Rostaf., Vers. Syst. Mycetozoen 14 (1873) (26 species)
Nom. cons., see Lado et al., Taxon 54(2). 545 (2005) and Gams, Taxon 54(3): 829 (2005).
Metatrichia Ing, Trans. Brit. Mycol. Soc. 47(1): 51 (1964) (6 species)
Oligonema Rostaf., Śluzowce monogr. 291 (1875) (7 species)
Prototrichia Rostaf., Śluzowce monogr. suppl. 38 (1876) (1 species)
Trichia Haller, Hist. stirp. Helv. 3: 114 (1768) (35 species)

Order **Physarales** T. Macbr., N. Amer. Slime-moulds, ed. 2, 22 (1922) [Physarida]
Fruiting bodies with granular or crystalline calcareous deposits on some of their structures, rarely with oil or wax. Columella present or absent. Capillitium present. Spores in mass brown to black. Plasmodium of the phaneroplasmodium type. Consisting of 3 families, 18 genera, and 408 species.

Family Didymiaceae Rostaf. ex Cooke, Contr. mycol. brit. 29 (1877) [Didymiidae]
Published by Rostafiński, Vers. Syst. Mycetozoen 12 (1873) as "tribus," first spelt correctly and as family by Cooke, Contr. mycol. brit. 29 (Jan.–Jul. 1877), also by Luerssen, Handb. syst. Bot. 1: 43 (Dec. 1877).
= Spumariaceae Rostaf. ex Cooke, Contr. mycol. brit. 44 (1877)
Published by Rostafiński, Vers. Syst. Mycetozoen 13 (1873) as "tribus," first spelt correctly and as family by Cooke, Contr. mycol. brit. 29 (Jan.–Jul. 1877), also by Luerssen, Handb. syst. Bot. 1: 43 (Dec. 1877).
= Diachaeaceae Locq., Syn. Gen. Fung. [1] (1972), nom. inval. Art. 39.1.
= Lepidodermataceae Locq., Syn. Gen. Fung. [1] (1972), as "Lepidodermaceae," nom. inval. Art. 39.1.
Fruiting bodies sessile or stalked, sporocarpic or plasmodiocarpic, rarely aethalioid, without oil or wax. Peridium with granular or crystalline calcareous deposits, sometimes united to an egg-shell like cover. Columella present or absent, sometimes with a pseudocolumella. Capillitium filiform, not calcareous, rarely with some intermixed calcareous nodes (7 genera, 193 species).
Diachea Fr., Syst. orb. veg. 143 (1825) (13 species)
Diderma Pers., Neues Mag. Bot. 1: 89 (1794) (80 species)
Didymium Schrad., Nov. gen. pl. 20 (1797) (87 species)
Lepidoderma de Bary, in Rostafiński, Vers. Syst. Mycetozoen 13 (1873) (9 species)
Mucilago Battarra, Fungi arimin. 76 (1755) (1 species)
First employed by Micheli (1729) but pre-Linnaean.
Physarina Höhn., Sitzungsber. Kaiserl. Akad. Wiss., Math.-Naturwiss. Cl. 118: 431 (1909) (3 species)
Trabrooksia H.W. Keller, Mycologia 72(2): 396 (1980) (1 species)

Family Elaeomyxaceae Hagelst. ex M.L. Farr and H.W. Keller, Mycologia 74 (5): 857 (1982) [Elaeomyxidae]

Based on Hagelstein, Mycologia 34(5): 594 (1942), nom. inval., Art. 39.1. Fruiting bodies stalked, sporangiate, with oil or wax in some parts of their structures. Peridium without calcareous deposits. Capillitium filiform, not calcareous (1 genus, 4 species).

Elaeomyxa Hagelst., Mycologia 34(5): 593 (1942) (4 species)

Family Physaraceae Chevall., Fl. gén. env. Paris 1: 332 (Aug. 2–4, 1826), as "Physareae" [Physaridae]

Nom. cons. prop., see Doweld, Taxon 64(6): 1313 (2015). Also mentioned by Rostafiński, Vers. Syst. Mycetozoen 9 (1873) but as "tribus."

= Fuliginaceae Link, Abh. Köningl. Akad. Wiss. Berlin 1824 (Phys. Kl.): 168 (Feb. 26, 1826), as "Fuligineae"

Nom. rej. prop., see Doweld, Taxon 64(6): 1313 (2015), used as tribe by Brogniart, Dict. 33: 550 (1824), also Chevallier, Fl. gén. env. Paris 1: 346 (Aug. 2–4, 1826) as "Fulgiaceae."

= Crateriaceae Corda, Icon. fung. 5: 58 (1842), nom. inval., Art. 32.1

= Cienkowskiaceae Rostaf. ex Cooke, Contr. mycol. brit. 10 (1877)

Published by Rostafiński, Vers. Syst. Mycetozoen 9 (1873) as "tribus," first spelt correctly and as family by Cooke, Contr. mycol. brit. 10 (Jan.–Jul. 1877), also by Luerssen, Handb. syst. Bot. 1: 42 (Dec. 1877)

= Badhamiaceae Locq., Syn. gen. fung. [1] (1972), nom. inval., Art. 39.1

= Leocarpaceae Locq., Syn. gen. fung. [1] (1972), nom. inval., Art. 39.1

= Trichamphoraceae Locq., Syn. gen. fung. [1] (1972), nom. inval., Art. 39.1

= Protophysaraceae A. Castillo, Illana, and G. Moreno, Mycol. Res. 102(7): 842 (1998)

= Kelleromyxaceae D. Erastova, M. Okun, Novozh., and Schnittler, Mycol. Progress 12(3): 605 (2013), nom. inval. Art. 42.

Authors cited as "D. Erastova, M. Okun, A.M. Fiore-Donno, Novozh., and Schnittler" but see erratum, Mycol. Progress 12(3): 615 (2013)

Fruiting bodies sessile or stalked, sporocarpic or plasmodiocarpic, rarely aethalioid, with calcareous deposits in some parts of their structure, without oil or wax. Peridium usually with granular calcareous deposits. Columella absent, rarely present, sometimes with a calcareous pseudocolumella. Capillitium a network of hyaline tubules connecting calcareous nodes, sometimes with branched threads or simple and unbranched peridial outgrowths (10 genera, 211 species).

Badhamia Berk., Trans. Linn. Soc. London 21(2): 153 (1853) (31 species)

Badhamiopsis T.E. Brooks and H.W. Keller, in Keller and Brooks, Mycologia 68 (4): 835 (1976) (3 species)

Craterium Trentep., in Roth, Catal. bot. 1: 224 (1797) (17 species)

Fuligo Haller, Hist. stirp. Helv. 3: 110 (1768) (10 species)

Kelleromyxa Eliasson, in Eliasson, Keller, and Schoknecht, Mycol. Res. 95(10): 1205 (1991) (1 species)

Erastova et al. (2013), taking into account the particular characteristics and the phylogeny of the only known species, *Kelleromyxa fimicola*, supported the erection of a monotypic family for this species.

Leocarpus Link, Ges. Naturf. Freunde Berlin Mag. Neuesten Entdeck. Gesammten Naturk. 3: 25 (1809) (1 species)

Physarella Peck, Bull. Torrey Bot. Club 9(5): 61 (1882) (1 species)

Physarum Pers., Neues Mag. Bot. 1: 88 (1794) (144 species)

Protophysarum M. Blackw. and Alexop., Mycologia 67(1): 33 (1975) (2 species)

One species of the genus (*Protophysarum balticum* Dörfelt & Schmidt) is a fossil found in Baltic amber (Dörfelt and Schmidt, 2006). Castillo et al. (1998) raised this genus to family rank, but the arguments do not seem consistent with creating a new family.

Willkommlangea Kuntze, Revis. gen. pl. 2: 875 (1891) (1 species)

Subclass **Stemonitomycetidae** I.K. Ross, Mycologia 65 (2): 483 (1973)

Cavalier-Smith, Microbiol. Rev. 57: 971 (1993), give an alternative orthography as "Stemonitia Ross, 1973 orthogr. emend.," also Locquin, Tax. Fung. 6 (1974), as "Stemonomycetidae," but no description provided.

Spores borne in mass within fruiting bodies of various types. Development of fruiting bodies epihypothallic, the plasmodial protoplasm rising externally along the developing stalk in stipitate forms. Fruiting bodies sporocarpic, aethalioid or pseudoaethalioid, stalked or sessile, without calcareous deposits. Peridium not continuous with stalk and hypothallus, persistent or evanescent. Stalk, when present, filled with fibers or hollow. Columella usually present. Capillitium of simple, branched, or anastomosing threads, arising from the columella or from the sporophore base. Assimilative stage an aphanoplasmodium. Consisting of 1 order, 1 family, 17 genera and 211 species.

Order **Stemonitidales** T. Macbr., N. Amer. Slime-moulds, ed. 2. 22,148 (1922), as "Stemonitales" [Stemonitida]

= Amaurochaetales Hagelst., Mycetozoa N. Amer. 9 (1944), nom. inval., Art. 38.1.

With the characters of the subclass. Consisting of 1 family, 17 genera and 211 species.

Leontyev (2015), has proposed raising the order to the rank of superorder and segregated into 3 orders, order Meridermatales, with family Meridermataceae, order Stemonitidales, comprising the families Stemonitidaceae and Comatrichaceae, and the more questionable order Lamprodermatales, comprising the families Lamprodermataceae, Physaraceae, Kelleromyxaceae, and Didymiaceae.

Family Stemonitidaceae Fr., Syst. Mycol. 3(1): 75 (1829), as "Stemonitei," sanctioning citation [Stemonitidae]

First employed by Fries, Syst. mycol. 3(1): 75 (1829) as "Stemonitei," sanctioning citation according to the Index Fungorum (http://www.indexfungorum.org/names/NamesRecord.asp?RecordID=81588). First spelt correctly by Rostafiński, Vers. Syst. Mycetozoen 6 (1873) but as "tribus," first published as family by Cooke, Contr. mycol. brit. 45 (Jan.–Jul. 1877), also by Luerssen, Handb. syst. Bot. 1: 42 (Dec. 1877), both as "Stemonitaceae."

= Enerthenemataceae Rostaf. ex Cooke, Contr. mycol. brit.: 51 (1877), as "Enerthenemaceae"

Published by Rostafiński, Vers. Syst. Mycetozoen 8 (1873) as "tribus," first spelt correctly and as family by Cooke, Contr. mycol. brit.: 51 (Jan.–Jul. 1877), also by Luerssen, Handb. syst. Bot. 1: 42 (Dec. 1877).

= Amaurochaetaceae Rostaf. ex Cooke, Contr. mycol. brit. 52 (Jan.–Jul. 1877), as "Amaurochetaceae"

Published by Rostafiński,, Vers. Syst. Mycetozoen 8 (1873) as "tribus," first spelt correctly and as family by Cooke, Contr. mycol. brit. 52 (Jan.–Jul. 1877), also by Luerssen, Handb. syst. Bot. 1: 42 (Dec. 1877).

= Brefeldiaceae Rostaf. ex Cooke, Contr. mycol. brit.: 52 (Jan.–Jul. 1877)

Published by Rostafiński, Vers. Syst. Mycetozoen 8 (1873) as "tribus," first spelt correctly and as family by Cooke, Contr. mycol. brit. 52 (Jan.–Jul. 1877), also by Luerssen, Handb. syst. Bot. 1: 42 (Dec. 1877).

= Raciborskiaceae Berl., in Berlese, De Toni, and Fischer, Syll. fung. 7: 400 (1888)

= Lamprodermataceae T. Macbr., N. Amer. Slime-moulds, ed. 2. 189 (1922), as "Lamprodermaceae"

= Collodermataceae G. Lister, in Lister, Monogr. mycetozoa, ed. 3. 128 (1925), as "Collodermaceae"

= Comatrichaceae Locq., Syn. gen. fung. [1] (1972), nom. inval., Art. 39.1

= Macbrideolaceae Locq., Syn. gen. fung. [1] (1972), nom. inval., Art. 39.1

= Symphytocarpaceae Locq., Syn. gen. fung. [1] (1972), nom. inval., Art. 39.1

With the characters of the order (17 genera, 211 species).

Amaurochaete Rostaf., Vers. Syst. Mycetozoen 8 (1873) (4 species)

Nom. cons., Art. 14, see Lado et al., Taxon 54(2): 543 (2005) and Gams, Taxon 54(3): 829 (2005).

Brefeldia Rostaf., Vers. Syst. Mycetozoen 8 (1873) (1 species)

Collaria Nann.-Bremek., Proc. Kon. Ned. Akad. Wetensch., C. 70(2): 208 (1967) (5 species)

Colloderma G. Lister, J. Bot. 48: 312 (1910) (4 species)

Comatricha Preuss, Linnaea 24: 140 (1851) (40 species)

Diacheopsis Meyl., Bull. Soc. Vaud. Sci. Nat. 57: 149 (1930) (16 species)

Enerthenema Bowman, Trans. Linn. Soc. London 16: 152 (1830) (4 species)

Lamproderma Rostaf., Vers. Syst. Mycetozoen 7 (1873) (49 species)

Leptoderma G. Lister, J. Bot. 51: 1 (1913) (2 species)

Macbrideola H.C. Gilbert, Stud. Nat. Hist. Iowa Univ. 16(2): 155 (1934) (18 species)

Meriderma Mar. Mey. and Poulain, in Poulain, Meyer, and Bozonnet, Myxomycètes 1: 551 (2011) (4 species)

Leontyev (2015), has proposed to create, for this genus, a new order "Meridermatales" and a new family "Meridermataceae."

Paradiachea Hertel, Dusenia 7: 349 (1956) (5 species)

Paradiacheopsis Hertel, Dusenia 5: 191 (1954) (9 species)

Stemonaria Nann.-Bremek., R. Sharma, and Y. Yamam., in Nannenga-Bremekamp, Yamamoto, and Sharma, Proc. Kon. Ned. Akad. Wetensch., C. 87(4): 450 (1984) (14 species)

Stemonitis Gled., Meth. fung. 140 (1753) (17 species)

A species of the genus, *Stemonitis splendens* Rostaf., has been found well preserved in Baltic amber from the Tertiary Period, of the Eocene Epoch (Domke, 1952; Keller, 2012).

Stemonitopsis (Nann.-Bremek.) Nann.-Bremek., Nederlandse Myxomyceten (Zutphen) 203 (1975) (10 species)

Symphytocarpus Ing and Nann.-Bremek., Proc. Kon. Ned. Akad. Wetensch., C. 70(2): 218 (1967) (9 species)

Incertae sedis (1 genus, 1 species)

Trichioides Novozh., Hoof and Jagers, Mycol. Progress 14(1018): 2 (2015) (1 species)

This genus shows morphological affinities with members of the order Trichiales, such as the tubular capillitial threads ornamented by spiral bands characteristic of the genus *Trichia*, and with members of the order Cribrariales, such as the spores brown in mass, smooth and with a paler thinner area, a character only seen in some species of the genus *Licea* (Novozhilov et al., 2015).

THE NEW ERA OF TAXONOMY: HOW NEW TECHNOLOGIES CAN HELP TO RESOLVE CLASSICAL TAXONOMIC PROBLEMS

As previously defined, taxonomy deals with the study of identifying, grouping, and naming of organisms, based on similarities in such things as structure and origin, according to their established natural relationships. Taxonomy is one of the oldest branches of biology and in its modern form dates back to the work of Linnaeus in the middle of the 18th century. Taxonomy is the most fundamental of life sciences and is becoming crucial in many aspects of life and society, such as the biodiversity management, public health or agriculture. From a practical point of view, taxonomy involves the organization of large amounts of information. But before the web, so much of this taxonomic information was difficult to find, and most of it was hidden in journals that are not easily accessible.

The Internet, with its great efficiency as a source of information and as a rapid and effective method of communication, has had a great positive impact on the work of taxonomists. Web-taxonomy, usually called e-taxonomy, takes advantage of the Internet as a medium to find information, to connect specialists or amateurs working on the same group of organisms, and to develop new tools that make taxonomy work more efficient. In the case of myxomycetes, it allows taxonomic information to be found easily and accessibly to be shared, taxonomic hypotheses to be debated. In addition, numerous tools are available to enable them to perform their work more quickly and efficiently. However, it is important not to underestimate the vast task of transferring biodiversity information to the web, not just moving data from paper to digital form, but also the institutional and social challenges of presenting an old science in a new way.

When a new species of myxomycete is discovered or a new rank is to be proposed, the next step is to publish the results in a journal or other permanent resource. This act has important scientific significance and must fulfill a series of laws and requirements of the ICN (McNeill et al., 2012). It is important to know that a name does not become valid until it is published in accordance with these requirements. These current mandatory requirements are:

- Providing a Latin or English description or diagnosis of the taxon, or by a reference to a previously and effectively published description or diagnosis.
- Selecting a type specimen to be deposited in a public herbarium or collection.
- Adding a clear indication of the rank of the taxon involved.
- Registering the taxon or the name in a recognized repository (Mycobank is recommended).

It is also recommended that the description is supplemented with an illustration showing the essential characters and that the author's name is added after the name of the new taxon even when the authorship of the publication is the same. A unique aspect of taxonomy, unlike all other branches of science, is that taxonomists need to refer to material published in the 19th and even the 18th century. Therefore, a taxonomist needs to be very familiar with the species he works on and also have an intimate knowledge of the often-complex literature on the group. This dependence on the literature is the price that has had to be paid for stability, and it has served the subject well for 250 years. However, the question remains as to whether this can be modified in light of modern means of communicating and information sharing, especially using the web.

To evaluate the subject, let us identify some problems:

- Taxonomic information is highly scattered and difficult to access, which is an impediment to people who are not experts in the field. In the case of the myxomycetes, we usually know the literature of our country or continent but ignore the rest. More than 5000 articles or books are known that contain data on the myxomycetes, a figure unknown for many specialists and which would have been difficult to handle without current computer technology.
- Taxonomic research requires access to an extensive and often very old literature that is available only in major institutions. For example, the book

Śluzowce Monografia by Rostafiński (1874–76), the first monograph of myxomycetes in a modern sense, is very rare and difficult to obtain, and only a few copies exist in specialized libraries. Gaining access to this information through virtual libraries, for example, can be of great benefit, as you can read many books remotely. This book is now available in several web pages, such as https://archive.org or www.europeana.eu.

- Printed literature is often a poor medium for taxonomy. For one thing, it is expensive, which in particular restricts the use of illustrations. For example, in the monographs of myxomycetes in the *North American Flora* (Martin, 1949) and *Flora Neotropica* (Farr, 1976), illustrations could have resolved some interpretations of the text. Fortunately, digital images have solved most of these problems, given their low cost and easy production and reproduction.
- The language of some publications, such as classic Polish, Chinese, Ciril-ic, or even Latin, are unfamiliar to many researchers. The monograph of Rostafiński, in classic Polish, or *De Nederlandse Myxomyceten* by Nannenga-Bremekamp (1974), which in Dutch, are some examples. Computer translation programs and optical character recognition (optical character reader) (OCR), offer an electronic transversion of images of typed, handwritten, or printed text into machine-encoded text, allowing a rough idea of these texts in a short time or almost instantaneously.
- There is frequently a lack of connection between taxonomists and the people using their taxonomic products. Users normally need to know the current consensus taxonomy of a particular group, but consensus is often difficult to achieve because, typically, there are multiple taxonomic hypotheses of the species concept in the literature. The genus *Physarum*, with more than 140 species accepted and more than 500 proposals of names to be analyzed, may serve as an example. Widely used software, such as databases, word processors, or spreadsheets allows handling of large volumes of data quickly and efficiently.
- All taxonomies are hypotheses and change as research progresses, but tracing how a particular name or classification evolved is often difficult and can lead to ambiguities as to what concept was being referred to at any particular time. The concept of the genus *Badhamia* or *Lamproderma* are two examples of this complexity. Programs that reveal the evolution and analyze chronological changes or consult via Internet digital images of herbarium labels and type material are other tools that could be developed by e-taxonomy.
- Taxonomy is basically a descriptive science, but in the descriptions of species the authors do not always use the same terminology or use terms in the same sense. Sometimes, terms are inaccurately used and in some cases contain a high dose of subjectivity. The result is that the descriptions in these cases are not easy to interpret and are ambiguous, thereby creating uncertainty regarding the concept of the species. Programs should be developed

to create standardization of terms, facilitating descriptions and comparison of taxa, and creating synoptic and dichotomous keys.

- In terms of nomenclature, the new technologies are also allowed. The ICN has introduced new rules to regulate the use of electronic documents. In Art. 29, 30, where it treats the conditions that must be met for the effective publication of a new taxon, the ICN already accepts and regulates, from January 1, 2012, the use and distribution of electronic material in Portable Document Format (PDF), facilitating the use of online publications. Also, in Art. 42, established as mandatory, to register in a recognized repository (the www.mycobank.org is recommended) all names of new taxa, new combinations, names at new ranks, or replacement names of fungi published on or after January 1, 2013, as well as the citation in the prologue of the identifier issued.
- Finally, it is often difficult for nontaxonomists to contribute to the taxonomic research program, although they sometimes have valuable observations and material to add. Fortunately, current research projects are more open and available to all people through web pages, Internet forums, social networks and other participatory media.

The rapid and unstoppable development of the computer and Internet of the last 2 decades has allowed a creation of an arsenal of new technological tools, designed to streamline and make more precise the work of the taxonomist of our century. There is even a proposal to try to create a single web-medium, called "unitary taxonomy" (Scoble, 2004), that would serve as a "one-stop shop," providing all the resources that a taxonomist needs for work on a particular group beyond the physical specimens themselves. It would also provide an interface to allow nonspecialists not only to gain access to information about a particular group but also to enable them to contribute information, such as new records, new images, and new evidences in their life cycle (e.g., plasmodial color). It would increase the number of users of taxonomy.

Some researchers, teams, projects, and governments are developing initiatives or programs at different levels to solve the problems raised earlier in this chapter, some of which are discussed in the following sections.

Virtual Libraries and Herbaria

These new generation libraries have produced a revolution in this matter and are a clear example of the impact of new technologies in our daily work. By accessing these tools the user can consult many old books for no higher cost than a fee for connecting to the Internet and a few seconds spent in the search. Prominent examples are The Biodiversity Heritage Library (www.biodiversitylibrary.org), which specializes in old books; the Digital Library (http://bibdigital.rjb.csic.es), developed by the Royal Botanic Garden of Madrid, which currently offers an open access to more than 1.6 million pages and more than 2900 titles of botanical books; and Gallica (http://gallica.bnf.fr) or Europeana (www.europeana.eu), more generalistic initiatives that offer free access to old periodicals, manuscripts, maps, images, and rare books. Many reference works for taxonomists, such as the Taxonomic Literature (TL) (http://www.sil.si.edu/DigitalCollections/tl-2/index.cfm), the Botanico Periodicum Huntianum (BPH) (http://fmhibd.library.cmu.edu/HIBD-DB/bpho/findrecords.php), or the Index Herbariorum (http://sweetgum.nybg.org/science/ih/), are also available on the web.

A virtual herbarium is a herbarium in digitalized form and can serve as a very useful tool for studying details of the labels or parts of the specimen without touching the latter, thus avoiding possible deterioration of the sample. Most important taxonomic institutions have ongoing programs of digitalization of their collections, with special attention to type material. Information about the programs, formats, and digitalized collections is available on the websites of the respective institutions.

Repositories of Biodiversity Information

Countries or groups of countries along with important institutions have seen the importance of these technological opportunities, and support several e-taxonomy initiatives designed to provide large volumes of information on the biodiversity of a region, country, continent or worldwide. The Global Biodiversity Information Facility (GBIF) (www.gbif.org) is a good example. GBIF is an international governmental initiative that enables free and open access to biodiversity data online. One can explore species, countries, or datasets and obtain, with only a click, large amounts of information. This includes nomenclatural information, where the type or authentic specimens are deposited, total number of collections available, a representation on a map of the records deposited in offical herbaria or collections, and the institutions that provide the information. This international and cooperative initiative provides information of more than 1.6 million species and more than 649 million occurrences of organisms around the world. Another similar initiative is the ITIS (Integrated Taxonomic Information Systems), which provides information from North America (United States, Canada, and Mexico). GBIF and ITIS are also partners of other e-taxonomy initiatives, such as Species 2000 (www.sp2000.org), which provides access to species names and associated data through distributed databases compiled by specialists, the Catalogue of Life (CoL) (www.catalogueoflife.org), a single integrated species checklist and taxonomic hierarchy, or the Encyclopedia of Life (EoL) (www.eol.org), a global access to knowledge about life on Earth.

Projects

Given the magnitude of the problems addressed in e-taxonomy, several collaborative projects have been developed in recent years. One is EU BON (European Biodiversity Observation Network) (http://eubon.eu). This complex project, in which 31 partners from 18 countries are involved, has among its objectives the establishment and adoption of new data standards and integration techniques, and harmonized data collections. The Virtual Biodiversity Research and Access Network for Taxonomy (ViBRANT) is a project funded by the European Union FP7 and running from December 2010 to 2013 to support the development of virtual research communities involved in biodiversity science. The goal is to provide a more integrated and effective framework for biodiversity data that has often been generated by taxonomists. ViBRANT provides a virtual research environment (Scratchpads), analytical services for users to build identification keys and phylogenetic trees, a publication platform for users to automatically compile biodiversity science manuscripts from their research database, a portal for users to publicly accessible biodiversity research information and literature, and training, helping research communities to use these tools and services. The Distributed European School of Taxonomy (DEST) (www.taxonomytraining.eu), which is using the Scratchpads biodiversity online platform (scratchpads.eu), is another project. The TDWG Taxonomic Database Working Group (TDWG) is one of the most active groups and their mission is the development of standards and guidelines for the recording and exchange of biological and biodiversity data. Some of their more common botanical standards are "Authors of Plant Names," "World Geographical Scheme for Recording Plant Distribution," "Index Herbariorum," or "Structured Descriptive Data." The Biodiversity Virtual e-Laboratory (BioVeL Project) (http://www.biovel.eu) supports scientists to carry out research on biodiversity by offering computerized tools (workflows) to process large amounts of data, as well as tools for designing and running workflows. LIFEWATCH (http://www.lifewatch.eu) is another European project to provide E-Science European Infrastructure for Biodiversity and Ecosystem Research, and places emphasis on environmental issues.

Banks of Data

Another extremely useful tool is represented by the databanks, which act as global taxonomic data repositories. Among them are MycoBank (http://www.mycobank.org) and Index Fungorum (http://www.indexfungorum.org), two online databases designed as a service by documenting mycological nomenclatural novelties (new names and combinations) and associated data, such as descriptions, illustrations, and status of the names. Also, NomenMyx (www.eumycetozoa.com), a database specifically dedicated to myxomycetes and related organisms (protostelids and dictyostelids), provides nomenclatural information on more than 4600 names employed in this group. Additionally, for old literature, it can be consulted for a pdf of the original pages in which the species were described as "new." To demonstrate the impact of these banks of data, in a year and a half almost 50,000 pages have been consulted on NomenMyx, which is a specialized web page, and more than 7,300 individuals from all over the world have visited this site. GenBank (http://www.ncbi.nlm.nih.gov/genbank/), another good example, is a genetic sequence database and an annotated collection of all publicly available DNA sequences. GenBank also provides information about genomes, metagenomes, revision history of a sequence, as well as many other databases and tools with the idea of offering the most up to date and comprehensive DNA sequence information. Morphobank (http://www.morphobank.org) is a web application with tools and archives for evolutionary research, specifically systematics, and is a project designed to build the Tree of Life with phenotypes.

Virtual Encyclopedias

Two of the more general initiatives in the whole web of life are the Encyclopedia of Life' (EoL) (www.eol.org) and "Discover Life" (www.discoverlife.org). EoL is a web page for every species, with information on classification, nomenclature, taxonomy, references, images, but also with IUCN red list status, molecular biology and genetics, maps, and comments. It is linked to other initiatives previously mentioned, such as Species 2000, ITIS, Catalogue of Life or GBIF, and compiles much of the information provided by these other specialized web pages. Discover Life is in the same line as EoL and has as its mission to assemble and share knowledge in order to improve education, health, agriculture, economic development, and conservation throughout the world. This makes the work of taxonomists, which is often silent and anonymous, more visible and useful to society.

Web Pages

Web pages are single, usually hypertext documents, suitable for the World Wide Web and identified by a unique URL (Uniform Resource Locator). This document can incorporate text, images, sounds, videos, links to other web pages, and endless possibilities that the mainstream media are not able to offer. Due to its versatility, the web pages have been chosen as a means of dissemination of many research projects to people interested in the myxomycetes. As an example, the web page developed by the Eumycetozoan Project (http://slimemold.uark.edu) could be mentioned. It is an excellent compilation of information about "slime molds" assembled at the University of Arkansas in the context of grants from the Planetary Biodiversity Inventories (PBI) and Partnerships for Enhancing Expertise in Taxonomy (PEET) programs of the National Science Foundation of the United States. On this web page, we can find educational materials about myxomycetes, as well as databases, a species image gallery, and reports of the expeditions made by the different members of the team in many parts of the world. This is probably the most complete web page dedicated to myxomycetes. Another important web page is MYXOTROPIC (www.myxotropic.org), produced at the Real Jardín Botánico (CSIC) of Madrid, which shows the results of this project that has as its objective the study of the myxomycetes in the Neotropics, one of the richest biogeographic zones of the planet, with many biodiversity hotspots. On this web page, with versions in English and Spanish, information about the project can be found, such as a brief history of the project, its objectives, research areas, digital images of species, pdfs of publications of the team and a data portal with extensive information on the distribution of these organisms in this biogeographical region. This makes this web page to be the most complete source of information on the myxomycetes of a particular region of the world, in this case the Neotropics. Other web pages, some developed by people interested in the myxomycetes, with complementary information about myxomycetes (videos, digital images of high quality, and biological information) are http://www.myxomycetes.net, http://myxo.be, http://henkeikin.org, http://greekmyxomycetes.blogspot.com.es, http://hiddenforest.co.nz/slime, http://mushroomobserver.org/species_list/show_species_list/728, http://www.myx.dk, http://www.physarumplus.org, http://bioweb.uwlax.edu/bio203/2010/renner_brad/index.htm.

Software

Bioinformatics is a discipline that has been very actively developed in recent years. Among its tasks is the development of software to formalize and streamline many of the tasks of taxonomists. Many of these software tools are free, offer excellent performance, and do a number of different things, such as to create automatically descriptions or keys to species (DELTA), to measure or compare structures (ImageJ, SigmaScan, Scion Image), to align or compare DNA sequences (BLAST), to produce, manipulate, or stack digital images (Photoshop, CombineZP, Helicon focus), to georeference localities of sampling and GIS (Google Earth, GDAL, DIVA-GIS) to elaborate distribution maps of the species (DIVA-GIS, ArcGIS), or to correlate climatic parameters with environmental requirements of species (Maxent). Many of these programs or tools have been designed by taxonomists themselves or by specialized engineers in bioinformatics in close cooperation with taxonomists. Also, one should not forget computer search engines (like Google) that enable us to track all kinds of information, from the most general to the most specific. Finding information on a scientific name on the network is no longer a problem.

As already presented, the applications of information technology are numerous and varied, from the accessibility of information to the development of standardizing terminology, standardizing descriptions, producing taxon description "kits," automatic key updates, morphological repositories, galleries of images, and graphics tools. The use of these tools seeks greater efficiency in the work of taxonomists, and new terms, such as e-taxonomy, web-taxonomy, cybertaxonomy, or bioinformatics are increasingly being used by those individuals who use or develop these new tools.

CHALLENGES IN TAXONOMY

The taxonomist of today can work at orders of magnitude faster, thanks to more efficient and collaborative modes of work. Governments and institutions are aware of the challenges and support programs or projects to improve infrastructure, to develop new tools, to implement new applications, ultimately, to launch new ideas and new objectives. As Wheeler (2008) reminds us, "Against formidable odds and with minimal funding, equipment, infrastructure, organization and encouragement, taxonomist have discovered, described and classified nearly 1.8 million species." Hobern et al. (2012) in their Global Biodiversity Informatics Outlook (GIBO) document identified four major focal areas of work in the digital age, and offered a framework for reaching a much deeper understanding of the world's biodiversity. Collaborative networks seem to be the fastest and most effective way to further develop the current taxonomy. Each day more taxonomists are attracted by the possibilities that e-taxonomy offers to work in one of the oldest branches of biology. The challenges that arise are many and exciting. Some of them are:

- To complete the digitalization of all the published literature on myxomycetes from 1753.
- To digitalize records associated with specimens in collections, herbaria, and fungaria.
- To produce a website containing a modern treatment of a taxonomic group.
- To create a consensus taxonomy without loss of alternative taxonomic hypotheses.
- To ascertain that the taxonomic content on the web is produced by taxonomists, corrected and up to date.
- To explore new tools using bioinformatic specialists.
- To enable e-taxonomy to link to other initiatives, such as global catalogues, the web of life initiative, bar-coding and other molecular sites, as well as to specimen-level databases.
- To work within the current relevant taxonomic codes, while exploring the practicalities and implications of taxonomic publication on the Internet.
- To store basic taxonomic data in such a way that it can be used by scientists and end-users.

These are just a few examples but can serve as encouragement to continue working on this intriguing group of organisms. This is a proposition of enormous audacity but one in which all taxonomists, researchers, and people interested in myxomycetes can contribute with their individual grain of sand.

ACKNOWLEDGMENTS

Carlos Lado thanks the Spanish Government grant CGL2014-52584P for economic support of the research. Uno Eliasson thanks the Royal Society of Arts and Sciences in Gothenburg for economic support of fieldwork and other research related work on myxomycetes in Sweden and abroad during several years. Both authors thank Carlos de Mier for his technical assistance with the images and micrographs.

REFERENCES

Adl, S.M., Simpson, A.G.B., Lane, C.E., Lukes, J., Bass, D., Bowser, S.S., Brown, M.W., Burki, F., Dunthorn, M., Hampl, V., Heiss, A., Hoppenrath, M., Lara, E., Le Gall, L., Lynn, D.H., McManus, H., Mitchell, E.A., Mozley-Stanridge, S.E., Parfrey, L.W., Pawlowski, J., Rueckert, S., Shadwick, L., Schoch, C.L., Smirnov, A., Spiegel, F.W., 2012. The revised classification of eukaryotes. J. Eukaryot. Microbiol. 59 (5), 429–493.

Alexopoulos, C.J., 1973. Myxomycetes. In: Ainsworth, G.C., Sparrow, F.K., Sussman, A.S. (Eds.), The Fungi: An Advanced Treatise. Academic Press, New York, NY, pp. 39–60.

Alexopoulos, C.J., 1978. The evolution of the taxonomy of the myxomycetes. In: Subramanian, C.V. (Ed.), In: Proceedings of the International Symposium on Taxonomy of Fungi Part I. University of Madras, pp. 1–8.

Batsch, A.J.G.C., 1783–1789. Elenchus Fungorum. In: Gebauer, J.J., (Ed.), Halle.

Blackwell, M., Alexopoulos, C.J., 1975. Taxonomic studies in the Myxomycetes IV: *Protophysarum phloiogenum*, a new genus and species of Physaraceae. Mycologia 67, 32–37.

Brefeld, O., 1869. *Dictyostelium mucoroides*: Ein neuer Organismus aus der Verwandschaft der Myxomyceten. Abhandlungen der Senckenbergischen Naturforschenden Gesellschaft Frankfurt 7, 85–107.

Castillo, A., Illana, C., Moreno, G., 1998. *Protophysarum phloiogenum* and a new family in the Physarales. Mycol. Res. 102 (7), 838–842.

Cavalier-Smith, T., 1998. A revised six-kingdom system of life. Biol. Rev. Camb. Philos. Soc. 73, 203–266.

Cavalier-Smith, T., 2013. Early evolution of eukaryote feeding modes, cell structural diversity, and classification of the protozoan phyla Loukozoa, Sulcozoa, and Choanozoa. Eur. J. Protistol. 49 (2), 115–178.

Clark, J., 1996. Mating systems of Myxomycetes. ICSEM2 Abstract volume: 41. Real Jardín Botánico, CSIC, Madrid

Clark, J., Haskins, E., 2002. Reproductive systems of *Comatricha laxa* and *Lamproderma arcyrionema*. ICSEM4 Abstract volume: 15. National Botanic Garden of Belgium, Meisse.

Clark, J., Haskins, E.F., 2014. Sporophore morphology and development in the myxomycetes: a review. Mycosphere 5, 153–170.

Clark, J., El Hage, N., Stephenson, S.L., 1999. Biosystematics of *Didymium squamulosum*. ICSEM3 Abstract volume: 66. USADA-ARS National Fungus Collection (BPI), Beltsville, Maryland.

Clark, J., Haskins, E., Stephenson, S.L., 2002a. Biosystematics of *Badhamia gracilis*. ICSEM4 Abstract volume: 16. National Botanic Garden of Belgium, Meisse.

Clark, J., Schnittler, M., Stephenson, S.L., 2002b. Biosystematics of *Arcyria cinerea*. ICSEM4 Abstract volume: 17. National Botanic Garden of Belgium, Meisse.

Cooke, M.C., 1877. The myxomycetes of the United States. Ann. Lye. Nat. Hist. N. Y. 11, 378–409.

Corda, A.K.J., 1842. Icones Fungorum hucusque cognitorum, vol. 5. Apud J.G. Calve, Pragae.

David, J.C., 2002. A preliminary catalogue of the names of fungi above the rank of order. Constancea 83, 1–30.

de Bary, A., 1858. Ueber die Myxomyceten. Bot. Ztg. 16, 357–358, 361–364, 365–369.

de Bary, A., 1859. Die Mycetozoen: Ein Beitrag zur Kenntnis der niedersten Thiere. Z. Wiss. Zool. 10, 88–175.

de Bary, A., 1864. Die Mycetozoen (Schleimpilze): Ein Beitrag zur Kenntnis der niedersten Organismen. Verlag Von Wilhelm Engelmann, Leipzig.

de Bary, A., 1884. Vergleichende Morphologie und Biologie der Pilze, Mycetozoen, und Bacterien. Verlag Von Wilhelm Engelman, Leipzig.

de Schweinitz, L.D., 1822. Synopsis fungorum Carolinae superioris secundum observationes Ludovici Davidis de Schweinitz. Schriften der Naturforschenden Gesellschaft zu Leipzig 1, 20–131.

de Schweinitz, L.D., 1832. Synopsis fungorum in America boreali media degentium: secundum observations. Trans. Am. Philos. Soc. 4, 141–316.

Domke, W., 1952. Der erste sichere Fund eines Myxomyceten im Baltischen Bernstein (Stemonitis splendens Rost. fa. succini fa. nov. foss.). Mitt. Geol. Staatsinst. Hamburg 21, 154–161.

Dörfelt, H., Schmidt, A.R., 2006. An archaic slime mould in Baltic amber. Palaeontology 49 (5), 1013–1017.

Dörfelt, H., Schmidt, A.R., Ullmann, P., Wunderlich, J., 2003. The oldest fossil myxogastroid slime mould. Mycol. Res. 107 (1), 123–126.

Doweld, A.B., 2015. (2392) Proposal to conserve the name Physaraceae against Fuliginaceae (myxomycetes). Taxon 64 (6), 1313.

Eliasson, U.H., 1977. Recent advances in the taxonomy of myxomycetes. Bot. Not. 130, 483–492.

Eliasson, U.H., 2017. Review and remarks on current generic delimitations in the myxomycetes, with special emphasis on Licea, Listerella, and Perichaena. Nova Hedwigia 104, 343–350.

Eliasson, U.H., Gilert, E., 1982. A SEM-study of *Listerella paradoxa* (myxomycetes). Nord. J. Bot. 2, 249–255.

Eliasson, U.H., Keller, H.W., Schoknecht, J.D., 1991. *Kelleromyxa*, a new generic name for *Licea fimicola* (myxomycetes). Mycol. Res. 95, 1201–1207.

Eliasson, U.H., Sjögren, J., Castillo, A., Moreno, G., 2010. Two species of the Stemonitales (myxomycetes) new to Sweden with remarks on generic delimitations. Boletín de la Sociedad Micológica de Madrid 34, 147–154.

Emoto, Y., 1977. The Myxomycetes of Japan. Sangyo Tosho Publishing Co., Tokyo.

Erastova, D.A., Okun, M.V., Fiore-Donno, A.M., Novozhilov, Y.K., Schnittler, M., 2013. Phylogenetic position of the enigmatic myxomycete genus *Kelleromyxa* revealed by SSU rDNA sequences. Mycol. Prog. 12, 599–608.

Estrada-Torres, A., Gaither, T.W., Miller, D.L., Lado, C., Keller, H.W., 2005. The myxomycete genus *Schenella*: morphological and DNA sequence evidence for synonymy with the gasteromycete genus *Pyrenogaster*. Mycologia 97, 139–149.

Farr, M.L., 1976. Myxomycetes. In: Rogerson, C.T. (Ed.), Flora Neotropica Monograph 16. New York Botanical Garden, New York, NY, pp. 1–305.

Farr, M.L., Alexopoulos, C.J., 1977. Validation of subclass Ceratiomyxomycetida and order Ceratiomyxales (class myxomycetes). Mycotaxon 6 (2), 213–214.

Fiore-Donno, A.M., Berney, C., Pawlowski, J., Baldauf, S.L., 2005. Higher-order phylogeny of plasmodial slime molds (Myxogastria) based on elongation factor 1-A and small subunit rRNA gene sequences. J. Eukaryot. Microbiol. 52 (3), 1–10.

Fiore-Donno, A.M., Clissman, F., Meyer, M., Schnittler, R.M., Cavalier-Smith, T., 2013. Two-gene phylogeny of bright-spored myxomycetes (slime moulds, Superorder Lucisporidia). PLoS One 8, e62586.

Fiore-Donno, A.M., Haskins, E.F., Pawlowski, J., Cavalier-Smith, T., 2009. *Semimorula liquescens* is a modified echinostelid myxomycete (Mycetozoa). Mycologia 101, 773–776.

Fiore-Donno, A.M., Kamono, A., Meyer, M., Schnittler, M., Fuku, M., Cavalier-Smith, T., 2012. 18s rRNA phylogenics of *Lamproderma* and allied genera (Stemonitales, Myxomycetes, Amoebozoa). PLoS One 7 (4), e35359.

Fiore-Donno, A.M., Meyer, M., Baldauf, S.L., Pawlowski, J., 2008. Evolution of dark-spored Myxomycetes (slime-molds): molecules versus morphology. Mol. Phylogenet. Evol. 46, 878–889.

Fiore-Donno, A.M., Nikolaev, S.I., Nelso, M., Pawlowski, J., Cavalier-Smith, T., Baldauf, S., 2010. Deep phylogeny and evolution of slime moulds (Mycetozoa). Protist 161, 55–70.
Fries, E.M., 1829. Systema mycologicum, sistens fungorum ordines, genera et species. In: Mauritii E. (Ed.), Greifswald, vol. 3. 67–199.
Gilbert, H.C., 1935. Critical events in the life history of *Ceratiomyxa*. Am. J. Bot. 22, 52–74.
Gilert, E., 1985. Ultrastructure of *Licea kleistobolus* (myxomycetes) and its bearing on the taxonomic affinity of the species. Nord. J. Bot. 5, 99–104.
Gilert, E., 1987. Morphology and ultrastructure of the plasmodiocarpous species *Licea variabilis* and *L. retiformis* (myxomycetes). Nord. J. Bot. 7, 569–575.
Gilert, E., 1990. On the identity of *Perichaena liceoides* (myxomycetes). Mycol. Res. 94, 698–704.
Gilert, E., 1997. Morphological and ultrastructural features in selected species of *Licea* (myxomycetes). Nord. J. Bot. 16, 515–547.
Hagelstein, R., 1944. The Mycetozoa of North America, Published by the author, Mineola, NY.
Härkönen, M., Sivonen, E., 2011. Limasienet. Botanical Museum, Finnish Museum of Natural History, Helsinki.
Haskins, E.F., Mcguinness, M.D., Berry, C.S., 1983. *Semimorula*: new genus with myxomycete and protostelid affinities. Mycologia 75, 153–158.
Haskins, E.F., Wrigley de Basanata, D., 2008. Methods of agar culture of myxomycetes: an overview. Rev. Mex. Micol. 27, 1–7.
Hobern, D., Apostolico, A., Arnaud, E., Bello, J.C., Canhos, D., Dubois, G., Field, D., García, E.A., Hardisty, A., Harrison, J., Heidorn, B., Krishtalka, L., Mata, E., Page, R., Parr, C., Price, J., Willoughby, S., 2012. Global Biodiversity Informatics Outlook. Delivering biodiversity knowledge in the information age. GBIF Secretariat. Available from: http://www.gbif.org/resource/80859
Hochgesand, E., Gottsberger, G., 1989. *Arcyriatella congregata*, a new genus and new species of the Trichiaceae (myxomycetes). Nova Hedwigia 48, 485–489.
Ing, B., 1999. The Myxomycetes of Britain and Ireland. Richmond Publishing Co., Slough, UK.
Jahn, E., 1928. Myxomycetes (Mycetozoa, Phytosarcodina, Schleimpilze, Pilztiere). In: Engler, A., Prantl, K. (Eds.), Die natürlichen Pflanzenfamilien, Vol. 2, Engelmann, Leipzig, pp. 304–339.
Keller, H.W., 2012. Myxomycete history and taxonomy: highlights from the past, present, and future. Mycotaxon 122, 369–387.
Keller, H.W., Braun, K.L., 1999. Myxomycetes of Ohio: Their Systematics, Biology, and Use in Teaching. Ohio Biological Survey, Columbus, OH.
Kowalski, D.T., 1966. New records of myxomycetes from California I. Madroño 18 (5), 140–142.
Kowalski, D.T., 1967. New records of myxomycetes from California II. Madroño 19 (2), 43–46.
Kowalski, D.T., 1973. New records of myxomycetes from California V. Madroño 22 (2), 97–100.
Kowalski, D.T., Curtis, D.H., 1968. New records of myxomycetes from California III. Madroño 19 (7), 246–249.
Kowalski, D.T., Curtis, D.H., 1970. New records of myxomycetes from California IV. Madroño 20 (7), 377–381.
Kretzschmar, M., Kuhnt, A., Bonkowski, M., Fiore-Donno, A.M., 2016. Phylogeny of the highly divergent Echinosteliales (Amoebozoa). J. Eukaryot. Microbiol. 63, 453–459.
Krzemieniewska, H., 1960. Sluzowce Polski na tle flory sluzowców Europejskich. Polska Akademia nauk Instytut Botaniki, Warszawa.
Lado, C., 2001. NOMENMYX. A nomenclatural taxabase of Myxomycetes. Cuadernos de trabajo de Flora micológica Ibérica 16, 1–221.
Lado, C., 2005–2016. An online nomenclatural information system of Eumycetozoa. Available from: http://www.nomen.eumycetozoa.com

Lado, C., Eliasson, U., Stephenson, S.L., Estrada-Torres, A., Schnittler, M., 2005. (1688–1691) Proposals to conserve the names Amaurochaete against Lachnobolus, Ceratiomyxa against Famintzinia, Cribraria Pers. against Cribraria Schrad. ex J.F. Gmel. and Hemitrichia against Hyporhamma (myxomycetes). Taxon 54 (2), 543–545.

Lado, C., Pando, F., 1998. (1340) Proposal to conserve the name Reticularia (myxomycetes) with a conserved type. Taxon 47, 453–454.

Lakhanpal, T.N., Mukerji, K.G., 1981. Taxonomy of the Indian Myxomycetes. Bibliotheca Mycologica 78, 1–531.

Leers, J.D., 1789. Flora herbornensis exhibens plantas circa Herbornam Nassoviarum crescentes. Impensis C.F. Himburgi, Berlin.

Leontyev, D.V., 2015. The prospects and perspectives of the phylogenetic system of myxomycetes (Myxogastrea). Ukr. Bot. J. 72 (2), 147–155.

Leontyev, D.V., 2016. The evolution of sporophore in Reticulariaceae (myxomycetes). Ukr. Bot. J. 73 (2), 178–184.

Leontyev, D.V., Schnittler, M., Moreno, G., Stephenson, S.L., Mittchell, D.W., Rojas, C., 2014. The genus Alwisia (myxomycetes) revalidated, with two species new to science. Mycologia 106 (3), 936–948.

Li, Y., Li, H.-Z., Wang, Q., Chen, S.-L., 2008a. Flora Fungorum Sinicorum. Myxomycetes I: Ceratiomyxales, Echinosteliales, Liceales, Trichiales. Science Press, Beijing.

Li, Y., Li, H.-Z., Wang, Q., Chen, S.-L., 2008b. Flora Fungorum Sinicorum. Myxomycetes II. Physarales, Stemonitales. Science Press, Beijing.

Link, J.H.F., 1833. Handbuch zur Erkennung der nutzbarsten und am häufigsten vorkommenden Gewächse 3, 405–422, 432–433, der Haude und Spenerschen Buchhandlung (S.J. Joseephy), Berlin.

Linnaeus, C., 1753. Species Plantarum. Impensis Laurentii Salvii, Holmiae [Stockholm].

Linnaeus, C., 1763. Species Plantarum, second ed. Impensis Direct, Laurentii Salvii, Holmiae [Stockholm].

Lister, A., 1894. A Monograph of the Mycetozoa. British Museum, London.

Lister, A., 1911. A Monograph of the Mycetozoa, second ed. British Museum, London, revised by G. Lister.

Lister, G., 1916. The life-history of Mycetozoa, with special reference to *Ceratiomyxa*. J. R. Microsc. Soc. 1916, 361–365.

Lister, A., 1925. A Monograph of the Mycetozoa, third ed. British Museum, London, revised by G. Lister.

Macbride, T.H., 1899. The North-American Slime-Moulds. Macmillan, London.

Macbride, T.H., 1922. The North-American Slime-Moulds, second ed. Macmillan, London.

Macbride, T.H., Martin, G.W., 1934. The Myxomycetes. Macmillan, New York, NY.

Martin, G.W., 1949. Class Myxomycetes. North American FloraNew York Botanical Garden, New York, NY, 1.

Martin, G.W., 1961. The systematic position of the Myxomycetes. Mycologia 52, 119–129, [1960].

Martin, G.W., 1966. The genera of myxomycetes. Stud. Nat. Hist. Iowa Univ. 20 (8), 3–32.

Martin, G.W., Alexopoulos, C.J., 1969. The Myxomycetes. University of Iowa Press, Iowa City.

Martin, G.W., Alexopoulos, C.J., Farr, M.L., 1983. The Genera of Myxomycetes. University of Iowa Press, Iowa City.

Massee, G., 1892. A Monograph of the Myxogastres. Methuen & Company, London.

McNeill, J., Barrie, F.R., Buck, W.R., Demouilin, V., Greuter, W., Hawksworth, D.L., Herendeen, P.S., Knapp, S., Marhold, K., Prado, J., Prud'Homme van Reine, W.F., Smith, G.F., Wiersema,

J.H., Turland, N.J., 2012. International Code of Nomenclature for algae, fungi, and plants (Melbourne Code). Regnum Vegetabile, A.R.G. Gantner Verlag KG, pp. 154.

Micheli, P.A., 1729. Nova plantarum genera iuxta Tournefortii methodum disposita. Typis Bernardi Paperinii, Florentiae.

Moreno, G., Singer, H., Sánchez, A., Illana, C., 2004. A critical study of some Stemonitales of North American herbaria and comparison with European nivicolous collections. Bol. Soc. Micol. Madrid 28, 21–41.

Morgan, A.P., 1893a. The myxomycetes of the Miami Valley, Ohio I. J. Cincinnati Soc. Nat. Hist. 15, 127–143.

Morgan, A.P., 1893b. The myxomycetes of the Miami Valley, Ohio II. J. Cincinnati Soc. Nat. Hist. 16, 13–36.

Morgan, A.P., 1894. The myxomycetes of the Miami Valley, Ohio III. J. Cincinnati Soc. Nat. Hist. 16, 127–156.

Morgan, A.P., 1896. The myxomycetes of the Miami Valley, Ohio IV. J. Cincinnati Soc. Nat. Hist. 18, 36–45.

Morgan, A.P., 1900. The myxomycetes of the Miami Valley, Ohio V. J. Cincinnati Soc. Nat. Hist. 19, 147–166.

Nannenga-Bremekamp, N.E., 1967. Notes on myxomycetes. XII. A revision of the Stemonitales. Proc. Kon. Ned. Akad. Wet. Ser. C 70 (2), 201–216.

Nannenga-Bremekamp, N.E., 1974. De Nederlandse Myxomyceten. Biblioth. Kon. Nederl. Natuurhist. Ver. 18, 1–440.

Nannenga-Bremekamp, N.E., 1982. Notes on myxomycetes. XXI. The use of polarized light as an aid in the taxonomy of the Trichiales. Proc. Kon. Ned. Akad. Wet. Ser. C 85 (4), 541–562.

Nannenga-Bremekamp, N.E., 1985. Notes on myxomycetes XXII. Three new species, two new families and four new combinations. Proc. Kon. Ned. Akad. Wet. Ser. C 88 (1), 121–128.

Nannenga-Bremekamp, N.E., 1991. A Guide to Temperate Myxomycetes. Biopress Ltd., Bristol.

Nelson, R.K., Scheetz, R.W., 1975. Swarm cell ultrastructure in *Ceratiomyxa fruticulosa*. Mycologia 67, 733–740.

Nelson, R.K., Scheetz, R.W., 1976. Thread phase ultrastructure in *Ceratiomyxa fruticulosa*. Mycologia 68, 144–150.

Neubert, H., Nowotny, W., Baumann, K., 1993–2000. Die Myxomyceten Deutschlands und des angrenzenden Alpenraumes unter besonderer Berücksichtigung Österreichs. 3 vols. Karlheinz Baumann Verlag, Gomaringen.

Novozhilov, Y.K., 1993. Definitorium Fungorum Rossiae. Divisio Myxomycota. Fasc. 1. Classis Myxomycetes. Institutum Botanicum nomine V.L. Komarovii, Petropolis. Nauka.

Novozhilov, Y.K., Mitchell, D.W., Schnittler, M., 2003. Myxomycete biodiversity of the Colorado Plateau. Mycol. Prog. 2 (4), 243–258.

Novozhilov, Y.K., Schnittler, M., 2000. A new coprophilous species of Perichaena (Myxomycetes) from the Russian Artic (the Taimyr Peninsula and the Cruckchi Penninsula). Karstenia 40, 117–122.

Novozhilov, Y.K., van Hooff, H., Jagers, M., 2015. Trichioides iridescens, a new genus and new species of the Trichiaceae (myxomycetes). Mycol. Prog. 14 (1018), 2–7.

Novozhilov, Y.K., Zemlyanskaya, I.V., Schnittler, M., Stephenson, S.L., 2008. Two new species of Perichaena (Myxomycetes) from arid areas of Russia and Kazakhstan. Mycologia 100, 816–822.

Olive, L.S., 1970. The Mycetozoa: a revised classification. Bot. Rev. 36, 59–89.

Olive, L.S., 1975. The Mycetozoans. Academic Press, New York, NY.

Olive, L.S., Stoianovitch, C., 1966. A new two-spored species of *Cavostelium* (Protostelida). Mycologia 58, 440–451.

Panckow, T., 1654. Herbarium Portatile. In verlegung de Historis, Berlin.
Persoon, P.H., 1794. Neuer Versuch einer systematischen Eintheilung der Schwämme. Neues Mag. Bot. 1, 63–128.
Poulain, M., Meyer, M., Bozonnet, J., 2011. Les Myxomycètes. 2 vols. Fédération mycologique et botanique Dauphiné-Savoie, Sevrier.
Ride, W.D.L., 1999. International Code of Zoological Nomenclature (ICZN), fourth ed. The International Trust for Zoological Nomenclature. Natural History Museum, London.
Ross, I.K., 1973. The Stemonitomycetidae, a new subclass of myxomycetes. Mycologia 65 (2), 477–484.
Rostafiński, J., 1873. Versuch eines Systems der Mycetozoen. Inaugural-dissertation, Strassburg.
Rostafiński, J., 1874. Śluzowce (Mycetozoa) Monografia. Pamietn. Towarz. Nauk. Sci. Paryzu 5, 1–215.
Rostafiński, J., 1875. Śluzowce (Mycetozoa) Monografia. Pamietn. Towarz. Nauk. Sci. Paryzu 6, 216–432.
Rostafiński, J., 1876. Śluzowce (Mycetozoa) Monografia. Pamietn. Dodatek. Towarz. Nauk. Sci. 8, 1–43.
Ruggiero, M.A., Gordon, D.P., Orrell, T.M., Bailly, N., Bourgoin, T., Brusca, R.C., Cavalier-Smith, T., Guiry, M.D., Kirk, P.M., 2015. A higher level classification of all living organisms. PLoS One 10 (6), e0130114.
Scheetz, R.W., 1972. The ultrastructure of *Ceratiomyxa fruticulosa*. Mycologia 64 (1), 38–53.
Scheetz, R.W., Nelson, R.K., Carlson, E.C., 1980. Scanning electron microscopy of *Ceratiomyxa fruticulosa*. Can. J. Bot. 58, 392–400.
Schnittler, M., Stephenson, S.L., Novozhilov, Y., 2000. Ultrastructural studies of *Barbeyella minutissima* (myxomycetes). Karstenia 40, 159–166.
Schrader, H.A., 1797. Nova Genera Plantarum. Pars prima, Lipsiae.
Scoble, M.J., 2004. Networks and their role in e-taxonomy. In: Wheeler, Q.D. (Ed.), The New Taxonomy. CRC Press, Boca Raton, FL, pp. 19–30.
Schoknecht, J.D., Keller, H.W., 1989. Peridial calcification in the Myxomycetes. In: Crick, R.E. (Ed.), Origin, Evolution, and Modern Aspects of Biomineralization in Plants and Animals. Plenum Press, New York, pp. 455–488.
Shadwick, L.L., Spiegel, F.W., Shadwick, J.D.L., Brown, M.W., Silberman, J.D., 2009. Eumycetozoa = Amoebozoa?: SSUrDNA phylogeny of protosteloid slime molds and its significance for the Amoebozoan supergroup. PLoS One 4, e6754.
Shchepin, O., Novozhilov, Y., Schnittler, M., 2014. Nivicolous myxomycetes in agar culture: some results and open problems. Protistology 8 (2), 53–61.
Stephenson, S.L., 2003. Fungi of New Zealand. Myxomycetes of New Zealand, vol. 3. Fungal Diversity Press, Hong Kong.
van Tieghem, P. van, 1880. Sur quelques Myxomycètes a plasmodes agrégé. Bull. Soc. Bot. France 27, 317–322.
Walker, L.M., Stephenson, S.L., 2016. The species problem in Myxomycetes revisited. Protist 167, 329–338.
Wallroth, C.W.F., 1831–1833. Flora cryptogamica Germaniae, 2 parts. Norimbergae.
Wheeler, Q.D., 2008. The New Taxonomy. CRC Press, Boca Raton, FL.
Winter, G., 1880. Die Pilze Deutschlands, Oesterreichs und der Schweiz I. Abtheilung: Schizomyceten, Saccharomyceten und Basidiomyceten. Rabenh. Krypt.-Fl. ed. 2, 1(1).
Wrigley de Basanta, D., Lado, C., 2005. A taxonomic evaluation of the stipitate *Licea* species. Fungal Divers. 20, 261–314.
Wrigley de Basanta, D., Lado, C., García-Martín, J.M., Estrada-Torres, A., 2015. Didymium xerophilum, a new myxomycete from the tropical Andes. Mycologia 107 (1), 157–168.
Yamamoto, Y., 1998. The Myxomycete Biota of Japan. Toyo Shorin Publishing Co., Tokyo.

Chapter 8

Ecology and Distribution of Myxomycetes

Yuri K. Novozhilov*, Adam W. Rollins, Martin Schnittler[†]**
**Komarov Botanical Institute of the Russian Academy of Sciences, St. Petersburg, Russia; **Lincoln Memorial University, Harrogate, TN, United States; [†]Institute of Botany and Landscape Ecology, Ernst Moritz Arndt University Greifswald, Greifswald, Germany*

INTRODUCTION

General patterns of community structure for terrestrial macroorganisms (plants, animals, and macrofungi) are relatively well known. However, similar information on microorganisms remains rather fragmentary and has been limited, to a large extent, by methodical difficulties relating to carrying out inventories due to their small size, their high reproductive productivity and dispersal capacity, overall abundance, and certain unique aspects of the life cycles, such as extremely resistant resting stages and the presence of both sexual or asexual life cycles. Fortunately, the mathematical methods of community ecology, developed to analyze and compare diversity, species abundance patterns and interactions between organisms and their physical, chemical, and biological environments (Jongman et al., 1995; Krebs, 1999; Magurran, 2004) are amenable to studies involving the fruiting bodies produced by myxomycetes.

Synecological studies of assemblages of myxomycetes were rare and rather general in nature until the 1980s, when the first relatively comprehensive ecological studies were carried out. These included studies of the myxomycetes associated with tropical vegetation types in Brazil (Maimoni-Rodella and Gottsberger, 1980) and the upland forests of southwestern Virginia in the United States (Stephenson, 1988). Further studies focused as well on the autecology of myxomycetes, analyzing the occurrence of fruiting bodies in relation to environmental gradients, including the first studies of niche breadth and niche overlap (Rojas et al., 2008; Schnittler, 2001a). In contrast to the sometimes clearly visible fruiting bodies, the trophic stages of myxomycetes are microscopic and/or have a rather hidden life style; therefore, our knowledge of the ecological requirements of amoeboflagellates and plasmodia and their trophic interactions is still limited (Eliasson, 1977b; Madelin, 1984a; Shchepin et al., 2014; Sunhede, 1973).

Myxomycetes: Biology, Systematics, Biogeography, and Ecology. http://dx.doi.org/10.1016/B978-0-12-805089-7.00008-1

The current species concept for myxomycetes is based almost entirely upon morphological features of the fruiting bodies and their spores, which allow identification to the species level (Lister and Lister, 1925; Martin and Alexopoulos, 1969; Schnittler and Mitchell, 2000; Schnittler et al., 2012; Stephenson, 2011). Since trophic stages of myxomycetes (amoeboflagellates and plasmodia) display little or no meaningful taxonomic characters, studies of fruiting bodies from field collections or moist chamber cultures is usually the only way to tell species apart. Fruiting bodies can be collected, air-dried, and preserved in much the same manner as regular herbarium specimens of other groups of organisms, such as vascular plants, bryophytes, and macrofungi. Fruiting bodies or colonies of fruiting bodies provide a basic unit to estimate species richness and species abundance because they are extractable, identifiable, and quantifiable. In contrast to most other groups within the Amoebozoa, these features allow researchers to accumulate data on occurrence, ecology, and geography of myxomycetes. Examples of such data can be accessed by using such web sites as http://slimemold.uark.edu/ or http://www.gbif.org/

Approximately 60% of all morphospecies can be detected in the field with the naked eye or a simple hand lens. The corticolous species with minute fruiting bodies (e.g., species of *Licea* or *Echinostelium*) can reliably be detected only in the laboratory using the moist chamber culture technique and examination of substrate samples under a dissecting microscope. Only a very few species like *Echinostelium bisporum* require agar culture techniques and the use of a compound microscope, similar to the techniques developed for protosteloid amoebae (Schnittler et al., 2015b; Spiegel et al., 2004). It has been hypothesized that some species of myxomycetes do not form fruiting bodies and can be detected only by metagenomic analysis of environmental DNA (for a description of the first studies of this kind, see Clissmann et al., 2015; Fiore-Donno et al., 2016; Hoppe and Schnittler, 2015; Kamono et al., 2013). If these methods become more widely available and barcoding approaches, such as the first study of Feng and Schnittler (2017) become more common, this will result in databases that allow a reliable assignment of operational taxonomic units (OTUs) in relation to existing species. This would result in a tremendous expansion of our knowledge of myxomycete ecology. Until recently, ecological studies of myxomycetes were limited by the necessity to detect and identify fruiting bodies. This limitation must be kept in mind when examining methodological approaches and reports on the current state of knowledge about myxomycete ecology, which is the task of this chapter.

METHODS OF STUDY

Field Surveys

Field-based surveys of myxomycetes must be planned to account for seasonal and environmental variation across the study area being considered to avoid artifacts associated with phenological variation in sporulation activity. In the tropics, field surveys appear to be most productive during the rainy season. Tran et al. (2006) examined fruiting phenology across 5 study sites in northern Thailand and recovered 62 species during the rainy season but only 2 species during the dry season. In contrast, productivity in broadleaf deciduous forests in eastern North America seems to peak around August (Stephenson, 1988); in temperate Europe, the peak shifts toward autumn or late autumn (Schnittler and Novozhilov, 1998). High-latitude sites appear to be most productive during the narrow window associated with the growing season. Completely different is the phenology of nivicolous myxomycetes, with a peak fruiting period during the time of spring snow melt (Ronikier and Ronikier, 2009; Schnittler et al., 2015a).

For wood-inhabiting species, an intense survey conducted during the peak season (under typical conditions) can produce a representative species list of the most common taxa for a particular area, if all major microhabitats are systematically investigated. An experienced collector usually can locate 50–70 collections during a typical 6-h field day in a forested area in the temperate or boreal zone. As such, an intensive survey of approximately 12–16 days can produce a fairly complete species list in these zones. However, tropical areas require a much greater sampling effort. For example, a project carried out by the senior author in southern Vietnam required three surveys over as many years (representing 276 h of collecting) to produce a fairly complete species list (Novozhilov et al., 2017a).

Field surveys in arid regions are not very effective. In contrast, especially for winter-cold deserts lacking succulent plants, moist chambers are very productive. Almost all specimens found in the winter-cold deserts and steppes of Kazakhstan (Schnittler and Novozhilov, 2000), western Mongolia (Novozhilov and Schnittler, 2008), the Lower Volga River basin of Russia (Novozhilov et al., 2006) and the Tarim Basin in China (Schnittler et al., 2013) were obtained from moist chamber cultures.

Annual shifts in species abundance have been documented for many species. This is exemplified by data from the monsoon tropical forests of southern Vietnam where, in spite of comparable field efforts during visits across 3 years in the same season, the abundance of myxomycete fruiting bodies varied considerably (Novozhilov et al., 2017a). Abundance of the fruiting bodies of nivicolous myxomycetes in alpine and subalpine vegetation has also been documented to be extremely variable between years, ranging from nearly zero to hundreds of colonies (Schnittler et al., 2015a). For this reason, at least two independent field surveys separated in time are necessary to ensure the recording of all phenological groups.

Various designs and placement of study plots have been used for myxomycetes. The plots can be arranged systematically along transects or collecting can be carried in completely randomized plots. Opportunistic collecting without the use of plots has been used in some studies. A survey of the literature reveals that the number of transects and plots has varied widely across different studies. For example, Stephenson (1988), in a study of myxomycete ecology carried out in upland forest communities in southwestern Virginia, used 0.1 ha (20 × 50 m) permanent plots established within each of five different forest types. Each plot was (1) representative of the particular forest community with respect to both vegetation and site conditions, (2) consisted of a relatively homogenous unit of vegetation located in an area of essentially uniform topography, and (3) there was no evidence of appreciable recent disturbance by man or other causes. Rollins and Stephenson (2013), when surveying grasslands across the central United States, used two transects, which were established at each study site. Five samples were collected from each predetermined available microhabitat at approximately 5 m intervals along each transect. In the Maquipucuna Cloud Forest Reserve in Ecuador (Stephenson et al., 2004a), substrate samples for moist chamber cultures were collected along a transect ca. 200 m in length, which was established in each study area. Within a distance of 10 m, several substrate pieces of one type were collected and pooled to produce a sample of approximately 15–25 g weight. Sampling was repeated along the transect to obtain a series of substrate samples of each type. Schnittler (2001b), in his study of myxomycetes associated with liverworts in the Neotropics, collected substrate samples over a distance of approximately 10 m along a transect of 200–250 m in length and 6 m in width.

Novozhilov et al. (2013b), in a survey of nivicolous myxomycetes in the northwestern Caucasus Mountains, used a design consisting of 18 circular plots of ca. 50 m diameter arranged along an elevational transect, which were revisited for 4 years (Schnittler et al., 2015a). Every year the whole transect was examined for several days and every myxomycete fruiting was recorded, including those in a weathered or severely decayed condition. Rojas and Doss (2014), in ecological study of myxomycetes in Costa Rica, used four transects running south to north that were established perpendicular to the sharp ecotone between forested and nonforested patches in the study area and placed 100 m from each other. Along each transect, four circular plots 10 m in radius were established every 50 m. Two of these study points were located inside the forested habitat zones, one at the forest border, and one in the nonforested (pasture) habitat.

Moist Chamber Cultures

The moist chamber culture technique, first described by Gilbert and Martin (1933), can supplement the information obtained from collecting specimens that have fruited in the field under natural conditions and thus represents a major source of data for diversity studies in myxomycetes (Eliasson and Lundqvist, 1979; Härkönen, 1977, 1978; Härkönen and Ukkola, 2000; Stephenson, 1985; Wrigley de Basanta et al., 2002). First of all, this technique allows a better detection of species with minute fruiting bodies via the use of a dissecting microscope. Moreover, it helps to overcome biases that result from the sporadic and ephemeral nature of myxomycete fruiting. However, the technique itself has an inherent bias since it favors species with short developmental cycles and small fruiting bodies. Species with large compound fruiting bodies (e.g., species in the genera *Tubifera* or *Lycogala*) appear much more rarely in moist chambers than in the field. Nevertheless, the moist chamber culture technique is especially helpful to study ecological groups with very small fruiting bodies, such as corticolous (those species associated the bark surface of living trees) or coprophilous (fimicolous) myxomycetes (those species associated with the weathered dung of herbivorous animals), where fruiting bodies are rarely encountered in the field. The use of moist chamber cultures is particularly important for supplementing surveys in extreme environments, such as arid habitats, high-elevation montane sites, and high-latitude regions where fruiting bodies rarely develop under natural conditions.

The moist chamber culture technique is both simple to use and effective, requiring only Petri dishes, filter paper, and a good dissecting microscope. The technique does not require sterile conditions and can easily be modified to suit the researcher's situation. For example, if filter paper is not available, toilet paper works as well, and if Petri dishes are not available, they can be replaced by plastic containers (like those used for food products). For comparable results, the modifications should account (as best as possible) for the water holding capacity of the paper and the amount of substrate material placed in each container. Substrate samples are collected in the field and placed in small paper bags, transported to the laboratory, and allowed to air-dry.

Typically, 1–10 g of dry substrate material is placed in each of a series of 100 × 15 mm plastic Petri dishes in which the bottom has been lined with filter paper. Water is added to submerge the substrate material, which is then allowed to soak for approximately 24 h, at which time the pH of the supernatant is determined with a flat-plate electrode and the excess water is decanted. A dissecting microscope is used to monitor the cultures for the occurrence of myxomycete fruiting bodies. The monitoring schedule and total duration varies among published studies but cultures are generally maintained for 2–3 months under diffuse daylight at room temperature (typically 22–23°C).

When myxomycete fruiting bodies are observed, they are collected and preserved in the same manner as field specimens. A small piece of the substrate containing mature fruiting bodies is glued to a paper tray, which is placed into a small pasteboard box. Fruiting bodies of minute species are immediately preserved in polyvinyl lactophenol or glycerol gelatine when calcareous structures are present. The occurrence of one or more fruiting bodies of one species considered to have developed from a single plasmodium is recorded as one collection. Optionally, for statistical evaluations, the number of fruiting bodies in a moist chamber culture may be estimated.

Traditionally, the moist chamber culture technique has been used to examine the species associated with the microhabitats represented by (1) forest floor litter, including leaf litter, twigs, decaying conifer cones, seeds and fruits; (2) the bark surface of living trees and shrubs; and (3) various types of woody debris.

However, as ecological studies of myxomycetes have increased across the tropics, in arid areas, and in high-latitude regions, this set of substrates has been expanded to include a number of additional microhabitats, including aerial litter (dead but still attached plant parts above the ground), inflorescences of large tropical giant herbs (especially members of the family Heliconiaceae; Schnittler and Stephenson, 2002), the litter of fleshy herbaceous plant parts, such as shoots and leaves of *Musa* sp. and *Colocasia* sp. (Novozhilov et al., 2017a), living leaves of understory plants with a cover of leafy liverworts present (Schnittler, 2001b), the bark of lianas (Wrigley de Basanta et al., 2008), the litter of forbs (non-grass broadleaf plants) and grasses (Rollins and Stephenson, 2013), the litter of succulent plants (Estrada-Torres et al., 2009) and weathered dung of herbivorous mammals (e.g., camels, cows, horses, sheep, reindeer, moose, and various rodents) and birds, such as grouse (Eliasson and Keller, 1999), and soil (Stephenson et al., 2011). Examples of many of these microhabitats are illustrated in Figs. 8.1 and 8.2.

Despite its importance, the moist chamber technique does have some limitations. Since the technique creates nonnatural conditions, such as a long-lasting period of moisture, it may produce records of species from dormant stages that would not or very rarely have formed fruiting bodies under natural conditions. In addition, the limited amount of resources available in these microcosms does not appear to be sufficient for species with large compound fruiting bodies, since the latter rarely appear under such circumstances. Perhaps, for similar reasons lignicolous myxomycetes that are often abundant on coarse wood debris in the field rarely appear in moist chamber cultures. For example, in studies across southern Vietnam (Novozhilov et al., 2017a) and the Russian Far East (Novozhilov et al., 2017b), the ratio of records of *Hemitrichia serpula* from the field and in moist chamber cultures were 10:1 and 63:0, respectively. Although this kind of bias cannot be ruled out, comparisons of results obtained from moist chamber cultures and field surveys in intensively studied areas have not revealed such large discrepancies. Most species were found both in the field and in moist chamber cultures, but most were much more commonly recorded from one or the other.

Another limitation of the moist chamber culture technique is that it can be time intensive, especially when trying to quantify the number of fruiting bodies and searching for minute species. Myxomycetes, which produce minute fruiting bodies, account for almost one third of all described taxa among the corticolous myxomycetes and are not always easily detected even with a dissecting microscope (Mitchell, 1977, 1978). Even if the magnification of an excellent instrument potentially allows the detection of nearly every fruiting body, the field of view becomes very small and screening the whole surface of a Petri dish (about 60 cm^2 for standard dishes of 9 cm diameter) can be laborious. Therefore, the smallest myxomycetes such as *Echinostelium bisporum* (two-spored fruiting bodies of 18–25 μm total height) or *E. lunatum* (4–8 spores) usually escape detection by this method and can be recovered only in agar cultures if these are examined with a compound microscope.

FIGURE 8.1 Microhabitats for myxomycetes. (A) *Opuntia* sp., a specific microhabitat for succulenticolous myxomycetes in deserts. (B) Weathered dung of herbivorous animals. (C) Coarse woody debris of a conifer in a taiga forest. (D) Accumulation of leaf litter in a broadleaf deciduous forest. (E) Mosses in forest-tundra covered by immature fruiting bodies of *Fuligo muscorum*. (F) Colony of fruiting bodies of *Colloderma robustum* on a single patch of the liverowort *Mylia taylorii* (Hook.) Gray.

FIGURE 8.2 Additional microhabitats for myxomycetes. (A) David Mitchell collecting specimens of the bark of a juniper for use in moist chamber cultures. (B) Bark of lianas. (C) Decaying floral parts of *Heliconia* sp., which provide a microhabitat for myxomycetes. (D) Aerial litter of a bird's nest fern (Asplenium nidus s.l.) in a lowland monsoon forest of southern Vietnam.

Agar Cultures

The use of agar cultures represents the classic approach for the physiological and autecological studies of myxomycetes. The first spore-to-spore cultures of myxomycetes were reported by Lister (1901). At present, only about 100 myxomycete morphospecies (about 10% of morphospecies known worldwide) have been cultured from spore-to-spore on agar media (Haskins and Wrigley de Basanta, 2008; Liu et al., 2010). The method has numerous variations, but in general spores collected from mature fruiting bodies are sown on 0.75% water agar at pH 7.0. Parafilm is used to seal the Petri dishes, which are kept at room temperature (20–23°C) in diffuse light. Upon germination, amoeboflagellates are transferred to water agar (ranging from 1.5% to 5%), then to carrot agar (Indira, 1968) or to SM agar (1.5% agar on SM broth; Kerr and Sussman, 1958). Bacteria incidentally sown along with the spores serve as a food source. Crushed autoclave-sterilized oat flakes can also be used as an additional food source for plasmodia.

Although the percentage of positive cultures is lower when compared to moist chamber cultures, the agar culture method can be extremely effective for detecting minute myxomycetes (Schnittler et al., 2015b) and protostelids (Spiegel et al., 2004), due to the relatively weak, resource-poor agar that keeps competing fungal growth at a minimum. However, myxomycetes with larger fruiting bodies and/or colonies are usually absent in these cultures or develop only a few fruiting bodies.

Pieces of the substrate used in agar cultures need to be small (e.g., 1 × 2–4 mm) and comparatively tall (1–2 mm), since only the outline of myxomycetes can be detected with the use of a compound microscope. In addition, it is useful to arrange a fixed number of pieces in a line to assist the observer in maintaining orientation while checking the cultures. For example, Schnittler et al. (2015b) used 4–6 substrate pieces in 4 lines, giving a total of 120–200 mm outline to scan for the presence of myxomycetes. A modification of this method was recently used to examine the trophic stages of myxomycetes present in wood, utilizing half-strength corn meal agar overlaid with a dilute suspension of bacteria (*Aerobacter* sp.) and 2.0 mL of sterile water (Taylor et al., 2015). This study revealed the presence of the amoeboflagellates of myxomycetes in 47% of the wood slivers examined but found no evidence of plasmodia.

Agar cultures also appear to represent a promising method to study the myxomycetes associated with soil (Indira, 1968; Stephenson et al., 2011; Warcup, 1950). A modified version of the "Cavender method" (Cavender and Raper, 1965) has been used in many culture-based studies that have attempted to characterize and quantify myxomycete occurrence in the soil microhabitat (Feest, 1985, 1987; Kerr, 1994; Stephenson and Landolt, 1996). Feest and Madelin (1985) used culture plates prepared with half-strength corn meal agar inoculated with a suspension of washed cells of *Saccharomyces cerevisiae*. These studies commonly count plasmodium-forming units (PFUs), thus quantifying the occurrence of all groups of soil mycetozoans. It should be noted, however, that this method probably produces an underestimation (possibly several fold), since for sexual myxomycetes two cells are required to form a plasmodium, plasmodia can fuse, and most species cannot be isolated using culture-based methods. This approach also introduces a bias toward members of the Physarales, which are more readily grown in culture.

Metacommunity Analysis

In contrast to all other detection techniques, this is the only approach that is able to detect the trophic stages of myxomycetes. A study exclusively focusing on actively living protists in soil, employing a transcriptomic approach (RNA) revealed the Amoebozoa as one of the major terrestrial protist groups, with myxomycetes accounting for 25% of all ribotypes in the soil sample, thus representing the single largest component of total protozoan soil biodiversity (Urich et al., 2008). Comparatively easier to carry out are studies targeting the more stable DNA, although dormant stages (such as inactive spores) will be detected as well. Primer specificity is crucial; in a metagenomics study in which universal primers targeting the whole SSU were used (Lesaulnier et al., 2008), myxomycetes were absent from the investigated soils. This discrepancy is likely the result of the high divergence of SSU rDNA sequences in both size and sequence in different species of myxomycetes (Fiore-Donno et al., 2010). Even primers known to be nearly universal for the Amoebozoa often do not amplify myxomycetes (Cavalier-Smith et al., 2015). In addition, the SSU gene contains a large number of introns of varying lengths, which may take up to 70% of the sequence (Feng and Schnittler, 2015; Feng et al., 2016; Wikmark et al., 2007a,b). Given this situation, considerable effort has focused on the design of appropriate primers, in particular for DNA barcoding of species of myxomycetes (Fiore-Donno et al., 2005, 2010, 2013; Martin et al., 2003; Novozhilov et al., 2013b; Schnittler et al., 2017). Even with the appropriate primers, sequences resulting from PCR of environmental samples can be assigned to species only by comparison with sequences coming directly from fruiting bodies (Feng and Schnittler, 2017). Not surprisingly, the first studies employing ePCR (Clissmann et al., 2015; Fiore-Donno et al., 2016; Kamono et al., 2013) revealed a significant proportion of unknown sequences; these are clearly of myxomycete origin but do not cluster with sequences of known species. This proportion of "hidden diversity" continues to pose a serious methodological problem, and further research is needed to show to what extent the presence of these sequences can be explained by (1) technical errors (for instance, chimera formation), (2) lack of comparison sequences from fruiting bodies, or (3) or that they belonging to yet truly undiscovered taxa (possibly amoebal strains that have lost the ability to fruit without consequences for vegetative reproduction). This raises as well the question of the threshold for the assignment of sequences to OTUs, which may be derived from an analysis of existing sequences from fruiting bodies. A systematic barcoding approach is highly desirable, since for many morphospecies we do not yet have a sequence for even a single marker gene.

Data Evaluation for Diversity Studies in Myxomycetes

At the level of amoebal populations, at least the numbers of individual amoebae can be estimated. Until the publication of a method for the enumeration of myxomycetes in soil (Feest, 1987; Feest and Madelin, 1985; Madelin, 1990), no data on the numbers of myxomycete propagules in any substrate had been reported. In their papers, Feest and Madelin (1985) showed it was possible to obtain numerical data for soil using a unit (the plasmodium-forming unit, PFU) analogous to the clone-forming unit of dictyostelids used by Cavender and Kawabe (1989). Based on the ability to form plasmodia, which distinguishes myxomycete amoeboflagellates apart from free-living amoebae in soil, Feest and Madelin (1985) used the presence or absence of plasmodia as an unambiguous indicator of the presence of myxomycetes in a particular substrate. As such, this method permits us to estimate a total abundance of myxomycetes in soil but is time-consuming. Since amoeboflagellates or plasmodia usually cannot be determined to the species level, this approach does not work for estimating the abundance of separate species.

As such, virtually all data on myxomycete diversity is derived from fruiting body-based surveys, where fruiting bodies can be used as an abundance unit because they can be collected, identified to species and quantified. The great advantage of this approach is the simple way to collect specimens. Colonies of fruiting bodies simply can be dried, glued along with the substrate in small boxes, such as matchboxes and stored in the same manner as other types of herbarium specimens. Eliasson (1981) and Stephenson (1988) considered a "collection" as one or more colonies of fruiting bodies that originated from a single plasmodium. In these studies, colonies of fruiting bodies were regarded as separate collections if they were separated by a distance of at least 30 cm. With such approaches, we can apply the methods of community ecology, used in diversity studies of various groups of organisms, as well as for myxomycetes. However, one central term is difficult to define for myxomycetes—the individual as the basic unit to count abundances (Novozhilov et al., 2000). Two kinds of data have been widely used in diversity studies of myxomycetes. These are presence/absence data, leading to checklists, and abundance data, based on the number of records for a particular species (Schnittler, 2001a). For the latter approach, usually one or several colonies of fruiting bodies developing on the same piece of substrate is regarded as an "individual," taking into account that (1) plasmodia can fuse or separate into pieces and (2) that a piece of a phanero- or aphanoplasmodium usually segregates into a colony of several gregarious fruiting bodies.

Most commonly used is the second approach, which involves expressing species abundance as the number of records (colonies of fruiting bodies assumed to have developed independently from each other). Stephenson et al. (1993) introduced a simple scale (often abbreviated ACOR), based on the proportion of a species in the total number of records with R: rare (<0.5%); O: occasional (0.5%–1.5%); C: common (1.5%–3%); and A: abundant (>3%). This scale is widely used in many modern studies of myxomycetes and permits comparison of different data sets if roughly the same approach (method of collecting, selection of substrates for moist chamber cultures, etc.) has been used.

Since many species of myxomycetes produce compound fruiting bodies in which individual fruiting bodies are still apparent, a third approach (in addition to presence/absence of species and counting numbers of collections per species) would be appropriate for diversity studies—the number of fruiting bodies (or fruiting body-equivalents) per species. Considered from an ecological point of view, myxomycete fruiting bodies are not actively living stages but instead serve solely as spore dispensers. Therefore, estimating spore numbers per species is probably the most accurate measure of abundance. This approach was proposed to measure the spore numbers for species cultivated in moist chamber cultures by Schnittler (2001a). In this approach, the number of spores per fruiting body was estimated for each species, using average fruiting body size and shape (e.g., half-globose, globose, and cylindrical) to calculate fruiting body volume, and spore diameters when assuming the densest package of spheres. The dimension of these values was confirmed by counting the spore numbers of a few selected fruiting bodies with a counting chamber (as used for determination of erythrocyte levels). However, this method is too laborious for large-scale applications and has not been used for subsequent studies.

The most popular abundance estimation is the number of records per species; for studies with a moist chamber culture component, the occurrence of one species in one moist chamber culture constitutes a collection (record). Less commonly used is weighted abundance, calculated by dividing the absolute number of fruiting bodies recorded in a particular moist chamber/colony by the mean value from all moist chambers with this species present (Schnittler, 2001a; Stephenson et al., 2004a). Consequently, the sum of all weighted abundances for all cultures with a species is equal to the number of records for this species. This approach can help to determine substrate preferences for a particular species.

For diversity studies, a number of standards are now used by most researchers. These are that (1) common and easily recognizable species are sometimes only recorded instead of being collected, whereas for rare and not easily recognizable species every colony of fruiting bodies will be preserved as a herbarium specimen to verify determinations; (2) the genus *Ceratiomyxa* is included due to its ecological equivalence to the true myxomycetes; (3) all fruiting bodies which share the same substrate and are clustered together (and thus are likely to have developed from one plasmodium) are recognized as one colony and are registered as one record; (4) specimens have to be identified to the lowest possible taxonomic level according to standard monographs (Martin and Alexopoulos, 1969) and various original descriptions from the literature, utilizing a morphospecies concept; and (5) nomenclature used for species follows the standard list of Lado (2005–16).

Many studies have shown that the diversity of myxomycetes in a particular vegetation type can be determined only if all major substrate types are sampled for moist chamber cultures, and records from moist chamber and from the field often complement each other (Novozhilov et al., 2017a; Schnittler et al., 2002). This is especially important in arid ecosystems, where most to nearly all species could be detected only in moist chamber cultures (Novozhilov and Schnittler, 2008; Schnittler and Novozhilov, 2000).

If abundances of species are recorded, then individual-based species accumulation curves (SAC) based on the Hulbert formula can also be constructed using programs like EstimateS version 9.0 (Colwell, 2013), the Vegan package of R (Oksanen et al., 2016) or SPADE (Chao et al., 2016); nonparametric estimators for expected species richness could also be calculated (Chao et al., 2006). Most commonly used are the Chao1 (for abundance data) or the Chao2 (for replicated incidence data) estimators (Gotelli and Colwell, 2011). As a second approach, the SAC can be fitted with a hyperbolic function $y = ax/(b + x)$, where x is the number of samples, y represents the number of species recorded, and the parameter a refers to the maximum number of species to be expected (Unterseher et al., 2008).

To characterize assemblages of myxomycetes and to make comparisons among assemblages, in addition to species richness (S), evenness can be quantified and expressed with the respective indices (Magurran, 2004). Typically, assemblages of myxomycetes show low evenness values, since a few very common species are accompanied by many rare ones, even if the sampling effort is extensive (Novozhilov et al., 2017a; Stephenson, 1988). The most often used is the Shannon index $H' = -\Sigma p_i \ln p_i$, where p_i is the relative abundance (the proportion of the total number of individuals or records represented by the ith species) is a heterogeneity index, influenced by both species richness and evenness. Often used to describe evenness of a community (as species' dominance) is Simpson's dominance index $D = 1/\sum p_i^2$ (see the textbooks of Krebs, 1999; Magurran, 2004 for details). A simple indicator of overall taxonomic diversity is the species per genus (S/G) ratio as described by Stephenson et al. (1993).

Diversity comparisons between study sites (geographical regions, habitats, or substrates) can be carried out by calculating Fisher's α, Shannon H', and the first three numbers of Hill's diversity series (Hill, 1973; Morris et al., 2014). In addition, a rank-abundance distribution (Whittaker, 1965) of the species can be constructed and the most probable abundance distribution model can be determined by testing different models, following Wilson (1991). For example, these include the null (fits the MacArthur broken stick model), preemption (fits the geometric series or Motomura model), log-normal, Zipf, and Mandelbrot models. The goodness of fit is usually compared with a χ^2 test, using the deviations between the observed and expected number of species for the various classes. The Vegan package of R with the functions *renyi* and *radfit*, provides the necessary algorithms (Dagamac et al., 2017b).

To compare species overlap between assemblages of myxomycetes from different habitats or substrates, the adjusted incidence-based index Chao-Sørensen (Chao et al., 2006), computed with EstimateS or the Vegan package of R, can be used. To test if any two species occur more (association) or less (avoidance) often together in samples (moist chamber cultures) than would be expected by chance, the Cole (Cole, 1949) or Brave indices of interspecific association and its standard errors can be computed (Novozhilov et al., 2006). Both indices are based on a 2×2 contingency table for presence and absence of a pair of species in one moist chamber culture, ranging from -1 (the species never occur together in the same moist chamber) to 1 (the species always occur together in the same moist chamber); significance can be tested by a χ^2 test.

In summary, the ideal (but time-consuming) approach for an all-taxon inventory of myxomycetes would consist of a field survey component, moist chambers cultures with substrate samples, agar cultures with discrete substrate pieces for minute myxomycetes, and additional metagenomic techniques. Generating databases for marker genes by the less error-prone Sanger sequencing of fruiting bodies is a precondition for reliable identification and interpretation of sequences from environmental PCR. In addition, we need more knowledge about the genetic, morphological and ecological diversity of the various morphospecies (Stephenson, 2011).

Microhabitat Descriptions

Many ecological studies of myxomycetes focus on defining habitat and microhabitat parameters to describe ecological niches. Although myxomycete distribution is almost undoubtedly limited by macroclimate and habitat (especially vegetation) requirements of particular species, there is strong evidence that microhabitat availability is the factor acting most strongly on species distribution. The term "microhabitat," used herein *sensu* Stephenson (1988), refers to a small section of a habitat that is spatially homogeneous in both biotic and abiotic factors (e.g., a section of the trunk of a shrub or small, shaded patches of litter with relatively the same thickness and moisture level (Schnittler, 2001a).

Ordination techniques (multivariate statistics) are a valuable approach for exploring and visualizing possible effects of micro- and macroenvironmental variables, but the latter still need to be quantified. Often used methods (Jongman et al., 1995; Oksanen et al., 2016; Ter Braak, 1987; ter Braak and Verdonschot, 1995) include nonmetric multidimensional scaling ordination (NMS), principal component analysis (PCA), canonical correspondence analysis (CCA), and correspondence analysis (CA). Rojas and Stephenson (2007) used NMS with the program PC-ORD (McCune and Mefford, 2006), Sørensen distances and a Monte Carlo test in a study of myxomycetes in oak forests of Cerro Bellavista in Costa Rica. Various microenvironmental parameters, including pH, substrate moisture, and diameter, height above the ground and canopy openness were recorded. PCA based on correlations also was performed with the same set of data to evaluate the relative importance of the different variables to the structure and composition of the assemblage of species present and to evaluate similarities with the previous ordination. CCA was carried out to assess the relative importance of recorded environmental factors on the assemblages of myxomycetes present in arid regions in Eurasia and in tropics (Novozhilov and Schnittler, 2008; Schnittler, 2001b). CA was used to assess the distributional relationships and microhabitat preferences of quantitatively important species of myxomycetes associated with the three types of temperate grasslands (tall grass, mixed grass, and short grass) found across the western central United States (Rollins and Stephenson, 2013). A discriminant analysis ordination was carried out to evaluate compositional differences in the assemblages that existed for height (ground and aerial) and diameter (large and small) of lianas in northern Thailand (Ko Ko et al., 2010). Dagamac et al. (2017b) used Bray-Curtis distances and PERMANOVA, which was performed to test for the influence of region versus elevation, using the functions *hclust*, *metaMDS*, and *adonis* of R.

The output of all these methods depends largely on the way in which environmental parameters are recorded. In his study of myxomycete ecology in Kazakhstan deserts, Schnittler (2001a) successfully utilized a modular system to describe microhabitat parameters, which was subsequently adopted for use in other studies (Novozhilov and Schnittler, 2008; Novozhilov et al., 2006). The parameters included the type of substrate (e.g., bark, litter, wood, and weathered dung of herbivorous animals), texture of the bark subdivided in five texture groups according to its physical features and the type ground litter (leafy litter, grass litter, and herbaceous litter). Environmental parameters included sampling height above the ground, light intensity, wind exposure, water retention, and pH of the substrate. Light intensity was estimated using five categories ranging from complete darkness (e.g., under stones) over various degrees of shade to full sunshine. Wind exposure was described using a scale of four categories reaching from fully sheltered to strongly exposed. To estimate the water retention of bark, sticks about 8 cm long were removed from the trunks of all sampled species of shrubs. Two or sometimes three sticks with typical bark structure were collected within the diameter range used for substrate sampling. Their ends were covered with nail polish to prevent water from soaking through wood pores. Sticks were dried, weighed, and then watered for 2 h, simulating strong rainfall, and then weighed again. The differences between dry and wet weight and the diameter and length of the sticks were used to calculate water retention in mL/dm^2 bark surface.

In a study of the assemblages of myxomycetes associated with cloud forests of the Maquipucuna Cloud Forest Reserve in the western Andes of Ecuador (Schnittler et al., 2002; Stephenson et al., 2004a), all myxomycete substrates were classified into seven categories. These were the bark of living trees and shrubs; decaying, formerly solid wood in various stages of decay; decaying woody but soft plant parts with a small diameter (e.g., lianas and shoots of climbing members of the family Araceae in tropics); decaying corolla parts and bracts of inflorescences of giant herbs, with all samples obtained from living plants above ground; leafy litter, either aerial (dead but still attached plant parts), or from the forest floor; litter of fleshy herbaceous plant parts, such as shoots of members of the family Heliconiaceae; and epiphyllic liverworts on living, mostly leathery leaves of understory shrubs and trees.

Schnittler et al. (2010) recorded 14 environmental parameters, including microclimatic parameters, in a study of the ecology of a highly specialized assemblage of ravine myxomycetes from sandstone gorges of the Saxonian Switzerland region (Germany). These parameters were (1) height of the myxomycete colony above ground (where the soil starts, precision 10 cm, range 0–80 m); (2) height of the locality above the valley bottom (with the rivulet taken as baseline, precision 0.5 m, range 0–80 m); (3) height of the rocks above the colony (all rocks not covered by soil, precision 0.5 m, range 0–80 m); (4) percent horizon opening, as seen with a fisheye lens (precision 1%, range 1%–100%); (5) exposition of the rock surface (in degrees from North = 0 degrees, precision 5 degrees, range 0–360 degrees); (6) inclination of the rock surface (using a ruler and a compass with inclination scale, precision 2 degrees, range 0–180 degrees); and (7) pH (measured 3 times on three samples of bryophytes covered by myxomycetes using an Orion 610 pH meter with a flat surface electrode, precision 0.05 units, range 2.5–4.5 units).

In the relevé areas (20 cm × 20 cm), percent coverage was determined for (8) the extent of the bryophyte-covered rock surface, (9) the extent of the algae-covered rock surface and (10) the extent of bare rock; and finally, and (11–14) the coverage of the four most common bryophyte species (all precision 5%, range 0%–100%). A total of 127 small-scale vegetation relevés showed that the assemblage occurs only in deep and narrow gorges (mean horizon openness 4.9%) on nearly vertical rocks (mean inclination 79°), and preferentially on rocks with a northern exposure (42% of all relevés).

Climatic parameters are very important for nivicolous myxomycetes (Schnittler et al., 2015a). Data on temperature, snow cover, and growth limits for amoeboflatellates (Shchepin et al., 2014) indicate that nivicolous myxomycetes appear to be abundant predators of below-snow microbial communities. Most probably, microclimatic conditions rather than food shortages represent decisive factors determining the abundance of fruiting bodies. According to this study, abundant fruitings can be expected if (1) the first sharp frosts in autumn (<-5°C) occur after the first snowfalls, (2) a continuous snow cover persists until hard night frosts cease in spring, and (3) snow packs reach a depth of at least 30 cm to shelter the ground from hard winter frosts, ensuring a sufficiently long period of snow melt providing the most suitable temperatures for growth of amoeboflagellates.

Environmental Factors and Ecological Niche

A major challenge for researchers is determining which factors can constrain species distributions. The main theoretical concept applied in such studies is the ecological niche of species (Broennimann et al., 2012; Soberón, 2007). Historically, such studies of myxomycete ecology have focused primarily on differences in microhabitats (Stephenson, 1988). However, in spite of the long history of such studies, our knowledge about the ecological requirements of species of myxomycetes remains very fragmentary, with only a few taxa studied (Eliasson, 1977a; Sunhede, 1973). Currently, methods for quantifying the environmental niche and estimating niche differences typically rely on ordination techniques (Broennimann et al., 2012). In myxomycete studies, the traditional or classical approach has been to calculate values for niche breadth *sensu* Whittaker et al. (1973) with the formula $NB = 1/\sum P_{ij}$ (Novozhilov et al., 2006; Schnittler, 2001a; Stephenson, 1988, 1989). The sum refers to the number of states for the environmental parameter defining a niche dimension, with P_{ij} as the proportion of species *i* associated with state *j* divided by the total abundance of species *i* across all states (Feinsinger et al., 1981). Many measurements of ecologically and evolutionarily relevant factors are possible at the species and individual levels of organization in ecological studies of myxomycetes. The main bias for all such studies is that at present only data on the occurrence of fruiting bodies are available. As abundance measures, numbers of records of fruiting bodies or their total abundance are used. Due to the difficulties in identification and detection, studies of how the trophic stages (amoeboflagellates and plasmodia) vary in their requirements for and tolerances of these factors in the field are lacking.

Ecological requirements for niche dimensions of species have been documented in a study of myxomycetes in a winter-cold desert of western Kazakhstan (Schnittler, 2001a) and the Caspian lowland (Novozhilov et al., 2006). The parameters used as resource states were the type of substrate, height above the ground, light intensity, wind exposure, substrate water retention, pH, the type of bark texture, and the development time of fruiting bodies (the 5 observation days for moist chamber cultures were 2, 6, 11, 21, and 40). The pH value, ranging from 4.5 to 8.5, was subdivided into classes of 0.5, whereas resource states for the other parameters were defined as explained earlier. In the same manner, niche overlap was computed, using the symmetrical index $NO_{ik} = \sum P_{ij}P_{ik}/\text{sqrt}(\sum P_{ij}^2 \sum P_{ik}^2)$, with P_{ij} and P_{ik} as the proportions of the *i*th resource state by the *j*th and *k*th species, respectively (Colwell and Futuyma, 1971; Levins, 1968), as modified by Pianka (1973). Values for niche breadth and niche overlap range from 0 to 1.

The ecology of the three morphospecies of *Ceratiomyxa* was studied by Rojas et al. (2008) to evaluate both direct and multivariable responses of the species to their environment. As such, microenvironmental parameters associated with fruiting bodies were measured or determined directly in the field. These including pH, substrate moisture, substrate type, diameter or thickness of the substrate as an index of substrate mass, height above the ground, and canopy openness as a way to evaluate light availability. The first two parameters were measured with a calibrated pH meter and an electronic moisture meter when possible, whereas the other parameters were determined as described in Stephenson et al. (2004a). Light availability was assessed with a system utilizing five discrete categories and which was transformed into a standard percentage scale comparable to the readings obtained with the moisture meter. The categorical arrangement used for ecosystem types and forest subtypes was considered part of the microenvironmental setting. The study revealed a clear separation of niches between the two tropical species, which was interpreted as an indication of resource partitioning within the genus. The cosmopolitan and most variable *Ceratiomyxa fruticulosa* had the broadest niche of the three species. Collectively, the niche overlap value between *Ceratiomyxa morchella* and *Ceratiomyxa sphaerosperma*, when considered in conjunction with the results of multivariate CDA analysis, suggested that these two species are more specialized than *C. fruticulosa*.

Acidity of Substrate (pH)

Information available in the literature consistently indicates that substrate pH may represent one of the most influential factors affecting myxomycete assemblages. As such, this parameter may be considered as one of the major drivers of myxomycete distribution. Some species (e.g., *Arcyria cinerea*) are regularly reported from substrates with a wide pH range, whereas others (e.g., *Paradiacheopsis fimbriata*) seem to be restricted to narrower pH range. Most of the studies in which an effort has been made to access the influence of substrate acidity on myxomycete distribution have been directed toward corticolous species (Everhart et al., 2008; Härkönen, 1977; Härkönen and Vänskä, 2004; Kilgore et al., 2009; Snell and Keller, 2003).

It is well known that low substrate pH appears to be a limiting factor for the growth and development of many species of myxomycetes. In general, deciduous trees characterized by moderately acidic bark display a more diverse assemblage of corticolous myxomycetes than coniferous trees with highly acidic bark (Ndiritu et al., 2009; Snell and Keller, 2003). The results from an experiment with simulated acid rain and the bark of *Quercus petraea* revealed a trend of decreasing number of corticolous myxomycetes with decreasing pH (Wrigley de Basanta, 2004). It was reported that the abundance of myxomycetes was affected by the constant acidification of the substrates; however, fewer individuals within each species appeared at pH 3.0 and 4.0. The author hypothesized that at low pH values food resources may become very limited or no longer available. Novozhilov et al. (1999) reported that most species of myxomycetes encountered in the Taimyr Peninsula appeared to have a relatively wide pH tolerance but displayed different pH optima.

As a general observation, corticolous species seem to have a narrower pH tolerance than wood-inhabiting species. For example, *Comatricha nigra* was collected 30 times on the bark of living *Larix* and *Picea* but only once on the bark of *Duschekia fruticosa*, and *P. fimbriata* was found only on *Larix* in a study carried out by Novozhilov et al. (1999). It is likely that for many corticolous members of the *Physarales* and *Echinosteliales* that are adapted to higher pH values, this is a major limiting factor. If so, this would be an explanation as to why they are rare in boreal coniferous forests. In contrast, compared with the results from surveys carried out in other geographic regions, the biota of myxomycetes encountered in deserts of Eurasia is rather poor but one of the most distinctive (Novozhilov and Schnittler, 2008; Schnittler, 2001a). In addition to obvious features, such as absence of trees and succulent plants or the harsh, arid conditions, the high (7.5–8.0) pH of almost all of the substrates available appears to be a limiting factor (Schnittler and Novozhilov, 2000; Stephenson et al., 2000). The reasons seem to be the often salty soils and the actual plant substrates, especially the abundant members of the *Chenopodiaceae* with high Na^+ and ash contents. These high pH values may be a reason for the absence of most members of the Stemonitales, Liceales, and Trichales except for the genus *Perichaena*. The only exception is represented by the bark of *Tamarix* (pH 4.6–7.2, $n = 5$), which supports a different assemblage of myxomycetes (e.g., *Comatricha laxa*, *Comatricha pulchella*, *Arcyria minuta*, and *Licea biforis*). The very basic pH, as recorded for most substrates by Novozhilov and Schnittler (2008), probably accounts for the distinctive assemblage of myxomycetes found in deserts. Still higher pH values (8.7–10.4) were recorded from the dead pith of cacti from the Sonoran Desert (Blackwell and Gilbertson, 1984). Humid regions, especially those with coniferous trees, possess more substrates with low pH values, as preferred by most members of the Stemonitales or certain species of *Licea* (Stephenson, 1989; Stephenson et al., 2000). Only one member of the Stemonitales (*Macbrideola oblonga*) was common in deserts, and this was the only member of this taxonomic order with an apparent optimum for higher pH values. Similarly, certain species of *Perichaena* are probably an exception among the order Trichiales, where most of the species inhabit decaying wood with low pH values. Reports for *Perichaena chrysosperma* (Stephenson, 1989) seem to fit this picture as is also the case for the records of *P. corticalis* reported in the study of Blackwell and Gilbertson (1980).

In studies of the assemblages of myxomycetes associated with the inflorescences of large Neotropical herbs, Schnittler and Stephenson (2002) reported that at least two of the species (*Physarum compressum* and *Ph. didermoides*) most commonly encountered preferred substrates with a very basic pH (8–9). Interestingly, both species are commonly found on the weathered dung of herbivorous animals, which is also characterized by an alkaline pH (Novozhilov, personal communication). The actual substrates for myxomycetes in inflorescences are the rapidly decaying floral parts enclosed by the massive, still living bracts. It is likely that the residues of floral nectar or secretions from extrafloral nectaries provide the basic resource for a rapidly developing community of yeasts and bacteria, as well as dung. A high density of these organisms is suggested by the frequent occurrence of myxobacteria in the moist chamber cultures prepared with floral parts and dung.

It is interesting to note that the assemblages of acidophilic corticolous species inhabiting the bark of conifers are clearly different between boreal and tropical forests. For example, the most common corticolous species on the bark of *Pinus sylvestris* in the taiga is *P. fimbriata*, whereas in the tropical montane forests of Vietnam it is *Cribraria confusa* (Novozhilov et al., 2017a).

It should be noted that almost all of the aforementioned studies focused on the synecological and biogeographical aspects of myxomycete assemblages. There are very few autecological studies that have examined the influence of environmental factors on a single species. According to Rojas et al. (2008), *C. morchella* is strongly limited to decaying wood with a very acidic pH, a rather rare habitat in tropical forests. Without exception, the habitat of the more than 100 records obtained for this species was the decorticated, moderately to strongly decayed wood of large logs with pH values between 3.0 and 4.5. This precludes any confusion with *C. sphaerosperma*, the second tropical species of the genus, which inhabits exclusively substrates with a pH above 6.5 (up to 8), and most often is found on the decaying shells or husks of fruits. In the context of ecological and distributional studies of *Barbeyella minutissima* (Schnittler et al., 2000), it was shown that this species tends to be associated with the liverworts (*Nowellia* sp. and *Cephalozia* sp.), which often occur together as pioneer species on decorticated coniferous logs, with a pH optimum 4.6–5.2. It has been suggested that these bryophytes can well serve as indicator species for *Barbeyella*.

Even species belonging to one genus can have different pH ranges. For example, *Licea minima* is abundant on substrates with a pH of 4.0–5.0 (optimum pH 4.3; Stephenson, 1989). In contrast, *Licea succulenticola* occurs on succulent plants with a pH of 10.0 (Mosquera et al., 2003). In general, members of the Physarales prefer neutral to basic substrates such plant litter, whereas most members of the Trichiales and Cribrariales occur more commonly on acidic bark and the coarse woody debris of conifers (pH usually 3.5–5.5). Exceptions prove this rule, since five species of *Perichaena* were commonly found on substrates with a relatively high pH (6.5–6.8) in the deserts of the Caspian Lowland (Novozhilov et al., 2006).

Several published studies that have analyzed the results obtained from moist chamber cultures prepared with samples of ground litter and bark collected across tropical and boreal forests but also including deserts, have demonstrated a pattern of increasing species diversity with increasing pH values. For both litter and bark, a higher substrate pH tends to be positively correlated with higher species diversity. Low substrate pH appears to be a limiting factor for growth and development of most species of myxomycetes, which may simply result from the fact that most bacteria do not grow well under these conditions.

Microhabitats and Ecological Assemblages of Myxomycetes

A detailed knowledge of the small-scale distribution of species of myxomycetes across different microhabitats (substrates) is essential for understanding the composition of myxomycete assemblages and their functional roles in ecosystems. Numerous published studies of myxomycete diversity across different substrates have demonstrated that species composition varies considerably for different types of substrates (e.g., the bark of different species of trees). However, the set of species that is dominant for a certain sample of a particular substrate is rather limited. In most studies, substrate samples are typically characterized by one to four species, which comprise more than 30%–50% of the assemblage present (potential dominants). With the help of the moist chamber technique (Gilbert and Martin, 1933) these dominants are revealed rather easily and definitely, which provides an opportunity to carry out mass analyses and produce statistically sound results. Utilization of moist chamber cultures permits one to evaluate α-diversity and abundance as "frequency of species" and determine an assemblage of core species, especially for arid and high-latitude regions with extreme climates.

Ground Litter

Plant litter represents a major source of organic carbon in forest soils (Anichkin and Tiunov, 2011; Berg et al., 2001). The decomposition of plant litter is a complex process that involves mineralization and transformation of organic matter. Decomposition of plant litter is a key step in nutrient recycling. As most of the plant biomass-derived carbon in the forests is mineralized in the litter layer, an understanding of this process and the microorganisms involved is essential for the identification of factors that affect global carbon fluxes (Voříšková and Baldrian, 2013). As such, environmental heterogeneity and resource availability within this context have been pointed out as essential factors strongly affecting species richness (Yang et al., 2015).

The leaf litter microhabitat is heterogeneous and relatively complex, both chemically (containing various soluble substances, cellulose, and lignin) and structurally (consisting of a mixture of different types of decaying plant detritus, such as leaves, fruits, flowers, seeds, bark fragments, and twigs). This microhabitat is intimately associated with (and ultimately contributes to the formation of) the soil. Given its high level of heterogeneity, the leaf litter microhabitat could conceivably be quite productive for soil myxomycetes. More recently, certain components of the leaf litter microhabitat, such as fruits (Novozhilov et al., 2001, 2017a) and twigs (Stephenson et al., 2008), have been given some attention. Furthermore, it is generally accepted, although very little quantitative data exist, that amoeboflagellates are normal and sometimes quite abundant components of the soil microbiota and probably have a substantial role in nutrient cycling and the detrital food chain (Stephenson et al., 2011; Urich et al., 2008). Other than a single paper by Stephenson and Landolt (1996), little information exists with respect to the distributional patterns of myxomycetes within the leaf litter microhabitat. Rollins and Stephenson (2012) reported that there were inherent differences between different levels (or strata) of the forest floor leaf litter microhabitat. Each stratum was characterized by a distinct assemblage of species; moreover, both species richness and abundance varied throughout. The leaf litter microhabitat represents a heterogeneous and understudied microhabitat that can be very productive (at least in temperate forests) for myxomycetes. Even when species richness is relatively low for a given stratum, the total number of collections can be appreciable. A well-defined stratification of assemblages of myxomycetes may represent the effects of competition and/or other microenvironmental influences. The highly decomposed A-1 stratum and twigs are found to be exceptionally diverse for myxomycetes. The relatively high diversity of the twig stratum is probably not surprising, since Stephenson et al. (2008) reported more than 40 species associated with twigs in a series of studies carried out in many different localities around the world.

It seems to be a general trend in the tropics for litter is more diverse for myxomycetes than decaying wood and the bark of living trees (Schnittler and Stephenson, 2000; Stephenson et al., 2004a). Since fruiting bodies of many litter-inhabiting myxomycetes are easy to overlook and quickly destroyed by numerous predatory invertebrates, only field surveys carried out in combination with the moist chamber culture technique are suitable to recover the diversity of myxomycete assemblages, especially for litter. For example, in lowland monsoon forests of southern Vietnam (Novozhilov et al., 2017a), in spite of a high number of intensively studied plots, ground litter appeared poor in the field (22 species from 84 records, $H' = 2.44$, $D = 6.88$), compared to records obtained in moist chamber cultures (47 species from 272 records, $H' = 3.33$, $D = 20.03$). Samples from the leafy fraction of litter yielded more records and species than those from all other fractions collectively. In addition, moist chamber cultures were more productive for the leafy fraction. Species inhabiting the ground litter of herbaceous plants seem to be more specialized than myxomycetes which prefer general leaf litter and twigs.

Despite the low plant biomass of arid regions, litter represents an important microhabitat for myxomycetes. For example, in the steppe zone of Eurasia, the intrazonal arboreal communities of the steppe and desert zones provide all of the microhabitats for myxomycetes typically found in temperate forests (Novozhilov et al., 2006). It is interesting to note that most of the rare species obtained in the Caspian lowlands were collected in the field on woody debris and leaf litter in the intrazonal arboreal habitats located in gullies and riparian forests. In these areas, the conditions are rather similar to those in temperate deciduous forests. In grassland communities of North America, the greatest number of collections of myxomycetes was obtained from microhabitats represented by forbs, defined as broadleaf herbaceous plants (Rollins and Stephenson, 2013). The greatest species richness occurred in tall grasslands, whereas mixed grasslands and short grasslands produced an equal number of species. The difference in the numbers of collections obtained from forb litter and grass litter was striking, whereas the forb aerial litter and forb ground litter microhabitats were found to be the most similar.

In desert regions, litter plays an important role as a microhabitat for myxomycetes (Blackwell and Gilbertson, 1984; Lado et al., 2002; Novozhilov and Schnittler, 2008; Schnittler, 2001a; Schnittler and Novozhilov, 2000). It has been documented that plant litter mineralizes rather quickly in deserts (Dobrovol'skaya et al., 1997) due to the high abundance of hydrolytic bacteria, which consequently represent a source of food for myxomycetes. However, collectively ground litter was less productive than the bark of trees and in total was characterized by lower values of diversity and species richness in deserts than in forests. Among the species recorded from litter in temperate deserts, the dominant litter-inhabiting forms tend to be species *of Badhamia*, *Comatricha*, *Didymium*, *Echinostelium*, *Licea*, *Perichaena*, and *Physarum* (Novozhilov and Schnittler, 2008; Novozhilov et al., 2003, 2010; Schnittler et al., 2013).

Coarse Woody Debris

For many organisms, coarse woody debris represents an important factor with respect to their distribution (Davis et al., 2010; Harmon et al., 2004; Nordén et al., 2004; Ohlson et al., 1997; Samuelsson et al., 1994; Siitonen, 2001), and the myxomycetes are no exception. Results from studies carried out across different types of terrestrial ecosystems suggest that the species associated with coarse woody debris represent one of the main components of overall myxomycete diversity (Ing, 1994; Maimoni-Rodella and Gottsberger, 1980; Novozhilov et al., 2006; Rufino and Cavalcanti, 2007; Schnittler and Novozhilov, 1996; Stephenson et al., 2008; Takahashi, 2001, 2004; Takahashi and Harakon, 2012). Myxomycetes, especially species forming large (>1–2 cm) compound fruiting bodies (e.g., species of *Reticularia*, *Lycogala*, *Tubifera*, and *Fuligo*) have a strong preference for large pieces of coarse woody debris and are found mostly in the field, whereas the moist chamber culture technique is much less effective in recovering these species.

The majority of the data on substrate preferences of lignicolous myxomycetes has been derived from records of fruiting bodies appearing on the surface of logs. As such, we do not have a clear picture of whether or not the trophic stages of lignicolous myxomycetes live close to the surface or also occur within the logs, where microclimatic conditions should be more stable. Two studies (Ostrofsky and Shigo, 1981; Taylor et al., 2015) claim to have detected myxomycetes within solid wood; in both studies agar cultures were prepared. A recent metagenomics approach for the bright-spored clade (superorder Lucisporidia; Clissmann et al., 2015) revealed 260 SSU sequences of bright-spored myxomycetes that were assembled into 29 OTUs from wood cores of logs.

Wood has a wide range of chemical and physical characteristics (Samuelsson et al., 1994) and thus may provide a wide range of putative ecological niches for myxomycetes. Evidence comes from the well-known turnover of different species associated with different decay states (Takahashi and Hada, 2009). A study classifying coarse woody debris into four decay stages found the highest myxomycete diversity was associated with final stages of decay (Schnittler and Novozhilov, 1996). Takahashi and Hada (2009) suggested that the state of wood decay affects the species composition of the microbes that inhabit it and, in turn, the species composition of their predators, the myxomycetes. These authors report that *Stemonitis splendens* significantly preferred slightly decayed, rather than dry wood. Species of *Physarum* emerged mainly on slightly decomposed, fairly hard wood, but rapidly decreased in occurrence with the progression of decay. Conversely, five species in the Cribrariaceae and *Ceratiomyxa fruticulosa* var. *porioides* increased significantly with the progression of decay. *Lindbladia cribrarioides* was reported to prefer soft damp wood. Because wood decay is accompanied by increasing water content (Renvall, 1995), the moisture content of the wood depends on the state of decay. Consequently, Takahashi and Hada (2009) suggested that water retention determines as well the occurrence of lignicolous myxomycetes. In this study, the distribution of almost all species was attributed to a narrow range of moisture content, which was restricted to 60%–100% on average, except for *S. splendens*, which occurred on low-moisture wood.

Another very important factor in the occurrence of lignicolous species is the pH of the wood. Clissmann et al. (2015) found pH to be the main ecological parameter accounting for the distribution of OTUs on coarse woody debris. Logs with a pH range between 5.34 and 5.66 had the most diverse myxomycete assemblages, suggesting a link between pH and diversity. It is likely that pH and the decay stage of wood are not completely independent variables.

Bark of Living Woody Plants

The guild of corticolous myxomycetes (ca. 120 species), the members of which are associated with the bark surface of living trees, shrubs and lianas has been known for a long time (Gilbert and Martin, 1933; Keller and Brooks, 1973, 1976a,b; Lister, 1915; McHugh, 1998; Mitchell, 1978; Penfound, 1940). Most species in this guild possess minute fruiting bodies and can be detected reliably only through the use of the moist chamber culture technique (Härkönen, 1977, 1978; Keller and Brooks, 1977; Mitchell, 1977; Snell et al., 2003; Stephenson, 1985). The first time this technique was used in Scotland by W. Cran (Lister, 1938), but it became popular only after being described by Gilbert and Martin (1933). While studying epiphytic algae, these authors found that numerous fruiting bodies of *P. fimbriata* appeared on pieces of poplar bark placed in a moist chamber culture. Subsequent years of research have revealed a whole assemblage of species apparently adapted for survival on tree bark (Brooks et al., 1977; Keller, 1980; Keller and Brooks, 1976b; Rodríguez-Palma et al., 2002).

Early students doubted that the results obtained from bark cultures reflected a natural assemblage. Peterson (1952) suggested that the bark works only as a spore trap for myxomycetes but does not represent the real microhabitat for their trophic stages. Later experiments, covering bark pieces under a plastic film, confirmed that some species complete their entire life cycle from spore to fruiting bodies in bark fissures (Pendergrass, 1976). Many species having a protoplasmodium, especially members of the genera *Echinostelium* and *Licea*, are well known to be corticolous, developing rapidly on bark that dries out soon after a period of moist weather. Members of this group tend to develop fruiting bodies in a few days and possess protoplasmodia or minute aphanoplasmodia. Some authors differentiate between obligate and facultative bark-inhabiting species. For example, *Diachea arboricola* has been found only on the bark of living trees above a height of 3 m and is considered a true corticolous myxomycete (Keller et al., 2004). In contrast, although primarily corticolous, *Echinostelium minutum* is not uncommon on litter and dung. Although corticolous species are often abundant and widespread, they appear to be more specialized than other ecological groups of myxomycetes.

Bark texture, pH, water-holding capacity, tree height, and epiphyte load can be very important environmental factors for corticolous myxomycetes (Everhart and Keller, 2008; Everhart et al., 2008). Stephenson (1989) reported that the thick, furrowed, rough bark of trees could be effective at trapping myxomycete spores from the air, and in his study the former yielded more species than trees with a thin, smooth bark surface. In contrast, he found high species richness on *Betula alleghaniensis*, which has smooth bark. A similar trend was found for lianas, where the highest species richness and highest diversity index were recorded for smooth bark, followed by intermediate-furrowed bark and then rough bark (Ko Ko et al., 2010). The authors suggested that perhaps spores of some species of myxomycetes possess a high enough viscosity to allow them to adhere to smooth bark surfaces.

In arid regions, it has been reported (Novozhilov and Schnittler, 2008; Novozhilov et al., 2006; Schnittler and Novozhilov, 2000; Schnittler et al., 2013, 2015b) that the fissured and peeling bark of desert shrubs was characterized by the most distinctive assemblage of myxomycetes, with sharply contrasting species preferences and clearly associated common species, whereas smooth and smooth but breaking with age bark, and fibrous types of bark clustered together and displayed no clear patterns with respect to the occurrence of the more common species. The high percentage of positive bark cultures seemed to indicate that textured bark serves as a better spore trap than smooth bark. Deeply fissured bark provided as well a higher surface area per square centimeter and better water retention, therefore allowing a prolonged time window for bacterial growth and myxomycete development. However, Snell and Keller (2003) did not find a correlation between the water-holding capacity of bark and species richness and diversity of corticolous myxomycetes.

A few studies have investigated the assemblages of myxomycetes found at different heights on trees (Everhart et al., 2009; Snell and Keller, 2003; Wrigley de Basanta, 1998). These studies did not show any strong trends related to height in the canopy except for a single type of tree. There was a significant decrease in species richness of myxomycetes from the bottom to the top of *Platanus occidentalis* (Everhart et al., 2009).

Bark pH (see earlier) seems to be inversely correlated with species richness of corticolous myxomycetes. Assemblages of myxomycetes from coniferous trees with low pH differ, often strongly, from assemblages on deciduous broadleaf trees with a bark with a slightly acidic to neutral pH (Everhart et al., 2008; Schnittler et al., 2016; Snell et al., 2003).

The moss-covered logs (Ing, 1983) of Western Great Britain and Ireland (McHugh, 1998) are characterized by a high diversity of myxomycetes in spite of the dense mats of epiphytic organisms present upon the bark. Schnittler and Stephenson (2000) suggested that a high epiphyte cover seems to be one—although not the only—reason for low abundance and diversity of corticolous myxomycetes in the tropics. The most probable reasons for this phenomenon are an excess in rainfall, connected with the fact that the bark surface of tree trunks in closed-canopy cloud forests rarely dries out and is often has an extensive cover of liverworts and mosses present. Furthermore, the heavy tropical rains probably remove myxomycete plasmodia and spores from the bark surface. In addition to this mechanical effect, a leaching effect caused by the sheer amount of rainfall cannot be ruled out, with soluble nutrients and microorganisms being removed from the bark.

Overall, the general pattern for corticolous myxomycetes is an increase in diversity and percentage of positive moist chamber cultures from tropical forests (Ko Ko et al., 2010; Novozhilov et al., 2017a; Rojas et al., 2010; Schnittler and Stephenson, 2000; Stephenson et al., 2004b), to boreal forests (Novozhilov et al., 1999; Schnittler and Novozhilov, 1996; Schnittler et al., 2016), to deciduous broadleaf forests (Novozhilov et al., 2017b; Peterson, 1952; Stephenson, 1989), to Mediterranean forests (Pando, 1989; Pando and Lodo, 1987; Wrigley de Basanta, 1998) and ultimately to grasslands and deserts (Davison et al., 2008; Estrada-Torres et al., 2009; Novozhilov and Schnittler, 2008; Novozhilov et al., 2006, 2013a; Schnittler et al., 2013).

Aerial Microhabitats

Aerial litter consists of dead but still attached leaves of trees and other plant parts, fallen leaves trapped in tree branches, dead fruits, flowers, and woody remnants of dead twigs and lianas, which collectively make up a significant amount of dead biomass in forest ecosystems (Thomas and Packham, 2007). Assemblages of protosteloid amoebae on aerial litter (Aguilar et al., 2007; Moore and Stephenson, 2003; Shadwick et al., 2009; Stephenson et al., 1999; Zahn et al., 2014) have been more thoroughly characterized for this microhabitat than myxomycetes. Temperate deciduous forests with a seasonal variation in leaf-fall contain much less aerial litter than is the case for tropical forests. Aerial litter in the former consists mainly of woody remnants, which represent suitable microhabitats for lignicolous myxomycetes (Schnittler et al., 2006).

These authors estimated the total number of species in this microhabitat to vary between 42 and 45. The most common species were *A. cinerea*, *Perichaena depressa*, *P. chrysosperma*, *P. vermicularis*, *Physarum leucophaeum*, and *Stemonitis fusca*.

In tropical forests, myxomycetes often inhabit aerial microhabitats. For example, studies of such microhabitats in the Yucatan recorded 17 species (Stephenson et al., 2003), in Ecuador 11 species (Stephenson et al., 2004a), in Australia 11 species (Black et al., 2004), and in Vietnam 42 species (Novozhilov et al., 2017a). Results from recent studies in Brazil indicated that aerial litter above the tide level was the preferred microhabitat for myxomycetes in mangroves (Cavalcanti et al., 2016), and eight species were recorded. The most common species on aerial woody remnants and leaves in tropical forests appear to be *A. cinerea*, *Didymium squamulosum*, *Collaria arcyrionema*, and *Ph. compressum*. The composition of this assemblage differs from that found in temperate forests. In addition to woody remnants, aerial microhabitats in the tropics include foliicolous liverworts and lichens covering leaves of understory woody plants (Schnittler, 2001b; Stephenson et al., 2004a), inflorescence of giant herbaceous plants (Schnittler and Stephenson, 2002) and the aerial "canopy soil" (Stephenson and Landolt, 2015) associated with vascular and nonvascular epiphytes.

Lowland rain forests with a high annual rainfall seem to provide the best conditions for the growth of epiphyllous myxomycetes associated with the liverworts (mostly members of the family Lejeuneaceae) covering living leaves (Schnittler, 2001b; Schnittler and Stephenson, 2000). Compared with other microhabitats, such as leafy litter on the forest floor or the bark of living trees, assemblages of myxomycetes associated with epiphyllous liverworts appear to be rather species poor, and Schnittler (2001b) recorded only 11 species from this microhabitat. The average frequency of the three most common species (*A. cinerea, Didymium iridis*, and *D. squamulosum*) was surprisingly high. These species are likely to be regular inhabitants of liverwort-covered leaves. However, none of the myxomycetes found the study mentioned earlier seem to be specialized for living leaves as a microhabitat. Unusually small numbers of fruiting bodies found in the moist chamber cultures, along with the occurrence of atypically small fruiting bodies, suggest a microhabitat poor in nutrients. More studies, such as axenic culture of epiphyllous specimens of *Didymium* and moist chamber cultures of twigs with leaves kept alive by watering are necessary to understand more completely the ecology of epiphyllous myxomycetes.

The assemblage of myxomycetes associated with inflorescences is also rather poor in species (Schnittler and Stephenson, 2002). A comparison with the myxomycetes recorded from other litter substrates showed that more than half of the 13 more common species are ubiquists, occurring with about the same frequency or even more often on other litter substrates. Prominent examples are *D. squamulosum* and *D. iridis*, both very common on all kinds of aerial litter. Both species are several times more common on other litter substrates than on inflorescences.

The relatively low percentage of positive cultures suggests that the frequency of occurrence of myxomycetes in canopy soil is not especially high, but the numbers of plasmodia recorded from those cultures that were positive also suggests that these organisms can be relatively common in this microhabitat (Stephenson and Landolt, 2015). Myxomycete assemblages associated with aerial litter seem not to consist of specialized species. Studies carried out in grassland ecosystems found that the aerial litter of grasses and forbs represent a suitable microhabitat for many species, but of the 16 species recorded from this microhabitat, the two most common (*Badhamia melanospora* and *S. fusca*) were often found on ground litter and wood (Rollins and Stephenson, 2013). In contrast, 6 of the 13 species of myxomycetes from inflorescences of tropical herbs, including the two most common examples (*Ph. compressum* and *Ph. didermoides*) showed a clear preference for this microhabitat (Schnittler and Stephenson, 2002).

The canopy as myxomycete habitat was first investigated for broadleaf deciduous forests of the Appalachians with the single-rope technique, focusing on corticolous species associated the tree trunks (Keller et al., 2004; Snell et al., 2003). Another study used a canopy crane (Schnittler et al., 2006; Unterseher et al., 2005) to investigate small twigs from the upper- and outermost canopy (height of 30 m) for the occurrence of myxomycetes. Wood of dead twiglets attached to their still living parts seem to form appropriate "islands" in the canopy for many lignicolous myxomycetes. The species richness of lignicolous canopy myxomycetes (32 species from 146 cultures in this study) seems to be comparable with that of corticolous canopy myxomycetes recorded by Snell and Keller (2003) from the Great Smoky Mountains (48 species from more than 418 cultures). In contrast to the tropics, where many species occur almost exclusively in the canopy, most of the species detected in this study also may fruit on logs on the ground.

Tropical Lianas

At least a few lianas occur in most types of temperate forests, but this growth form is most common in tropical forests (Schnitzer and Bongers, 2002). Relying on the supportive structure of trees to reach the light, lianas typically produce a soft wood with wide vessels. This results in high water retention for its woody debris and thus makes this microhabitat very suitable for myxomycetes (Wrigley de Basanta et al., 2008). In addition, the bark and woody remnants of lianas have near-neutral pH values (5.4–8.5) that are suitable for many species of myxomycetes. This microhabitat was investigated for Neotropical forests (Lado et al., 2003; McHugh, 2005; Schnittler et al., 2002; Wrigley de Basanta et al., 2008), forests in Northern Queensland in Australia (Black et al., 2004) and Southeastern Asia (Ko Ko et al., 2010). *A. cinerea*, *Diderma hemisphaericum*, *D. squamulosum*, *Physarum pusillum*, and *S. fusca* var. *nigrescens* appear to be among the more consistently abundant and widespread members of the assemblage of myxomycetes associated with lianas.

Weathered Dung of Herbivorous Animals

The dung of herbivorous animals consists of the macerated and undigested remains of plants plus vast quantities of bacteria and animal waste products. The nature of herbivore dung depends on the efficiency of the digestive tract, which, in turn, depends on the animal's digestive anatomy and its microflora. Most studies of coprophilous organisms have focused on fungi (Angel and Wicklow, 1975; Dickinson and Underhay, 1977; Dix and Webster, 1995; Krug et al., 2002; Nyberg and Persoon, 2002) or beetles (Hanski and Cambefort, 1991). Coprophilous (or fimicolous) myxomycetes have been known for a long time (Eliasson and Lundqvist, 1979; Krug et al., 2004; Marchal, 1895); however, the ecology of this group is not well studied. There are approximately 115 species of myxomycetes that have been recorded from dung (Eliasson, 2013), and most of these commonly occur in other microhabitats and thus their occurrence on dung is merely coincidental. However, compared to most other ecological assemblages, at least some coprophilous myxomycetes show a high level of specialization (Eliasson and Keller, 1999). Among the species most commonly encountered on dung are *Badhamia apiculospora*, *B. spinispora*, *Licea alexopouli*, *Perichaena liceoides*, *P. luteola*, and *Kelleromyxa fimicola*.

Most records for this group have been obtained from moist chamber culture studies in regions with a monsoon or semiarid climate (Bezerra et al., 2008; Chung and Liu, 1995, 1996; García-Zorrón, 1977), whereas in the humid tropics dung often does not last long enough because of rapid degradation by arthropods. To a lesser extent, the dung of herbivorous animals and birds which feed on seeds, buds or other plant parts may accumulate in temperate and boreal forests (Eliasson and Lundqvist, 1979), as well as in tundra (Novozhilov and Schnittler, 2000; Novozhilov et al., 1999), where coprophilous myxomycetes have been documented as well. In general, this substrate is not very productive (Stephenson, 1989), with deserts and grasslands representing the most noteworthy exceptions. The maximum diversity and abundance of coprophilous myxomycetes seems to be associated with cool and dry climates, where the prevailing conditions slow down the decomposition of dung (Novozhilov and Schnittler, 2008; Novozhilov et al., 2006, 2010; Schnittler and Novozhilov, 2000; Schnittler et al., 2015b).

Considering animal species, the dung of domestic herbivores, such as the cow and horse has been found to be more productive than that from wild animals. Eliasson (2013) reviewed many aspects of the diversity and ecology of coprophilous myxomycetes and emphasized the need for special studies to elucidate possible adaptations related to the passage of myxomycete spores through the digestive system of a herbivore. Key features seem to be the ultrastructure of spore walls and spore germination. Thick-walled spores, like these found in *L. alexopouli* (Blackwell, 1974), *K. fimicola* (Eliasson et al., 1991; Erastova et al., 2013), and *Trichia brunnea* (Eliasson and Keller, 1999), indicate that spores taken up by herbivores can remain alive after passage though the digestive system.

Soil

It is generally assumed that soil protozoans (Bonkowski, 2004; Sherman, 1916) and bacterivorous nematodes (Kudrin et al., 2015) largely control the growth of bacteria in soils. As mentioned earlier, myxomycetes may represent the largest component of total protozoan soil biodiversity, and species diversity can be very high. Currently, most studies of soil myxomycetes have provided only estimates of the abundance of trophic stages or propagules and could not identify the species involved because of the problem of identifying amoeboflagellates to species. According to these estimations, the density of the amoeboflagellate cells of myxomycetes reaches hundreds of thousand cells per gram dry soil (Feest, 1985). Approaches employing environmental PCR do not allow for overall estimates of amoeboflagellate density but have revealed a high genetic diversity of the myxomycete assemblages present in soil (Fiore-Donno et al., 2016). However, despite these efforts, our understanding of soil myxomycete diversity and functioning in ecosystems remains poor. Stephenson and Feest (2012) reviewed all that is currently known about the diversity and ecology of soil myxomycetes. Most species associated with soil fruit well on leaf litter. Most of the sequences from dark-spored myxomycetes observed from soil by molecular methods were assigned to *Lamproderma* spp. and *Didymium* spp. (Fiore-Donno et al., 2016). Tiunov et al. (2015) suggested that a significant difference in δ13C and δ15N values shows that prey types and/or basal food sources may differ among different species of myxomycetes. This suggestion remains speculative, as there is no detailed information on the feeding mode of amoeboflagellates or plasmodia in natural conditions.

Aquatic Habitats

Despite the fact that myxomycetes are typically considered as terrestrial organisms, the plasmodia and amoeboflagellates of some species have been observed to survive under water. *Didymium aquatile*, one of the few species described from aquatic environments, was found in a stream in the tropical forests of Brazil as submerged fruiting bodies (Gottsberger and Nannenga-Bremekamp, 1971). Some experiments using aquaria and moist chamber cultures indicate that terrestrial species may fruit in submerged conditions. An early report described a species of myxomycete (probably *Didymium difforme* or *Physarum cinereum*) fruiting on plant roots under water (Ward, 1886). Subsequently, the ability to grow under water has been reported for *Physarum gyrosum*, *Ph. album*, *Fuligo cinerea*, *Fuligo septica* (Parker, 1946), *Didymium nigripes* (Kappel, 1992; Müller et al., 2008), *D. iridis* (Keller and Braun, 1999), and *Diderma effusum* (Tamayama and Keller, 2013). In addition, myxomycetes have been recorded from moist chamber cultures of aquatic substrates (Bodyagin and Barsukova, 2009; Lindley et al., 2007; Shearer and Crane, 1986), although direct evidence that the entire life cycle of these species takes place in the water was not obtained. More studies, especially those employing environmental PCR and/or cultures to reveal the presence of myxomycetes in bodies of water, are required.

Effect of Disturbance Events on Assemblages of Myxomycetes

Disturbance effects on communities of larger organisms, such as plants and animals are well documented, but much less attention has been directed toward microorganisms despite their importance in biogeochemical processes through decomposition and other mechanisms (Carreño-Rocabado et al., 2012; Ohlson et al., 1997; Sun et al., 2015). The impact of different types of disturbance events on myxomycetes is limited to a very small number of studies, including wildfire (Adamonytė et al., 2016), flooding (Novozhilov et al., 2017a), fertilization by bird colonies (Adamonytė et al., 2013), and various anthropogenic factors studied by Härkönen (1977), Härkönen and Vänskä (2004), and Wrigley de Basanta (2000). For example, Adamonytė et al. (2016) reported that in the sites they studied only crown fire plots showed a clear chronosequence of postfire myxomycete assemblages. The impact of fire was to promote the establishment and/or fruiting of species of myxomycetes which are rare in similar unburned sites or are usually confined to other types of forests and substrates.

Limited data from field studies indicate that forest structure seems to be a driver of differences in species composition and abundance of the myxomycetes associated with aerial litter, ground litter, and soil (Rojas and Stephenson, 2013). Rojas and Valverde (2015) reported more narrow niches and a higher level of coexistence for myxomycetes in dry compared to wet forests. Data from Ethiopia indicate that human activity (collecting of fuelwood) reduces species diversity of lignicolous myxomycetes (Dagamac et al., 2017a). As such, it remains unclear to what extent myxomycete assemblages change in response to changes of climate, vegetation disturbance, and landscape cover alterations.

Phenology of Sporulation (Seasonality of Myxomycetes)

Since many species of myxomycetes display pronounced fruiting peaks which correspond to seasons, at least two field surveys in a year are often necessary to document the species inventory for a region (Schnittler and Novozhilov, 1996). Since this poses logistical problems for field work, many published checklists from singular surveys can be expected to have a large bias for certain phenological groups. Only a few surveys have investigated temporal changes in the fruiting of myxomycetes (Blackwell and Gilbertson, 1984; Eliasson, 1981; Kazunari, 2010; Ko Ko et al., 2011; Stephenson, 1988; Stojanowska, 1980a,b; Takahashi, 1995; Takahashi and Hada, 2008; Tran et al., 2006). As expected, the highest fluctuations in fruiting activity were found for regions with a pronounced dry season, such as the Pacific coast of Costa Rica (Stephenson et al., 2004b) or regions of southeastern Asia with a monsoon climate (Ko Ko et al., 2011), where species richness and diversity were higher for the warm-wet season than for the cool-dry season. Some members of the order Stemonitales were encountered only during the warm-wet season, whereas members of the order Physarales appear to be more common during the cool-dry season. Different groups of myxomycetes with different fruiting activity have also been reported in temperate deciduous (Schnittler et al., 2010; Stephenson, 1988; Stojanowska, 1980b) and boreal forests (Eliasson, 1981; Schnittler and Novozhilov, 1996; Vlasenko and Novozhilov, 2011).

Even within one ecological group the phenology of species can differ. Novozhilov and Schnittler (1997) classified the cryophilous litter-inhabiting myxomycetes on the Kola Peninsula into two subgroups, based on phenology and habitat requirements. The first subgroup contains the "true" nivicolous species (*Diacheopsis effusa*, *Diderma niveum*, and species of *Lepidoderma* and *Lamproderma* with the exception of *Lamproderma sauteri*). The second subgroup of species was more cryophilous, growing predominantly in summer under cool and wet conditions on litter, especially in shady woodlands. *Ph. cinereum*, *Didymium deplanatum*, *D. dubium*, and perhaps *Trichia alpina* and *L. sauteri* can be placed in this subgroup. The most prominent example of seasonal fruiting has been documented for *B. minutissima* (Schnittler et al., 2000). In southern regions *Barbeyella* fruits in the winter, whereas in more northern regions peak fruiting occurs in September to early October (eastern North America) or mid-October (Germany). Observations from the German Alps (Schnittler and Novozhilov, 1998) and the Leningrad region of Russia (Novozhilov pers. comm.) provided evidence that fruiting bodies of *Barbeyella* can develop at temperatures between 0 and 10°C.

Relationships With Other Organisms

Myxomycetes are frequently associated with insects, especially beetles (Coleoptera). Stephenson and Stempen (1994) along with Dudka and Romanenko (2006) reviewed the cooccurrences of myxomycetes and beetles recorded since the end of the nineteenth century, and listed the beetle families (Rhizodidae, Leiodidae, Staphylinidae, Clambidae, Eucinetidae, Sphindidae, Cerylonidae, Latridiidae) most frequently involved in these associations. The numerous and repeated records of some beetles on certain species of myxomycetes at least suggests a positive association. Some beetles feed not only on spores but also on plasmodia (Lawrence and Newton, 1980; Wheeler, 1980). The term myxomycetophagy has been proposed to describe the phenomenon that beetles feed on slime molds (Newton and Stephenson, 1990). These authors discussed the extent to which members of the most specialized family (Latridiidae) are obligate or facultative feeders on myxomycetes. Newton and Stephenson (1990) considered their new species of *Dienerella* from India to be a facultative feeder. Results from the Ukraine suggest that *Latridius hirtus*, *Enicmus rugosus*, and *E. fungicola* may be obligate feeders (Dudka and Romanenko, 2006). In particular, *L. hirtus* has been shown to be capable of completing its life cycle with only myxomycetes as a food supply. However, *Corticarina truncatella* is more likely to be facultative, since records of this species on myxomycetes are rather rare, and it is usually found in forest floor litter, on hay and other plant remains. Furthermore, gut content analysis shows that this species feeds on fungal spores and on those of myxomycetes.

It seems that some species of myxomycetes may be closely associated with bryophytes and blue-green algae (Stephenson and Studlar, 1985). Schnittler and Novozhilov (1996), while working in Karelia, found large colonies of *Colloderma oculatum* and *Lepidoderma tigrinum* on the layers of mosses situated on rocks. There was no wood available as an alternative substrate, and the only conclusion was that some species of myxomycetes are well-adapted to living together with mosses. Ing (1983) described a myxomycete association in similar microhabitats in England but with an almost completely different set of species. The ecology of this system was characterized by Schnittler et al. (2010), who suggested that algae may be a possible food source for the plasmodia. *B. minutissima* and *C. oculatum* have a remarkable preference for decorticated logs coated with unicellular algae, which form gelatinous layers. Evidence for a stable association of these myxomycetes with algae was presented by Schnittler and Novozhilov (1998). Bacteria apparently represent the main food resource for both trophic stages in the myxomycete life cycle, but plasmodia are also known to feed upon yeasts, algae (including cyanobacteria), and fungal spores and hyphae (Amewowor and Madelin, 1991; Kalyanasundaram, 2004; Madelin, 1984a,b; Stephenson and Stempen, 1994).

CONCLUSIONS

In spite of a number of serious efforts that have been carried out during the last few decades, we still must acknowledge that there are major gaps in our understanding of the roles that myxomycetes play in ecological systems and their relationships with other organisms. The major reason is the inaccessibility of their tropic stages, and most of the available data are based on collections of fruiting bodies and probably show only (in a literal sense) the tip of the iceberg, which may represent only a minor fraction of the real diversity of myxomycetes in different habitats. Advanced molecular methods (environmental PCR connected with metagenomic approaches together with attempts at barcoding) may help to visualize the real diversity of their trophic stages. Analyses of stable isotopes may help to elucidate the position of myxomycetes in the food web if these data can be successfully related to the stable isotope composition of soil bacteria. For the same reasons, we still lack precise information on the true niches of species of myxomycetes. For example, we know almost nothing about the horizontal and vertical differentiation between populations of amoeboflagellates in ground litter and soil. Large-scale studies may help to determine the crucial biotic and abiotic factors determining species richness and diversity of myxomycetes in major ecosystems of the world. As a precondition for such macroecological studies, the gaps which exist for poorly studied regions (e.g., Africa, Southern Asia, Siberia, and the Far East) need to be filled by intense regional surveys. Experiments using microcosms and different manipulation methods (Altermatt et al., 2015) also may help to understand causal niche differentiation and speciation in myxomycetes (Drake and Kramer, 2012).

ACKNOWLEDGMENTS

Yuri Novozhilov acknowledges the programs "Biodiversity and spatial structure of assemblages of fungi and myxomycetes in natural and anthropogenic ecosystems" № 01201255604 and "Mycobiota of southern Vietnam" № 01201255603 (Komarov Botanical Institute RAS) for support of his research. All three coauthors are especially grateful to Dr. Roland McHugh (Dublin, Ireland), Dr. Grazina Adamonytė (Vilnius, Lithuania), and Dr. Bruce Ing (Ullapool, Great Britain) for helpful proofreading and their comments relating to this manuscript.

REFERENCES

Adamonytė, G., Iršėnaitė, R., Motiejūnaitė, J., Taraškevičius, R., Matulevičiūtė, D., 2013. Myxomycetes in a forest affected by great cormorant colony: a case study in Western Lithuania. Fungal Divers. 59, 131–146.

Adamonytė, G., Motiejūnaitė, J., Iršėnaitė, R., 2016. Crown fire and surface fire: effects on myxomycetes inhabiting pine plantations. Sci. Total Environ. 572, 1431–1439.

Aguilar, M., Lado, C., Spiegel, F.W., 2007. Protostelids from deciduous forests: first data from southwestern Europe. Mycol. Res. 111, 863–872.

Altermatt, F., Fronhofer, E.A., Garnier, A., Giometto, A., Hammes, F., Klecka, J., Legrand, D., Maechler, E., Massie, T.M., Pennekamp, F., Plebani, M., Pontarp, M., Schtickzelle, N., Thuillier, V., Petchey, O.L., 2015. Big answers from small worlds: a user's guide for protist microcosms as a model system in ecology and evolution. Methods Ecol. Evol. 6, 218–231.

Amewowor, D.H.A.K., Madelin, M.F., 1991. Numbers of myxomycetes and associated microorganisms in the root zones of cabbage (Brassica-Oleracea) and broad bean (*Vicia faba*) in field plots. FEMS Microbiol. Ecol. 86, 69–82.

Angel, K., Wicklow, D.T., 1975. Relationships between coprophilous fungi and fecal substrates in a Colorado grassland. Mycologia 67, 63–74.

Anichkin, A.E., Tiunov, A.V., 2011. Dynamics of plant litter decomposition. In: Tiunov, A.V. (Ed.), Structure and Functions of Soil Communities of a Monsoon Tropical Forest (Cat Tien National Park, Southern Vietnam). KMK Scientific Press, Moscow, pp. 188–206 [In Russian with English summary.].

Berg, B., McClaugherty, C., Santo, A.V.D., Johnson, D., 2001. Humus buildup in boreal forests: effects of litter fall and its N concentration. Can. J. For. Res. 31, 988–998.

Bezerra, M.F.A., Silva, W.T.M., Cavalcanti, L.H., 2008. Coprophilous myxomycetes of Brazil: first report. Rev. Mex. Micol. 27, 29–37.

Black, D.R., Stephenson, S.L., Pearce, C.A., 2004. Myxomycetes associated with the aerial litter microhabitat in tropical forests of Northern Queensland, Australia. Syst. Geogr. Plants 74, 129–132.

Blackwell, M., 1974. A new species of *Licea* (myxomycetes). Proc. Iowa Acad. Sci. 81, 6.

Blackwell, M., Gilbertson, R.L., 1980. Sonoran Desert myxomycetes. Mycotaxon 11, 139–149.

Blackwell, M.M., Gilbertson, R.L., 1984. Distribution and sporulation phenology of myxomycetes in the Sonoran Desert of Arizona. Microb. Ecol. 10, 369–377.

Bodyagin, V.V., Barsukova, T.N., 2009. Myxomycetes registered in basins of Moscow and Moscow oblast. Mikol. Fitopatol. 43, 281–283, [in Russian, with English summary].

Bonkowski, M., 2004. Protozoa and plant growth: the microbial loop in soil revisited. New Phytol. 162, 617–631.

Broennimann, O., Fitzpatrick, M.C., Pearman, P.B., Petitpierre, B., Pellissier, L., Yoccoz, N.G., Thuiller, W., Fortin, M.-J., Randin, C., Zimmermann, N.E., Graham, C.H., Guisan, A., 2012. Measuring ecological niche overlap from occurrence and spatial environmental data. Global Ecol. Biogeogr. 21, 481–497.

Brooks, T.E., Keller, H.W., Chassain, M., 1977. Corticolous myxomycetes VI: a new species of *Diderma*. Mycologia 69, 179–184.

Carreño-Rocabado, G., Peña-Claros, M., Bongers, F., Alarcón, A., Licona, J.-C., Poorter, L., 2012. Effects of disturbance intensity on species and functional diversity in a tropical forest. J. Ecol. 100, 1453–1463.

Cavalcanti, L.d.H., Damasceno, G., Costa, A.A.A., Bezerra, A.C.C., 2016. Myxomycetes in Brazilian mangroves: species associated with *Avicennia nitida*, *Laguncularia racemosa* and *Rhizophora mangle*. Mar. Biodivers. Rec. 9, 1–7.

Cavalier-Smith, T., Fiore-Donno, A.M., Chao, E., Kudryavtsev, A., Berney, C., Snell, E.A., Lewis, R., 2015. Multigene phylogeny resolves deep branching of Amoebozoa. Mol. Phylogenet. Evol. 83, 293–304.

Cavender, J.C., Kawabe, K., 1989. Cellular slime moulds of Japan. I. Distribution and biogeographical considerations. Mycologia 81, 683–691.

Cavender, J.C., Raper, K.B., 1965. The Acrasiae in nature II. Forest soil as a primary habitat. Am. J. Bot. 52, 297.

Chao, A., Chazdon, R.L., Colwell, R.K., Shen, T.J., 2006. Abundance-based similarity indices and their estimation when there are unseen species in samples. Biometrics 62, 361–371.

Chao, A., Shen, T.-J., Ma, K.H., Hsieh, T.C., 2016. User's Guide for Program SPADE (Species Prediction And Diversity Estimation). Available from: http://chao.stat.nthu.edu.tw/

Chung, C.H., Liu, C.H., 1995. First report of fimicolous myxomycetes from Taiwan. Fungal Sci. 10, 33–35.

Chung, C.H., Liu, C.H., 1996. More fimicolous myxomycetes from Taiwan. Taiwania 41, 259–264.

Clissmann, F., Fiore-Donno, A.M., Hoppe, B., Krüger, D., Kahl, T., Unterseher, M., Schnittler, M., 2015. First insight into dead wood protistan diversity: a molecular sampling of bright-spored myxomycetes (Amoebozoa, slime-moulds) in decaying beech logs. FEMS Microbiol. Ecol. 91, doi: 10.1093/femsec/fiv050.

Cole, L., 1949. The measurement of interspecific association. Ecology 30, 411–424.

Colwell, R.K., 2013. EstimateS: Statistical Estimation of Species Richness and Shared Species from Samples. Version 9. User's Guide and Application. Available from: http://purl.oclc.org/estimates

Colwell, R.K., Futuyma, D.J., 1971. On the measurement of niche breadth and overlap. Ecology 52, 567–576.

Dagamac, N.H.A., Hoffmann, M., Novozhilov, Y.K., Schnittler, M., 2017a. Myxomycetes from the highlands of Ethiopia. Nova Hedwigia 104 (1–3), 11–127.

Dagamac, N.H.A., Novozhilov, Y.K., Stephenson, S.L., Lado, C., Rojas, C., dela Cruz, T.E.E., Unterseher, M., Schnittler, M., 2017b. Biogeographical assessment of myxomycete assemblages from Neotropical and Asian Paleotropical forests. J. Biogeogr., doi: 10.1111/jbi.12985.

Davis, J.C., Castleberry, S.B., Kilgo, J.C., 2010. Influence of coarse woody debris on the soricid community in southeastern Coastal Plain pine stands. J. Mammal. 91, 993–999.

Davison, E.M., Davison, P.J.N., Brims, M.H., 2008. Moist chamber and field collections of myxomycetes from the Northern Simpson desert, Australia. Aust. Mycol. 27, 129–135.

Dickinson, C.H., Underhay, V.H.S., 1977. Growth of fungi in cattle dung. Trans. Br. Mycol. Soc. 69, 473–477.

Dix, N.J., Webster, J., 1995. Coprophilous fungi. In: Dix, N.J., Webster, J. (Eds.), Fungal Ecology. Springer, Dordrecht, The Netherlands, pp. 203–224.

Dobrovol'skaya, T.G., Chernov, I.Y., Zvyagintsev, D.G., 1997. Characterizing the structure of bacterial communities. Microbiologia 66, 408–414.

Drake, J., Kramer, A., 2012. Mechanistic analogy: how microcosms explain nature. Theor. Ecol. 5, 433–444.

Dudka, I.A., Romanenko, E.A., 2006. Co-existence and interaction between myxomycetes and other organisms in shared niches. Acta Mycol. 41, 99–112.

Eliasson, U.H., 1977a. Ecological notes on *Amaurochaete* Rost. (myxomycetes). Bot. Notiser 129, 419–425.

Eliasson, U.H., 1977b. Recent advances in the taxonomy of myxomycetes. Bot. Notiser 130, 483–492.

Eliasson, U.H., 1981. Patterns of occurence of myxomycetes in a spruce forest in South Sweden. Holarct. Ecol. 4, 20–31.

Eliasson, U., 2013. Coprophilous myxomycetes: recent advances and future research directions. Fungal Divers. 59, 85–90.

Eliasson, U., Keller, H.W., 1999. Coprophilous myxomycetes: updated summary, key to species, and taxonomic observations on *Trichia brunnea*, *Arcyria elaterensis*, and *Arcyria stipata*. Karstenia 39, 1–10.

Eliasson, U., Keller, H.W., Schoknecht, J.D., 1991. *Kelleromyxa*, a new generic name for *Licea fimicola* (Myxomycetes). Mycol. Res. 95, 1201–1207.

Eliasson, U.H., Lundqvist, N., 1979. Fimicolous myxomycetes. Bot. Notiser 132, 551–568.

Erastova, D.A., Okun, M.V., Novozhilov, Y.K., Schnittler, M., 2013. Phylogenetic position of the enigmatic myxomycete genus *Kelleromyxa* revealed by SSU rDNA sequences. Mycol. Res. 12, 599–608.

Estrada-Torres, A., Wrigley de Basanta, D., Conde, E., Lado, C., 2009. Myxomycetes associated with dryland ecosystems of the Tehuacán-Cuicatlán Valley Biosphere Reserve. Mexico Fungal Divers. 36, 17–56.

Everhart, S.E., Ely, J.S., Keller, H.W., 2009. Evaluation of tree canopy epiphytes and bark characteristics associated with the presence of corticolous myxomycetes. Botany 87, 509–517.

Everhart, S.E., Keller, H.W., 2008. Life history strategies of corticolous myxomycetes: the life cycle, plasmodial types, fruiting bodies, and taxonomic orders. Fungal Divers. 29, 1–16.

Everhart, S.E., Keller, H.W., Ely, J.S., 2008. Influence of bark pH on the occurrence and distribution of tree canopy myxomycete species. Mycologia 100, 191–204.

Feest, A., 1985. Numerical abundance of myxomycetes (myxogastrids) in soil in the West of England. Microb. Ecol. 31, 353–360.

Feest, A., 1987. The quantitative ecology of soil mycetozoa. Prog. Protistol. 2, 331–361.

Feest, A.M., Madelin, M.F., 1985. A method for the enumeration of myxomycetes in soil and its application to a wide range of soils. FEMS Microbiol. Lett. 31, 103–109.

Feinsinger, P., Spears, E.E., Poole, R.W., 1981. A simple measure of niche breadth. Ecology 62 (1), 27–32.

Feng, Y., Klahr, A., Janik, P., Ronikier, A., Hoppe, T., Novozhilov, Y.K., Schnittler, M., 2016. What an intron may tell: several sexual biospecies coexist in *Meriderma* spp. (myxomycetes). Protist 167, 234–253.

Feng, Y., Schnittler, M., 2015. Sex or no sex? Group I introns and independent marker genes reveal the existence of three sexual but reproductively isolated biospecies in *Trichia varia* (Myxomycetes). Org. Divers. Evol. 15, 631–650.

Feng, Y., Schnittler, M., 2017. Molecular or morphological species? Myxomycete diversity in a deciduous forest in northeastern Germany. Nova Hedwigia 104 (1–3), 359–380.

Fiore-Donno, A.-M., Berney, C., Pawlowski, J., Baldauf, S.L., 2005. Higher-order phylogeny of plasmodial slime molds (Myxogastria) based on elongation factor 1-A and small subunit rRNA gene sequences. J. Eukaryot. Microbiol. 52, 1–10.

Fiore-Donno, A.M., Clissmann, F., Meyer, M., Schnittler, M., Cavalier-Smith, T., 2013. Two-gene phylogeny of bright-spored myxomycetes (slime moulds, superorder Lucisporidia). PLoS One 8, e62586.

Fiore-Donno, A.M., Nikolaev, S.I., Nelson, M., Pawlowski, J., Cavalier-Smith, T., Baldauf, S.L., 2010. Deep phylogeny and evolution of slime moulds (Mycetozoa). Protist 161, 55–70.

Fiore-Donno, A.M., Weinert, J., Wubet, T., Bonkowski, M., 2016. Metacommunity analysis of amoeboid protists in grassland soils. Sci. Rep. 6, 19068.

García-Zorrón, N., 1977. Mixomicetos coprofilos del Uruguay. Rev. Biol. Uruguay 5, 47–50.

Gilbert, H.C., Martin, G.W., 1933. Myxomycetes found on the bark of living trees. Univ. Iowa Stud. Nat. Hist. 15, 3–8.

Gotelli, N.J., Colwell, R.K., 2011. Chapter IV: estimating species richness. In: Magurran, A.E., McGill, B.J. (Eds.), Biological Diversity Frontiers in Measurement and Assessment. Oxford University Press, New York, NY, pp. 39–54.

Gottsberger, G., Nannenga-Bremekamp, N.E., 1971. A new species of *Didymium* from Brazil Proceedings Koninklijke Nederlandse Akademie van Wetenschappen. Ser. C. Proc. Koninklijke Nederlandse Akademie van Wetenschappen. Ser. C 74, 264–268.

Hanski, I., Cambefort, Y., 1991. Dung Beetle Ecology. Princeton University Press, Princeton, NJ.

Härkönen, M., 1977. Corticolous myxomycetes in three different habitats in southern Finland. Karstenia 17, 19–32.

Härkönen, M., 1978. Corticolous myxomycetes in Northern Finland and Norway. Ann. Bot. Fennici 15, 32–37.

Härkönen, M., Ukkola, T., 2000. Conclusions on myxomycetes compiled of twenty-five years from 4793 moist chamber cultures. Stapfia 73, 105–112.

Härkönen, M., Vänskä, H., 2004. Corticolous myxomycetes and lichens in the Botanical Garden in Helsinki, Finland: a comparison after decades of recovering from air pollution. Syst. Geogr. Plants 74, 183–187.

Harmon, M.E., Franklin, J.F., Swanson, F.J., Sollins, P., Gregory, S.V., Lattin, J.D., Anderson, N.H., Cline, S.P., Aumen, N.G., Sedell, J.R., Lienkaemper, G.W., Cromack, K.J., Cummins, K.W., 2004. Ecology of coarse woody debris in temperate ecosystems. Adv. Ecol. Res. 34, 59–234.

Haskins, E.F., Wrigley de Basanta, D., 2008. Methods of agar culture of myxomycetes: an overview. Rev. Mex. Micol. 27, 1–7.

Hill, M.O., 1973. Diversity and eveness: a unifying notation and its consequences. Ecology 54, 427–432.

Hoppe, T., Schnittler, M., 2015. Characterization of myxomycetes in two different soils by TRFLP analysis of partial 18S rRNA gene sequences. Mycosphere 6, 216–227.

Indira, P.U., 1968. Some Slime moulds from Southern India. IX. Distribution, habitat and variation. J. Indian Bot. Soc. 67, 330–341.

Ing, B., 1983. A ravine association of myxomycetes. J. Biogeogr. 10, 299.

Ing, B., 1994. The phytosociology of myxomycetes. New Phytol. 126, 175–201.

Jongman, R.H.G., Ter Braak, C.J.F., van Tongeren, O.F.R., 1995. Data Analysis in Community and Landscape Ecology. Cambridge University Press, Cambridge.

Kalyanasundaram, I., 2004. A positive ecological role for tropical myxomycetes in association with bacteria. Syst. Geogr. Plants 74, 239–242.

Kamono, A., Meyer, M., Cavalier-Smith, T., Fukui, M., Fiore-Donno, A.-M., 2013. Exploring slime mould diversity in high-altitude forests and grasslands by environmental RNA analysis. FEMS Microbiol. Ecol. 84, 98–109.

Kappel, T., 1992. An aquarium myxomycetes: *Didymium nigripes*. Mycologist 6, 106–107.

Kazunari, T., 2010. Succession in myxomycete communities on dead *Pinus densiflora* wood in a secondary forest in southwestern Japan. Ecol. Res. 25 (5), 995–1006.

Keller, H.W., 1980. Corticolous myxomycetes VIII: *Trabrooksia*, a new genus. Mycologia 72, 395–403.

Keller, H.W., Braun, K., 1999. Myxomycetes of Ohio: Their Systematics, Biology and Use in Teaching. Ohio Biological Survey, Columbus.

Keller, H.W., Brooks, T.E., 1973. Corticolous myxomycetes I: two new species of *Didymium*. Mycologia 65, 286–294.

Keller, H.W., Brooks, T.E., 1976a. Corticolous myxomycetes V: observations on the genus *Echinostelium*. Mycologia 68, 1204–1220.

Keller, H.W., Brooks, T.E., 1976b. Corticolous myxomycetes: IV. *Badhamiopsis*, a new genus for *Badhamia ainoae*. Mycologia 68, 834–841.

Keller, H.W., Brooks, T.E., 1977. Corticolous myxomycetes VII: contribution toward a monograph of *Licea*, five new species. Mycologia 69, 667–684.

Keller, H.W., Skrabal, M., Eliasson, U., Gaither, T., 2004. Tree canopy biodiversity in the Great Smoky Mountains National Park: ecological and development observations of a new myxomycete species of *Diachea*. Mycologia 96, 537–547.

Kerr, S.J., 1994. Frequency of recovery of myxomycetes from soils of the Northern United States. Can. J. Bot. 72, 771–778.

Kerr, N.S., Sussman, M., 1958. Clonal development of the true slime mould, *Didymium nigripes*. J. Gen. Microbiol. 19, 173–177.

Kilgore, C.M., Keller, H.W., Ely, J.S., 2009. Aerial reproductive structures of vascular plants as a microhabitat for myxomycetes. Mycologia 101, 305–319.

Ko Ko, T.W., Stephenson, S.L., Hyde, K.D., Lumyong, S., 2011. Influence of seasonality on the occurrence of myxomycetes. Chiang Mai J. Sci. 38, 71–84.

Ko Ko, T.W., Stephenson, S.L., Hyde, K.H., Rojas, C., Lumyong, S., 2010. Patterns of occurrence of myxomycetes on lianas. Fungal Ecol. 3, 302–310.

Krebs, C.J., 1999. Ecological Methodology. Addison-Welsey Educational Publishers, Menlo Park, CA.

Krug, J.C., Benny, G.L., Keller, H.W., 2002. Coprophilous Fungi. Smithsonian Institution Press, Washington, DC.

Krug, J.C., Benny, G.L., Keller, H.W., 2004. Coprophilous fungi. In: Mueller, G.M., Bills, G.F., Foster, M.S. (Eds.), Biodiversity of Fungi. Inventory and Monitoring Methods. Elsevier Academic Press, Burlington, pp. 467–499.

Kudrin, A.A., Tsurikov, S.M., Tiunov, A.V., 2015. Trophic position of microbivorous and predatory soil nematodes in a boreal forest as indicated by stable isotope analysis. Soil Biol. Biochem. 86, 193–200.

Lado, C., 2005–2016. An on-line nomenclatural information system of Eumycetozoa. Available from: http://www.nomen.eumycetozoa.com

Lado, C., Estrada Torres, A., Ramirez, M., Conde, E., 2002. A study of the succulenticolous myxomycetes from arid zones of Mexico. Scripta Bot. Belg. 22, 58.

Lado, C., Estrada-Torres, A., Stephenson, S.L., Wrigley de Basanta, D., Schnittler, M., 2003. Biodiversity assessment of myxomycetes from two tropical forest reserves in Mexico. Fungal Divers. 12, 67–110.

Lawrence, J.F., Newton, A.F., 1980. Coleoptera associated with the fruiting bodies of slime molds (myxomycetes). Coelopt. Bull. 34, 129–143.

Lesaulnier, C., Papamichail, D., McCorkle, S., Ollivier, B., Skiena, S., Taghavi, S., Zak, D., van der Lelie, D., 2008. Elevated atmospheric CO_2 affects soil microbial diversity associated with trembling aspen. Environ. Microbiol. 10, 926–941.

Levins, R., 1968. Evolution in Changing Environments: Some Theoretical Explorations. Princeton University Press, Princeton, NJ.

Lindley, L.A., Stephenson, S.L., Spiegel, F.W., 2007. Protostelids and myxomycetes isolated from aquatic habitats. Mycologia 99, 504–509.

Lister, A., 1901. On the cultivation of Mycetozoa from spores. J. Bot. 39, 5–8.

Lister, G., 1915. Japanese mycetozoa. Trans. Br. Mycol. Soc. 5, 67–84.

Lister, G., 1938. The Revd. William Cran and his scientific work. J. Bot. 76, 319–327.

Lister, A., Lister, G., 1925. A Monograph of the Mycetozoa: A Descriptive Catalogue of the Species in the Herbarium of the British Museum, third ed. Trustees of the British Museum, London.

Liu, P., Wang, Q., Li, Y., 2010. Spore-to-spore agar culture of the myxomycete *Physarum globuliferum*. Arch. Microbiol. 192, 97–101.

Madelin, M.F., 1984a. Myxomycete data of ecological significance. Trans. Br. Mycol. Soc. 83, 1–19.

Madelin, M.F., 1984b. Myxomycetes,microorganisms and animals: a model of diversity in animal-microbial ineractions. In: Anderson, J.M., Rayner, A.D.M., Walton, D.W.H. (Eds.), Invertebrate-Microbial Interactions. Cambridge University Press, Cambridge, pp. 1–33.

Madelin, M.F., 1990. Methods for studying the ecology and population dynamics of soil myxomycetes. Methods Microbiol. 22, 405–416.

Magurran, A.E., 2004. Measuring Biological Diversity. Blackwell, Malden.

Maimoni-Rodella, R.C., Gottsberger, G., 1980. Myxomycetes from the forest and the Cerrado vegetation in Botucatu, Brazil: a comparative ecological study. Nova Hedwigia 34, 207–245.

Marchal, E., 1895. Champignons coprophile de Belgique. VII. Bull. Soc. R. Bot. Belg. 34, 125–149.

Martin, G.W., Alexopoulos, C.J., 1969. The Myxomycetes. University of Iowa Press, Iowa City.

Martin, M.P., Lado, C., Johansen, S., 2003. Primers are designed for amplification and direct sequencing of ITS region of rDNA from myxomycetes. Mycologia 95, 474–479.

McCune, B., Mefford, M.J., 2006. PC-ORD. Multivariate Analysis of Ecological Data, Version 4.0. MjM Software Design, Gleneden Beach, OR.

McHugh, R., 1998. Corticolous myxomycetes from Glen Mhuire, Co. Wicklow. Mycologist 12, 166–168.

McHugh, R., 2005. Moist chamber culture and field collections of myxomycetes from Ecuador. Mycotaxon 92, 107–118.

Mitchell, D.W., 1977. The bark myxomycetes—their collection, culture and identification. School Sci. Rev. 58, 444–455.

Mitchell, D.W., 1978. A key to the corticolous myxomycetes part II. Bull. Br. Mycol. Soc. 12, 90–107.

Moore, D.L., Stephenson, S.L., 2003. Microhabitat distribution of protostelids in a tropical wet forest in Costa Rica. Mycologia 95, 11–18.

Morris, E.K., Caruso, T., Buscot, F., Fischer, M., Hancock, C., Maier, T.S., Meiners, T., Müller, C., Obermaier, E., Prati, D., Socher, S.A., Sonnemann, I., Wäschke, N., Wubet, T., Wurst, S., Rillig, M.C., 2014. Choosing and using diversity indices: insights for ecological applications from the German Biodiversity Exploratories. Ecol. Evol. 4, 3514–3524.

Mosquera, J., Lado, C., Estrada-Torres, A., Beltrán Tejera, C., 2003. Description and culture of a new Myxomycete, *Licea succulenticola*. Anales Jard. Bot. Madrid 60, 3–10.

Müller, H.T., Hoppe, T., Ferchen, T., 2008. Notizen über einen aquatischen Schleimpilz. DATZ 61, 80–81.

Ndiritu, G.G., Spiegel, F.W., Stephenson, S.L., 2009. Distribution and ecology of the assemblages of myxomycetes associated with major vegetation types in Big Bend National Park. Fungal Ecol. 2, 168–183.

Newton, A.F.S., Stephenson, S.L., 1990. A beetle/slime mold assemblange from Northern India (Coleoptera; myxomycetes). Orient. Insects 24, 197–218.

Nordén, B., Ryberg, M., Götmark, F., Olausson, B., 2004. Relative importance of coarse and fine woody debris for the diversity of wood-inhabiting fungi in temperate broadleaf forests. Biol. Conserv. 117, 1–10.

Novozhilov, Y.K., Erastova, D.A., Shchepin, O.N., Schnittler, M., Aleksandrova, A.V., Popov, E.S., Kuznetsov, A.N., 2017a. Myxomycetes associated with monsoon lowland tropical forests in southern Vietnam. Nova Hedwigia 104 (1–3), 143–182.

Novozhilov, Y.K., Mitchell, D.W., Schnittler, M., 2003. Myxomycete biodiversity of the Colorado Plateau. Mycol. Prog. 2, 243–258.

Novozhilov, Y.K., Okun, M.V., Erastova, D.A., Shchepin, O.N., Zemlyanskaya, I.V., García-Carvajal, E., Schnittler, M., 2013a. Description, culture and phylogenetic position of a new xerotolerant species of *Physarum*. Mycologia 105, 1535–1546.

Novozhilov, Y.K., Schnittler, M., 1997. Nivicole myxomycetes of the Khibine mountains (Kola peninsula). Nordic J. Bot. 16, 549–561.

Novozhilov, Y.K., Schnittler, M., 2000. A new coprophilous species of *Perichaena* (myxomycetes) from the Russian Arctic (the Taimyr Peninsula and the Chukchi Peninsula). Karstenia 40, 117–122.

Novozhilov, Y.K., Schnittler, M., 2008. Myxomycete diversity and ecology in arid regions of the Great Lake Basin of western Mongolia. Fungal Divers. 30, 97–119.

Novozhilov, Y.K., Schnittler, M., Erastova, D.A., Okun, M.V., Schepin, O.N., Heinrich, E., 2013b. Diversity of nivicolous myxomycetes of the Teberda State Biosphere Reserve (Northwestern Caucasus, Russia). Fungal Divers. 59, 109–130.

Novozhilov, Y.K., Schnittler, M., Erastova, D.A., Shchepin, O.N., 2017b. Myxomycetes of the Sikhote-Alin State Nature Biosphere Reserve (Far East, Russia). Nova Hedwigia 104 (1–3), 183–209.

Novozhilov, Y.K., Schnittler, M., Rollins, A., Stephenson, S.L., 2001. Myxomycetes in different forest types in Puerto Rico. Mycotaxon 77, 285–299.

Novozhilov, Y.K., Schnittler, M., Stephenson, S.L., 1999. Myxomycetes of the Taimyr Peninsula (north-central Siberia). Karstenia 39, 77–97.

Novozhilov, Y.K., Schnittler, M., Vlasenko, A.V., Fefelov, K.A., 2010. Myxomycete diversity of the Altay Mts. (southwestern Siberia, Russia). Mycotaxon 111, 91–94.

Novozhilov, Y.K., Schnittler, M., Zemlianskaia, I.V., Fefelov, K.A., 2000. Biodiversity of plasmodial slime moulds (Myxogastria): measurement and interpretation. Protistology 1, 161–178.

Novozhilov, Y.K., Zemlyanskaya, I., Schnittler, M., Stephenson, S.L., 2006. Myxomycete diversity and ecology in the arid regions of the Lower Volga River Basin (Russia). Fungal Divers. 23, 193–241.

Nyberg, A., Persoon, I.-L., 2002. Habitat differences of coprophilous fungi on moose dung. Mycol. Res. 106, 1360–1366.

Ohlson, M., Söderström, L., Hörnberg, G., Zackrisson, O., 1997. Habitat qualities versus long-term continuity as determinants of biodiversity in boreal old-growth swamp forests. Biol. Conserv. 81, 221–231.

Oksanen, J., Blanchet, F.G., Friendly, M., Kindt, R., Legendre, P., McGlinn, D., Minchin, P.R., O'Hara, R.B., Simpson, G.L., Solymos, P., Stevens, M.H.H., Szoecs, E., Wagner, H., 2016. Community Ecology Package "Vegan," 2.4-0 ed. Available from: https://cran.r-project.org, https://github.com/vegandevs/vegan

Ostrofsky, A., Shigo, A.L., 1981. A myxomycete isolated from discolored wood of living red maple. Mycologia 73, 997–1000.

Pando, F., 1989. Un estudio sobre los myxomycetes corticolas de la isla de Mallorca. Anales Jard. Bot. Madrid 46, 181–188.

Pando, F., Lodo, C., 1987. Myxomycetes corticicolas ibericos, I: especies sobre Juniperus thurifera. Bol. Soc. Micol. Madrid 11, 203–212.

Parker, H., 1946. Studies in the nutrition of some aquatic Myxomycetes. J. Elisha Mitchell Sci. Soc. 62, 231–247.

Pendergrass, L., 1976. Further studies on corticolous myxomycetes from within the city limits of Atlanta, Georgia. Department of Biology, Atlanta University, Atlanta, p. 136.

Penfound, W.T., 1940. A note concerning the relation between drainage pattern, bark conditions, and the distribution of corticolous bryophytes. Bryologist 43, 169–170.

Peterson, J.E., 1952. Myxomycetes developed on bark of living trees in moist chamber culture. Master thesis, Michigan State College, East Lansing, MI, p. 104.

Pianka, E.R., 1973. The structure of lizard communities. Ann. Rev. Ecol. Syst. 4, 53–74.

Renvall, P., 1995. Community structure and dynamics of wood-rotting Basidiomycetes on decomposing conifer trunks in northern Finland. Karstenia 35, 1–51.

Rodríguez-Palma, M., Varela-Garcia, A., Lado, C., 2002. Corticolous myxomycetes associated with four tree species in Mexico. Mycotaxon 81, 345–355.

Rojas, C., Doss, R.G., 2014. Does habitat loss affect tropical myxomycetes. Mycosphere 5, 692–700.

Rojas, C., Stephenson, S.L., 2007. Distribution and ecology of myxomycetes in the high-elevation oak forests of Cerro Bellavista, Costa Rica. Mycologia 99, 534–543.

Rojas, C., Stephenson, S., 2013. Effect of forest disturbance on myxomycete assemblages in the southwestern Peruvian Amazon. Fungal Divers. 59, 45–53.

Rojas, C., Schnittler, M., Biffi, D., Stephenson, S.L., 2008. Microhabitat and niche separation in species of *Ceratiomyxa*. Mycologia 100, 843–850.

Rojas, C., Valverde, R., 2015. Ecological patterns of lignicolous myxomycetes from two different forest types in Costa Rica. Nova Hedwigia 101, 21–34.

Rojas, C., Valverde, R., Stephenson, S.L., Vargas, M.J., 2010. Ecological patterns of Costa Rican myxomycetes. Fungal Ecol. 3, 139–147.

Rollins, A.W., Stephenson, S.L., 2012. Myxogastrid distribution within the leaf litter microhabitat. Mycosphere 3, 543–549.

Rollins, A., Stephenson, S., 2013. Myxomycetes associated with grasslands of the western central United States. Fungal Divers. 59, 147–158.

Ronikier, A., Ronikier, M., 2009. How "alpine" are nivicolous myxomycetes? A worldwide assessment of altitudinal distribution. Mycologia 101, 1–16.

Rufino, M.U.L., Cavalcanti, L.H., 2007. Alterations in the lignicolous myxomycete biota over two decades at the Dois Irmãos Ecologic State Reserve, Recife, Pernambuco, Brazil. Fungal Divers. 24, 159–171.

Samuelsson, J., Gustafsson, L., Ingelög, T., 1994. Dying and Dead Trees: A Review of Their Importance for Biodiversity. Art Databanken, Upsala.

Schnittler, M., 2001a. Ecology of Myxomycetes of a winter-cold desert in western Kazakhstan. Mycologia 93, 653–669.

Schnittler, M., 2001b. Foliicolous liverworts as a microhabitat for neotropical myxomycetes. Nova Hedwigia 72, 259–270.

Schnittler, M., Dagamac, N.H.A., Sauke, M., Wilmking, M., Buras, A., Ahlgrimm, S., Eusemann, P., 2016. Ecological factors limiting occurrence of corticolous myxomycetes—a case study from Alaska. Fungal Ecol. 21, 16–23.

Schnittler, M., Erastova, D.A., Shchepin, O.N., Heinrich, E., Novozhilov, Y.K., 2015a. Four years in the Caucasus—observations on the ecology of nivicolous myxomycetes. Fungal Ecol. 14, 105–115.

Schnittler, M., Lado, C., Stephenson, S.L., 2002. Rapid biodiversity assessment of a tropical myxomycete assemblage—Maquipucuna Cloud Forest Reserve, Ecuador. Fungal Divers. 9, 135–167.

Schnittler, M., Mitchell, D.W., 2000. Species diversity in Myxomycetes based on the morphological species concept: a critical examination. Stapfia 73, 55–61.

Schnittler, M., Novozhilov, Y.K., 1996. The myxomycetes of boreal woodlands in Russian northern Karelia: a preliminary report. Karstenia 36, 19–40.

Schnittler, M., Novozhilov, Y.K., 1998. Late-autumn myxomycetes of the northern Ammergauer Alps. Nova Hedwigia 66, 205–222.

Schnittler, M., Novozhilov, Y.K., 2000. Myxomycetes of the winter-cold desert in western Kazakhstan. Mycotaxon 74, 267–285.

Schnittler, M., Novozhilov, Y.K., Carvajal, E., Spiegel, F.W., 2013. Myxomycete diversity in the Tarim basin and eastern Tian-Shan, Xinjiang Prov., China. Fungal Divers. 59, 91–108.

Schnittler, M., Novozhilov, Y.K., Romeralo, M., Brown, M., Spiegel, F.W., 2012. Fruit body-forming protists: myxomycetes and myxomycete-like organisms (Acrasia, Eumycetozoa). In: Frey, W. (Ed.), Engler's Syllabus of Plant Families, Part 1/1. Blue-Green Algae, Myxomycetes and Myxomycete-Like Organisms, Phytoparasitic Protists, Heterotrophic Heterokontobionta and Fungi. thirteenth ed. Bornträger, Stuttgart, pp. 40–88.

Schnittler, M., Novozhilov, Y.K., Shadwick, J.D.L., Spiegel, F.W., García-Carvajal, E., König, P., 2015b. What substrate cultures can reveal: myxomycetes and myxomycete-like organisms from the Sultanate of Oman. Mycosphere 6, 356–384.

Schnittler, M., Shchepin, O.N., Dagamac, N.H.A., Borg Dahl, M., Novozhilov, Yu, K., 2017. Barcoding myxomycetes with molecular markers: challenges and opportunities. Nova Hedwigia 104 (1–3), 323–341.

Schnittler, M., Stephenson, S.L., 2000. Myxomycete biodiversity in four different forest types in Costa Rica. Mycologia 92, 626–637.

Schnittler, M., Stephenson, S.L., 2002. Inflorescences of neotropical herbs as a newly discovered microhabitat for myxomycetes. Mycologia 94, 6–20.

Schnittler, M., Stephenson, S.L., Novozhilov, Y.K., 2000. Ecology and world distribution of *Barbeyella minutissima* (Myxomycetes). Mycol. Res. 104, 1518–1523.

Schnittler, M., Unterseher, M., Pfeiffer, T., Novozhilov, Y.K., Fiore-Donno, A.-M., 2010. Ecology of sandstone ravine myxomycetes from Saxonian Switzerland (Germany). Nova Hedwigia 90, 277–302.

Schnittler, M., Unterseher, M., Tesmer, J., 2006. Species richness and ecological characterization of myxomycetes and myxomycete-like organisms in the canopy of a temperate deciduous forest. Mycologia 98, 223–232.

Schnitzer, S.A., Bongers, F., 2002. The ecology of lianas and their role in forests. Trends Ecol. Evol. 17, 223–230.

Shadwick, J.D., Stephenson, S.L., Spiegel, F.W., 2009. Distribution and ecology of protostelids in Great Smoky Mountains National Park. Mycologia 101, 320–328.

Shchepin, O.N., Novozhilov, Y.K., Schnittler, M., 2014. Nivicolous myxomycetes in agar culture: some results and open problems. Protistology 8, 53–61.

Shearer, C.A., Crane, J.L., 1986. Illinois fungi XII. Fungi and myxomycetes from wood and leaves submerged in southern Illinois swamps. Mycotaxon 25, 527.

Sherman, J.M., 1916. Studies on soil protozoa and their relation to the bacterial flora. II. J. Bacteriol. 1, 165–185.

Siitonen, J., 2001. Forest management, coarse woody debris and saproxylic organisms: fennoscandian boreal forests as an example. Ecol. Bull. 49, 11–41.

Snell, K.L., Keller, H.W., 2003. Vertical distribution and assemblages of corticolous myxomycetes on fire tree species in the Great Smoky Mountains National Park. Mycologia 95, 565–576.

Snell, K.L., Keller, H.W., Eliasson, U.H., 2003. Tree canopy myxomycetes and new records from ground sites in the Great Smoky Mountains National Park. Castanea 68, 97–108.

Soberón, J., 2007. Grinnellian and Eltonian niches and geographic distributions of species. Ecol. Lett. 10, 1115–1123.

Spiegel, F.W., Stephenson, S.L., Keller, H.W., Moore, D.L., Cavender, J.C., 2004. Mycetozoans. In: Mueller, G.M., Bills, G.F., Foster, M.S. (Eds.), Biodiversity of Fungi: Inventory and Monitoring Methods. Elsevier Academic Press, Amsterdam, pp. 547–576.

Stephenson, S.L., 1985. Slime molds in the laboratory II: moist chamber culture. Am. Biol. Teach. 47, 487–489.

Stephenson, S.L., 1988. Distribution and ecology of myxomycetes in temperate forests I. Patterns of occurence in the upland forests of southwestern Virginia. Can. J. Bot. 66, 2187–2207.

Stephenson, S.L., 1989. Distribution and ecology of myxomycetes in temperate forests. II. Patterns of occurence on bark surface of living trees, leaf litter, and dung. Mycologia 81, 608–621.

Stephenson, S.L., 2011. From morphological to molecular: studies of myxomycetes since the publication of the Martin and Alexopoulos (1969) monograph. Fungal Divers. 50, 21–34.

Stephenson, S.L., Estrada-Torres, A., Schnittler, M., Lado, C., Wrigley de Basanta, D., Ogata, N., 2003. Distribution and ecology of myxomycetes in the forests of Yucatan. In: Gomez-Pompa, A., Allen, M.F., Fedick, S.L., Jiménez-Osornio, J.J. (Eds.), The Lowland Maya Area. Three Millenia at the Human-Wildland Interface. Food Products Press, New York, NY, pp. 241–259.

Stephenson, S.L., Feest, A., 2012. Ecology of soil eumycetozoans. Acta Protozool. 51, 201–208.

Stephenson, S.L., Fiore-Donno, A.M., Schnittler, M., 2011. Myxomycetes in soil. Soil Biol. Biochem. 43, 2237–2242.

Stephenson, S.L., Kalyanasundaram, I., Lakhanpal, T.N., 1993. A comparative biogeographical study of myxomycetes in the mid-Appalachians of eastern North America and two regions of India. J. Biogeogr. 20, 645–657.

Stephenson, S.L., Landolt, J.C., 1996. The vertical distribution of dictyostelids and myxomycetes in the soil/litter microhabitat. Nova Hedwigia 62, 105–117.

Stephenson, S.L., Landolt, J., 2015. Occurrence of myxomycetes in the aerial "canopy soil" microhabitat. Mycosphere 6, 74–77.

Stephenson, S.L., Landolt, J.C., Moore, D.L., 1999. Protostelids, dictyostelids, and myxomycetes in the litter microhabitat of the Luquillo Experimental Forest, Puerto Rico. Mycol. Res. 103, 209–214.

Stephenson, S.L., Novozhilov, Y.K., Schnittler, M., 2000. Distribution and ecology of myxomycetes in high-latitude regions of the northern hemisphere. J. Biogeogr. 27, 741–754.

Stephenson, S.L., Schnittler, M., Lado, C., 2004a. Ecological characterization of a tropical myxomycete assemblage—Maquipucuna Cloud Forest Reserve. Ecuador. Mycol. 96, 488–497.

Stephenson, S.L., Schnittler, M., Lado, C., Estrada-Torres, A., Wrigley de Basanta, D., Landolt, J.C., Novozhilov, Y.K., Clark, J., Moore, D.L., Spiegel, F.W., 2004b. Studies of neotropical mycetozoans. Syst. Geogr. Plants 74, 87–108.

Stephenson, S.L., Stempen, H., 1994. Myxomycetes: A Handbook of Slime Molds. Timber Press, Portland, OR.

Stephenson, S.L., Studlar, S.M., 1985. Myxomycetes fruiting upon bryophytes: coincidence or preference. J. Bryol. 13, 537.

Stephenson, S.L., Urban, L., Rojas, C., McDonald, M., 2008. Myxomycetes associated with woody twigs. Rev. Mex. Micol. 27, 21–28.

Stojanowska, W., 1980a. Comparison of myxomycetes of the forest in Skarszyn and of the beech reserve in Muszkowice. Acta Mycol. 16, 221, [in Polish, with English summary].

Stojanowska, W., 1980b. The seasonal variation of myxomycetes flora in the Muszkowicki Las Bukowy reserve (Lower Silesia). Fragm. Flor. Geobot. 26, 103–113, [in Polish, with English summary].

Sun, H., Santalahtia, M., Pumpanen, J., Köster, K., Berninger, F., Raffaello, T., Jumpponen, A., Asiegbu, F.O., Heinonsalo, J., 2015. Fungal community shifts in structure and function across a boreal forest fire chronosequence. Appl. Environ. Microbiol. 81, 7869–7880.

Sunhede, S., 1973. Studies in myxomycetes. I. On the growth and ecology of the myxomycete Relicularia lycoperdon. Svensk Bot. Tidskr. 67, 172–176.

Takahashi, K., 1995. An ecology of myxomycetes in secondary forests of coastal areas in Okayama Prefecture—regarding substratum and seasonal changes of myxomycetes. Bull. Okayama Pref. Nat. Conserv. Center 3, 23–31.

Takahashi, K., 2001. Ocurrence of lignicolous myxomycetes and their association with the decaying state of coniferous wood in sub-alpine forests of Central Japan. Asahi Bull. 22, 3–15.

Takahashi, K., 2004. Occurence of myxomycetes on different decay states of decidouos broadleaf and coniferous wood in a natural temperate forest at southwest of Japan. Syst. Geogr. Plants 74, 133–142.

Takahashi, K., Hada, Y., 2008. Seasonal change in the species composition and dominant species of myxomycete communities on dead wood of *Pinus densiflora* in a warm temperate forest. Hikobia 15, 145–154.

Takahashi, K., Hada, Y., 2009. Distribution of myxomycetes on coarse woody debris of *Pinus densiflora* at different decay stages in secondary forests of western Japan. Mycoscience 50, 253–260.

Takahashi, K., Harakon, Y., 2012. Comparison of wood-inhabiting myxomycetes in subalpine and montane coniferous forests in the Yatsugatake Mountains of Central Japan. J. Plant Res. 125, 327–337.

Tamayama, M., Keller, H.W., 2013. Aquatic myxomycetes. FUNGI 6, 18–24.

Taylor, K.M., Feest, A., Stephenson, S.L., 2015. The occurrence of myxomycetes in wood. Fungal Ecol. 17, 179–182.

Ter Braak, C.J.F., 1987. The analysis of vegetation-environment relationships by canonical correspondence analysis. Vegetatio 69, 69–77.

ter Braak, C.J.F., Verdonschot, P.E.M., 1995. Canonical correspondence analysis and related multivariate methods in aquatic ecology. Aquat. Sci. 57, 255–289.

Thomas, P., Packham, J., 2007. Ecology of Woodlands and Forests. Description, Dynamics and Diversity. Cambridge University Press, New York, NY.

Tiunov, A.V., Semenina, E.S., Aleksandrova, A.V., Tsurikov, S.M., Anichkin, A.E., Novozhilov, Y.K., 2015. Stable isotope composition (δ13C and δ15N values) of slime molds: placing bacterivorous soil protozoans in the food web context. Rapid Commun. Mass Spectrom. 29, 1–8.

Tran, H.T.M., Stephenson, S.L., Hyde, K.D., Mongkolporn, O., 2006. Distribution and occurrence of myxomycetes in tropical forests of northern Thailand. Fungal Divers. 22, 227–242.

Unterseher, M., Otto, P., Morawetz, W., 2005. Species richness and substrate specificity of lignicolous fungi in the canopy of a temperate, mixed deciduous forest. Mycol. Prog. 4, 117–132.

Unterseher, M., Schnittler, M., Dormann, C., Sickert, A., 2008. Application of species richness estimators for the assessment of fungal diversity. FEMS Microbiol. Lett. 282, 205–213.

Urich, T., Lanzén, A., Qi, J., Huson, D.H., Schleper, C., Schuster, S.C., 2008. Simultaneous assessment of soil microbial community structure and function through analysis of the metatranscriptome. PLoS One, e2527.

Vlasenko, A.V., Novozhilov, Y.K., 2011. Phenological features of myxomycetes in pine forests of the right-bank of the Upper Ob' River. Plant World of Asian Russia 2, 3–9, [in Russian].

Voříšková, J., Baldrian, P., 2013. Fungal community on decomposing leaf litter undergoes rapid successional changes. ISME J. 7, 477–486.

Warcup, J., 1950. The soil plate method for the isolation of fungi from soil. Nature 166, 117–118.

Ward, M.H., 1886. The morphology and physiology of an aquatic myxomycete. Q. J. Microsc. Sci. 24, 64–86.

Wheeler, Q.D., 1980. Studies in neotropical slime mold/beetle relationships, part I: natural history and description of a new species of *Anisotoma* from Panama (Coleoptera: Leiodidae). Proc. Entomol. Soc. Wash. 82, 493–498.

Whittaker, R.H., 1965. Dominance and diversity in plant communities. Science 147, 250–260.

Whittaker, R.H., Levin, S.A., Root, R.B., 1973. Niche, habitat, and ecotope. Am. Nat. 107, 321–338.

Wikmark, O.G., Haugen, P., Haugli, K., Johansen, S.D., 2007a. Obligatory group I introns with unusual features at positions 1949 and 2449 in nuclear LSU rDNA of *Didymiaceae* myxomycetes. Mol. Phylogenet. Evol. 43, 596–604.

Wikmark, O.G., Haugen, P., Lundblad, E.W., Haugli, K., Johansen, S.D., 2007b. The molecular evolution and structural organization of group I introns at position 1389 in nuclear small subunit rDNA of myxomycetes. J. Eukaryot. Microbiol. 54, 49–56.

Wilson, J.B., 1991. Methods for fitting dominance/diversity curves. J. Veg. Sci. 2, 35–46.

Wrigley de Basanta, D., 1998. Myxomycetes from the bark of the evergreen oak *Quercus ilex*. Anales Jar. Bot. Madrid 56, 3–14.

Wrigley de Basanta, D., 2000. Acid deposition in Madrid and corticolous myxomycetes. Stapfia 73, 113.

Wrigley de Basanta, D., 2004. The effect of simulated acid rain on corticolous myxomycetes. Syst. Geogr. Plants 74, 175–181.

Wrigley de Basanta, D., Lado, C., Stephenson, S.L., Estrada-Torres, A., 2002. Myxomycetes from moist chamber cultures of neotropical substrates. In: Rammeloo, J., Bogaerts, A. (Eds.), Fourth International Congress on Systematics and Ecology of Myxomycetes. National Botanic Garden, Brussels, p. 100.

Wrigley de Basanta, D., Stephenson, S.L., Lado, C., Estrada-Torres, A., Nieves-Rivera, A.M., 2008. Lianas as a microhabitat for myxomycetes in tropical forests. Fungal Divers. 28, 109–125.

Yang, Z., Liu, X., Zhou, M., Ai, D., Wang, G., Wang, Y., Chu, C., Lundholm, J.T., 2015. The effect of environmental heterogeneity on species richness depends on community position along the environmental gradient. Sci. Rep. 5, 15723.

Zahn, G., Stephenson, S.L., Spiegel, F.W., 2014. Ecological distribution of protosteloid amoebae in New Zealand. PeerJ 2, e296.

Chapter 9

Biogeographical Patterns in Myxomycetes

Martin Schnittler*, Nikki Heherson A. Dagamac*, Yuri K. Novozhilov**
**Institute of Botany and Landscape Ecology, Ernst Moritz Arndt University Greifswald, Greifswald, Germany; **Komarov Botanical Institute of the Russian Academy of Sciences, St. Petersburg, Russia*

INTRODUCTION

Only a very few major lineages of life, namely the Metazoa (animals), Embryophyta (terrestrial plants), and Eumycota (some Ascomycetes and Basidiomycetes) evolved into truly multicellular, macroscopic organisms. Myxomycetes are a significant exception among protists due to their noncellular fruiting bodies, which are not formed out of a growth process but by reformation of the second vegetative stage, the *plasmodium*. Their mostly macroscopic fruiting bodies are the precondition to colonize habitat islands, characterized by a locally higher density of other microorganisms serving as prey, by spores (Schnittler and Tesmer, 2008). Understanding dispersal ecology and speciation patterns in the group is the key to understand and describe species distribution and diversity patterns.

Spore Dispersal as a Key Feature for Myxomycetes

Among protists, myxomycetes are unique for their multinucleate and macroscopic plasmodia but not in dispersing by spores. Stalked fruiting bodies that release airborne spores evolved independently in several protistean groups, characterized by similarities in their life style (i.e., predators of other microbes). These groups can be seen as an ecological guild (myxomycetes and myxomycete-like organisms, Schnittler et al., 2006). If the myxobacteria (Reichenbach, 1993) are considered, this would even include a group of prokaryotes. All Eumycetozoans [dictyostelids (Romeralo et al., 2010, 2011; Swanson et al., 2002)], the isolated genus *Ceratiomyxa* (Fiore-Donno et al., 2010), and the myxomycetes (Schnittler et al., 2012; Stephenson and Schnittler, 2016) form primarily stalked fruiting bodies. Quite heterogenous and paraphyletic protistean groups dispersing via spores are the protosteliid (Spiegel et al., 2004), and acrasiid amoebae

Myxomycetes: Biology, Systematics, Biogeography, and Ecology. http://dx.doi.org/10.1016/B978-0-12-805089-7.00009-3

(Brown et al., 2012). Surprisingly, even a genus within Ciliata [*Sorogena* (Bardele et al., 1991)] develops stalked fruiting bodies that release spores. These cases of parallel evolution can be seen as evidence that the terrestrial life style required an efficient means of dispersal, since metabolically active cells are easily transported by water but not by air due to the danger of desiccation. The obvious solution are spores, being simultaneously dormant and transport stages. Spores in the guild range from 1 to 3 μm (prokaryotic myxobacteria with a much smaller genome) over (3−)5–8 μm in dictyostelids, (4)7–12(−20) μm in myxomycetes to ca. 35 μm in *Sorogena*, where the spores transport a micro- and a macronucleus as it is typical for ciliates.

Spore size is the first critical parameter for dispersal, determining largely the terminal (sedimentation) velocity of spores in still air. For myxomycetes, figures between 1 mm/s (spore size ca. 7 μm) and 3 mm/s (12 μm) were reported by Tesmer and Schnittler (2007) and Schnittler et al. (2006). These values correlate well with figures calculated employing Stokes' law (predicting the terminal velocity of small spherical bodies), and if this holds true, spore size should be under strong selective pressure, since a spore of *Trichia varia* (mean diameter 13.0 μm) falls about four times faster than one of *Stemonitis axifera* (mean diameter 6.3 μm). Consequently, species with smaller spores should achieve higher dispersal abilities.

The second critical parameter for dispersal is the starting point of a spore, determined mostly by the ability of myxomycete plasmodia to react positively phototropic as soon as the point of no return for the conversion of the plasmodial biomass into fruiting bodies is reached. As such, aerial habitats, such as the tree canopy (Schnittler et al., 2006; Snell et al., 2003) are especially effective for myxomycete dispersal. Since the trophic stages of myxomycetes, amoeboflatellates, and plasmodia, both require liquid water for active growth and movement, fruiting bodies often develop on substrates covered by a film of water. Taking this into consideration, the lengths of 0.5–3 mm typically achieved by the stalks of myxomycete fruiting bodies make a difference, since they elevate the spore case above the water film, allowing the spores to dry out and become airborne. Coinciding with these considerations, stalked fruiting bodies are an ancestral character for myxomycetes (Fiore-Donno et al., 2010). Stalk length, taking away resources from the developing *plasmodium*, seems to be under selective pressure; widely distributed species of myxomycetes seem to develop longer stalks in tropical regions (Stephenson et al., 2004, 2008). In periodically dry environments, such as the bark surface of living trees, in addition to stalked species with a short development time, a second strategy can be seen—species developing more slowly with more robust, sessile fruiting bodies, which open in dry periods (Schnittler, 2001).

An additional dispersal strategy includes animal vectors, often invertebrates. Among the Coleoptera, the genera *Anisotoma* and *Agathidium* are specialized in feeding on myxomycete fruiting bodies (Blackwell et al., 1982; Wheeler, 1984). Other recorded observations of animals dispersing myxomycete spores include Diptera (Buxton, 1954), Collembola (Chassain, 1973), acarids (Keller and Smith, 1978), tartigrades (Kylin, 1991), various earthworms (Murray et al., 1985), slugs (Keller and Snell, 2002), and nematodes (Kilgore and Keller, 2008). In addition, birds (Sutherland, 1979), as well as lizards (Townsend et al., 2005) have been observed to disperse myxomycetes. Dispersal of spores by water cannot be ruled out, since experiments have confirmed observations that myxomycetes may fruit and produce viable spores in submerged conditions (Lindley et al., 2007; Vlasenko et al., 2016).

Whereas most species of myxomycete develop colonies of stalked but separated fruiting bodies, about 10% of all species develop compound fruiting bodies, which can reach a considerable size (in *Brefeldia maxima* up to 1m). Some of these fruiting bodies mimic puffballs (raindrop ballists; Dixon, 1963), and like some gasteromycetes the spores of these genera (*Lycogala*, *Reticularia*, *Tubifera*) possess a reticulum of elevated ridges, making them extremely hydrophobic (Hoppe and Schwippert, 2014). Species with compound fruiting bodies rely more on animal vectors (*Fuligo* and *Symphytocarpus*) seem to have less hydrophobic spores, ornamented by warts or spines. Smaller compound fruiting bodies may be fascicle-stalked (*Arcyria cinerea* var. *digitata* and *Alwisia bombarda*), whereas large compound fruiting bodies (exceeding 2 cm diam.) are always sessile (*Fuligo* spp. and *Tubifera* spp.). The transition from solitary to compound fruiting bodies occurred in several lineages independently (Leontyev et al., 2015).

The third parameter important for dispersal is spore ornamentation, enhancing the hydrophobicity of spores. A high hydrophobicity will enable spores to become airborne again, even from wet surfaces, although no quantitative data are yet available for this secondary dispersal. Only a very few species of myxomycetes possess completely smooth spores. Noteworthy are some coprophilous species, which fruit on herbivore dung. Their thick-walled spores seem to be adapted to passage through the digestive tract of herbivorous animals (Eliasson and Lundqvist, 1979).

Height of release, spore size, and ornamentation seem to be the three decisive parameters for long-distance dispersal of spores. Myxomycete spores have been detected in the air (Kamono et al., 2009), but no quantitative dispersal models have yet been developed for myxomycetes. From experiments with fungal spores (Penet et al., 2014) and pollen (Ottewell et al., 2012; Robledo-Arnuncio and Gil, 2005) of comparable sizes, we can assume successful long-distance dispersal but with leptokurtic dispersal curves (having a long "tail"), which enable myxomycetes to reach distant habitat islands with a sufficiently high density of microbes. This was indirectly confirmed by surveys on remote islands, which revealed relatively diverse myxomycete assemblages [Macquarie Island, south of Tasmania (Stephenson et al., 2007); La Reunion (Adamonytė et al., 2011); Papua New Guinea (Kylin et al., 2013); Bohol Island (Macabago et al., 2017); and the occurrence of myxomycetes in the aerial woody debris and "canopy soil" (Schnittler et al., 2006; Stephenson and Landolt, 2015) microhabitats. As shown by model calculations, the successful colonization of such habitat islands depends on the reproductive system of a species (Schnittler and Tesmer, 2008). Homothallic strains [presumably asexual (Clark and Haskins, 2013)] need only a single spore to colonize a new habitat, whereas heterothallic (sexual) strains require two compatible spores. Since in addition to the formation of fruiting bodies, the predatory life style is another joint character of the guild of myxomycetes and myxomycete-like organisms, all of the groups of organisms mentioned at the beginning of this text must be assumed to be locally distributed (i.e., confined to habitat islands, such as decaying logs, accumulations of litter, or other decaying plant material that provide abundant microbial prey).

MYXOMYCETE BIOGEOGRAPHY—WHAT CAN WE SEE?

Due to their comparatively conspicuous fruiting bodies, myxomycetes were "discovered" much earlier than most other protistean groups, attracting the attention of naturalists, including Linnaeus (1753) who described some species as miniature puffballs—in an ecological sense, he was not terribly wrong. The first major monograph of the group (Rostafinsky, 1875) recognized a significant proportion of the ca. 945 morphospecies known today (Lado, 2005). Therefore, we can look back on a history of nearly 200 years of records of fruiting bodies, which has resulted in a significant body of data on myxomycete distribution.

However, it should be noted that this entire body of data is based only on records of the fruiting bodies, not amoeboflagellates or plasmodia as the active life forms. In this respect, we are like a gardener who wants to know the distribution of apple trees in an orchard, but he can see only the apples, not the trees. Small and nonfruiting trees under suboptimum conditions remain invisible for him. Therefore, we must assume that we see nothing more than the "tip of the iceberg"—myxomycete populations growing under optimum conditions that successfully fruit. This might be especially true for soil myxomycetes [see Stephenson et al. (2011) for a review]. Amoeboflagellates of myxomycetes were found to be the most abundant group of protists in a study based on ePCR with mRNA sequences (Urich et al., 2008), but fruiting bodies of myxomycetes on bare soil are rarely reported; accumulations of litter with a higher number of nutrients and thus microbes seem to be required for formation of fruiting bodies. In addition, it cannot be ruled out that nonfruiting strains of myxomycetes do exist in nature, because a loss of functionality in a single gene in the complex ontogenesis of fruiting bodies may halt spore formation and thus severely interrupt dispersal abilities, similar to the loss of a stalk in the sessile *Semimorula liquescens*, where molecular data suggest a close relationship to *Echinostelium*, a genus forming long-stalked fruiting bodies (Fiore-Donno et al., 2009). There is no reason to assume that nonfruiting strains cannot persist indefinitely as populations of amoeboflatellates in soil, even if they lost the ability for long-distance dispersal via spores. As such, we can expect that a species may often have a larger range than indicated by records of fruiting bodies. This has been shown to be the case for nivicolous myxomycetes, a peculiar group of soil inhabiting myxomycetes, which fruit in spring near melting snow banks (Schnittler et al., 2015a). Fruitings are abundant only in alpine situations (Ronikier and Ronikier, 2009), which usually means mountains with a long-lasting contiguous snow cover. However, two studies based on ePCR from the lowlands of Germany (Fiore-Donno et al., 2016) and the lowlands of northwestern Russia (Shchepin, Novozhilov pers. commun.) detected sequences of several nivicolous species in areas where their fruiting bodies have never been found. As such, we must assume that a particular species of myxomycete may be more widely distributed than indicated by records of fruiting bodies alone.

However, a second line of thinking points in the opposite direction. The current species concept in myxomycetes is a morphological one, at least with respect to our data on species distribution. Another concept, though several decades old, found never its way into diversity research, since it relies on amoeboflagellate compatibility. Early experiments, carried out mostly with cultivable members of the Physarales, demonstrated that a morphospecies might include strains that are incompatible to each other, forming separate biospecies (Collins, 1979). In addition, asexual (presumably diploid) strains may exist (Clark and Haskins, 2013). Therefore, one morphospecies may include several cryptic biospecies. Verifying this for a given morphospecies requires laborious cultivation, but only a minor fraction of all myxomycetes (mostly members of the Physarales) have been cultivated successfully from spore to spore. The different reproductive options and its consequences for species concepts are discussed in Feng et al. (2016) and Walker and Stephenson (2016).

Molecular studies may be a tool for independent verification of the morphospecies concept. In contrast to pathogenic protists, where molecular research was spurred by health arguments [e.g., *Entamoeba* (García et al., 2014)], myxomycetes are neither pathogenic nor of significant economic importance. Therefore, molecular tools for species differentiation are still at an early state of development, and the species concept in the group is entirely based on the morphology of the fruiting bodies. The first phylogenetic studies (Fiore-Donno et al., 2005, 2008, 2010, 2011, 2012, 2013) did not only challenge the classical taxonomic concept of the group but also enabled researchers to develop barcoding markers (Schnittler et al., 2017). As in other protists, the 18S rRNA gene (SSU), which is usually a multicopy gene (Torres-Machorro et al., 2010), is the most promising candidate for such a barcode (Adl et al., 2014). The first studies employing barcodes [*Badhamia melanospora* (Aguilar et al., 2014) and nivicolous myxomycetes (Novozhilov et al., 2013b)] always found several ribotypes per morphospecies. This was confirmed by other studies that used additional markers, which enabled these authors to demonstrate cryptic speciation [*T. varia* (Feng and Schnittler, 2015), *Meriderma* spp. (Feng et al., 2016), and *Hemitrichia serpula* (Dagamac et al., 2017c)]. In addition, a first survey of wood inhabiting, bright-spored myxomycetes (Feng and Schnittler, 2017) estimated a relation between morphospecies and ribotypes of 1:2 to 1:10, which points to a considerable amount of cryptic speciation. In other words, the current morphospecies concept is very likely underestimating diversity, in spite of the steep increase in newly described morphospecies within the past few decades (Schnittler and Mitchell, 2000). In addition, two studies tracing the worldwide distribution of a morphospecies based on the distribution of its ribotypes (Aguilar et al., 2014; Dagamac et al., 2017c) found ribotypes, constituting putative cryptic species, to be more limited in distribution than the morphospecies as a whole. This provides evidence for a contrasting hypothesis that many morphospecies may in fact be complexes of cryptic species showing more limited distribution patterns than the morphospecies as a whole.

Biogeographic Hypotheses About Myxomycete Distribution

The considerations mentioned earlier have to be taken into account for the discussion of two controversial hypotheses that are used to explain biogeographic patterns in myxomycetes. Similar to the situation in other protists, the Baas-Becking model of ubiquity (Bass and Boenigk, 2011; Finlay, 2002; Finlay and Clarke, 1999) was opposed by the moderate endemicity model (Foissner, 1999). The first model argues that the ability of protists to have almost unlimited dispersal due to their small cell (spore) sizes causes them to be found everywhere (Finlay et al., 2004); geographical barriers should not limit distribution (Finlay, 2002). In this case, only habitat suitability will limit the occurrence of a protist species in a certain region. The second model claims that some protists may be cosmopolitan but others show geographically restricted distribution patterns (Foissner, 2006) and still others may even be endemic for a particular locality (Cotterill et al., 2008; Martiny et al., 2006). These models attenuate the tenet of "everything is everywhere" by proposing that (1) the abundances and thus the migration rates are rather low for the majority of species, (2) extinction rates are moderate, and (3) the proportion of the global species pool found locally is moderate. Indeed, for ciliates, a group usually not dispersing by spores, about 30% of the species were found to be regionally endemic (Foissner, 2008). Thus far, local endemicity has not been reported for myxomycetes, and even surveys on remote archipelagos have not revealed endemic taxa at the morphospecies level [Galapagos Islands (Eliasson and Nannenga-Bremekamp, 1983); the Hawaiian archipelago (Eliasson, 1991); Palawan Island (Pecundo et al., 2017)]. Regional endemicity may occur in some species. Possible examples include species recently described from deserts, such as *Physarum pseudonotabile* in Central Asian deserts (Novozhilov and Schnittler, 2008; Novozhilov et al., 2013a), or *Perichaena calongei, Didymium infundibuliforme*, and *Physarum atacamense* in South America (Araujo et al., 2015; Lado et al., 2009; Wrigley de Basanta et al., 2009, 2012). More systematic surveys are needed to determine if these species truly represent cases of regional endemicity or simply appear so due to a lack of data from other regions.

Whereas already rare dispersal events are sufficient to colonize new and remote localities, regular dispersal is necessary to ensure a minimum level of gene flow between local populations to prevent regional speciation. The first available studies employing molecular methods to investigate geographic separation among populations of myxomycetes at transcontinental scales revealed different patterns. For *Didymium difforme* [mtDNA (Winsett and Stephenson, 2011)] and *D. squamulosum* [ITS (Winsett and Stephenson, 2008)] no or only weak evidence for geographic separation was found. Marker resolution may be a question. A study investigating *T. varia* throughout Eurasia with three different markers [SSU, EF1A, COI (Feng and Schnittler, 2015)] found cryptic speciation but with no apparent geographic limitation for these species. However, the investigated region was limited to Eurasia. In contrast, studies of *Badhamia melanospora* [SSU (Aguilar et al., 2014)] and *Hemitrichia serpula* [SSU, EF1A (Dagamac et al., 2017c)], based on specimens collected in different parts of the world, revealed clear evidence for geographical limitation of certain genotypes. In such cases, low gene flow and/or mutations in mating-type genes governing amoeboflagellate compatibility may facilitate the evolution of regionally endemic species.

To settle this discussion, more population genetic studies of clear cut and widely occurring species of myxomycetes will be needed, which could determine the intensity of gene flow between local populations. This gene flow may be just strong enough to occupy all possible habitats within geological areas, allowing a morphospecies to appear cosmopolitan, but not strong enough to break the dominance of locally adapted genotypes, which may or may not evolve into cryptic species that become regional endemic.

Understanding which of the two models can be best applied to myxomycetes is crucial to estimate the overall diversity of the group. If all or most species are cosmopolitan in distribution, the global diversity of the group should be comparatively low. However, if indeed a significant proportion of species of myxomycetes display restricted distribution patterns and these are not simply caused by the patchiness of suitable habitats, their global diversity may be higher (Mitchell and Meisterfeld, 2005). Whereas the postulated broader distribution of amoeboflagellates in comparison to fruiting populations would support the ubiquity model, the discovery of cryptic species with a more restricted population than the respective morphospecies would comply better with the moderate endemicity model.

TWO-HUNDRED YEARS OF FRUITING BODY-BASED DIVERSITY RESEARCH IN MYXOMYCETES

The first comprehensive monograph of the group (Rostafinsky, 1875) listed a significant part of the ca. 1000 species currently described for the myxomycetes (Lado, 2005), although the number of species per year described as new to science seems to steadily increase. As such, for many morphospecies a considerable body of data exists, relating to collections of fruiting bodies from all over the world. However, as typical for organisms with hidden life styles, systematic surveys for myxomycetes have always been carried out by only a few specialists. As indicated by monographic treatments, the first studied regions were Europe (Lister, 1894, 1911, 1925; Rostafinsky, 1875) and Eastern North America (Hagelstein, 1944; Martin and Alexopoulos, 1969; Massee, 1892). Systematic surveys carried out within the past three decades have added a considerable number of studies of the Neotropics (Lado et al., 2003; Rojas et al., 2010, 2012a,b, 2013; Schnittler et al., 2002). Rather well studied are the deserts of western South America (Estrada-Torres et al., 2009; Lado et al., 2011, 2013, 2014, 2016, 2017; Wrigley de Basanta et al., 2010, 2013) and Central Asia (Novozhilov et al., 2006, 2009; Novozhilov and Schnittler, 2008; Schnittler and Novozhilov, 2000; Schnittler et al., 2013). Only recently have systematic studies been carried out in the Paleotropics, particularly in the Southeast Asian region (Alfaro et al., 2015; Dagamac et al., 2010, 2011, 2012, 2014, 2015a,b,c; Ko Ko et al., 2010b, 2012, 2013; Macabago et al., 2010, 2012, 2016 ; Novozhilov et al., 2017a; Rea-Maminta et al., 2015; Tran et al., 2006, 2008), whereas most parts of tropical Africa remain to be covered. In spite of this geographically unevenly distributed study intensity, the available information makes it possible to compile worldwide distribution maps for particular species.

At least for the formation of fruiting bodies, temperature seems to be an important factor. A significant proportion of the described species of myxomycetes appears to be restricted to either temperate or tropical zones (Stephenson and Stempen, 1994). For instance, two species of *Ceratiomyxa* (*C. morchella* and *C. sphaerosperma*) have been found exclusively in the tropics (Rojas et al., 2008, Fig. 9.1). In contrast, the patchy but worldwide range of *Barbeyella*

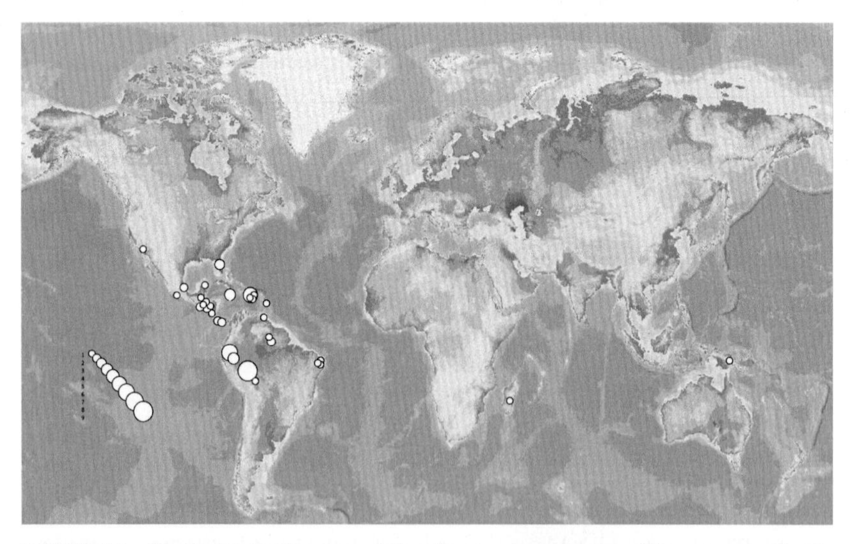

FIGURE 9.1 World distribution map of *Ceratiomyxa morchella* providing an example of a largely tropical species with restricted distribution. The size of the circles indicates the number of records per investigation site.

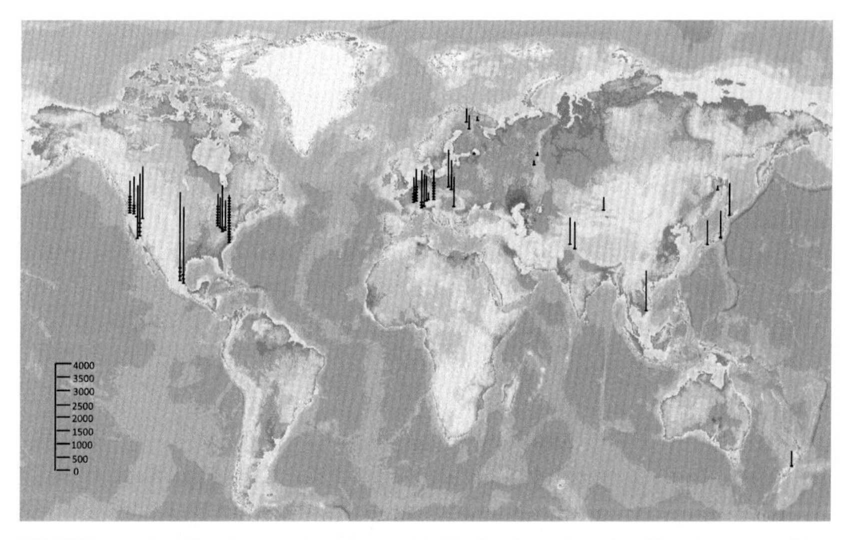

FIGURE 9.2 Needle plot showing the world distribution of *Barbeyella minutissima* from fructification records (updated from Schnittler et al., 2000). Each needle arises from the collecting site, its length indicates elevation, and the number of ticks the number of records for this side (see scale at the lower left of the figure).

minutissima, a specialist of temperate coniferous forests (Schnittler et al., 2000), corresponds mostly to the temperate zone of the planet, with some occurrences in high mountains of tropical zones (Fig. 9.2). These observations support the ubiquity model, because all regions with suitable microhabitats seem to be covered.

On a regional scale, such patterns are now well documented. For example, Rojas et al. (2012b) compared the myxomycete communities associated with highland areas along a transect that extended from North America to Central America. They showed that the similarity of species composition in these highland areas decreased with decreasing latitude (from Mexico to Costa Rica). Moreover, the similarity of species composition in these highland areas, when compared with the data available from the temperate zones (in this case the United States) decreased as well.

A severe problem for biogeographical studies is the apparent rarity of the fruiting bodies of many species of myxomycetes. Even surveys represented by a large number of specimens show a high proportion of singletons [e.g., Novozhilov et al. (2017a): Southern Vietnam, 1136 records, 107 taxa, 23% represented by singletons and Novozhilov et al. (2017b): Russian Far East, 3280 records, 161 taxa, 21% as singletons]. Schnittler and Mitchell (2000) estimated that more than half of the 446 species of myxomycetes for which data were available were represented by fewer than three collections thus far, which makes up for a significant proportion of the species known for the group, especially of the newly described species. Biogeographical studies carried in the Americas found that species abundances could be predicted on the basis of forest structure

(Rojas et al., 2011) but this pattern may as well have been caused by historical-geographical events (Estrada-Torres et al., 2013).

Another problem relates to the different methods used. For example, many species with minute fruiting bodies can be detected only by placing samples of substrates in moist chamber cultures (Gilbert and Martin, 1933), which have been used only in a fraction of all surveys, due to the intense amount of labor required. Whereas in forested areas most species can be observed directly in the field (Novozhilov et al., 2017b), in deserts without a high proportion of succulent plants, nearly 100% of the species present can be detected only with the moist chamber culture technique (Schnittler et al., 2015b). Most often, both approaches complement each other and are equally needed (Dagamac et al., 2015a,c; Novozhilov et al., 2017a,b; Schnittler and Stephenson, 2000).

Best studied are the myxomycetes of the north temperate zones, and these seem to be among the regions with the highest morphospecies diversity. For Germany, with a long history of myxomycete research, 373 species have been recorded (Schnittler et al., 2011). Japan and eastern North America, with their predominately broadleaf forests, seem to have an even higher species diversity, although precise numbers are difficult to obtain (Yamamoto, 1998). Temperate zones also have the highest diversity of corticolous and follicolous species, which can be detected best with the moist chamber culture technique (Gilbert and Martin, 1933; Goad and Stephenson, 2013; Härkönen, 1981). Members of this group, comprising ca. 120 species (Mitchell, 2004) with mostly very small fruiting bodies, inhabit the bark surface of living trees and show two life strategies (Schnittler, 2001). These are (1) stalked species with a short development time and minute, usually stalked fruiting bodies with evanescent peridia (*Echinostelium* spp. and *Macbrideola* spp.), and (2) sessile species with a longer development time and robust fruiting bodies, usually covered by a thicker peridium. A special, yet species-poor ecological group found within the temperate zone are bryophilous myxomycetes (Schnittler et al., 2010), which are adapted to moss layers in permanently humid gorges covering porous rocks or decaying logs.

In comparison, our current knowledge relating to the diversity of tropical myxomycetes seems to defy the traditional concept (as latitude increases, species richness decreases) that applies to many groups of macroorganisms. Systematic studies in tropical regions point toward a lower species richness than for temperate regions, especially for southern temperate zones. For example, the comparatively well studied country of Costa Rica has 225 myxomycete species recorded (Rojas et al., 2010, 2015; Schnittler et al., 2002). Although many tropical regions are still not systematically studied, the few systematic surveys carried out in tropical forests point seem to support this same trend [e.g., Ecuador: Yasuni, 86 species (Lado et al., 2017); southern Vietnam, Cat Tien, 107 species (Novozhilov et al., 2017a); and the Philippine archipelago, 158 species (Dagamac and dela Cruz, 2015; Macabago et al., 2017)]. Among the countries in the Neotropical zone were myxomycete are relatively well studied [a comprehensive checklist for this region lists a total of 431 taxa, (Lado and Wrigley de Basanta, 2008)] Mexico, which has a high proportion of arid regions along high volcanoes with coniferous forests, contributes the largest number of species (323). Even with this still fragmentary knowledge, we can conclude that at least the steep increase in species numbers toward tropical regions, very well known for vascular plants or insects, is not a pattern shared by myxomycetes.

The most likely reason for this discrepancy is the moisture regime. Two studies along a gradient of increasing elevation and moisture revealed that in the tropical regions being considered [volcano Cacao, Costa Rica (Schnittler and Stephenson, 2002) and slopes of the western Andes, Maquipucuna, Ecuador (Schnittler et al., 2002)] showed that species richness and also the proportion of moist chamber cultures positive for myxomycetes decreased with elevation. The lowest figures were always found for the tropical mountains, which are characterized by continuously wet forests at the highest elevations. This is matched by the observation that in contrast to forests of the temperate zones, aerial litter is more diverse than litter from the forest floor (Alfaro et al., 2015; Stephenson et al., 2008). In addition, there are unique aerial microhabitats in the tropics, such as inflorescences of large monocotyledonous forbs with high substrate pH [especially Zingiberales (Schnittler and Stephenson, 2002), dead lianas (Coelho and Stephenson, 2012; Ko Ko et al., 2010a; Wrigley de Basanta et al., 2008), or epiphytic liverworts (Schnittler, 2000)]. A few species, such as *Physarum didermoides*, *Ph. superbum* or *Ph. compressum* occur very regularly in aerial microhabitats. Further investigations are needed to elucidate the reasons for the apparent lower myxomycete diversity in the wet tropics. It is possible that species may be fruit less abundantly because desiccation as a trigger of fruiting body formation is absent or fruiting bodies may decompose very rapidly and thus appear to be less abundant when this is not actually the case.

DIVERSITY AND SPECIES COMPOSITION OF MYXOMYCETE COMMUNITIES FROM MAJOR ECOSYSTEMS OF THE WORLD

Tundra, Forest-Tundra and Subantarctic Habitats

As to be expected from the scarce and less diverse cover of vegetation, zonal arctic, subarctic and subantarctic ecosystems are rather species-poor for myxomycetes. Stephenson et al. (2000), analyzing data from ca. 2000 specimens collected from the tundra and forest-tundra of Alaska (Stephenson and Laursen, 1998), Iceland (Götzsche, 1990), Greenland (Götzsche, 1989), and Russian Arctic regions (Novozhilov and Schnittler, 1997; Novozhilov et al., 1999), recorded a total of 150 species, but only 33 of these were widely distributed enough to be regarded as species that are regularly associated with high-latitude regions. Lesser known are the myxomycetes of subarctic regions (Arambarri, 1973; Arambarri and Spinedi, 1989; Wrigley de Basanta et al., 2010). Two main factors are likely to limit myxomycete distribution toward the extreme high latitudes of both hemispheres. These are substrate availability and climate. Some lineages may be better suited than others. For example, Wrigley de Basanta et al. (2010) found that members of the order Trichiales dominated in a survey from Patagonia, whereas members of the order Physarales are more common in temperate and tropical South America.

Due to the inherent resistance of the three dormant stages (microcysts, sclerotia, and spores) in the life cycle of a myxomycete, low winter temperatures would seem to be a relatively unimportant factor. Indeed, the Taimyr Peninsula, with the most extreme winter temperatures of all the high-latitude areas intensively studied, was not only rich in species but also supported *Cribraria violacea*, whose distribution is apparently centered on submeridional to tropical regions. A possible explanation may be the rather high mean summer temperature connected with the highly continental climate, since for approximately one month temperatures are high enough to allow myxomycetes to complete their life cycle. Interestingly, even species developing large, compact fruiting bodies occur this far north. Examples are *Lycogala epidendrum*, *Mucilago crustacea*, and *Enteridium splendens* var. *juranum*. Two of these three species are wood inhabitants, whose distribution is limited by the presence of larger logs. In contrast, *M. crustacea* is frequently observed in pure tundra regions, emerging from thin mats of raw humus and litter, sometimes covered with bryophytes and lichens (Stephenson and Laursen, 1993). Seemingly, the latter species is not limited by the availability of a certain substrate and may accept a wide range of food organisms. Most obligate corticolous members of the *Physarales* and *Echinosteliales* are adapted to near-neutral substrates; therefore, they are relatively rare in coniferous communities of forest tundra, where tree bark is more acidic [pH 4.5–3.0 (Novozhilov et al., 1999)]. A few acidophilic species [*Echinostelium brooksii* and *Paradiacheopsis fimbriata*, Taimyr Peninsula (Novozhilov et al., 1999) and *Paradiacheopsis solitaria*, Alaska (Schnittler et al., 2016)] conform to this pattern and appear to be consistent inhabitants of boreal coniferous forests. As a litter-inhabiting species, *Leocarpus fragilis* spreads beyond the timberline into arctic tundra, where this species frequently occurs on the bark of shrubs that provides a higher pH (Schnittler et al., 2016).

Still understudied are the communities of nivicolous myxomycetes associated with subarctic and arctic mountains. Most of the data has originated from the Scandinavian Mountains (Schinner, 1983) and the Khibine Mountains [Kola peninsula (Erastova et al., 2017; Novozhilov and Schnittler, 1997)]. Only three of 32 species (*Diderma alpinum*, *Diderma niveum*, and *Physarum albescens*) were found to be widely distributed in the Khibine Mountains, where species richness and diversity decreases from subalpine crooked-stem birch-mountain ash forests to alpine mountain tundra (Erastova et al., 2017).

Boreal Forests (Taiga) and Subalpine/Montane Coniferous Forests

Taiga (coniferous boreal forests, including montane coniferous forests) represents the world's largest biome, and myxomycetes associated with this type of ecosystem are rather well studied [Alaska (Schnittler et al., 2016; Stephenson, 2004; Stephenson and Laursen, 1998); Scandinavia (Eliasson and Strid, 1976; Härkönen, 1977; Schinner, 1983); Russia (Kosheleva et al., 2008; Novozhilov and Fefelov, 2000; Schnittler and Novozhilov, 1996; Stephenson, 2004; Vlasenko and Novozhilov, 2011); German Alps (Schnittler and Novozhilov, 1998); Japan (Takahashi and Hada, 2012; Takahashi and Harakon, 2012)]. The most common taxa in these biomes are members of the Cribrariaceae, Trichiaceae and Stemonitidaceae on decaying conifer wood with an acidic pH. Typical species include *S. axifera*, *Licea minima*, *Comatricha elegans*, *C. nigra*, and *Cribraria* spp. In addition, a second species assemblage (*Lepidoderma tigrinum* and *Lamproderma columbinum* but especially *Colloderma oculatum* and *B. minutissima*) can be called bryophilous, since they are associated with layers of bryophytes on decorticated coniferous logs (Schnittler and Novozhilov, 1996, 1998) and rocks (Schnittler et al., 2010). Food organisms may include unicellular algae, which form gelatinous layers on these substrates. This association would explain the late peak of fruiting—cool nights in late autumn characterized by an extended dewfall, thus keeping the bryophyte layer continuously wet for some weeks and thereby allowing algal growth. Moist, *Sphagnum*-rich spruce woodlands are typical for taiga forests and have a species-poor assemblage of specialized species, such as *Didymium melanospermum, Physarum virescens*, *Ph. confertum*, and *Fuligo muscorum*, which typically fruit in the shelter of moss tussocks. Not a lot is known about *Sphagnum* bogs, but at least three species (*Badhamia lilacina*, *Diderma simplex*, and *Symphytocarpus trechisporus*) seem to be specialized for this microhabitat and occur even in raised bogs. Nivicolous myxomycetes are found in low-elevation mountains (Ronikier and Ronikier, 2009; Ronikier et al., 2008). Commonly encountered species are *Diderma niveum*, *Lepidoderma chailletii* and *L. ovoideum*, whereas in lowland coniferous forests *Meriderma carestiae* and *D. niveum* may occur (Erastova and Novozhilov, 2015).

Temperate Forests

Temperate deciduous forests support the most diverse and abundant myxomycete communities in the world (Ing, 1994; Novozhilov et al., 2017b; Stephenson et al., 2001; Takahashi, 2004). Especially rich appear to be broadleaf deciduous forests, occurring in climates with a summer peak in rainfall. Examples include eastern North America [Great Smoky Mts. National Park (Stephenson et al., 2001)] or the Manchurian-Japanese mixed broadleaf forests of northeast Asia [Sikhote-Alin Biosphere Reserve (Novozhilov et al., 2017b)]. A number of microhabitats with specialized myxomycete assemblages have been described during the past decade, including bark and decaying twigs in the forest canopy (Schnittler et al., 2006; Snell et al., 2003) or ravine myxomycetes (Ing, 1994), which exhibit a preference for bryophyte covers on rocks (see earlier). The latter are known from the British Islands (Ing, 1983) and Germany (Schnittler et al., 2010). Both species diversity and abundance in temperate forests tend to achieve their maximum near the wetter end of the moisture gradient (Rollins and Stephenson, 2011).

Mediterranean Forests, Woodlands, Scrub

Mediterranean forests and scrublands are characterized by hot and dry summers, while winters tend to be cool and moist. These two types of vegetation occur between 30 and 40 degree northern and southern latitudes on the westward sides of continents and include five regions. These are the Mediterranean, south central and southwestern Australia, the fynbos of southern Africa, the Chilean matorral, and the Mediterranean ecoregions of California with scrubland vegetation (chaparral). Well studied are the species-rich communities of myxomycetes around the Mediterranean Sea (Binyamini, 1997; Härkönen, 1988; Lado, 1994) and the Californian scrubland (Critchfield and Demaree, 1991; Estrada-Torres et al., 2009). Less complete information is available about southern Africa (Ndiritu et al., 2009c), the Chilean matorral (Lado et al., 2013), and scrub communities of Australia (Mitchell, 1995). High moisture gradients at small scales together with the regular fluctuations in moisture are the most likely reason for the high diversity, especially for corticolous species. Many new taxa have been described in the past few decades (Pando, 1997), and these show some similarities with communities from desert areas (Schnittler and Novozhilov, 2000). Preferred substrates for myxomycetes are *Olea europaea*, *Juniperus* spp. (Schnittler et al., 2015a), and *Quercus* spp., which support various minute species [*Echinostelium*, *Licea*, and *Macbrideola* spp. (Pando and Lado, 1990; Wrigley de Basanta, 1998)]. In addition, a large supply of slowly decaying leaf litter from sclerophyllous shrubs supports many litter-inhabiting species, such as *Diderma asteroides* and *Physarum brunneolum.*

Tropical and Subtropical Forests

Stephenson et al. (2004) summarized the body of information available on Neotropical myxomycetes and suggested that three major trends seem to exist. First, myxomycete species richness and abundance appear to be lower in tropical forests when compared to temperate forests. Second, both abundance and richness of myxomycetes decrease with increasing moisture (Schnittler and Stephenson, 2000). Third, some microhabitats with no equivalents in temperate regions support distinct assemblages of myxomycetes. Some species (e.g., *A. bombarda* and *C. sphaerosperma*), which require high temperature for development, seem to be restricted largely if not exclusively to tropical climates.

Somewhat lagging behind are research efforts directed toward the Paleotropics, especially in Africa. A comprehensive checklist by Ndiritu et al. (2009c), listed a total of only 294 species represented by 49 genera reported from 31 African countries, most of which are anecdotal in nature.

Some relatively well-studied localities in Africa include Ethiopia (Dagamac et al., 2017a), Tanzania (Ukkola et al., 1996), and Kenya (Ndiritu et al., 2009a). The survey with the highest number of species (124) reported was carried out in Madagascar, once again a region with a significant proportion of highlands and arid regions, with one species (*Perichaena madagascariensis*) reported as new to science (Wrigley de Basanta et al., 2013). Better studied areas are the Southeast Asian Paleotropics, with surveys carried out in Singapore [92 species (Rosing et al., 2011)], Myanmar [67 species (Ko Ko et al., 2013)], Vietnam [107 species (Novozhilov et al., 2017a)], Laos [44 species (Ko Ko et al., 2012)], the Philippines [158 species (Dagamac and dela Cruz, 2015; Macabago et al., 2017)], and Thailand [132 species (Ko Ko et al., 2010b; Tran et al., 2008)]. Factors, such as seasonality (Dagamac et al., 2012; Ko Ko et al., 2011), disturbance (Dagamac et al., 2015a; Rea-Maminta et al., 2015), and litter heterogeneity (Alfaro et al., 2015; Tran et al., 2006) have been investigated to explain the occurrence and distribution of myxomycetes in the region. In the past decade, a number of new species have been described from the region, including *Comatrichia spinispora* (Novozhilov and Mitchell, 2014), *Craterium retisporum* (Moreno et al., 2009), *Cribraria tecta* (Hooff, 2009), and *Perichaena echinolophospora* (Novozhilov and Stephenson, 2015).

For tropical regions, the moist chamber culture technique seems to provide both lower species diversity and fewer fruiting bodies per species found, following a gradient of habitat aridity from deserts (where moist chamber cultures work best) to temperate zones to tropical zones. As a consequence, the litter microhabitat, when examined by moist chamber cultures, yields a considerably lower species richness than what would be expected for the same microhabitat in temperate regions (Stephenson et al., 1999).

In addition, a clear pattern of decreasing species richness with increasing elevation and moisture has been noted in several studies [field collections from Puerto Rico (Novozhilov et al., 2000), a moisture gradient along a volcano in Costa Rica (Schnittler and Stephenson, 2000) and the western slopes of the Andes (Schnittler et al., 2002)]. Most likely, continuous moisture does not trigger the formation of fruiting bodies, and strong rainfall events seem to wash away amoeboflagellates and/or plasmodia. Evidence for the latter hypothesis comes from the observation that the proportion of species with phaneroplasmodia (the most robust of the types of plasmodia produced by myxomycetes) increases with increasing elevation and moisture. In addition, species richness of bark-inhabiting myxomycetes has been found to be negatively correlated with the amount of epiphyte coverage and moisture. Further support comes from aerial microhabitats, which in the tropics support a higher diversity and abundance of myxomycetes when compared to ground microhabitats. Presumably, since these microhabitats are not in contact with the forest floor, they have a better chance of drying out after rainfall events in the wet tropics.

High elevation tropical forests have remained mostly unstudied. Studies of high-elevation *Quercus* forests in Costa Rica revealed a myxomycete community strongly differing from those communities characteristic of lowland tropical forests but displaying similarities (both taxonomically and ecologically) to assemblages associated with temperate forests (Rojas and Stephenson, 2007; Rojas et al., 2010). The authors reported several species, most notably *L. fragilis*, that tend to be largely absent from tropical regions but are characteristic of temperate regions. A similar trend has been reported for mountain forests of southern Vietnam, which have a vegetation dominated by families with many temperate tree genera from the Fagaceae, Magnoliaceae and Pinaceae), where *B. minutissima* and *L. columbinum* were present (Novozhilov, pers. comm.).

Asian Paleotropical forests seem to be richer in species of myxomycete than Neotropical forests (Dagamac et al., 2017b) and some species seem to be more common in the Asian Paleotropics, like *Physarum echinosporum* (Fig. 9.3). The presence of conifers (*Pinus* spp.) with acidic bark and wood in the mountains of Vietnam and Thailand may contribute to this pattern. Several species (e.g., *B. minutissima, E. brooksii, E. colliculosum, L. columbinum, Licea kleistobolus, Lindbladia tubulina, Paradiacheopsis rigida*, and *Trichia persimilis*) that are known to be common in the temperate zones have been reported from these mountain forests.

As noted earlier in this chapter, it has been suggested that moist tropical forests may receive too much rain to be conducive to the successful completion of the myxomycete life cycle, which appears to be best suited to alternating wet and dry periods. This hypothesis seems to be supported by reports that aerial microhabitats, in the tropics, support a greater diversity and abundance of myxomycetes when compared to ground microhabitats (as already noted

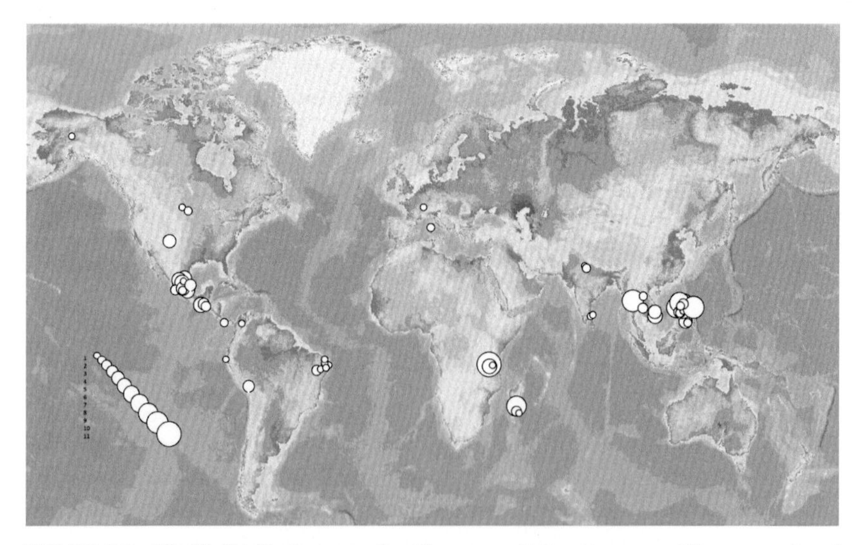

FIGURE 9.3 World distribution map for *Physarum echinosporum* providing examples of largely tropical species with restricted distribution. The size of the circles indicates the number of records per investigation site.

but also see Chapter 8). Specific microhabitats for tropical myxomycetes include inflorescences of monocotyledonous herbs with an extremely high pH (Schnittler et al., 2002), living leaves in the forest understory that have been overgrown with liverworts (Schnittler, 2000), and both living and decaying lianas (Wrigley de Basanta et al., 2008), once again a substrate characterized by rather high pH values.

Steppes and Prairies

At a first glance, due to a lack of woody debris, grasslands would not seem to provide good habitats for myxomycetes, despite evidence from studies showing that amoeboflagellates are more common in grassland soils than in forest soils (Feest and Madelin, 1988). However, studies carried out in the steppe regions of Russia (Novozhilov et al., 2006, 2010), Mongolia (Novozhilov and Schnittler, 2008), and the Midwestern United States (Rollins and Stephenson, 2013) reported a surprisingly high diversity of myxomycetes. Recently, Fiore-Donno et al. (2016) analyzed soil samples from a temperate grassland in Germany by ePCR and found that the most abundant OTUs belonged to the genera *Lamproderma* and *Didymium*. A special assemblage of species in grasslands often abundantly grazed by herbivores, such as horses and cattle includes coprophilous species (Eliasson and Lundqvist, 1979). Typical examples include *Kelleromyxa fimicola* and *Perichaena liceoides* associated with dung and *D. squamulosum*, *Echinostelium minutum*, *Fuligo cinerea*, and *Ph. pseudonotabile* (Novozhilov et al., 2013a), which are associated with grass litter.

At the grassland-forest transition zone, species diversity tends to increase toward the latter. Looking at the species/genus ratio, Rollins and Stephenson (2013) found a trend of increasing taxonomic diversity moving eastward from short to tallgrass prairie. In the Russian Altay (Novozhilov et al., 2010), a pronounced trend of increasing species richness was found when moving from dry steppe (6 species, $H' = 1.6$) over dark coniferous taiga and secondary mixed aspen and birch forests (99 species, $H' = 4.1$) to mixed forests (116 species, $H' = 4.2$); diversity decreased again toward the forest-steppe zone (65 species, $H' = 3.7$). Overall species dominance in the treeless dry steppe was found to be higher ($D = 0.26$) when compared to the forest steppe, where lignicolous myxomycetes occur in forest islands near rivulets ($D = 0.05$). As would be expected, the occurrence of woody debris causes pronounced differences between the assemblagesof myxomycetes associated with open grasslands and adjacent gallery forest (Rollins and Stephenson, 2013).

Subalpine and Alpine Grasslands

Subalpine meadows and alpine grasslands occur above the timberline in the high mountains, mostly in regions with a temperate climate. Climatic conditions (low temperatures and heavy snow accumulation) limit the period of vegetative growth to a few months but such areas still support rich communities of tall forbs, such as can be observed in the northwestern Caucasus (Onipchenko, 2004) or the Southern Alps of New Zealand (Stephenson, pers. observ.). This type of vegetation supports a rich assemblage of snowbank (nivicolous) myxomycetes, with about 100 species having been described since the pioneering studies of Meylan (Kowalski, 1975). Nivicolous myxomycetes are best studied in the northern hemisphere (Lado, 2004; Lado and Ronikier, 2008, 2009; Meylan, 1914; Moreno et al., 2005; Ronikier and Ronikier, 2009; Singer et al., 2005; Stephenson and Shadwick, 2009). The soil-inhabiting nivicolous myxomycetes are not strictly alpine (Ronikier and Ronikier, 2009) but occur as well in boreal lowland forests and in low-elevation mountains (Erastova and Novozhilov, 2015; Müller, 2002; Ronikier et al., 2008; Tamayama, 2000; Yajima et al., 2006). As already suggested in an initial paper by Schinner et al. (1990), the prevailing conditions determine the occurrence of myxomycetes. Since amoeboflagellates are susceptible to frost, early snowfalls and a long, contiguous snow cover, providing stable temperatures around zero degrees under the snow, are crucial for the formation of fruiting bodies (Schnittler et al., 2015a; Shchepin et al., 2014). These studies have suggested that nivicolous myxomycetes are important predators in the microbial communities that exist beneath the snow.

Deserts and Other Arid Areas

Since myxomycetes need water or substrates covered by a film of water for their active life, one might not expect these organisms to occur in deserts. However, numerous studies from arid ecosystems, often employing the moist chamber culture technique, have revealed astounding myxomycete diversity in arid regions. A series of surveys throughout Middle and Central Asia, extending from the Caspian Lowlands (Novozhilov et al., 2006), western Kazakhstan (Schnittler and Novozhilov, 2000), across the inner mountain basins of the Russian Altay (Novozhilov et al., 2010) to Mongolia (Novozhilov and Schnittler, 2008) to the Chinese province of Xinjiang (Schnittler et al., 2013) have explored the diversity of myxomycetes in winter-cold steppes and deserts. More limited data exist for Australian deserts (Davison et al., 2008; S.L. Stephenson, unpub. data) and the southwestern United States (Blackwell and Gilbertson,1980, 1984; Evenson, 1961; Ndiritu et al., 2009b; Novozhilov et al., 2003). A rich body of data also exists for the deserts of Central and South America [arid regions of Mexico (Estrada-Torres et al., 2009) and Chile (Lado et al., 2007)]. Published species lists from other deserts throughout the world include the Sinai Peninsula (Ramon, 1968), Morocco (Yamni and Meyer, 2008), Oman (Schnittler et al., 2015b), and a few reports from the Sahara (Faurel et al., 1965).

Although the lack of moisture in desert environments undoubtedly places severe constraints on the growth and development of myxomycetes, two strategies have evolved that allow myxomycetes to utilize the few suitable microhabitats. First, minute, usually stalked, species develop rapidly from protoplasmodia or very small aphanoplasmodia and thus can benefit from occasional rainfall events. Second, species forming sessile fruiting bodies with a hard-shelled peridium seem to be able to withstand repeated phases of drought during development (Schnittler, 2001). Eventually, the peridium dehisces to release the spores when the substrate on which the fruiting bodies occur dries out completely. In addition, special microhaibtats, most prominently succulent plants, support a distinctive assemblage of myxomycete assemblages. One prominent example is *B. melanospora* (Aguilar et al., 2014). In response to moisture fluctuations, these species may repeatedly switch between actively feeding plasmodia and dormant sclerotia (Estrada-Torres et al., 2009). The three dormant stages (spores, microcysts, and sclerotia) are the key for myxomycete survival in deserts. Blackwell and Gilbertson (1984) reported that myxomycete sclerotia incubated at 70°C still had significant survival rates, although survival differed between the species examined.

Many substrates for desert myxomycetes are characterized by a much higher pH in comparison to most substrates from other environments (Schnittler, 2001). These observations may explain the high number of apparent specialists, limited to particular microhabitats or even specific life forms of vascular plants [such as the succulenticolous myxomycetes (Estrada-Torres et al., 2009)]. A remarkably high number of species have been described from material collected in deserts, and many of these are not yet known from any other type of ecosystem (Blackwell and Gilbertson, 1980; Lado et al., 2007; Mosquera et al., 2000; Novozhilov et al., 2008, 2013a; Novozhilov and Zemlyanskaya, 2006; Wrigley de Basanta et al., 2009, 2010).

Coastal Habitats and Mangroves

Only a few studies directed toward the myxomycete communities associated with coastal habitats and mangroves have been carried out, although coastal ecosystems encompass a broad range of habitat types, such as dune grass vegetation (Howard, 1948), mangrove swamps (Cavalcanti et al., 2014, 2016; Kohlmeyer, 1969), shingle beaches (Ing, 1967), or marshes and coastal forests (Eliasson, 1971). The most important environmental factors shaping myxomycete diversity and distribution in coastal habitats may be strong wind, sea level oscillations (high salinity), temporarily high temperatures for coastlines in the tropics and subtropics, anaerobic soils and the absence of forest vegetation. Ing (1994), reviewing myxomycete diversity in sea shore communities, stated that there are no exclusively marine myxomycetes known, although species common in adjacent woodland communities have been found on dune grasses and herbaceous litter, as well as on driftwood accumulated on the sea shore (Hagelstein, 1930). Ing (1968) noted that in areas where woodland is scarce, driftwood may be a valuable reservoir for common lignicolous species. In addition, some rare species may be common in these habitats. For example, on *Cladonia* spp. in dry slacks, the very rare *Listerella paradoxa* has been found in Scotland; elsewhere, it is known only from lichen heath on shingle or moorland. *Diacheopsis mitchellii* was described on the basis of material collected on *Cladonia* in dune systems in southeast England (Ing, 1994). Adamonytė et al. (2013) studied a Great Cormorant (*Phalacrocorax carbo*) colony in Lithuania, obtaining in moist chamber cultures of various kinds of substrates such rare species as *Arcyria leiocarpa*, *Badhamia apiculospora*, and *Comatricha mirabilis*. Reports of myxomycetes from mangrove belts come from Brazil (Cavalcanti et al., 2014, 2016; Damasceno et al., 2011). Overall diversity is low, and regular inundation by the sea appears to inhibit fruiting body development. Currently, 31 species have been recorded from mangrove forests (Cavalcanti et al., 2016); the most common are *A. cinerea*, *A. denudata*, *Collaria arcyrionema* and *Stemonitis fusca*.

Agricultural and Urban Habitats

Ing (1994) noted that temperate grasslands are rather poor in myxomycetes, although *Badhamia foliicola*, *M. crustacea*, and (especially) *Physarum cinereum* often occur in residential lawns. Saunders and Saunders (1900), who examined piles of rotting straw near cornfields, reported that different species of myxomycetes tended to be associated with different parts of the pile. *Fuligo cinerea*, *Physarum didermoides*, and *Ph. pusillum* were found in association with the dry outer parts of the pile, whereas *Didymium difforme* and *D. vaccinum* were consistently associated with the bottom of the pile, where water retention was greatest. Myxomycetes are not pathogenic to plants, although they occasionally cause indirect injury. This occurs when they cover and shade plant tissues and inhibit photosynthesis.

Recently, *Comatricha pulchella* and *Fuligo septica* were recorded on cultivated *Dendrobium candidum* and apparently affected the growth of this plant (Tu et al., 2016). Well known is the occurrence of *F. septica* on the wood chips used as organic mulch in gardening. Myxomycetes are common in parks and gardens, where they can use natural as well as artificial substrates. Ing (1994) mentioned the studies of Brãndză (1924), where the author constructed "nurseries" (e.g., piles of branches, sawdust, leaves, manure, and waste paper) to attract myxomycetes. In this way, he recorded 33 species, including nine new to Romania. Corticolous species are not rare in parks of cities but may suffer from acid pollution (Härkönen and Vänskä, 2004; Wrigley de Basanta, 2000).

Interestingly, the low diversity of fruiting bodies as based on field observations does not agree well with the abundance of amoeboflagellates and plasmodia in the soil, as shown by culturing (Feest, 1987) and ePCR (Fiore-Donno et al., 2016). Much more work is needed to correlate the occurrence of myxomycete fruiting bodies with the real diversity of these organisms in artificial habitats. Without doubt, artificial habitats can attract specialists that are rare or absent in natural vegetation. For instance, the acidotolerant corticolous *Cribraria confusa* is very common in artificial plantations and gardens in southern Vietnam on the acidic bark of *Pinus kesiya* and *Anacardium occidentale* but rare in natural forests (Novozhilov et al., 2017a).

Heathlands

With its low and dense shrub cover warming up quickly, Ing (1994) indicated that some species of myxomycete may be associated with the litter and decaying twiglets of shrubs, such as *Calluna vulgaris*. Candidate species may include *Listerella paradoxa* on lichens and *Diderma simplex* and *Fuligo muscorum* on litter and old stems of *Calluna* and *Erica* (Santesson, 1948).

Savannahs and Semi-Arid Grasslands

The few studies carried out in semi-arid habitats include those of the cerrado in Brazil (Maimoni-Rodella and Gottsberger, 1980) and the African savannas and tropical dry forests (Dagamac et al., 2017a; Härkönen, 1981). Most common are litter-inhabiting species. A number of rarely recorded species were recorded from dry forests in Ethiopia (Dagamac et al., 2017a), including *Didymium saturnus, Metatrichia floripara, Perichaena areolata*, and *Physarina echinospora.* Moreover, there is some evidence that similar to its unique flora, the east African mountain ranges harbor a diverse and distinctive assemblage of myxomycetes. Of particular interest as substrates for myxomycetes are the hollow decaying trunks of giant tree-like lobelias. Samples collected from similar and placed in moist chamber cultures, Rammeloo (1975a,b) described several new species of myxomycete in the 1960s.

ACKNOWLEDGMENTS

Y. Novozhilov acknowledges support from several grants from RFBR (Russian Foundation of Basic Research, 15-29-02622) for field work summarized in this chapter.

REFERENCES

Adamonytė, G., Iršėnaitė, R., Motiejūnaitė, J., Matulevičiūtė, D., Taraškevičius, R., 2013. Myxomycetes in a forest affected by great cormorant colony: a case study in Western Lithuania. Fungal Divers. 59, 131–146.

Adamonytė, G., Stephenson, S.L., Michaud, A., Seraoui, E.H., Meyer, M., Novozhilov, Y.K., Krivomaz, T., 2011. Myxomycete species diversity on the island of La Réunion (Indian Ocean). Nova Hedwigia 92, 523–549.

Adl, S.M., Habura, A., Eglit, Y., 2014. Amplification primers of SSU rDNA for soil protists. Soil Biol. Biochem. 69, 328–342.

Aguilar, M., Fiore-Donno, A.M., Lado, C., Cavalier-Smith, T., 2014. Using environmental niche models to test the 'everything is everywhere' hypothesis for *Badhamia*. ISME J. 8, 737–745.

Alfaro, J.R.D., Alcayde, D.L.I.M., Agbulos, J.B., Dagamac, N.H.A., dela Cruz, T.E.E., 2015. The occurrence of myxomycetes from a lowland montane forest and agricultural plantations of Negros Occidental, Western Visayas, Philippines. Fine Focus 1, 7–20.

Arambarri, A.M., 1973. Myxomycetes de Tierra del Fuego I. Especies nuevas y criticas del genero *Diderma* (Didymiaceae). Bol. Soc. Argent. Botán. 15, 175–182.

Arambarri, A.M., Spinedi, H.A., 1989. Antarctic Myxomycetes [Mixomicetes antárticos]. Instituto Antártico Argentino, Buenos Aires. Contribución 365, pp. 12.

Araujo, J.C., Lado, C., Xavier-Santos, S., 2015. *Perichaena calongei* (Trichiales): a new record of Myxomycetes from Brazil. Curr. Res. Environ. Appl. Mycol. 5, 352–356.

Bardele, C.F., Foissner, W., Blanton, R.L., 1991. Morphology, morphogenesis and systematic position of the sorocarp forming ciliate *Sorogena stoianovitchae* Bradbury & Olive, 1980. J. Protozool. 38, 7–17.

Bass, D., Boenigk, J., 2011. Everything is everywhere: a twenty-first century de-/reconstruction with respect to protists. In: Fontaneto, D. (Ed.), Biogeography of Microscopic Organisms: Is Everything Small Everywhere?. Cambridge University Press, Cambridge, pp. 88–110.

Binyamini, N., 1997. Myxomycetes from Israel IV. Mycoscience 38, 87–89.

Blackwell, M., Gilbertson, R.L., 1980. *Didymium eremophilum*: a new myxomycete from the Sonoran Desert. Mycologia 72, 791–797.

Blackwell, M., Gilbertson, R.L., 1984. Distribution and sporulation phenology of myxomycetes in the Sonoran desert of Arizona. Microb. Ecol. 10, 369–377.

Blackwell, M.R., Laman, T.G., Gilbertson, R., 1982. Spore dispersal in *Fuligo septica* (Myxomycetes) by lathridiid beetles. Mycotaxon 14, 58–60.

Brãndzã, M., 1924. Sur l'apparition des Myxomycetes dans le Bugey. Bull. Fed. Centr. Hist. Mykol. 3, 13–20.

Brown, M.W., Silberman, J.D., Spiegel, F.W., 2012. A contemporary evaluation of the acrasids (Acrasidae, Heterolobosea, Excavata). Eur. J. Protistol. 48, 103–123.

Buxton, P.A., 1954. British diptera associated with Fungi. 2. Diptera bred from myxomycetes. Proc. Roy. Entomol. Soc. 29, 163–171.

Cavalcanti, L.H., Damasceno, G., Bezerra, A.C.C., Costa, A.A.A., 2014. Mangrove myxomycetes: species occurring on *Conocarpus erectus* L. (Combretaceae). Sydowia 66, 183–190.

Cavalcanti, L.H., Damasceno, G., Costa, A.A.A., Bezerra, A.C.C., 2016. Myxomycetes in Brazilian mangroves: species associated with *Avicennia nitida*, *Laguncularia racemosa* and *Rhizophora mangle*. Mar. Biodiv. Rec. 9.

Chassain, M., 1973. Capture d'un insecte collembole par deux Myxomycètes. Doc. Mycol. 8, 37–38.

Clark, J., Haskins, E.F., 2013. The nuclear reproductive cycle in the myxomycetes: a review. Mycosphere 4, 233–248.

Coelho, I.L., Stephenson, S.L., 2012. Myxomycetes associated with pipevine, a temperate liana. Mycosphere 3, 245–249.

Collins, O.N.R., 1979. Myxomycete biosystematics: some recent developments and future research opportunities. Bot. Rev. 45, 145–201.

Cotterill, F.P.D., Al-Rasheid, K., Foissner, W., 2008. Conservation of protists: Is it needed at all? Biodiv. Conserv. 17, 427–443.

Critchfield, R.L., Demaree, R.S., 1991. Annotated checklist of California Myxomycetes. Madrõna 38, pp. 45–56.

Dagamac, N.H.A., dela Cruz, T.E.E., 2015. Myxomycete research in the Philippines: updates and opportunities. Mycosphere 6, 784–795.

Dagamac, N.H.A., dela Cruz, T.E.E., Pangilinan, M.V.B., Stephenson, S.L., 2011. List of species collected and interactive database of myxomycetes (plasmodial slime molds) for Mt. Arayat National Park, Pampanga, Philippines. Mycosphere 2, 449–455.

Dagamac, N.H.A., dela Cruz, T.E.E., Rea-Maninta, M.A.D., Aril-dela Cruz, J.V., Schnittler, M., 2015a. Rapid assessment of myxomycete diversity in Bicol Peninsula. Nova Hedwigia 100, 31–46.

Dagamac, N.H.A., Hoffman, M., Novozhilov, Y.K., Schnittler, M., 2017a. Myxomycetes from the highlands of Ethiopia. Nova Hedwigia 104, 111–127.

Dagamac, N.H.A., Leontyev, D.V., dela Cruz, T.E.E., 2010. Corticolous myxomycetes associated with *Samonea samans* (Jacq.) Merr. collected from different sites in Luzon Island, Philippines. Philippine Biota 43, 2–15.

Dagamac, N.H.A., Novozhilov, Y.K., Stephenson, S.L., Lado, C., Rojas, C., dela Cruz, T.E.E., Unterseher, M., Schnittler, M., 2017b. Biogeographical assessment of myxomycete assemblages from Neotropical and Asian Paleotropical forests. J. Biogeogr. 44(7), 1524–1536.

Dagamac, N.H.A., Rea-Maminta, M.A.D., dela Cruz, T.E.E., 2015b. Plasmodial slime molds of a tropical karst forest, Quezon National Park, the Philippines. Pacific Sci. 69, 407–418.

Dagamac, N.H.A., Rea-Maminta, M.A.D., Batungbacal, N.S., Jung, S.H., Bulang, C.R.T., Cayago, A.G.R., dela Cruz, T.E.E., 2015c. Diversity of plasmodial slime molds (myxomycetes) on coastal, mountain, and community forests of Puerto Galera, Oriental Mindoro, Philippines. J. Asia Pac. Biodiv. 8, 322–329.

Dagamac, N.H.A., Rojas, C., Novozhilov, Y.K., Moreno, G.H., Schlueter, R., Schnittler, M., 2017c. Speciation in progress? A phylogeographic study among populations of *Hemitrichia serpula* (Myxomycetes). PLoS One 12 (4), e0174825.

Dagamac, N.H.A., Stephenson, S.L., dela Cruz, T.E.E., 2012. Occurrence, distribution and diversity of myxomycetes (plasmodial slime molds) along two transects in Mt. Arayat National Park, Pampanga, Philippines. Mycology 3, 119–126.

Dagamac, N.H.A., Stephenson, S.L., dela Cruz, T.E.E., 2014. The occurrence of litter myxomycetes at different elevations in Mt. Arayat, National Park, Pampanga, Philippines. Nova Hedwigia 98, 187–196.

Damasceno, G., Tenorio, J.C.G., Cavalcanti, L.H., 2011. Stemonitaceae (Myxomycetes) in Brazilian mangroves. Sydowia 63, 9–22.

Davison, E.M., Davison, P.J.N., Brims, M.H., 2008. Moist chamber and field collections of myxomycetes from the northern Simpson Desert, Australia. Aust. Mycol. 27, 129–135.

Dixon, P.A., 1963. Spore liberation by water drops in some myxomycetes. Trans. Br. Mycol. Soc. 46, 615–619.

Eliasson, U., 1971. A collection of myxomycetes from the Galapagos Islands. Svensk Botan. Tids. 65, 105–111.

Eliasson, U., 1991. The myxomycete biota of the Hawaiian Islands. Mycol. Res. 95, 257–267.

Eliasson, U., Lundqvist, N., 1979. Fimicolous myxomycetes. Botan. Notiser 132, 551–568.

Eliasson, U., Nannenga-Bremekamp, N.E., 1983. Myxomycetes from the Scalesia forest, Galapagos Islands. Proc. Konin. Neder. Akad. Weten. 86, 148–153.

Eliasson, U., Strid, Å., 1976. Wood-inhabiting fungi of alder forests in north-central Scandinavia. 3. Myxomycetes. Botan. Notiser 129, 267–272.

Erastova, D.A., Novozhilov, Y.K., 2015. Nivicolous myxomycetes of the lowland landscapes of the Northwest of Russia. Mikol. Fitopatol. 49, 9–18.

Erastova, D.A., Novozhilov, Y.K., Schnittler, M., 2017. Nivicolous myxomycetes of the Khibiny Mountains, Kola Peninsula, Russia. Nova Hedwigia 104, 85–100.

Estrada-Torres, A., Wrigley de Basanta, D., Conde, E., Lado, C., 2009. Myxomycetes associated with dryland ecosystems of the Tehuacán-Cuicatlán Valley Biosphere Reserve, Mexico. Fungal Diver. 36, 17–56.

Estrada-Torres, A., Wrigley de Basanta, D., Lado, C., 2013. Biogeographic patterns of the myxomycete biota of the Americas using a parsimony analysis of endemicity. Fungal Diver. 59, 159–177.

Evenson, A.E., 1961. A preliminary report of the Myxomycetes of southern Arizona. Mycologia 53, 137–144.

Faurel, L., Feldmann, J., Schotter, G., 1965. Catalogue des Myxomycétes de l'Afrique du Nord. Bull. Soc. Hist. Nat. l'Afrique Nord 55, 7–39.

Feest, A., 1987. The quantitative ecology of soil Mycetozoa. Prog. Protist. 2, 331–361.

Feest, A., Madelin, M.F., 1988. Seasonal population changes of myxomycetes and associated organism in four woodland soils. FEMS Microbiol. Ecol. 53, 133–140.

Feng, Y., Klahr, A., Janik, P., Ronikier, A., Hoppe, T., Novozhilov, Y.K., Schnittler, M., 2016. What an intron may tell: several sexual biospecies coexist in *Meriderma* spp. (Myxomycetes). Protist 167, 234–253.

Feng, Y., Schnittler, M., 2015. Sex or no sex? Independent marker genes and group I introns reveal the existence of three sexual but reproductively isolated biospecies in *Trichia varia* (Myxomycetes). Organ. Diver. Evol. 15, 631–650.

Feng, Y., Schnittler, M., 2017. Molecular or morphological species? Myxomycete diversity in a deciduous forest in northeastern Germany. Nova Hedwigia 104, 359–382.

Finlay, B.J., 2002. Global dispersal of free-living microbial eukaryotic species. Science 296, 1061–1063.

Finlay, B.J., Clarke, K.J., 1999. Ubiquitous dispersal of microbial species. Nature 400, 828.

Finlay, B.J., Esteban, G.F., Fenchel, T., 2004. Protist diversity is different? Protist 155, 15–22.

Fiore-Donno, A.M., Berney, C., Pawlowski, J., Baldauf, S.L., 2005. Higher-order phylogeny of plasmodial slime molds (Myxogastria) based on elongation factor 1-A and small subunit rRNA gene sequences. J. Eukaryot. Microbiol. 52, 201–210.

Fiore-Donno, A.M., Clissmann, F., Meyer, M., Schnittler, M., Cavalier-Smith, T., 2013. Two-gene phylogeny of bright-spored myxomycetes (slime moulds, superorder Lucisporidia). PLoS One 8, e62586.

Fiore-Donno, A.M., Haskin, E.F., Pawlowski, J., Cavalier-Smith, T., 2009. *Semimorula liquescens* is a modified echinostelid myxomycete (Mycetozoa). Mycologia 101, 773–776.

Fiore-Donno, A.M., Kamono, A., Meyer, M., Schnittler, M., Fukui, M., Cavalier-Smith, T., 2012. 18S rDNA phylogeny of *Lamproderma* and allied genera (Stemonitales, Myxomycetes, Amoebozoa). PLoS One 7, e35359.

Fiore-Donno, A.M., Meyer, M., Baldauf, S.L., Pawlowski, J., 2008. Evolution of dark-spored myxomycetes (slime-molds): molecules versus morphology. Mol. Phylog. Evol. 46, 878–889.

Fiore-Donno, A.M., Nikolaev, S.I., Nelson, M., Pawlowski, J., Cavalier-Smith, T., Baldauf, S.L., 2010. Deep phylogeny and evolution of slime moulds (Mycetozoa). Protist 161, 55–70.

Fiore-Donno, A.M., Novozhilov, Y.K., Meyer, M., Schnittler, M., 2011. Genetic structure of two protist species (Myxogastria, Amoebozoa) suggests asexual reproduction in sexual amoebae. PLoS One 6, e22872.

Fiore-Donno, A.M., Weinert, J., Wubet, T., Bonkowski, M., 2016. Metacommunity analysis of amoeboid protists in grassland soils. Sci. Rep. 6, 19068.

Foissner, W., 1999. Protist diversity: estimates of the near imponderable. Protist 150, 363–368.

Foissner, W., 2006. Biogeography and dispersal of micro-organisms: a review emphasizing protists. Acta Protozool. 45, 111–136.

Foissner, W., 2008. Protist diversity and distribution: some basic considerations. Biodiv. Conserv. 17, 235–242.

García, G., Ramos, F., Perez, R.G., Yañez, J., Estrada, M.S., Mendoza, L.H., Martinez-Hernandez, F., Gaytan, P., 2014. Molecular epidemiology and genetic diversity of *Entamoeba* species in a chelonian collection. J. Med. Microbiol. 63, 271–283.

Gilbert, H.C., Martin, G.W., 1933. Myxomycetes found on the bark of living trees. Univ. Iowa Stud. Nat. Hist. 15, 3–8.

Goad, A.E., Stephenson, S.L., 2013. Myxomycetes appearing in moist chamber cultures on four different types of dead leaves. Mycosphere 4, 707–712.

Götzsche, H.F., 1989. Myxomycetes from Greenland. Opera Botan. 100, 93–104.
Götzsche, H.F., 1990. Notes on Icelandic myxomycetes. Acta Botan. Islandica 10, 3–21.
Hagelstein, R., 1930. Mycetozoa from Jones Beach State Park. Mycologia 22, 256–262.
Hagelstein, R., 1944. Mycetozoa of North America. Published by the author, Mineola, New York.
Härkönen, M., 1977. Corticolous myxomycetes in three different habitats in southern Finland. Karstenia 17, 19–32.
Härkönen, M., 1981. Myxomycetes developed on litter of common Finnish trees in moist chamber cultures. Nord. J. Bot. 1, 791–794.
Härkönen, M., 1988. Some additions to the knowledge of Turkish myxomycetes. Karstenia 27, 1–7.
Härkönen, M., Vänskä, H., 2004. Corticicolous myxomycetes and lichens in the Botanical Garden in Helsinki, Finland: a comparison after decades of recovering from air pollution. Syst. Geogr. Plants 74, 183–187.
Hooff, J.P.M., 2009. *Cribraria tecta*, a new myxomycete from Vietnam. Bol. Soc. Micol. Madrid 33, 129–136.
Hoppe, T., Schwippert, W.W., 2014. Hydrophobicity of myxomycete spores: an undescribed aspect of spore ornamentation. Mycosphere 5, 601–606.
Howard, H.J., 1948. The mycetozoa of sand-dunes and marshland. Southeast. Nat. 53, 26–30.
Ing, B., 1967. Myxomycetes as sources of food for other organisms. Proc. S. Lond. Entomol. Nat. Hist. Soc. 1967, 18–23.
Ing, B., 1968. A census catalogue of british myxomycetes. Br. Mycol. Soc. Foray Comm.
Ing, B., 1983. A ravine association of myxomycetes. J. Biogeograph. 10, 299–306.
Ing, B., 1994. The phytosociology of myxomycetes. New Phytol. 126, 175–201.
Kamono, A., Kojima, H., Matsumoto, J., Kawamura, K., Fukui, M., 2009. Airborne myxomycete spores: detection using molecular techniques. Naturwissenschaften 96, 147–151.
Keller, H.W., Smith, D.M., 1978. Dissemination of myxomycete spores through the feeding activities (ingestion-defecation) of an acarid mite. Mycologia 70, 1239–1241.
Keller, H.W., Snell, K.L., 2002. Feeding activities of slugs on myxomycetes and fungi. Mycologia 94, 757–760.
Kilgore, C.M., Keller, H.W., 2008. Interactions between myxomycete plasmodia and nematodes. Inoculum 59, 1–3.
Ko Ko, T.W., Rosing, W.C., Ko Ko, Z.Z.W., Stephenson, S.L., 2013. Myxomycetes of Myanmar. Sydowia 65, 267–276.
Ko Ko, T.W., Stephenson, S.L., Hyde, K.D., Lumyong, S., 2011. Influence of seasonality on the occurrence of myxomycetes. Chiang Mai J. Sci. 38, 71–84.
Ko Ko, T.W., Stephenson, S.L., Hyde, K.D., Rojas, C., Lumyong, S., 2010a. Patterns of occurrence of myxomycetes on lianas. Fungal Ecol. 3, 302–310.
Ko Ko, T.W., Tran, H.T.M., Clayton, M.E., Stephenson, S.L., 2012. First records of myxomycetes in Laos. Nova Hedwigia 96, 73–81.
Ko Ko, T.W., Tran, H.T.M., Stephenson, S.L., Mitchell, D.W., Rojas, C., Hyde, K.D., Lumyong, S., 2010b. Myxomycetes of Thailand. Sydowia 62, 243–260.
Kohlmeyer, J., 1969. Ecological notes on fungi in mangrove forests. Trans. Br. Mycol. Soc. 53, 237–250.
Kosheleva, A.P., Novozhilov, Y.K., Schnittler, M., 2008. Myxomycete diversity of the state reserve “Stolby” (southeastern Siberia, Russia). Fungal Diver. 31, 45–62.
Kowalski, D.T., 1975. The myxomycete taxa described by Charles Meylan. Mycologia 67, 448–494.

Kylin, J.H., 1991. On the feeding habits of a tardigrade: selective foraging on myxomycetes. Mycologist 5, 54–55.

Kylin, H., Mitchell, D.W., Seraoui, E.H., Buyck, B., 2013. Myxomycetes from Papua New Guinea and New Caledonia. Fungal Diver. 59, 33–44.

Lado, C., 1994. A checklist of myxomycetes of Mediterranean countries. Mycotaxon 52, 117–185.

Lado, C., 2004. Nivicolous myxomycetes of the Iberian Peninsula: considerations on species richness and ecological requirements. Syst. Geogr. Plants 74, 143–157.

Lado, C., 2005–16. An online nomenclatural information system of Eumycetozoa. Available from: http://www.nomen.eumycetozoa.com

Lado, C., Estrada-Torres, A., Stephenson, S.L., Wrigley de Basanta, D., Schnittler, M., 2003. Biodiversity assessment of myxomycetes from two tropical forest reserves in Mexico. Fungal Diver. 12, 67–110.

Lado, C., Estrada-Torres, A., Stephenson, S.L., 2007. Myxomycetes collected in the first phase of a north-south transect of Chile. Fungal Diver. 25, 81–101.

Lado, C., Estrada-Torres, A., Wrigley de Basanta, D., Schnittler, M., Stephenson, S.L., 2017. A rapid biodiversity assessment of myxomycetes from a primary tropical moist forest of the Amazon basin in Ecuador. Nova Hedwigia 104, 293–321.

Lado, C., Ronikier, A., 2008. Nivicolous myxomycetes from the Pyrenees: notes on the taxonomy and species diversity. Part 1. Physarales and Trichiales. Nova Hedwigia 87, 337–360.

Lado, C., Ronikier, A., 2009. Nivicolous myxomycetes from the Pyrenees: notes on the taxonomy and species diversity. Part 2. Stemonitales. Nova Hedwigia 89, 131–145.

Lado, C., Wrigley de Basanta, D., 2008. A review of Neotropical myxomycetes (1828–2008). Anal. Jard. Bot. Madrid 65, 211–254.

Lado, C., Wrigley de Basanta, D., Estrada-Torres, A., 2011. Biodiversity of Myxomycetes from the Monte desert of Argentina. Anal. Jard. Bot. Madrid 68, 61–95.

Lado, C., Wrigley de Basanta, D., Estrada-Torres, A., García-Carvajal, E., 2014. Myxomycete diversity of the Patagonian Steppe and bordering areas in Argentina. Anal. Jard. Bot. Madrid 71, e0006.

Lado, C., Wrigley de Basanta, D., Estrada-Torres, A., García-Carvajal, E., Aguilar, M., Hernández, J.C., 2009. Description of a new species of *Perichaena* (Myxomycetes) from arid areas of Argentina. Anal. Madrid 66, 63–70.

Lado, C., Wrigley de Basanta, D., Estrada-Torres, A., Stephenson, S.L., 2013. The biodiversity of myxomycetes in Central Chile. Fungal Diver. 59, 3–32.

Lado, C., Wrigley de Basanta, D., Estrada-Torres, A., Stephenson, S.L., 2016. Myxomycete diversity in the coastal desert of Peru with emphasis on the lomas formation. Anal. Jard. Bot. Madrid 73, e032.

Leontyev, D.V., Schnittler, M., Stephenson, S.L., 2015. A critical revision of the *Tubifera ferruginosa* complex. Mycologia 107, 959–985.

Lindley, L.A., Stephenson, S.L., Spiegel, F.W., 2007. Protostelids and myxomycetes isolated from aquatic habitats. Mycologia 99, 504–509.

Linnaeus, C., 1753. Species Plantarum. 2 vols. Salvius, Stockholm. Facsimile edition, 1957–1959, Ray Society, London.

Lister, A., 1894. A Mongraph of the Mycetozoa, third ed. British Museum of Natural History, London, (Revised by G. Lister).

Lister, A., 1911. A Mongraph of the Mycetozoa. British Museum of Natural History, London, (Revised by G. Lister).

Lister, A., 1925. A Mongraph of the Mycetozoa. British Museum of Natural History, London, (Revised by G. Lister).

Macabago, S.A.B., Dagamac, N.H.A., dela Cruz, T.E.E., 2010. Diversity and distribution of plasmodial myxomycetes (slime molds) from La Mesa Ecopark, Quezon City, Philippines. Biotropia 17, 51–61.

Macabago, S.A.B., Dagamac, N.H.A., dela Cruz, T.E.E., Stephenson, S.L., 2017. Implications of the role of dispersal on the occurrence of litter-inhabiting myxomycetes in different vegetation types after a disturbance. Nova Hedwigia 104(1–3): 221–236.

Macabago, S.A.B., dela Cruz, T.E.E., Stephenson, S.L., 2012. First records of myxomycetes from Lubang Island, Occidental Mindoro, Philippines. Sydowia 64, 109–118.

Macabago, S.A.B., Stephenson, S.L., dela Cruz, T.E.E., 2016. Diversity and distribution of myxomycetes in coastal and mountain forests of Lubang Island, Occidental Mindoro, Philippines. Mycosphere 7, 18–29.

Maimoni-Rodella, R., Gottsberger, G., 1980. Myxomycetes from the forest and the Cerrado vegetation in Botucatu, Brazil: a comparative ecological study. Nova Hedwigia 34, 207–245.

Martin, G., Alexopoulos, C.J., 1969. The Myxomycetes. University of Iowa Press, Iowa.

Martiny, H.J.B., Bohannan, B.J.M., Brown, J.H., Colwell, R.K., Fuhrman, J.A., Green, J.L., Horner-Devine, M.C., Kane, M., Krumins, J.A., Kuske, C.R., Morin, P.J., Naeem, S., Øvreås, L., Reysenbach, A.L., Smith, V.H., Staley, J.T., 2006. Microbial biogeography: putting microorganisms on the map. Nat. Rev. Microbiol. 4, 102–112.

Massee, G., 1892. A Monograph of the Myxogasteres. Methuen and Company, London.

Meylan, C., 1914. Remarques sur quelques espèces nivales de Myxomycètes. Bull. Soc. Vaud. Sci. Nat. 50, 1–14.

Mitchell, D.W., 1995. The myxomycota of Australia. Nova Hedwigia 60, 269–295.

Mitchell, D.W., 2004. A key to the corticolous Myxomycota. Syst. Geogr. Plants 74, 261–285.

Mitchell, E.A.D., Meisterfeld, R., 2005. Taxonomic confusion blurs the debate on cosmopolitanism versus local endemism of free-living protists. Protist 156, 263–267.

Moreno, G., Singer, H., Illana, C., 2005. A nivicolous species of *Lamproderma* from Japan. Bol. Soc. Micol. Madrid 29, 135–142.

Moreno, G., Mitchell, D.W., Stephenson, S.L., dela Cruz, T.E.E., 2009. A new species of *Craterium* (Myxomycetes) with reticulate spores. Bol. Soc. Micol. Madrid 33, 175–200.

Mosquera, J., Lado, C., Beltran-Tejera, E., 2000. Morphology and ecology of *Didymium subreticulosporum*. Mycologia 92, 978–983.

Müller, H., 2002. Beitrag zur Kenntnis und Verbreitung nivicoler Myxomyceten im Thüringer Wald. Zeit. Mykol. 68, 199–208.

Murray, P.M., Feest, A., Madelin, M.F., 1985. The numbers of viable myxomycete cells in the alimentary tract of earthworms and in earthworm casts. Bot. J. Linnean Soc. 91, 359–366.

Ndiritu, G.G., Spiegel, F.W., Stephenson, S.L., 2009a. Rapid biodiversity assessment of myxomycetes in two regions of Kenya. Sydowia 61, 287–319.

Ndiritu, G.G., Spiegel, F.W., Stephenson, S.L., 2009b. Distribution and ecology of the assemblages of myxomycetes associated with major vegetation types in Big Bend National Park, USA. Fungal Ecol. 2, 168–183.

Ndiritu, G.G., Winsett, K.E., Spiegel, F.W., Stephenson, S.L., 2009c. A checklist of African myxomycetes. Mycotaxon 107, 353–356.

Novozhilov, Y.K., Fefelov, K.A., 2000. An annotated checklist of the myxomycetes of Sverdlovsk region, West Siberian Lowland, Russia. Micol. Fitopatol. 35, 41–52.

Novozhilov, Y.K., Erastova, D.A., Shchepin, O.N., Schnittler, M., Aleksandrova, A.V., Popov, E.S., Kuznetsov, A.N., 2017a. Myxomycetes associated with monsoon lowland tall tropical forests in southern Vietnam. Nova Hedwigia 104 (1–3), 143–182.

Novozhilov, Y.K., Mitchell, D.W., 2014. A new species of *Comatricha* (Myxomycetes) from southern Vietnam. Nov. Sist. Nizsh. Rast. 48, 188–195.

Novozhilov, Y.K., Mitchell, D.W., Schnittler, M., 2003. Myxomycete biodiversity of the Colorado Plateau. Mycol. Prog. 2, 243–258.

Novozhilov, Y.K., Okun, M.V., Erastova, D.A., Shchepin, O.N., Zemlyanskaya, I.V., García-Carvajal, E., Schnittler, M., 2013a. Description, culture and phylogenetic position of a new xerotolerant species of *Physarum*. Mycologia 105, 1535–1546.

Novozhilov, Y.K., Schnittler, M., 1997. Nivicole myxomycetes of the Khibine mountains (Kola peninsula). Nordic J. Bot. 16, 549–561.

Novozhilov, Y.K., Schnittler, M., 2008. Myxomycete diversity and ecology in arid regions of the Great Lake Basin of western Mongolia. Fungal Diver. 30, 97–119.

Novozhilov, Y.K., Schnittler, M., Erastova, D.A., Okun, M.V., Schepin, O.N., Heinrich, E., 2013b. Diversity of nivicolous myxomycetes of the Teberda State Biosphere Reserve (Northwestern Caucasus, Russia). Fungal Diver. 59, 109–130.

Novozhilov, Y.K., Schnittler, M., Erastova, D.A., Schepin, O.N., 2017b. Myxomycetes of the Sikhote-Alin State Nature Biosphere Reserve (Far East, Russia). Nova Hedwigia 104 (1–3), 183–209.

Novozhilov, Y.K., Schnittler, M., Stephenson, S.L., 1999. Myxomycetes of the Taimyr Peninsula (north-central Siberia). Karstenia 39, 77–97.

Novozhilov, Y.K., Schnittler, M., Vlasenko, A.V., Fefelov, K.A., 2009. Myxomycete diversity of the Chuyskaya depression (Altay, Russia). Mikol. Fitopatol. 43, 522–534.

Novozhilov, Y.K., Schnittler, M., Vlasenko, A.V., Fevelov, K.A., 2010. Myxomycete diversity of the Altay Mountains (southwestern Siberia, Russia). Mycotaxon 111, 91–94.

Novozhilov, Y.K., Schnittler, M., Rollins, A.W., Stephenson, S.L., 2000. Myxomycetes from different forest types in Puerto Rico. Mycotaxon 77, 285–299.

Novozhilov, Y.K., Stephenson, S.L., 2015. A new species of *Perichaena* (Myxomycetes) with reticulate spores from southern Vietnam. Mycologia 107, 137–141.

Novozhilov, Y.K., Zemlyanskaya, I.V., 2006. A new species of *Didymium* (Myxomycetes) with reticulate spores. Mycotaxon 96, 147–150.

Novozhilov, Y.K., Zemlianskaia, I.V., Schnittler, M., Stephenson, S.L., 2006. Myxomycete diversity and ecology in the arid regions of the Lower Volga River Basin (Russia). Fungal Diver. 23, 193–241.

Novozhilov, Y.K., Zemlyanskaya, I.V., Schnittler, M., Stephenson, S.L., 2008. Two new species of *Perichaena* (Myxomycetes) from arid areas of Russia and Kazakhstan. Mycologia 100, 816–822.

Onipchenko, V.G., 2004. Alpine Ecosystems in the Northwestern Caucasus. Kluwer, Dordrecht, NL.

Ottewell, K., Grey, E., Castillo, F., Karubian, J., 2012. The pollen dispersal kernel and mating system of an insect-pollinated tropical palm, *Oenocarpus bataua*. Heredity 109, 332–339.

Pando, F., 1997. Bases corológicas de flora mycologica Iberica. NUMS, 1224–1411, Cuad. Trab. Fl. Micol. lbér. 12, 111–112.

Pando, F., Lado, C., 1990. A survey of the corticolous myxomycetes in Peninsular Spain and Balearte Islands. Nova Hedwigia 50, 127–137.

Pecundo, M.H., Dagamac, N.H.A., Stephenson, S.L., dela Cruz, T.E.E., 2017. First myxomycete survey in the limestone forest of Puerto Princesa Subterranean River National Park, Palawan, Philippines. Nova Hedwigia 104(1–3): 129–141.

Penet, L., Guyader, S., Pétro, D., Salles, M., Bussière, F., 2014. Direct splash dispersal prevails over indirect and subsequent spread during rains in *Colletotrichum gloeosporioides* infecting yams. PLoS One 9, e115757.

Rammeloo, J., 1975a. Structure of the epispore in the Stemonitales (Myxomycetes) as seen with the scanning electron microscope. Bull. Jard. Bot. Nat. Belgique 45, 301–306.

Rammeloo, J., 1975b. Structure of the epispore in the Trichiaceae (Trichiales, Myxomycetes), as seen with the scanning electron microscope. Bull. Jard. Bot. Nat. Belgique 107, 353–359.

Ramon, E., 1968. Myxomycetes of Israel. Isr. J. Botany 17, 207–211.

Rea-Maminta, M.A.D., Dagamac, N.H.A., Huyop, F.Z., Wahab, R.A., dela Cruz, T.E.E., 2015. Comparative diversity and heavy metal bisorption of myxomycetes from forest patches on ultramafic and volcanic soils. Chem. Ecol. 31, 741–753.

Reichenbach, H., 1993. Biology of the Myxobacteria: Ecology and Taxonomy. American Society of Microbiology, Washington DC.

Robledo-Arnuncio, J.J., Gil, L., 2005. Patterns of pollen dispersal in a small population of *Pinus sylvestris* L. revealed by total-exclusion paternity analysis. Heredity 94, 13–22.

Rojas, C., Biffi, D., Stephenson, S.L., Schnittler, M., 2008. Microhabitat and niche separation in species of *Ceratiomyxa*. Mycologia 100, 843–850.

Rojas, C., Herrera, N., Stephenson, S.L., 2012a. An update on the myxomycete biota (Amoebozoa: Myxogastria) of Colombia. Check List 8, 617–619.

Rojas, C., Morales, R.E., Calderón, I., Clerc, P., 2013. First records of myxomycetes from El Salvador. Mycosphere 4, 1042–1051.

Rojas, C., Schnittler, M., Stephenson, S.L., 2010. A review of the Costa Rican myxomycetes (Amebozoa). Brenesia 73–74, 39–57.

Rojas, C., Stephenson, S.L., 2007. Distribution and ecology of myxomycetes in the high-elevation oak forests of Cerro Bellavista, Costa Rica. Mycologia 99, 534–543.

Rojas, C., Stephenson, S.L., Huxel, G., 2011. Macroecology of high elevation myxomycete assemblages in the northern Neotropics. Mycol. Prog. 10, 423–437.

Rojas, C., Stephenson, S.L., Valverde, R., Estrada-Torres, A., 2012b. A biogeographical evaluation of high-elevation myxomycete assemblages in the northern Neotropics. Fungal Ecol. 5, 99–113.

Rojas, C., Valverde, R., Stephenson, S.L., 2015. New additions to the myxobiota of Costa Rica. Mycosphere 6, 709–715.

Rollins, A.W., Stephenson, S.L., 2011. Global distribution and ecology of myxomycetes. Curr. Top. Plant Biol. 12, 1–14.

Rollins, A.W., Stephenson, S.L., 2013. Myxomycetes associated with grasslands of the western central United States. Fungal Diver. 59, 147–158.

Romeralo, M., Cavender, J.C., Landolt, J.C., Stephenson, S.L., Baldauf, S.L., 2011. An expanded phylogeny of social amoebas (Dictyostelia) shows increasing diversity and new morphological patterns. BMC Evol. Biol. 11, 84.

Romeralo, M., Spiegel, F.W., Baldauf, S.L., 2010. A fully resolved phylogeny of the social amoebas (Dictyostelia) based on combined SSU and ITS rDNA sequences. Protist 161, 539–548.

Ronikier, A., Ronikier, M., 2009. How 'alpinc' are nivicolous myxomycetes? A worldwide assessment of altitudinal distribution. Mycologia 101, 1–16.

Ronikier, A., Ronikier, M., Drozdowicz, A., 2008. Diversity of nivicolous myxomycetes in the Gorce mountains—a low-elevation massif of the western Carpathians. Mycotaxon 103, 337–352.

Rosing, W.C., Mitchell, D.W., Moreno, G., Stephenson, S.L., 2011. Addition to the myxomycetes of Singapore. Pacific Sci. 65, 391–400.

Rostafinsky, J.T., 1875. Sluzowce monografia. Pamietn. Towarz. Nauk. Sci. Paryzu 6, 216–432.

Santesson, R., 1948. *Listerella paradoxa* Jahn och *Orcadella singularis* (Jahn) nov. comb., tvá für Sverige nya myxomyceter. Svensk Bot. Tids. 42, 42–50.

Saunders, J.E., Saunders, E., 1900. The habitats of the Mycetozoa. Trans. Hertfordshire Nat. Hist. Soc. 10, 169–172.

Schinner, F., 1983. Myxomycetes aus dem Gebiet des Torne Träsk (Abisko) in Schwedisch Lappland. Sydowia. Ann. Mycol. 36, 269–276.

Schinner, F., Kobilansky, C., Kolm, H., 1990. Ein Beitrag zur Ökologie der Myxomyceten. Beit. Kenn. Pilze Mittel. 5, 15–18.

Schnittler, M., 2000. Foliicolous liverworts as a microhabitat for Neotropical Myxomycetes. Nova Hedwigia 72, 259–270.

Schnittler, M., 2001. Ecology of Myxomycetes of a winter-cold desert in western Kazakhstan. Mycologia 93, 653–669.

Schnittler, M., Dagamac, N.H.A., Sauke, M., Wilmking, M., Buras, A., Ahlgrimm, S., Eusemann, P., 2016. Ecological factors limiting occurrence of corticolous myxomycetes: a case study from Alaska. Fungal Ecol. 21, 16–23.

Schnittler, M., Erastova, D.A., Shchepin, O.N., Heinrich, E., Novozhilov, Y.K., 2015a. Four years in the Caucasus: observations on the ecology of nivicolous myxomycetes. Fungal Ecol. 14, 105–115.

Schnittler, M., Kummer, V., Kuhnt, A., Krieglsteiner, L., Flatau, L., Müller, M., 2011. Rote Liste und Gesamtartenliste der Schleimpilze (Myxomycetes) Deutschlands. Schr. R. Veget. 70, 125–234.

Schnittler, M., Lado, C., Stephenson, S.L., 2002. Rapid biodiversity assessment of a tropical myxomycete assemblage: Maquipucuna Cloud Forest Reserve, Ecuador. Fungal Diver. 9, 135–167.

Schnittler, M., Mitchell, D.W., 2000. Species diversity in Myxomycetes based on the morphological species concept: a critical examination. Stapfia 73, 55–61.

Schnittler, M., Novozhilov, Y.K., 1996. The myxomycetes of boreal woodlands in Russian northern Karelia: a preliminary report. Karstenia 36, 19–40.

Schnittler, M., Novozhilov, Y.K., 1998. Late-autumn myxomycetes of the Northern Ammergauer Alps. Nova Hedwigia 66, 205–222.

Schnittler, M., Novozhilov, Y.K., 2000. Myxomycetes of the winter-cold desert in western Kazakhstan. Mycotaxon 74, 267–285.

Schnittler, M., Novozhilov, Y.K., Carvajal, E., Spiegel, F.W., 2013. Myxomycete diversity in the Tarim basin and eastern Tian-Shan, Xinjiang province, China. Fungal Diver. 59, 91–108.

Schnittler, M., Novozhilov, Y.K., Romeralo, M., Brown, M., Spiegel, F.W., 2012. Myxomycetes and myxomycete-like organisms. In: Frey, W. (Ed.), Englers Syllabus of Plant Families, 4, Bornträger, Stuttgart, pp. 12–88.

Schnittler, M., Novozhilov, Y.K., Shadwick, J.D.L., Spiegel, F.W., García-Carvajal, E., König, P., 2015b. What substrate cultures can reveal: myxomycetes and myxomycete-like organisms from the Sultanate of Oman. Mycosphere 6, 356–384.

Schnittler, M., Shchepin, O., Dagamac, N.H.A., Borg Dahl, M., Novozhilov, Y.K. 2017. Barcoding myxomycetes with molecular markers: challenges and opportunities. Nova Hedwigia 104, 183–209.

Schnittler, M., Stephenson, S.L., 2000. Myxomycete biodiversity in four different forest types in Costa Rica. Mycologia 92, 626–637.

Schnittler, M., Stephenson, S.L., 2002. Inflorescences of Neotropical herbs as a newly discovered microhabitat for myxomycetes. Mycologia 94, 6–20.

Schnittler, M., Stephenson, S.L., Novozhilov, Y.K., 2000. Ecology and world distribution of *Barbeyella minutissima* (Myxomycetes). Mycol. Res. 104, 1518–1523.

Schnittler, M., Tesmer, J., 2008. A habitat colonisation model for spore-dispersed organisms: does it work with eumycetozoans? Mycol. Res. 112, 697–707.

Schnittler, M., Unterseher, M., Pfeiffer, T., Novozhilov, Y.K., Fiore-Donno, A.M., 2010. Ecology of sandstone ravine myxomycetes from Saxonian Switzerland (Germany). Nova Hedwigia 90, 227–302.

Schnittler, M., Unterseher, M., Tesmer, J., 2006. Species richness and ecological characterization of myxomycetes and myxomycete-like organisms in the canopy of a temperate deciduous forest. Mycologia 98, 223–232.

Shchepin, O., Novozhilov, Y.K., Schnittler, M., 2014. Nivicolous myxomycetes in agar culture: some results and open problems. Protistology 8, 53–61.

Singer, H., Moreno, G., Illana, C., 2005. A taxonomic review on the nivicolous species described by Kowalski. I. Order Stemonitales. Mycol. Prog. 4, 3–10.

Snell, K.L., Keller, H.W., Eliasson, U.H., 2003. Tree canopy myxomycetes and new records from ground sites in the Great Smoky Mountains National Park. Castanea 68, 97–108.

Spiegel, F.W., Stephenson, S.L., Keller, H.W., Moore, D.L., Cavender, J.C., 2004. Mycetozoans. In: Mueller, G.M., Bills, G.F., Foster, M.S. (Eds.), Biodiversity of Fungi. Inventory and Monitoring Methods. Elsevier Academic Press, Amsterdam, pp. 547–576, Part IId.

Stephenson, S.L., 2004. Distribution and ecology of myxomycetes in southern Appalachian subalpine coniferous forests. Mem. NY Botan. Gard. 89, 203–212.

Stephenson, S.L., Fiore-Donno, A.M., Schnittler, M., 2011. Myxomycetes in soil. Soil Biol. Biochem. 43, 2237–2242.

Stephenson, S.L., Landolt, J., 2015. Occurrence of myxomycetes in the aerial "canopy soil" microhabitat. Mycosphere 6, 74–77.

Stephenson, S.L., Landolt, J.C., Moore, D.L., 1999. Protostelids, dictyostelids, and myxomycetes in the litter microhabitat of the Luquillo Experimental Forest, Puerto Rico. Mycol. Res. 103, 209–214.

Stephenson, S.L., Laursen, G.A., 1993. A preliminary report on the distribution and ecology of myxomycetes in Alaskan tundra. Bibl. Mycol. 150, 251–257.

Stephenson, S.L., Laursen, G.A., 1998. Myxomycetes from Alaska. Nova Hedwigia 66, 425–434.

Stephenson, S.L., Laursen, G.A., Seppelt, R.D., 2007. Myxomycetes of subantarctic Macquarie Island. Aust. J. Botany 55, 439–449.

Stephenson, S.L., Novozhilov, Y.K., Schnittler, M., 2000. Distribution and ecology of myxomycetes in high-latitude regions of the northern hemisphere. J. Biogeograph. 27, 741–754.

Stephenson, S.L., Schnittler, M., 2016. Myxomycetes. In: Archibald, J.M., Simpson, A.G.B., Slamovits, C.H. (Eds.), Handbook of the Protists. Springer, New York.

Stephenson, S.L., Schnittler, M., Lado, C., Estrada-Torres, A., Wrigley de Basanta, D., Landolt, J.C., Novozhilov, Y.K., Clark, J., Moore, D.L., Spiegel, F.W., 2004. Studies of neotropical mycetozoans. Syst. Geogr. Plants 74, 87–108.

Stephenson, S.L., Schnittler, M., Mitchell, D.W., Novozhilov, Y.K., 2001. Myxomycetes of the Great Smoky Mountains National Park. Mycotaxon 78, 1–15.

Stephenson, S.L., Schnttler, M., Novozhilov, Y.K., 2008. Myxomycete diversity and distribution from the fossil record to the present. Biodiv. Conserv. 17, 285–301.

Stephenson, S.L., Shadwick, J.D.L., 2009. Nivicolous myxomycetes from alpine areas of southeastern Australia. Aust. J. Bot. 57, 116–122.

Stephenson, S.L., Stempen, H., 1994. Myxomycetes: A handbook of slime molds. Timber Press Inc., USA.

Sutherland, J.B., 1979. Gray jay feeding on slime mold. Murrelet 60, 122.

Swanson, A.R., Spiegel, F.W., Cavender, J.C., 2002. Taxonomy, slime molds, and the questions we ask. Mycologia 94, 968–979.

Takahashi, K., 2004. Distribution of myxomycetes on different decay states of deciduous broadleaf and coniferous wood in a natural temperate forest in the southwest of Japan. Syst. Geogr. Plants 74, 133–142.

Takahashi, K., Hada, Y., 2012. Seasonal occurrence and distribution of myxomycetes on different types of leaf litter in a warm temperate forest of western Japan. Mycoscience 53, 245–255.

Takahashi, K., Harakon, Y., 2012. Comparison of wood-inhabiting myxomycetes in subalpine and montane coniferous forests in the Yatsugatake Mountains of Central Japan. J. Plant Res. 125, 327–337.

Tamayama, M., 2000. Nivicolous taxa of the myxomycetes in Japan. Stapfia 73, 121–129.

Tesmer, J., Schnittler, M., 2007. Sedimentation velocity of myxomycete spores. Mycol. Prog. 6, 229–234.

Torres-Machorro, A.L., Hernández, R., Cevallos, A.M., López-Villaseñor, I., 2010. Ribosomal RNA genes in eukaryotic microorganisms: witnesses of phylogeny? FEMS Microbiol. Rev. 34, 59–86.

Townsend, J.H., Aldrich, H.C., Wilson, L.D., McCranie, J.R., 2005. First report of sporangia of a myxomycete (*Physarum pusillum*) on the body of a living animal, the lizard *Corytophanes cristatus*. Mycologia 97, 346–348.

Tran, H.T.M., Stephenson, S.L., Hyde, K.D., Mongkolporn, O., 2006. Distribution and occurrence of myxomycetes in tropical forests of northern Thailand. Fungal Diver. 22, 227–242.

Tran, H.T.M., Stephenson, S.L., Hyde, K.D., Mongkolporn, O., 2008. Distribution and occurrence of myxomycetes on agricultural ground litter and forest floor litter in Thailand. Mycologia 100, 181–190.

Tu, Y., Xiao, F., Zhang, J., Lu, Q., Wang, L., 2016. Two "myxomycete diseases" occurred on cultivated fields of *Dendrobium candidum*. J. Zhejiang Univ. 42, 137–142.

Ukkola, T., Häarkönen, M., Saarimäki, T., 1996. Tanzanian myxomycetes: second survey. Karstenia 36, 51–77.

Urich, T., Lanzén, A., Qi, J., Huson, D.H., Schleper, C., Schuster, S.C., 2008. Simultaneous assessment of soil microbial community structure and function through analysis of the metatranscriptome. PLoS One 3, e2527.

Vlasenko, A.V., Novozhilov, Y.K., 2011. Myxomycetes of pine forests of the right-bank part of the Upper Ob River. Mikol. Fitopatol. 45, 1–13.

Vlasenko, A.V., Novozhilov, Y.K., Shchepin, O.N., Vlasenko, V.A., 2016. Hydrochory as certain mode of distribution of myxomycetes along floodlands in south of Western Siberia. Mikol. Fitopatol. 50, 14–23, [In Russian with English abstract].

Walker, L.M., Stephenson, S.L., 2016. The species problem in myxomycetes revisited. Protist 167 (4), 319–338.

Wheeler, Q.D., 1984. Associations of beetles with slime molds: ecological patterns in the *Anisotomini* (Leiodidae). Bull. Entomol. Soc. Am. 39, 14–18.

Winsett, K.E., Stephenson, S.L., 2008. Using ITS sequences to assess intraspecific genetic relationships among geographically separated collections of the myxomycete *Didymium squamulosum*. Rev. Mex. Micol. 27, 59–65.

Winsett, K.E., Stephenson, S.L., 2011. Global distribution and molecular diversity of *Didymium difforme*. Mycosphere 2, 135–146.

Wrigley de Basanta, D., 1998. Myxomycetes from the bark of the evergreen oak *Quercus ilex*. Anal. Jard. Bot. Madrid 56, 3–14.

Wrigley de Basanta, D., 2000. Acid deposition in Madrid and corticolous myxomycetes. Stapfia 73, 113–120.

Wrigley de Basanta, D., Lado, C., Estrada-Torres, A., 2012. Description and life cycle of a new *Physarum* (Myxomycetes) from the Atacama Desert in Chile. Mycologia 104 (5), 1206–1212.

Wrigley de Basanta, D., Lado, C., Estrada-Torres, A., Stephenson, S.L., 2009. Description and life cycle of a new *Didymium* (Myxomycetes) from arid areas of Argentina and Chile. Mycologia 101, 707–716.

Wrigley de Basanta, D., Lado, C., Estrada-Torres, A., Stephenson, S.L., 2010. Biodiversity of myxomycetes in subantarctic forests of Patagonia and Tierra del Fuego, Argentina. Nova Hedwigia 90, 45–79.

Wrigley de Basanta, D., Lado, C., Estrada-Torres, A., Stephenson, S.L., 2013. Biodiversity studies of myxomycetes in Madagascar. Fungal Diver. 59, 55–83.

Wrigley de Basanta, D., Stephenson, S.L., Lado, C., Estrada-Torres, A., Nieves-Rivera, A.M., 2008. Lianas as a microhabitat for myxomycetes in tropical forests. Fungal Diver. 28, 109–125.

Yajima, Y., Nishikawa, T., Yamamoto, Y., 2006. Studies on the myxomycetes of Hikkaido, Japan (II). Nivicolous species of Mount Asahidake in the Taisetsu Mountains. Rep. Taiset. Inst. Sci. 40, 53–57.

Yamamoto, Y., 1998. The myxomycete biota of Japan. Toyo Shorin Publishing Co., Ltd, Tokyo, Japan.

Yamni, K., Meyer, M., 2008. Myxomycetes succulenticoles du sud-ouest morocain. Bol. Soc. Micol. Madrid 32, 121–125.

FURTHER READING

Beltran, E., Lado, C., Barrera, J., Gonzalez, E., 2004. Myxomycete diversity in the Laurel forest of Garajonay National Park (Canary Islands, Spain). Syst. Geogra. Plants 74, 159–173.

Castillo, A., Illana, C., Moreno, G., 1998. *Protophysarum phloiogenum* and a new family in the Physarales. Mycol. Res. 102, 838–842.

Lado, C., Mosquera, J., Beltrán-Tejera, E., 1999. *Cribraria zonatispora*, development of a new myxomycete with unique spores. Mycologia 91, 157–165.

Ronikier, A., Lado, C., 2015. Nivicolous Stemonitales (Myxomycetes) from the Austral Andes. Analysis of morphological variability, distribution and phenology as a first step towards testing the large-scale coherence of species and biogeographical properties. Mycologia 107, 258–283.

Chapter 10

Techniques for Recording and Isolating Myxomycetes

Diana Wrigley de Basanta*, Arturo Estrada-Torres**
**Royal Botanic Garden (CSIC), Madrid, Spain; **Behavioural Biology Centre, The Autonomous University of Tlaxcala, Tlaxcala, Mexico*

INTRODUCTION

The first myxomycete ever recorded was reported in "Herbarium Portatile" by Panckow (1654), where a collection of what is now known as *Lycogala epidendrum* was illustrated and entitled "fungi cito crescentes" (literally "mushroom of rapid growth"). There is also a brief note on its rapid development in the volume, which is currently held at the Natural History Museum in London, and was examined by the first author in 2006. Lister (1918) also reported records of what is thought to be the same species, published by the English naturalist John Ray in 1696. A detailed history of the study of these organisms can be found in Chapter 2.

Since then, recording myxomycetes has been carried out for a variety of reasons, for pleasure, interest, and scientific study, from random collecting during a foray to more structured work. Many people have observed, searched for and found myxomycetes, setting up personal collections or adding to official herbaria in many countries. Historically, as well as currently, collectors have sometimes sent samples and *exsiccata* to known experts, who have exchanged these with individuals from all around the world. For example, a large number of samples were sent to Lister and his daughter, Gulielma, in the United Kingdom, including some from the Emperor Hirohito of Japan, a myxomycete enthusiast, who in 1928 hand-carried samples of *Hemitrichia imperialis* from Japan to London for study by the Listers. The samples still exist in the herbarium of the Natural History Museum in London (Lado and Wrigley de Basanta, in press). This work, to collect and record myxomycetes, has traditionally been done by mycologists, because these mostly visible microorganisms develop in habitats similar to those in which many different kinds of fungi are found. Myxomycetes can be recorded and isolated from the field where they are produced under natural conditions, or from laboratory culture by several different methods.

Myxomycetes: Biology, Systematics, Biogeography, and Ecology. http://dx.doi.org/10.1016/B978-0-12-805089-7.00010-X

FIELD COLLECTIONS

Myxomycetes are mainly recognized, recorded and/or collected in the sporulating or fruiting body stage of the life cycle, and current field collecting techniques normally involve fairly straightforward protocols, such as those outlined by Stephenson and Stempen (1994) or Ing (1999). The myxomycetes feed as unicellular stages (amoeboflagellates) or plasmodia, primarily on bacteria and other organic matter, in or on decaying plant material. They emerge from the substrate to produce the fruiting bodies, and these are often found on the surfaces of plant parts or remains from which they can be easily removed. The first step involved in collecting is to determine and select an area, habitat (Fig. 10.1A) or microhabitat to study, and then the most common method is to search all the known plant substrates or substrate groups where these organisms may be expected to develop, in a systematic manner and usually for a specified time. For example, in the survey of the coastal desert of Peru (Lado et al., 2016), four researchers collected in each specific locality for an hour. The localities were along predetermined North–South and West–East transects at intervals of approximately 25–50 km, to cover the maximum possible terrain in this very extensive belt of arid landscape. Schnittler et al. (2002) used 200 m transects at each study site and collected substrates for moist chamber cultures at 10 m intervals along these transects. In a 4-year study by Eliasson (1981), a marked route, in a study area 160 m × 120 m, was surveyed 2–3 times a week. Rodríguez-Palma (2003) established nine sampling stations in the fir forest on the summit of the Malintzi Volcano in Mexico. The sampling units were three fallen logs in each, and the myxomycete fruiting bodies on these were collected once a week in the rainy season (June–September) and every 2 weeks in the dry cold season (October–May). Transects can also be used in more detailed protocols where specific measured plots are set up, as they were for a very comprehensive biodiversity inventory in Costa Rica (Rossman et al., 1998).

Within the habitat or microhabitat, the substrate type determines the method used in searching for fruiting bodies, but it generally involves the use of a pocketknife or pruning clippers to remove small pieces of substrate supporting the fruiting bodies. Other useful equipment includes a good hand lens (with at least 10× magnification) to detect smaller specimens and a GPS receiver to determine the geographic coordinates of the sample and the elevation. These data are essential not only for confirmation or repetition of results, but also for any studies assessing accurate geographic distribution of species. A camera is also very valuable for recording and documenting the general habitat and the microhabitat of the specimen collected, and for later confirmation of substrate plant identity. A portable plant specimen dryer is sometimes necessary in very humid environments or seasons. Some places to examine within the habitat would include all surfaces of logs (Fig. 10.1C), other pieces of dead wood and old tree trunks, particularly in cracks or crevices, but mainly the lower surface, especially if the substrate is exposed to the sun, twigs, leaf litter that has accumulated, for

FIGURES 10.1 Habitats or microhabitats where myxomycetes can be found. (A) Forests, such as this one in Madagascar. (B) Leaves of rosette plants, such as *Puya* spp. (C) On logs, like this collection of a *Stemonitis*. (D) On the remains of cacti like this *B. melanospora.* Photo: Adam W. Rollins. (E) On leaf litter under a shrub, such as this *Zygophyllum stapfii* in Namibia. (F) At the edge of melting snow.

instance, under a shrub (Fig. 10.1E) or tree, and especially if there is some but not too much moisture present. Lifting the top drier layer with care will often reveal fruiting bodies on the slightly moist, sometimes compacted lower layer, and turning over individual leaves and examining the underside is also useful. Bark on living trees will sometimes have observable fruiting bodies in the field, although this substrate is often taken for later culture in the laboratory, as discussed later. In certain habitats, such as the drylands of the Americas, the inside of decaying cacti stems, or cladodes (Fig. 10.1D), produce many myxomycetes, and the extent of the decay is favored by different species. In addition, the bases of the dead leaves of rosette plants, such as *Puya* spp. (Fig. 10.1B) or *Agave* spp., especially in areas between the more decayed softer leaves, are productive substrates (Estrada-Torres et al., 2009) where fruiting bodies can last for several months in spite of the surrounding dryness. Stout gardening or hedger's gloves are necessary for field collecting of these and some other substrates, although sometimes even these will not fully protect one against spines, trichomes, and glochids. Gloves are particularly essential in tropical and subtropical areas where it is not uncommon to disturb scorpions, snakes, poisonous spiders and poisonous plants while searching leaf litter, logs or the remains of succulent plants.

Ing (1994), in his paper on the phytosociology of myxomycetes, discussed more than 15 potential substrate types where these organisms may be found and points out that the surface on which they fruit can sometimes be different from where they carry out their feeding stages as amoeboflagellates or plasmodia, as is the case for certain foliicolous species.

As the myxomycete fruiting bodies are produced ephemerally, it is necessary to carry out sampling over time and with a variety of methods. The sporulation phenology of different species of myxomycetes from trees in Mediterranean woodlands, for instance, was discussed in detail by Lado (1993), and showed distinct differences in the months of maximum appearance, dependent on available water and limited by temperature extremes. The author concluded that the sporulation phenology in these woodlands was also different from studies carried out at different latitudes. In work by Stephenson (1988) in forests of southwestern Virginia in the United States, the optimum time for the appearance of myxomycete fruiting bodies was July, a poor collecting time in the Mediterranean woodlands. In Sweden, on the other hand, there are species that appear in the spring, others in the summer or autumn and still others from May to October (Eliasson, 1981). Therefore, the optimum time for collecting myxomycetes in the field varies considerably according to the geographic area and the species sought.

The patterns of temperature and available moisture are the main limiting factors determining their appearance, but the range of tolerance of these parameters varies for different species and many species can be recorded from areas of extreme climatic conditions, for instance, from the edge of melting snow (Fig. 10.1F), as indicated in several studies (Kowalski, 1971; Lado et al., 2005; Ronikier and Ronikier, 2009). This ecological group of nivicolous species can appear in large and conspicuous fruitings on ground litter and the lower stems of living shrubs. In such situations, temperatures can be just above freezing when covered with snow, even if the air temperatures are lower, due to the insulating effect of the snow itself, which also buffers the abrupt day and night temperature fluctuation (Lado, 2004). This author included, in a detailed account of nivicolous myxomycetes in Spain and Portugal, valuable comments on their ecological requirements. He indicated that the abundance of myxomycetes in the high mountains, coinciding with the snow melt, can mean that four or five different species could be found in an area as small as a square meter (Lado, 2004). The nivicolous myxomycetes, as well as others, have been keyed and beautifully illustrated in the volumes by Poulain et al. (2011). The ecology of this group of myxomycetes in a biosphere reserve in Russia was investigated by Schnittler et al. (2015), who recorded more than 1000 fruitings in their survey over a period of 4 years. Using a transect system of 50 m circular plots, weather data and data loggers to monitor conditions beneath the snow. These authors found that observed fluctuations in the abundance of nivivolous myxomycetes from year to year are most probably due to climatic factors, such as insufficient snow depth or sharp frost when there is little or no snow cover (Schnittler et al., 2015). Ronikier and Lado (2015) discussed differences between myxomycete assemblages and the optimum times for the appearance of nivicolous species between different geographic regions. They suggested that the appearance of strictly nivicolous species, outside the normal snow-melt period, may indicate some special climatic conditions restricted to the southern hemisphere, since some species have been collected there even in the austral summer (Lado et al., 2013; Stephenson et al., 2007; Wrigley de Basanta et al., 2010).

When recording field collections, it is necessary to decide what is a collection or record of a species. A plasmodium could be considered to be an individual, but some species of myxomycetes produce just one aethalium from a plasmodium and others up to hundreds of fruiting bodies, all from one plasmodium. Smaller species may have many protoplasmodia, all producing an individual fruiting body but with many close together and at the same time.

As Eliasson (1981) pointed out, the use of determinants is arbitrary, but a reasonable practical compromise in the field is to consider fruiting bodies of the same species, that are at least 30 cm apart, as different collections. Stephenson (1988) also used this definition of a collection in a field study of myxomycetes in the upland forests of southwestern Virginia. The method used by Feest and Madelin (1985a), to quantify the myxomycetes present in the soil, established the number of plasmodium-forming units (PFU) present, but this method does not measure separate species and depends on the species capability of developing under culture conditions. A simple method of approximating numbers to find the relative abundance of a species was suggested by Stephenson et al. (1993) and is often used in surveys. It is a scale of the proportion of a species in the total number of collections and goes from A (abundant >3% of all collections) to R (rare, 0.5% of all collections).

Any specimens collected in the field are usually carried back in boxes. Some collectors use a box lined with cork and secure the specimens with pins (Ing, 1999), but others place them in plastic boxes of the type used to store nails and screws or fishing bait (Spiegel et al., 2007). Specimens can also be mounted with glue into cardboard boxes, like matchboxes, which is particularly advisable for transporting small delicate specimen. Specimens can be air-dried in open containers once back in either the laboratory or a working area (Fig. 10.2A) or the plant drier mentioned earlier can be used in more humid environments. Specimens are glued onto cardboard boxes, or card trays placed in boxes that then serve for storage in the herbarium or a private collection.

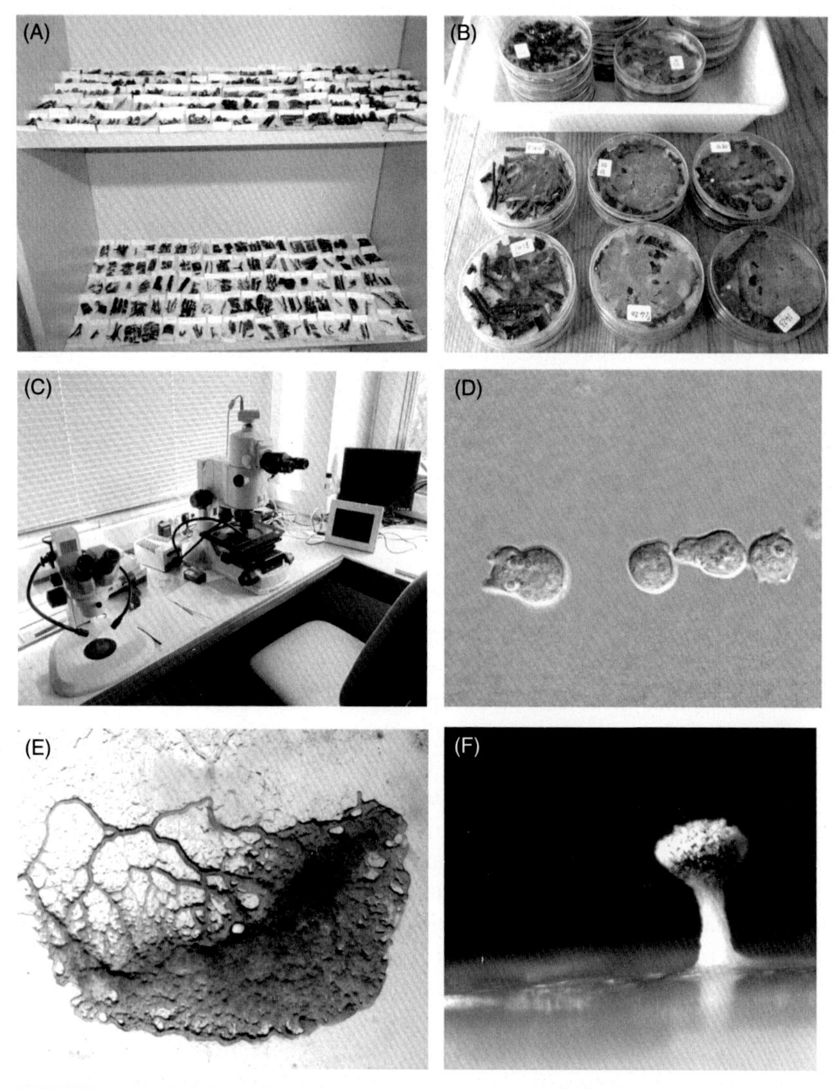

FIGURES 10.2 Laboratory culture. (A) Field collections air-drying. (B) Moist chamber cultures. (C) Laboratory equipment for observing agar cultures. (D) Myxamoebae. (E) A phaneroplasmodium. (F) A sporocarp forming on agar.

These cardboard boxes often come in flat packs, which are easy to transport to the collecting locality, where they can be readily assembled for use. Specimen are usually glued on the top of the box so that they are never separated from the label. Some useful tools at this point are fine forceps, a hand lens, a good portable light source, scissors, and pruning clippers for cleaning and separating the specimens. The label should include the genus and species (to the extent that can be determined), the unique collection number, the substrate, the country, province, geographic coordinates, elevation, date of collection, and the name of the collector. Some researchers include other data, such as the type of vegetation of the area and data on the microhabitat. If material is to be collected in a country other than one's own, certain permits are probably required, although the procedures required to obtain such a permit in each country vary widely. It is invaluable for these formalities to count on the cooperation of scientists working in the country involved. In addition, many substrate plants could be subject to the Cites convention and require special collecting and export permits (https://www.cites.org), and since 2014 probably come under the Nagoya protocol, an international treaty that prohibits the collection of plant material without an agreement with host countries and sharing of benefits (https://www.cbd.int/abs).

Field Collections: Discussion of Advantages and Constraints

The advantages of collecting specimens of myxomycetes in the field are many. First, the researcher can be sure that the particular species has developed there under natural conditions, and therefore the morphology and development, especially in large collections, will be reliable. Field-collected specimens are frequently present in large fruitings consisting of many individual fruiting bodies, thus showing a range of morphological characters and providing sufficient material to send duplicates to other herbaria. This is particularly important when the collections are used to describe a new species, and it allows for a more complete description, avoiding narrow and imprecise species concepts. Certain species will be collected only in the field, as they have not yet been obtained in laboratory culture. These include larger species of the genera *Fuligo*, *Lycogola*, or *Tubifera*, for instance. Some lignicolous species from decaying wood, like certain species of *Cribraria*, can also be collected only in the field. A very important benefit of collecting in the field is time. Even fairly comprehensive biodiversity field surveys can be completed in a few weeks (Wrigley de Basanta et al., 2013), and the field results are immediate, although any subsequent culture work takes longer, as discussed later. Several hours are sufficient to survey a particular vegetation type in any given locality (Novozhilov et al., 2000) or even 1 h each at multiple-staged collecting sites (Lado et al., 2014) along a transect.

The main constraints of field collecting are the effects of temperature, and water availability on the myxomycetes, and the ability of the investigator to coincide with the sporulation phenology of the maximum number of species in an area. As discussed earlier, taxa, such as *Ceratiomyxa* spp., can appear at the beginning of the local rainy season, whereas others appear only at the end of the rains. The latter is the case for many species of *Diderma*. This makes it essential to sample at different times of the year or in different years, but this can be difficult or too costly for remote regions and areas where access is a problem.

The cost of field collecting is a very real constraint, which can run into literally thousands of dollars for a 3-week collecting trip. Further problems can be encountered with the conditions that exist at the collecting sites, such as high elevations or extreme temperatures, as is the case in the arid areas of the Americas that we have been studying for the last 15 years (http://www.myxotropic.org). In addition, native animal or plant hazards, or inaccessible substrates, such as up on a rock face, on a remote island, or in the middle of a desert can complicate collecting. Many small species cannot be detected in the field and thus may be underrepresented in surveys that are based exclusively on field collections.

Time can also be a major constraint, as well as an advantage, depending on the type of survey. An estimate by Novozhilov et al. (2000) suggested 2–4 months for a regional study, but this will depend of course on the size of the region, on the objectives and the thoroughness of the study. It will often take repeated surveys throughout the year to coincide with seasonality of all species, and ideally repeated annual visits at different times to properly survey an area. In arid areas, samples can be old and withered and in humid areas frequently moldy. This can render difficult or impossible, the correct identification of species, in spite of all efforts to collect at a favorable season, because this is often difficult to determine. Some material can be destroyed in transit from the field, particularly small delicate species, and other organisms, such as beetles, springtails, or mites, can inadvertently be carried with them.

ISOLATION OF MYXOMYCETES BY LABORATORY CULTURE

Moist Chamber Cultures

A method of obtaining myxomycetes in addition to, or instead of, field collected specimens, is the use in the laboratory of what are known as moist chamber cultures (Fig. 10.2B). The use of the moist chamber culture method of isolating myxomycetes is often attributed to Gilbert and Martin (1933), who used the technique to allow algae to develop but then found it to be a valuable method for isolating myxomycetes. They described the method of removing the bark from various aspects of trees and placing it in culture and added a list of species isolated from their cultures. However, long before this, Lister (1894) described the development of plasmodia in a moist chamber from material gathered in the field and his daughter (Lister, 1918) stated how he had cultivated *Badhamia utricularis* at home and set up "glass boxes" where he gave the myxomycete plasmodia a "variety of food." In the first edition of his monograph, Lister

(1894) gives details of various experiments relating to feeding and nuclear division of different stages in the life history of these organisms. In a later paper (Lister, 1901), he explained how to culture them from spores.

Collection of material for later moist chamber culture involves several stages. The most meaningful use of this technique, when carrying out a biodiversity survey of large areas, is in conjunction with field collecting as described earlier. At field localities, suitable material for culture can be extracted at the same time as the collections encountered fruited in situ. This means that the two types of collections are complimentary to each other and often the same species will occur in both. However, this is not always the case, as the conditions in moist chamber cultures are obviously different from the natural conditions at the time of collection. The samples can be bark of living trees (Everhart et al., 2008, 2009; Scarborough et al., 2009), leaf litter, specific plant litter, for example, dead pieces of the epidermis or interior tissue of cacti or other succulent plants, dried inflorescences (Schnittler and Stephenson, 2002), on bryophytes in ravines (Ing, 1983), dry fruits, droppings of herbivorous animals (Eliasson, 2013; Eliasson and Lundquist, 1979), lianas (Wrigley de Basanta et al., 2008), twigs (Stephenson et al., 2008), small pieces of wood, grasses (Rollins and Stephenson, 2013), or other plant remains. As in field collecting, sometimes very specific protocols are necessary for collecting the substrate. For example, the bark of living trees must be carefully removed with a knife or fine chisel and care must be taken to avoid lifting phloem or inner tissue from the tree (Ing, 1999; Mitchell, 1980; Stephenson, 1989). In a study of the aerial reproductive structures of five representative plant families (Kilgore et al., 2009) a special rope-climbing technique was used for collection of aerial parts from the tops of tall trees. This technique, developed by Keller (2004) has been used to collect myxomycetes both in the field and for substrate collection for culturing corticolous myxomycetes. Other types of substrates, such as soils (Clarholm, 1981; Feest, 1987), have slightly different collecting protocols, specific to the substrate being collected.

The moist chamber culture technique can be used in many other types of study, too. For example, it has been used to study one specific substrate over a range of parameters, such as bark productivity versus tree height (Keller, 2004; Scarborough et al., 2009), acid rain effects on a specific type of tree bark (Wrigley de Basanta, 2000), to investigate the myxobiota over a given area (Kosheleva et al., 2008; Ukkola et al., 1996), or to examine a family of substrates to see the different productivity of each (Estrada-Torres et al., 2009), to name just a few. The technique can also be used for quantitative studies to improve our understanding of the biodiversity and ecology of myxomycetes, such as those discussed by Novozhilov et al. (2000).

Unless some special technique is required, the collected substrate samples, are placed in strong paper bags and allowed to air dry naturally. In very moist places, such as tropical rain forests, it can be difficult to dry samples and a plant drier can be useful in these situations or even a domestic hair dryer (with a cool setting and located at some distance) could be used as a last resort. Each sample should be labeled with the locality, name, type of substrate and date of collection. One example would be to use a three-letter code for the country, the year, and then sequential numbers for the locality, such as ARG 08-16, a sample collected at locality 16 in 2008 in Argentina. In this case, the country code is followed by the collector's numbering system, the name of the substrate and the date (e.g., dwb 1524, remains of *Oreocereus trolli*, xx-xx-2008) on each substrate bag. The preparation of material in the field by bagging, drying, and numbering could be the same for any of the studies mentioned earlier.

Once back in the laboratory, the series of samples are usually placed in sterile 9 cm plastic Petri dishes on Whatman number one filter paper (Fig. 10.2B), although other containers can be used, and some investigators use other types of absorbent paper (Schnittler et al., 2013). The objective is to maintain a chamber of constantly slow-drying moisture for which the Petri dish is perfect, as its lid allows for a minimal circulation of air. The samples should not normally exceed a covering over the bottom of the dish, as too much overlap of material makes examination difficult, and a dish packed too full will encourage the passage of mites from dish to dish and the proliferation of molds. The substrate in each dish is completely soaked with water on the first day and approximately 24 h later, after measuring the pH of the water, excess water is pipetted off, leaving the substrate moist, but not wet. The water used can range from double filtered or distilled sterile water, to cooled boiled tap water depending on the type of culture and the objectives of the study. Some authors correct the pH of distilled water with KOH or HCl, so that any change in pH is a reflection of the influence of the substrate used (Everhart et al., 2009; Scarborough et al., 2009).

The Petri dishes are examined with a stereomicroscope for fruiting bodies or plasmodia, and it is useful to examine each dish before wetting the substrate, because occasionally very small myxomycetes brought in from the field, that had been undetected, can be found. The frequency of examination depends to a large degree on the will and time of the investigator. Many tiny myxomycetes will appear as soon as 24 h after wetting, so it is a good practice to examine the dishes daily, or at least every 2 days, in the first week with a good dissecting microscope. After that, every 2 or 3 days can be enough, depending on the substrate (Wrigley de Basanta et al., 2009). Different observation protocols are used by other investigators, such as the specific days used by Schnittler et al. (2002) who used 5 observations on days 6, 17, 47, 81, and 109 after the culture was started, or intervals, such as Snell and Keller (2003) who observed their cultures every 7–10 days.

The succession of species that appear can sometimes follow a sequence of first members of the Echinosteliales, then members of the Liceales and finally members of the Trichiales; however, this can be completely different depending on the substrate and on whether the myxomycete was present on or in the substrate as a spore, a microcyst, or a sclerotium. In addition, there can be two cycles of a species in one culture but this seems to be very rare. Frequently, fruiting bodies of the same species will appear in the same culture over a variable incubation period. For instance, fruiting bodies of *Perichaena calongei*, a recently described species on dead leaf bases of the bromeliad species *Puya*, appeared in the same moist chamber culture over a period of 16–40 days, although the mean incubation time for the cultures in which the species appeared was 25 days (Lado et al., 2009), and it was impossible to know if the fruiting bodies were the product of one or more plasmodia. Probably, the species of *Echinostelium* that appear in huge numbers, and some of the *Licea* species, are each from individual protoplasmodia, but since it is not possible yet to ascertain this, most investigators consider any fruiting bodies of the same species produced in the same culture to be a single collection (Stephenson, 1989; Wrigley de Basanta et al., 2009).

Once myxomycete fruiting bodies appear, they need to be dried slowly or sometimes they will be malformed and difficult or impossible to identify. They can be carefully removed using very fine forceps, insect pin tips (for Echinosteliales), or an implement that we have found particularly useful is a disposable microscalpel of the type used in eye surgery (e.g., Amo knives by Allergan Medical Optics). The fruiting bodies are transferred to a dry and labeled Petri dish, to complete the drying process for at least 24 more hours. From here they can be mounted on cardboard trays and placed in cardboard boxes or even matchboxes, numbered with a unique specimen number and labeled with all the collecting data. Cultures can be maintained for up to 3 months, after which very few, if any, myxomycetes will develop. During this time, it is sometimes necessary to add small amounts of water, to rehydrate the culture if it is drying out too quickly, but this should never be to the point of overwatering, as the substrate should always be moist but not wet. A good guide is to maintain the filter paper in a moist condition.

Cultures should be placed in normal diffuse daylight, for instance, near a North-facing window in the northern hemisphere, but not exposed to direct sunlight, in a 12 h light–dark regime, as near as possible. Room temperature of approximately 20°C is sufficient. Various other microorganisms or even larger larvae, springtails, mites, and many different fungi from pin molds to small agarics can grow in cultures, but they can be discouraged by placing the culture in a refrigerator or heating with a microscope lamp, but ultimately if overrun, the culture must be sealed and autoclaved. This is what should also be done, before the disposal of the finished cultures, to prevent any contamination of the environment. Generally, the sooner the substrates are placed in culture the better the results (Ing, 1999), up to a year or more, but this depends a lot on the environment from which they come from. We have obtained good results from substrates cultured up to 3 years after they were collected.

One of the microenvironmental conditions, both in the field and in laboratory culture, that has been most consistently shown to affect myxomycetes, is the pH. Smart (1937) observed that the optimum pH for spore germination was between pH 4.5–7.0, and Gray (1939) noted that a low pH favored the chance of fruiting in *Physarum polycephalum*. Variations in substrate pH, caused by acidification of the environment by acid deposition, industrial pollutants or even excessive animal excreta will affect the number, variety, and assemblage of myxomycetes found, either in the field or in culture (Adamonytė et al., 2013; Everhart et al., 2008; Härkönen, 1977, 1978; Stephenson, 1989; Wrigley de Basanta, 2000, 2004). For example, Härkönen (1978) reported a decrease in the number and variety of myxomycetes cultured on tree bark from the south of Finland, compared with bark from the north, where the bark of the same tree species was on average less acidic. In addition, she reported a reduction in the species variety from an urban air-polluted park compared to a virgin forest (Härkönen, 1977). Later, she found that air improvement was related to an increase in the myxomycetes recovered (Härkönen and Vänscä, 2004). In a study using simulated acid rain, Wrigley de Basanta (2004) found that corticolous myxomycetes were adversely affected by the constant acidification of tree bark, in spite of some buffering by the bark itself. Both the number and species richness were affected and some species were found to be more toxitolerant than others. Wrigley de Basanta's results suggested the potential of myxomycetes as bioindicators of the corticolous habitat. Molecular sampling done by Clissmann et al. (2015) also confirmed that the pH was the main environmental variable determining myxomycete distribution in their study.

Under a variety of adverse conditions, such as desiccation, lack of food, or other unfavorable changes to the microenvironment, the plasmodium can become a hardened resistant sclerotium. This occurs frequently in moist chamber cultures but also in the field, and it is a critical life strategy for long-term resistance for the plasmodial stage of myxomycetes, and for incidental dispersal. The type of plasmodium dictates how the sclerotium is formed (Martin and Alexopoulos, 1969). In myxomycetes that produce a phaneroplasmodium, the protoplasm becomes aggregated in small spherules or macrocysts inside a horny protective cover. Brandza (1928) described conditions of their formation, and how sclerotium formation in calcareous myxomycetes could be total or partial, depending on the developmental stage of the plasmodium when the conditions became unfavorable. The encysted protoplasm survives with less water, and a greater lipid content, than a normal protoplasm (Olive, 1975), but if the desiccation is too fast, the sclerotium is not viable (Brandza, 1928). The sclerotia formed by aphanoplasmodia are separate microscopic droplets formed from the plasmodial veins, and protoplasmodia encyst entirely into one sclerotium (Martin and Alexopoulos, 1969). The encystment can be reversed when conditions become favorable once more, sometimes after many years. The sudden appearance of active plasmodia (Gray and Alexopoulos, 1968) or fruiting bodies that form within 24 h in moist chamber cultures, are most probably explained by the fact that the sclerotia were already formed and upon rehydration produced plasmodia (Martin and Alexopoulos, 1969). Myxomycetes can be collected in the field in the form of sclerotia and be transferred to either agar or moist chamber cultures back in the laboratory and allowed to complete their development and reveal their identity. The authors have activated several species in this manner. These methods of isolating myxomycetes by moist chamber culture are those that have been used by the authors, but there are many variations used by others that also produce good results.

Cold Moist Chamber Cultures

A variation on the methods of moist chamber culture has been to attempt to isolate nivicolous myxomycetes that had not been very successfully cultured before by the traditional methods. The very preliminary results of the attempts are outlined later. Since in the field these organisms appear near the edge of melting snow and require rather specific conditions to sporulate (Lado, 2004), an effort to simulate similar conditions in culture was made by the first author. A newly described nivicolous species, *Lamproderma argenteobruneum*, was cultured in the cold (5°C) and dark (Ronikier et al., 2010) instead of the usual moist chamber culture conditions of around 20°C and in diffuse light. Spores from a 13-year-old field collection were sprinkled onto small pieces of a pteridophyte, the substrate of the type collection of the new species. These pieces of substrate were placed between two moist filter paper discs in a sterile Petri dish and incubated at 5°C in the dark. After 3 months, six fruiting bodies of the species were obtained and found to be morphologically the same as the original collection.

Since then, the method has been repeated various times, but the results have been limited (Table 10.1). Some of the substrates were collected during a biodiversity survey in regions of Patagonia, Argentina, in November 2009, where field collections of nivicolous myxomycetes were made. The substrates

TABLE 10.1 Nivicolous Myxomycetes Obtained in Moist Chamber Cultures

Species	Substrate	Source of substrate	Culture temperature(s) (°C)
Lamproderma argenteobruneum	Pteridophyte stem	The Pyrenees, Spain	5
Lepidoderma sp.	Grass blades	The Patagonian Andes, Argentina	5
Lamproderma sp.	Bark	The Patagonian Andes, Argentina	5
Didymium sp.	*Alnus* sp. twigs	The Alps, France	5
Licea minima[a]	*Nothofagus* sp. bark Various substrates	The Patagonian Andes, Argentina —	5 20
Trichia sordida	*Berberis* sp. bark *Nothofagus* sp. twigs	Mendoza Argentina Rio Negro Argentina	5 20
Perichaena megaspora	Litter of an Asteraceae	Mendoza Argentina	20

[a]*Not strictly nivicolous.*

(bark, twigs, and grasses) were air dried in situ and sealed in paper bags. They were put into cold culture within a month of collection. Other substrates were removed from exsiccata of nivicolous collections from Europe, kindly provided by Marianne Meyer. For some of the cultures, the method was varied slightly to allow the culture dishes to "warm up" to simulate the days of sunshine, before snowmelt, where the temperature beneath the snow cover may be raised substantially for a short time in the middle of the day (Lado, 2004), as solar radiation passes through the snow. During the last month, these cultures were taken out of the cold during the day, and exposed to external temperatures not exceeding 16°C, and then returned to the cold. Two of these cultures did produce myxomycetes, suggesting a possible modification of the method in the future, but it is not known whether they appeared in response to the warming up or not. Two other cultures had visible sclerotia at the beginning of the experiment. One of these almost immediately produced active plasmodia that remained active throughout the culture period, but never produced fruiting bodies.

The cold moist chambers thus far have produced six species of myxomycetes that were mature and thus could be identified, at least to genus. In addition, several abnormal fruitings, plasmodia, and sclerotia were produced, giving a total productivity, in terms some evidence of myxomycetes, of more than 30% (Wrigley de Basanta, unpublished data). Another report of the moist chamber culture of a nivicolous myxomycete was published by Marx (1998), in which fruiting bodies of *Trichia sordida* were obtained after nearly 2 months incubating a substrate between 12 and 16°C. These results indicate how time-consuming and difficult it is to obtain even a few nivicolous myxomycetes in moist chamber cultures, but the fact that some of them have shown positive results probably indicates that perfecting this method of isolation, with all the benefits it would bring, is a question of finding the precise conditions for this group. Schnittler et al. (2015) have made a recent effort to further characterize the exact ecological conditions for nivicolous myxomycetes in the field by using data loggers, and Shchepin et al. (2014) investigated some of the ecological conditions in agar culture of nivicolous myxomycetes as discussed later. It is to be hoped that these results will encourage further investigations that will enable this group to be successfully cultured in the laboratory.

Moist Chambers: Discussion of Advantages and Constraints

Moist chamber culture technique has been found to substantially increase the yield of field collections in all surveys and to give a more complete picture of the real biodiversity of a particular locality. It is a method for obtaining relatively easy, inexpensive, reproducible, comparative results from different habitats, substrates, conditions or areas. Detection of very small species is especially difficult in the field and is one of the great advantages of laboratory culture, because some tiny species of myxomycetes, such as those from the genera *Licea*, *Echinostelium*, or *Macbrideola*, can be isolated using these techniques. In addition, other species can be obtained that may be thought to be rare, since collecting them in the field is dependent on the serendipity of coinciding with their optimum phenology of fruiting. Even new species can be obtained from moist chamber cultures (Lado et al., 2009; Wrigley de Basanta et al., 2013), although wherever possible characters should be confirmed with field-collected specimens as well.

In the table of an overview of some of the results from arid and semiarid surveys in the Neotropics carried out by our team (Table 10.2), moist chamber culture collections made up at least 22% of the total collections. In terms of the numbers of species, about a quarter of all the species came only from moist chamber cultures. These, of course, include the smaller species of *Licea*, *Macbrideola*, and *Echinostelium* but are not restricted to them, and the cultures produced many other species, and many large well-developed collections. More than 10% of the localities sampled in these surveys produced no myxomycetes in the field, only in culture, and for the Patagonian steppe this number rose to 38%. In addition, a test of independence on the collections made in a study in Peru indicated that the proportion of collections from different orders depended on the method of collection, whether from the field or from moist chamber cultures (Lado et al., 2016). These data support the belief that the method is useful in giving a more complete picture of the real biodiversity of any locality. This is especially true in arid areas where the hostile environment leaves a paucity of field collections and substrate material, due to the limiting factors of the lack of water, intense radiation, wind and temperature extremes of heat in the day and cold at night.

New plant substrates can be identified using moist chamber cultures, and a substantial increase made in the information on the geographic distribution of species and biodiversity. It is also possible, by means of moist chamber cultures, to produce collections in excellent condition with closed intact fruiting bodies, when these are often battered or spoiled and weathered in the field. In addition, it is possible to observe some stages of the life cycle of the myxomycetes directly during these cultures, which would be difficult or impossible in the field. For example, the development of the plasmodium is sometimes an important character for identification, or the time a species spends as a plasmodium before fruiting. For instance, in some moist chamber cultures plasmodia can be observed to go through a cycle of forming resistant sclerotia, reforming plasmodia then forming sclerotia again, and sometimes never forming fruiting bodies. In litter samples processed by Stephenson et al. (2000), 5%–10% of plasmodia in the moist culture chambers did not fruit. This is possibly due to lack of optimum conditions in a culture for sporulation, but it is also possible that this behavior happens in the field under frequent or rapidly changing environmental conditions, as discussed earlier.

TABLE 10.2 Results From Field (Fc) and Moist Chamber Culture (Mc) Collections in Arid Areas of the Neotropics

Study area	Tehuacan-Cuicatlan Mexico[a]	Central Chile[b]	Monte Argentina	Patagonian Steppe Argentina	Peru coastal desert
References	Estrada-Torres et al. (2009)	Lado et al. (2013)	Lado et al. (2011)	Lado et al. (2014)	Lado et al. (2016)
Collections total	1200	633	594	1134	723
Species total	104	110	72	133	78
Field collections	454	491	372	820	500
Moist chamber collections	746[a]	142	222	314	223
Species Fc only	38	69	28	73	33
Species Mc only	36[a]; 35%	24; 22%	19; 26%	36; 27%	20; 26%
Species both field and Mc	31	17	25	24	25
Localities with only Mc coll.	NA	14/108; 13%	11/105; 11%	63/168; 38%	19/96; 20%

[a]*Includes a special series of 250 Mc of cacti.*
[b]*Includes other vegetation in nonarid areas.*

Observations of sclerotia and determining conditions leading to these dormant stages are an added possibility with moist chamber cultures and other types of laboratory cultures. Cultures can also be used to determine the morphogenesis of the fruiting body and to confirm the stability of the phenotypic characters by comparison of specimens produced in moist chambers with field-collected specimens, as they were, for example, in the description of *Cribraria zonatispora* (Lado et al., 1999) or *Physarum atacamense* (Wrigley de Basanta et al., 2012). A further advantage of collecting myxomycetes in laboratory culture is that it becomes possible to note incubation times and the substrate pH over the course of the cultures. Observations of interactions with other organisms in the cultures can be made, too, such as those noted in the many reports of mites, beetles and fly larvae associated with myxomycete fruiting bodies, plasmodia, and spores (Dudka and Romanenko, 2006; Stephenson and Stempen, 1994). The systematic comparison of substrates or environments (Novozhilov et al., 2000) can be another advantage of the technique and in some studies carried out at the minimum cost of setting up the cultures, although in others studies, obtaining the substrates can be as costly as field collecting.

The constraints of this method of obtaining myxomycetes include the production of possible variations in morphology as a result of the culture conditions or the expression of different assemblages in the cultures done by different investigators, due to differences in culture protocols and controls. Not all species will appear, or form fruiting bodies in a moist chamber culture, as mentioned earlier, with myxomycetes cycling between the stages as plasmodia and sclerotia and never fruiting. Some will not even form plasmodia, as is the case for various lignicolous species. We have also found there can be a high specificity to certain plants or plant groups, as in the case of *Licea eremophila*. This species developed on *Puya* spp. but never in other cultures of a similar bromeliad genus *Hechtia* with similar morphology but a different phylogeny (Wrigley de Basanta et al., 2010). The necessity of some very specific biotic or abiotic element(s) from the substrate plant for successful culture also points to a possible constraint of the method.

Setting up and observing moist chamber cultures is very time-consuming, particularly as they have to be examined for a period of months. An estimate was made by Novozhilov et al. (2000) that it takes 30 min to examine each 9 cm Petri dish culture, but in our experience, it can often take up to 3 times as long, especially when minute specimens have to be lifted and mounted. This obviously limits the number of cultures that can be processed by one person at any one time. Species could appear that may not complete their development in the field, as a result of spores or microcysts present on the substrate developing under the more constant conditions of the culture. In addition, as pointed out by Clark and Haskins (2010a), a bias could be tipped toward the nonheterothallic isolates, as they would produce more individuals in the culture. The loss of cultures to mold or insect contamination is another constraint, as mentioned earlier.

Agar Cultures

Cultivation of myxomycetes from spores dates back well over 100 years. Lister (1901) cultured *B. utricularis* from spores to fruiting bodies. He did this by germinating the spores in water and using scalded slices of the fungus *Stereum hirsutum* as a medium for plasmodial growth. He encouraged sporulation by adding "well-washed thick sticks, with the bark on them" to the *Stereum* pieces with the plasmodia. In the same paper, he explained how he also cultured *Didymium comatum* from spores on scalded absorbent paper with boiled cress seeds (Lister, 1901). Many of the early reports of culture were on organic substrates to which spores or plasmodia were added (Martin, 1940), as in the two examples mentioned earlier. But Alexopoulos (1969) stressed how essential it was to be able to culture myxomycetes with consistent reproducible results and so to be able to compare them under laboratory conditions to fully understand their taxonomy. He explained the significance of experimental work to confirm the stability of characters and to understand the development of different species. However, it has become clear that some species are easier than others to culture, and even to the present day only a small percentage of species have completed their life cycle on agar (Clark, 1995; Haskins and Wrigley de Basanta, 2008). As agar culture is a subset of solid media culture, and agar may not always be the best solidifying agent, others could be tried. For instance, gelatin culture could be attempted, although this is not always a good idea due to the fact that many types of bacteria liquify gelatin, but other material such silica gel could be tried. A Gelrite medium, used to culture fern gametophytes, has been used with the culture of *Echinostelium corynophorum* and abundant sporulation and even growth of the amoeboflagellate phase were observed (E.F. Haskins, personal communication).

The basic protocols of agar culture always begin with the selection of the material to culture. An important precaution to take when selecting myxomycetes from field collections is to choose fully matured specimens with apparently well-developed spores. Once selected, these should be kept from contamination by transporting them from the field in a separate container from other specimens. We have found that a sterile PCR tube is perfect for this, but small boxes or paper bags will also suffice. Samples to be used for culture should not be fumigated or refrigerated, protocols that are often used before field-collected specimens are placed in a herbarium, but after drying, they should be kept at temperatures as close to field conditions as possible. When using collections from moist chamber cultures for agar culture, care should be taken also to select properly matured specimens, free from contaminants, such as fungal hyphae, and kept separately from the rest of the collections to avoid cross-contamination with spores from other species. It is both surprising and disappointing to get to the end of months of agar culture and subculture and find that a totally different species has developed because other spores were there as contaminants.

Many researchers have found difficulties germinating or culturing myxomycete spores. Although some species are reported to have germinated after

32 years (Smith, 1929) or even longer (Gray and Alexopoulos, 1968), different species have different requirements for germination, and are affected by the treatment they receive before they are put in culture. Smith (1929) showed that variations occurred in spore germination even within the same collection, irrespective of the age of the spores and concluded that the maturity of the spores at the time of collection was an evident cause. Alexopoulos (1960) concurred that many factors that have to do with the spores themselves were responsible for the differential ability to germinate, such as the fruiting bodies from which they were taken, maturity and specific conditions of maturing. In addition, Clark and Collins (1976) suggested that differences at the chromosome level or other variations in DNA could lead to sterile spores through the lack of correct meiosis and thus cause unexpected low viability of spores from some fruiting bodies.

However, spore germination has been completed for many species. Gray and Alexopoulos (1968) listed 134 species of myxomycetes for which spore germination has been reported by themselves or other investigators, including some typical small corticolous myxomycetes like *Macrideola decapillata* and species of *Echinostelium* and *Licea*. Germination can be achieved in a variety of ways, the simplest of which is placing spores in a hanging drop of sterile distilled or tap water (Smart, 1937). Smart also used diluted sterilized decoctions of the natural substrates of the myxomycetes, to compare the effect of the solutions on germination. He found that the nutrient solutions definitely promoted germination of the spores. Alternate freezing and thawing or wetting and drying have been suggested as influencing the ability of spores to germinate (Gilbert, 1929). Osmosis and enzyme action (Gilbert, 1928a) have also been suggested as mechanisms involved in the actual breakage of the spore wall. Gray and Alexopoulos (1968) gave a most detailed summary of all the early work done on germination, and the conditions of germination and amoeboflagellate physiology have recently been reviewed by Clark and Haskins (2016).

The methods we have used are detailed in Haskins and Wrigley de Basanta (2008), where tips on media preparation, cleaning cultures, and stasis can also be found. The method used is to invert a 9 cm Petri dish of sterile 0.75% water agar and draw a cross with a marker on the base to divide the dish into quarters. A small circle is then drawn in each quarter on the bottom of the dish and then, using aseptic technique, spores are placed with a fine flame-sterilized needle into each circle. For very small species like *Licea* or *Echinostelium*, this is best done under a dissecting microscope on a dark background so that the spores can be seen on the agar surface. Plates are sealed with parafilm and kept at room temperature (21–23°C). Since germination can take from hours to days, depending on the species (Gray and Alexopoulos, 1968), the use of the soft water-agar plates provides moisture to avoid the spores drying before germination. Usually, bacteria and yeasts, isolated as contaminants of the spores themselves, will begin to grow among the spores and will form food for the amoeboflagellates on germination. It has been suggested that the presence of these food organisms stimulates germination in some species, although species have been known to germinate in distilled water in a shorter time than can be explained by any bacteria present (Gilbert, 1928b). Addition of drops of sterile nutrient solution, made from 25 g of substrate remains in 1.0 L of water, such as that used in the culture of *P. atacamense*, has been of particular benefit in the germination and culture of some species. However, it is not known whether this may have been due to stimulation of bacterial growth or to some abiotic factor in the substrate itself. The benefit of some element(s) from the substrate plant in the form of the sterile nutrient medium indicates a particularly close association between the myxomycete and its substrate plant (Wrigley de Basanta et al., 2012). Germination can be observed closely by inoculating a thin agar block on a culture slide with spores in the manner described by Spiegel et al. (2007).

Once many amoeboflagellates (Fig. 10.2D) are visible, a flame-sterilized 3 mm spatula can be used to cut out a small square of agar supporting the amoeboflagellates, and this is placed on a clean sterile agar plate. As the amoeboflagellate culture grows, subcultures can be made until plasmodium formation (Fig. 10.2E) or sporulation (Fig. 10.2F). It is best if the subsequent subculture plates are of slowly increasing nutrient concentration and increasingly drier, starting with a very dilute malt/yeast agar (wMY) and increasing to half strength cornmeal agar (CM/2). Cultures of some myxomycetes have been found to thrive with the addition of drops of sterile substrate decoction or a dilute bacteria (*Escherichia coli*) or yeast (*Cryptococcus laurentii*) culture (Haskins and Wrigley de Basanta, 2008). Viewing plasmodia, particularly aphanoplasmodia and protoplasmodia, can be made much easier by using a microscope with differential interference contrast (DIC), such as the Nikon SMZ 1600 shown in Fig. 10.2C.

Complete in vitro cultivation of myxomycetes has had mixed results and has not been completed for many species (Clark, 1995; Gray and Alexopoulos, 1968; Haskins and Wrigley de Basanta, 2008). It has been most successfully done with larger myxomycetes, notably from the Physarales (Clark and Collins, 1976). Clark (1995) listed a total of 98 species that had completed the life cycle on agar, or other media, and six more were pointed out in the methods overview by Haskins and Wrigley de Basanta (2008). In the years since that publication, at least a further seven species have completed their life cycle on agar or other media (Phate and Mishra, 2014; Wrigley de Basanta et al., 2008 , 2009, 2010, 2011, 2012). However, most of these methods have involved at least two-membered cultures, since no attempt has been made to exclude food organisms from the culture. In fact, it is probable that the success of the cultures from spore to spore was because of the growth of natural microorganisms isolated at the time of germination, and even encouraged to grow by the addition of sterile substrate extract (Wrigley de Basanta et al., 2012). In a recent paper, Liu et al. (2010) described the spore-to-spore culture of *Physarum globuliferum* on cornmeal agar without adding bacteria or other microorganisms, but the authors did not comment on the possible growth of incidental microorganisms usually carried on the spores themselves nor any method of cleaning the subsequent culture of other organisms as discussed in Clark (1995) and Haskins and Wrigley de Basanta (2008).

Axenic cultures are clearly the optimum method for comparative studies of myxomycete development, but they are exceptionally difficult for many species (Gray and Alexopoulos, 1968; Haskins, 2008). Daniel and Rusch (1961) maintained *P. polycephalum* for more than 4 years on a soluble medium without other organisms, and since then many species have been grown in axenic culture, but most investigators have found difficulty in maintaining cultures for long without bacteria. Dead bacteria and fungi have sometimes been successfully used (Cohen, 1939) and cleaning cultures that have been started with bacteria, especially cleaning plasmodia, are methods that appear in the literature (Balaji et al., 1999).

A liquid culture medium was used by Taylor and Mallette (1976) to grow *P. gyrosum,* with the bacterium *E. coli* as a food source. However, many liquid media containing extracts, such as serum or liver infusions have been able to support even large cultures of a few species over long periods of time, although some have consisted only of amoeboflagellates or plasmodia and did not involve the complete cycle on these media (Brewer et al., 1964; Goodman, 1972; Haskins, 1970).

An isolation method for using amoeboflagellates, especially for testing reproductive systems, is setting up clonal populations, either from single spores or by selecting single amoeboflagellates from germination plates or slants (Clark, 1995; Clark and Haskins, 2010b, 2011). Some agar cultures of nivicolous myxomycetes have been reported (Kowalski, 1971), many of which involved only the germination of spores to produce amoeboflagellates or at most plasmodia (Schnittler et al., 2015; Shchepin et al., 2014; Wikmark et al., 2007). Methods for culturing myxomycetes are listed on the Internet, especially the website for those interested in working with *P. polycephalum* (http://www.physarumplus.org), which also gives some useful links to other information. The Eumycetozoa website has downloadable handbooks of culture and isolation methods (http://slimemold.uark.edu/educationframe.htm).

Agar Cultures: Discussion of Advantages and Constraints

When agar cultures of myxomycetes are successful, an enormous amount of information becomes available. With these cultures, it has become possible to document the entire life cycle of many species. Apart from increasing the number of collections of a given species, the technique can also be used to track changes in gene expression during life cycles and to determine the stability of phenotypic characters in the fruiting body, essential information for solving taxonomic problems (Alexopoulos, 1969). Observations of the type of germination can help to determine or confirm certain groups. Sometimes, it is possible to discern what kind of amoeboflagellate results from germination or the number of amoeboflagellates and their size, as an indication of the number of postmeiosis mitotic divisions in the spore (Clark and Haskins, 2015, 2016).

Amoeboflagellates can be observed in cultures and their growth, and sometimes feeding preferences, monitored, and other important metabolic by-products can be investigated and even mass-produced (Balaji et al., 1999; Brewer et al., 1964; Keller and Everhart, 2010). The conditions leading to the formation and excystment of microcysts can be noted and monitored in this type of culture. The presence of mating types, homothallism, or apomyxis can be determined (Clark, 1995; Clark and Haskins, 2011) and early plasmodia and their growth patterns observed. The type of plasmodium formed, especially when it is tiny and hyaline and thus impossible to view elsewhere, can also be determined. Fruiting body morphogenesis can be observed directly and details learned as to when lime is deposited and where, and conditions for the formation of resistant sclerotia can be monitored. Cultures can also provide additional material from spores, amoeboflagellates, plasmodia, or fruiting bodies for DNA sequences to construct phylogenies. Current research using myxomycetes in mazes or robotics would be impossible without methods of maintaining laboratory cultures (Adamatzky and Jones, 2008; Dimonte et al., 2014).

The major constraints to the techniques involved in the use agar or other types of laboratory culture are the difficulties encountered in determining the precise requirements of a given species to germinate or grow in artificial culture. This is very time-consuming and fraught with disappointments and in addition requires some rather more sophisticated and costly laboratory equipment than moist chamber cultures. Since, as mentioned earlier, barely 10% of the known species have ever been cultured from spore to spore, it appears possible that some species will simply not grow in the laboratory. In an exhaustive study of *Didymium squamulosum*, from a genus thought to be among those that can be readily cultured, less than 1% of 94 spore-to-spore cultures were successful (Winsett, 2011). In our own experience, more than 500 agar culture attempts have resulted in fewer than 10 species with completed life cycles. Some of the problems, as mentioned earlier, may be due to unknown culture conditions but sterility of the strain can occur and possibly does so in nature (Clark and Haskins, 2010a). Contamination of the cultures with other protists, bacteria, and fungi is a major constraint and causes loss of cultures, and time- and cost inefficiency.

Environmental Sequencing

Based on the isolation techniques discussed thus far that depend on the presence of fruiting bodies, many species of myxomycetes have been considered to be cosmopolitan. However, data on their distribution are at best incomplete, since forms that do not produce fruiting bodies exist in soils (Stephenson and Feest, 2012), in fresh water, where these organisms were not thought to exist (Fiore-Donno et al., 2010), and even as endocommensals of sea urchins (Dykova et al., 2007). As such, it seems that they are more abundant and widely distributed than suggested by inventories based on the recovery and study of the fruiting body stage of the life cycle alone (Fiore-Donno et al., 2016; Kamono et al., 2013; Stephenson et al., 2011), and other techniques, independent of the observation of this life stage, could be important to better understand the ecology of these organisms (Ko Ko et al., 2009; Stephenson et al., 2011). Molecular techniques provide a tool to reveal the hidden diversity of those species that rarely, or perhaps never, produce fruiting bodies and are therefore absent from current inventories.

Although detail of these techniques can be found in earlier chapters, they are also techniques for isolating myxomycetes, and so a brief summary is included here. The basic protocols involve extracting, amplifying and sequencing DNA or RNA from an environmental sample using the polymerase chain reaction (PCR) method. As Stephenson et al. (2011) pointed out, there have been two approaches used to date. The first is massive sequencing of all the RNA or DNA present in an environmental sample, without the use of specific primers, by metagenomics techniques. The second is using specifically designed primers to detect only certain groups of myxomycetes in the sample. In both cases the resulting sequences are compared to bioinformatic databases. The first method was used by Urich et al. (2008) and showed that eumycetozoans were the largest group of protists in the soil samples they examined by mass sequencing. The second approach is the one most commonly employed at present, using the partial nuclear small subunit ribosomal RNA gene (SSU rRNA) to obtain the DNA of this region (Clissmann et al., 2015; Fiore-Donno et al., 2016; Hoppe and Schnittler, 2015; Kamono and Fukui, 2006; Kamono et al., 2009a; Ko Ko et al., 2009), or the RNA product from this gene and using the reverse transcription method (RT-PCR) to obtain the complementary DNA (Kamono et al., 2009b, 2013).

This second method of investigation was used to study soils in the pioneering work carried out by Kamono and Fukui (2006), who separated the DNA amplicons they obtained by denaturing gradient gel electrophoresis (DGGE), a technique that enables fragments of DNA of similar size to be separated by differences in their nucleotide sequences. The DNA was then reamplified and the products sequenced. The same basic protocols were also used on soil samples by Kamono et al. (2009a), and by Ko Ko et al. (2009) to study leaf litter. Kamono et al. (2009b) modified the procedure to avoid the problems caused by the presence of introns in the SSU rDNA gene. They used different combinations of primers and varied the PCR cycle temperature to improve the specificity of product detection in a seminested reverse transcriptase (RT-PCR) approach. The products were then separated by DGGE, reamplified, and sequenced. Kamono et al. (2013) used this procedure to study soils from woodlands and grasslands at high elevations in three different parts of the world.

Clissmann et al. (2015) used a different approach to study the bright-spored myxomycetes in decaying wood, cloning the amplified DNA fragments from the substrate in a plasmid vector and providing a tool for future environmental studies. In the largest environmental sampling study carried out to date,

Fiore-Donno et al. (2016) used a similar technique to Clissmann et al. (2015) but to study dark-spored myxomycetes. Hoppe and Schnittler (2015) adapted a molecular technique called Terminal Restriction Fragment Length Polymorphism (TRFLP) to compare whole myxomycete communities based on genomic DNA extracted from the soil.

The studies by Kamono et al. (2013), Clissmann et al. (2015), and Fiore-Donno et al. (2016) have demonstrated the presence of a greater diversity of myxomycetes than what is reflected by inventories dependant on recovering fruiting bodies, although often the sequences they recovered could not be identified to the species level. Kamono et al. (2013) also observed marked differences between similar ecological areas that were geographically separated from each other, supporting the idea that many species of myxomycetes are restricted to certain areas and therefore could be considered to be strictly endemic, as has been suggested in previous studies (Estrada-Torres et al., 2013). The results from Fiore-Donno et al. (2016) indicated that the dark-spored communities were not only highly diverse and widely distributed but were not randomly distributed in the soils examined and showed a dominance of the genera *Lamproderma* and *Didymium*.

Environmental Sequencing: Discussion of Advantages and Constraints

Using molecular techniques to carry out environmental sampling has enormous potential for revealing the hidden diversity of myxomycetes (Clissmann et al., 2015; Fiore-Donno et al., 2016; Kamono et al., 2013). This would contribute to a greater understanding of the true ecology and distribution of myxomycetes and an increase in recording from specific localities and substrates, which results in more complete inventories. The techniques, based on sampling nucleic acids, afford the considerable advantage that they can be used for any stage in the life cycle of the organisms, from spores, amoeboflagellates, or plasmodia to the resistant stages of microcysts or sclerotia. In addition, almost any substrate can be used to sample the DNA, not only those used traditionally to isolate fruiting bodies as discussed earlier, but also soil, airborne spores, water, or other organisms, where other methods have had little success. Another advantage of environmental sampling is that it provides a method of generating comparable data on the relative abundance and assemblages of myxomycetes in different areas, and the samples can be collected in any season. The techniques can be used to great advantage in conjunction with traditional field and culture isolation (Wrigley de Basanta et al., 2015) to confirm the presence of certain species that may produce irregular fruitings or none at all.

The greatest disadvantage of this type of environmental sampling is that only a fraction of known species have been sequenced thus far, and so the majority of sequences obtained do not coincide with those in the databases. It is therefore very important to continue to incorporate sequences of more species, but these should be backed by careful taxonomic study. It has become evident that many morphospecies are actually complexes of genetically distinct units (Aguilar et al., 2014; Feng et al., 2016; Fiore-Donno et al., 2011), so it is even possible that incorrect names are ascribed to sequences deposited in the databases. In addition, some amoeboflagellates may no longer be able to sporulate and have become soil amoebae causing even greater taxonomic problems, as pointed out by Clark and Haskins (2016).

Another problem to consider with environmental sampling of nucleic acids is that the technique does not distinguish the active phases of the life cycle, such as amoeboflagellates or plasmodia, from the resting stages, such as spores, microcysts, or sclerotia, so results from dormant, or even dead, myxomycetes could be contributing to the total diversity being revealed (Kamono et al., 2013). In addition, none of the advances used in characterization of myxomycete assemblages by environmental sampling using molecular techniques have considered all groups of myxomycetes and there is a distinct bias in such studies toward the dark-spored genera.

CONCLUSIONS

The myxomycetes have interested scientists and other collectors for several centuries, and these organisms have been recorded and or isolated from every terrestrial ecosystem and recovered from every plant type and vegetation type investigated. As an essential and major component of biotic communities in every biome, their isolation and study is of evident importance. They can be recovered directly from the environment in which they develop, or be encouraged to develop under a variety of contrived but controlled conditions in the laboratory. In this chapter, an effort has been made to summarize the techniques and recording protocols currently in use, and to indicate some of the advantages and constraints of each.

ACKNOWLEDGMENTS

We thank Edward Haskins, Carlos Lado, Alba Mónica Montiel, and Christine Priestley for valuable suggestions and comments. Some of the research was supported by the Spanish Government grant CGL2014-52584P.

REFERENCES

Adamatzky, A., Jones, J., 2008. Towards *Physarum* robots: computing and manipulating on water surface. J. Bionic Eng. 5, 348–357.

Adamonytė, G., Reda Iršėnaitė, R., Motiejūnaitė, J., Taraškevičius, R., Matulevičiūtė, D., 2013. Myxomycetes in a forest affected by great cormorant colony: a case study in Western Lithuania. Fungal Divers. 59, 131–146.

Aguilar, M., Fiore-Donno, A.M., Lado, C., Cavalier-Smith, T., 2014. Using environmental niche models to test the "everything is everywhere" hypothesis for *Badhamia*. ISME J. 8, 737–745.

Alexopoulos, C.J., 1960. Morphology and laboratory cultivation of *Echinostelium minutum*. Am. J. Bot. 47, 37–43.

Alexopoulos, C.J., 1969. The experimental approach to the taxonomy of the myxomycetes. Mycologia 61, 219–239.

Balaji, S., Sujatha, A., Kalyanasundaram, I., 1999. A simple rapid procedure for obtaining axenic cultures of myxomycete plasmodia. Can. J. Microbiol. 45, 865–870.

Brandza, M., 1928. Observations sur quelques sclérotes de Myxomycètes Calcarées. Le Botaniste iv–v, 117–146.

Brewer, E.N., Kuraishi, S., Garver, J.C., Strong, F.M., 1964. Mass culture of a slime mold, *Physarum polycephalum*. Appl. Microbiol. 12, 161–164.

Clarholm, M., 1981. Protozoan grazing of bacteria in soil-impact and importance. Microb. Ecol. 7, 343–350.

Clark, J., 1995. Myxomycete reproductive systems: additional information. Mycologia 87, 779–786.

Clark, J., Collins, O.R., 1976. Studies on the mating systems of eleven species of myxomycetes. Am. J. Bot. 63, 783–789.

Clark, J., Haskins, E.F., 2010a. Reproductive systems in the myxomycetes: a review. Mycosphere 1, 337–353.

Clark, J., Haskins, E.F., 2010b. Culture and reproductive systems of 11 species of Mycetozoans. Mycologia 96, 36–40.

Clark, J., Haskins, E.F., 2011. Principles and protocols for genetical study of myxomycete reproductive systems and plasmodial coalescence. Mycosphere 2, 487–496.

Clark, J., Haskins, E.F., 2015. Myxomycete plasmodial biology: a review. Mycosphere 6, 643–658.

Clark, J., Haskins, E.F., 2016. Mycosphere essays 3. Myxomycete spore and amoeboflagellate biology: a review. Mycosphere 7, 86–101.

Clissmann, F., Fiore-Donno, A.M., Hoppe, B., Krüger, D., Kahl, T., Unterseher, M., Schnittler, M., 2015. First insight into dead wood protistan diversity: a molecular sampling of bright-spored myxomycetes (Amoebozoa, slime-moulds) in decaying beech logs. FEMS Microbiol. Ecol. 91, 1–8.

Cohen, A.L., 1939. Nutrition of the myxomycetes. I. Pure culture and two-membered culture of myxomycete plasmodia. Bot. Gaz. 101, 243–274.

Daniel, J.W., Rusch, H.P., 1961. The pure culture of *Physarum polycephalum* on a partially defined soluble medium. J. Gen. Microbiol. 25, 47–59.

Dimonte, A., Cifarelli, A., Berzina, T., Chiesi, V., Ferro, P., Besagni, T., Albertini, F., Adamatzky, A., Erokhin, V., 2014. Magnetic nanoparticles-loaded *Physarum polycephalum*: directed growth and particles distribution. Interdiscip. Sci. Comput. Life Sci. 6, 1–9.

Dudka, I.O., Romanenko, K.O., 2006. Co-existence and interaction between myxomycetes and other organisms in shared niches. Acta Mycol. 41, 99–112.

Dykova, I., Lom, J., Dvořáková, H., Pecková, H., Fiala, I., 2007. *Didymium*-like myxogastrids (class Mycetozoa) as endocommensals of sea urchins (*Sphaerechinus granularis*). Folia Parasitol. 54, 1–12.

Eliasson, U., 1981. Patterns of occurrence of myxomycetes in a spruce forest in south Sweden. Holarct. Ecol. 4, 20–31.

Eliasson, U., 2013. Coprophilous myxomycetes: recent advances and future research directions. Fungal Divers. 59, 85–90.

Eliasson, U., Lundquist, N., 1979. Fimicolous myxomycetes. Bot. Not. 132, 551–568.

Estrada-Torres, A., Wrigley de Basanta, D., Conde, E., Lado, C., 2009. Myxomycetes associated with dryland ecosystems of the Tehuacán-Cuicatlán Valley Biosphere Reserve, Mexico. Fungal Divers. 36, 17–56.

Estrada-Torres, A., Wrigley de Basanta, D., Lado, C., 2013. Biogeographic patterns of the myxomycete biota of the Americas using a parsimony analysis of endemicity. Fungal Divers. 59, 159–177.

Everhart, S.E., Ely, J.S., Keller, H.W., 2009. Evaluation of tree canopy epiphytes and bark characteristics associated with the presence of corticolous myxomycetes. Botany 87, 509–517.

Everhart, S.E., Keller, H.W., Ely, J.S., 2008. Influence of bark pH on the occurrence and distribution of tree canopy myxomycete species. Mycologia 100, 191–204.

Feest, A., 1987. The quantitative ecology of soil Mycetozoa. Prog. Protistol. 2, 331–361.

Feest, A., Madelin, M.F., 1985a. A method for the enumeration of myxomycetes in soils and its application to a wide range of soils. FEMS Microbiol. Ecol. 31, 103–109.

Feng, Y., Klahr, A., Janik, P., Ronikier, A., Hoppe, T., Novozhilov, Y.K., Schnittler, M., 2016. What an intron may tell: several sexual biospecies coexist in *Meriderma* spp. (myxomycetes). Protist 167, 234–253.

Fiore-Donno, A.M., Kamono, A., Chao, E.E., Fukui, M., Cavalier-Smith, T., 2010. Invalidation of *Hyperamoeba* by transferring its species to other genera of Myxogastria. J. Eukaryot. Microbiol. 57, 189–196.

Fiore-Donno, A.M., Novozhilov, Y.K., Meyer, M., Schnittler, M., 2011. Genetic structure of two protist species (Myxogastria, Amoebozoa) suggests asexual reproduction in sexual amoebae. PLoS One 6, e22872.

Fiore-Donno, A.M., Weinert, J., Wubet, T., Bonkowski, M., 2016. Metacommunity analysis of amoeboid protists in grassland soils. Sci. Rep. 6, 19068.

Gilbert, F.A., 1928a. Factors influencing the germination of myxomycetous spores. Am. J. Bot. 16, 280–286.

Gilbert, F.A., 1928b. A study of the method of spore germination in myxomycetes. Am. J. Bot. 15, 345–352.

Gilbert, F.A., 1929. Spore germination in the myxomycetes: a comparative study of spore germination by Families. Am. J. Bot. 16, 421–432.

Gilbert, H.C., Martin, G.W., 1933. Myxomycetes found on the bark of living trees. Univ. Iowa Stud. Nat. Hist. 15, 3–8.

Goodman, E.M., 1972. Axenic culture of myxamoebae of the myxomycete *Physarum polycephalum*. J. Bacteriol. III, 242–247.

Gray, W.D., 1939. The relation of pH and temperature to the fruiting of *Physarum polycephalum*. Am. J. Bot. 26, 709–714.

Gray, W.D., Alexopoulos, C.J., 1968. The Biology of the Myxomycetes. Ronald Press, New York, NY.

Härkönen, M., 1977. Corticolous myxomycetes in three different habitats in southern Finland. Karstenia 17, 19–32.

Härkönen, M., 1978. On corticolous myxomycetes in Northern Finland and Norway. Annal. Bot. Fenn. 15, 32–37.

Härkönen, M., Vänscä, H., 2004. Corticolous myxomycetes and Lichens in the Botanical Garden in Helsinki, Finland. Syst. Geogr. Plants 74, 183–187.

Haskins, E.F., 1970. Axenic culture of myxamoebae of the myxomycete *Echinostelium minutum*. Can. J. Bot. 48, 663–664.

Haskins, E.F., 2008. Useful methods for bringing field collections of myxomycetes into agar culture. Available from: http://slimemold.uark.edu/educationframe.htm

Haskins, E.F., Wrigley de Basanta, D., 2008. Methods of agar culture of myxomycetes: an overview. Rev. Mex. Micol. 27, 1–7.

Hoppe, T., Schnittler, M., 2015. Characterization of myxomycetes in two different soils by TRFLP—analysis of partial 18S rRNA gene sequences. Mycosphere 6, 216–227.

Ing, B., 1983. A ravine association of myxomycetes. J. Biogeogr. 10, 299–306.

Ing, B., 1994. The phytosociology of myxomycetes. New Phytol. 126, 175–201.

Ing, B., 1999. The Myxomycetes of Britain and Ireland. Richmond Publishing Co., Slough.

Kamono, A., Fukui, F., 2006. Rapid PCR-based method for detection and differentiation of *Didymiaceae* and *Physaraceae* (myxomycetes) in environmental samples. J. Microbiol. Methods 67, 496–506.

Kamono, A., Kojima, H., Matsumoto, J., Kawamura, K., Fukui, M., 2009a. Airborne myxomycete spores: detection using molecular techniques. Naturwissenschaften 96, 147–151.

Kamono, A., Matsumoto, J., Kojima, H., Fukui, M., 2009b. Characterization of myxomycete communities in soil by reverse transcription polymerase chain reaction (RT-PCR)-based method. Soil Biol. Biochem. 41, 1324–1330.

Kamono, A., Meyer, M., Cavalier-Smith, T., Fukui, M., Fiore-Donno, A.M., 2013. Exploring slime mould diversity in high-altitude forests and grasslands by environmental RNA analysis. FEMS Microbiol. Ecol. 84, 98–109.

Keller, H.W., 2004. Tree canopy biodiversity: student research experiences in Great Smoky Mountains National Park. Syst. Geogr. Plants 74, 47–65.

Keller, H.W., Everhart, S.E., 2010. The importance of myxomycetes in biological research and teaching. Fungi 3, 13–27.

Kilgore, C.M., Keller, H.W., Ely, J.S., 2009. Aerial reproductive structures of vascular plants as a microhabitat for myxomycetes. Mycologia 101, 303–319.

Ko Ko, T.W., Stephenson, S.L., Jeewon, R., Lumyong, S., Hyde, K.D., 2009. Molecular diversity of myxomycetes associated with decaying wood and forest floor leaf litter. Mycologia 101, 592–598.

Kosheleva, A.P., Novozhilov, Y.K., Schnittler, M., 2008. Myxomycete diversity of the state reserve "Stolby" (south-eastern Siberia, Russia). Fungal Divers. 31, 45–62.

Kowalski, D.T., 1971. The genus *Lepidoderma*. Mycologia 63, 491–516.

Lado, C., 1993. Myxomycetes of Mediterranean woodlands. In: Pegler, D.N., Boddy, L., Ing, B., Kirk, P.M. (Eds.), Fungi of Europe: Investigation, Recording and Conservation. Royal Botanic Gardens, Kew, pp. 93–114.

Lado, C., 2004. Nivicolous myxomycetes of the Iberian Peninsula: considerations on species richness and ecological requirements. Syst. Geogr. Plants 74, 143–157.

Lado, C., Wrigley de Basanta, D. Typification of the myxomycete taxa described by the Listers and preserved at the Natural History Museum, London (BM), (in press).

Lado, C., Mosquera, J., Beltrán Tejera, E., 1999. *Cribraria zonatispora*, development of a new myxomycete with unique spores. Mycologia 91, 157–165.

Lado, C., Ronikier, A., Ronikier, M., Drozdowicz, A., 2005. Nivicolous myxomycetes from the Sierra de Gredos. Nova Hedwigia 81, 371–394.

Lado, C., Wrigley de Basanta, D., Estrada-Torres, A., 2011. biodiversity of myxomycetes from the Monte Desert of Argentina. Anales del Jardin Botanico de Madrid 68 (1), 61–95.

Lado, C., Wrigley de Basanta, D., Estrada-Torres, A., García-Carvajal, E., 2014. Myxomycete diversity of the Patagonian Steppe and bordering areas in Argentina. Anales Jard. Bot. Madrid 71 (e006), 1–35.

Lado, C., Wrigley de Basanta, D., Estrada-Torres, A., García-Carvajal, E., Aguilar, M., Hernández, J.C., 2009. Description of a new species of *Perichaena* (myxomycetes) from arid areas of Argentina. Anales Jard. Bot. 66 (S1), 63–70.

Lado, C., Wrigley de Basanta, D., Estrada-Torres, A., Stephenson, S.L., 2013. The biodiversity of myxomycetes in central Chile. Fungal Divers. 59, 3–32.

Lado, C., Wrigley de Basanta, D., Estrada-Torres, A., Stephenson, S.L., 2016. Myxomycete diversity in the coastal desert of Peru with emphasis on the lomas formations. Anales Jard. Bot. Madrid 73 (e032), 1–27.

Lister, A., 1894. A Monograph of the Mycetozoa: Being a Descriptive Catalogue of Species in the Herbarium of the British Museum. Longman & Co., London.

Lister, A., 1901. On the cultivation of Mycetozoa from spores. J. Bot. 39, 5–8.

Lister, G., 1918. The Mycetozoa: a short history of their study in Britain; an account of their habitats generally; and a list of species recorded from Essex. Essex Naturalist [Essex field club special memoirs]. Mycetozoa 6, 1–54.

Liu, P., Wang, Q., Li, Y., 2010. Spore-to-spore agar culture of the myxomycete *Physarum globuliferum*. Arch. Microbiol. 192, 97–101.

Martin, G.W., 1940. The myxomycetes. Bot. Rev. 6, 356–388.

Martin, G.W., Alexopoulos, C.J., 1969. The Myxomycetes. University of Iowa Press, Iowa City.

Marx, H., 1998. *Trichia sordida* in Feuchtkammer—ein deutscher Erstnachweis—und andere seltene Myxomyceten aus Thüringen. Boletus 22, 107–111.

Mitchell, D.W., 1980. A Key to the Corticolous Myxomycetes. British Mycological Society, Cambridge.

Novozhilov, Y.K., Schnittler, M., Zemlianskaia, I.V., Fefelov, K.A., 2000. Biodiversity of plasmodial slime moulds (Myxogastria): measurement and interpretation. Protistology 1, 161–178.

Olive, L.S., 1975. The Mycetozoans. Academic Press, London.

Panckow T., 1654. Herbarium Portatile. Berlin.

Phate, P., Mishra, R.L., 2014. Culture of *Hemitrichia serpula* on wide range of agar medium. Int. J. Curr. Microbiol. Appl. Sci. 3, 480–488.

Poulain, M., Meyer, M., Bozonnet, J., 2011. Les Myxomycètes. Fédération Mycologique et Botanique Dauphiné-Savoie, Sevrier.

Rodríguez-Palma, M.M., 2003. Estudio monográfico de los Myxomycetes del estado de Tlaxcala. Tesis de doctorado, Facultad de Ciencias, UNAM.

Rollins, A.W., Stephenson, S.L., 2013. Myxomycetes associated with grasslands of the western central United States. Fungal Divers. 59, 147–158.

Ronikier, A., Lado, C., 2015. Nivicolous stemonitales from the Austral Andes: analysis of morphological variability, distribution and phenology as a first step toward testing the large-scale coherence of species and biogeographical properties. Mycologia 107, 258–283.

Ronikier, A., Lado, C., Meyer, M., Wrigley de Basanta, D., 2010. Two new species of nivicolous *Lamproderma* (myxomycetes) from the mountains of Europe and America. Mycologia 102, 718–728.

Ronikier, A., Ronikier, M., 2009. How 'alpine' are nivicolous myxomycetes? A worldwide assessment of altitudinal distribution. Mycologia 101, 1–16.

Rossman, A.Y., Tulloss, R.E., O'Dell, T.E., Thorn, R.G., 1998. Protocols for an All Taxa Biodiversity Inventory of Fungi in a Costa Rican Conservation Area. Parkway, Boone, NC.

Scarborough, A.R., Keller, H.W., Ely, J.S., 2009. Species assemblages of tree canopy myxomycetes related to bark pH. Castanea 74, 93–104.

Schnittler, M., Erastova, D.A., Shchepin, O.N., Heinricha, E., Novozhilov, Y.K., 2015. Four years in the Caucasus: observations on the ecology of nivicolous myxomycetes. Fungal Ecol. 14, 105–115.

Schnittler, M., Lado, C., Stephenson, S.L., 2002. Rapid biodiversity assessment of a tropical myxomycete assemblage—Maquipucuna Cloud Forest Reserve, Ecuador. Fungal Divers. 9, 135–167.

Schnittler, M., Novozhilov, Y.K., Carvajal, E., Spiegel, F.W., 2013. Myxomycete diversity in the Tarim basin and eastern Tian-Shan, Xinjiang Province, China. Fungal Divers. 59, 91–108.

Schnittler, M., Stephenson, S.L., 2002. Inflorescences as a new microhabitat for myxomycetes. Mycologia 94, 6–20.

Shchepin, O., Novozhilov, Y., Schnittler, M., 2014. Nivicolous myxomycetes in agar culture: some results and open problems. Protistology 8, 53–61.

Smart, R.F., 1937. Influence of certain external factors on spore germination in the myxomycetes. Am. J. Bot. 24, 145–159.

Smith, E.C., 1929. The longevity of myxomycete spores. Mycologia 21, 321–323.

Snell, K.L., Keller, H.W., 2003. Vertical distribution and assemblages of corticolous myxomycetes on five tree species in the Great Smoky Mountains National Park. Mycologia 95, 565–576.

Spiegel, F.W., Haskins, E.F., Cavender, J.C., Landolt, J.C., Lindley-Settlemyre, L.A., Edwards, S.M., Nderitu, G., Shadwick, J.D., 2007. A beginner's guide to isolating and culturing eumycetozoans. Available from: http://slimemold.uark.edu/educationframe.htm

Stephenson, S.L., 1988. Distribution and ecology of myxomycetes in temperate forests I. Patterns of occurrence in the upland forests of south-western Virginia. Can. J. Bot. 66, 2187–2207.

Stephenson, S.L., 1989. Distribution and ecology of myxomycetes in temperate forests II. Patterns of occurrence on bark surface of living trees, leaf litter, and dung. Mycologia 81, 608–621.

Stephenson, S.L., Feest, A., 2012. Ecology of soil eumycetozoans. Acta Protozool. 51, 201–208.

Stephenson, S.L., Fiore-Donno, A.M., Schnittler, M., 2011. Myxomycetes in soil. Soil Biol. Biochem. 43, 2237–2242.

Stephenson, S.L., Kalyanasundaram, I., Lakhanpal, T.N., 1993. A comparative biogeographical study of myxomycetes in the mid-Appalachians of eastern North America and two regions of India. J. Biogeogr. 20, 645–657.

Stephenson, S.L., Laursen, G.A., Seppelt, R.D., 2007. Myxomycetes of subantarctic Macquarie Island. Aust. J. Bot. 55, 439–449.

Stephenson, S.L., Novozhilov, Y.K., Schnittler, M., 2000. Distribution and ecology of myxomycetes in high-latitude regions of the northern hemisphere. J. Biogeogr 27, 741–754.

Stephenson, S.L., Stempen, H., 1994. Myxomycetes: A Handbook of Slime Molds. Timber Press, Portland, OR.

Stephenson, S.L., Urban, L.A., Rojas, R., McDonald, M.S., 2008. Myxomycetes associated with woody twigs. Rev. Mex. Micol. 27, 21–28.

Taylor, R.L., Mallette, M.F., 1976. Growth of *Physarum gyrosum* on agar plates and in liquid culture. Antimicrob. Agents Chemother. 10 (4), 613–617.

Ukkola, T., Härkönen, M., Saarimáki, T., 1996. Tanzanian myxomycetes: second survey. Karstenia 36, 51–77.

Urich, T., Lanzén, A., Qi, J., Huson, D.H., Schleper, C., Schuster, S.C., 2008. Simultaneous assessment of soil microbial community structure and function through analysis of the meta-transcriptome. PLoS One 3, e2527.

Wikmark, O.G., Huagen, P., Lundblad, E.W., Haugli, K., Johansen, S.D., 2007. The molecular evolution and structural organization of group I introns at position 1389 in nuclear small subunit rDNA of myxomycetes. J Eukaryot. Microbiol. 54, 49–56.

Winsett, K.E., 2011. Intraspecific variation in response to spore-to-spore cultivation in the myxomycete, *Didymium squamulosum*. Mycosphere 2, 555–564.

Wrigley de Basanta, D., 2000. Acid deposition in Madrid and corticolous myxomycetes. Stapfia 73 (155), 113–120.

Wrigley de Basanta, D., 2004. The effect of simulated acid rain on corticolous myxomycetes. Syst. Geogr. Plants 74, 175–181.

Wrigley de Basanta, D., Lado, C., Estrada-Torres, A., 2010. *Licea eremophila*, a new myxomycete from arid areas of South America. Mycologia 102, 1185–1192.

Wrigley de Basanta, D., Lado, C., Estrada-Torres, A., 2011. Spore to spore culture of *Didymium operculatum*, a new myxomycete from the Atacama Desert of Chile. Mycologia 103, 895–903.

Wrigley de Basanta, D., Lado, C., Estrada-Torres, A., 2012. Description and life cycle of a new *Physarum* (myxomycetes) from the Atacama Desert in Chile. Mycologia 104, 1206–1212.

Wrigley de Basanta, D., Lado, C., Estrada-Torres, A., Stephenson, S.L., 2009. Description and life cycle of a new *Didymium* (myxomycetes) from arid areas of Argentina and Chile. Mycologia 101, 707–716.
Wrigley de Basanta, D., Lado, C., Estrada-Torres, A., Stephenson, S.L., 2013. Biodiversity studies of myxomycetes in Madagascar. Fungal Divers. 59, 55–83.
Wrigley de Basanta, D., Lado, C., García-Martín, J.M., Estrada-Torres, A., 2015. *Didymium xerophilum*, a new myxomycete from the tropical Andes. Mycologia 107, 157–168.
Wrigley de Basanta, D., Stephenson, S.L., Lado, C., Estrada-Torres, A., Nieves-Rivera, A.M., 2008. Lianas as a microhabitat for myxomycetes in tropical forest. Fungal Divers. 28, 109–125.

FURTHER READING

Feest, A., Madelin, M.F., 1985b. Numerical abundance of myxomycetes (Myxogastrids) in soils in the West of England. FEMS Microbiol. Ecol. 1, 353–360.
Keller, H.W., Skrabal, M., Eliasson, U.H., Gaither, T.W., 2004. Tree canopy biodiversity in the Great Smoky Mountains National Park: ecological and developmental observations of a new myxomycete species of *Diachea*. Mycologia 96, 537–547.
Rojas, C., Stephenson, S.L., Valverde, R., Estrada-Torres, A., 2012. A biogeographical valuation of high-elevation myxomycete assemblages in the northern Neotropics. Fungal Ecol. 5, 99–113.

Chapter 11

Uses and Potential: Summary of the Biomedical and Engineering Applications of Myxomycetes in the 21st Century

Hanh T.M. Tran*, Andrew Adamatzky**
*Ho Chi Minh International University, Ho Chi Minh City, Vietnam; **The Unconventional Computing Centre, University of the West of England, Bristol, United Kingdom

PART A: BIOACTIVE COMPOUNDS FROM MYXOMYCETES

Hanh T.M. Tran

Bioactive compounds from myxomycetes were intensively studied during the period from the 1950s to the 1970s, but then such studies were largely discontinued until the early part of the 21st century. The biggest challenge in determining the nature of the bioactive compounds derived from myxomycetes has been the availability of samples to process. The number of fruiting bodies of a particular species appearing in the field is usually relatively limited except for a few species (e.g., *Fuligo septica* and *Lycogala epidendrum*) that achieve considerable size, and the plasmodia of most species of myxomycete are difficult to culture. A number of bioactive compounds, including those with antimicrobial and anticancer properties, have been isolated from the fruiting bodies and plasmodia of myxomycetes (Dembitsky et al., 2005). However, only a small number of these have been tested to evaluate their usefulness to humans. In addition to their antimicrobial and anticancer activities, 3,4-dihydroxyphenylalanine (L-DOPA), a neurotransmitter precursor used in the treatment of Parkinson's disease, has been recorded from *Stemonitis herbatica*. The potential for commercialization of this compound has been suggested by Loganathan (1998).

Among the myxomycetes, the plasmodium of *Physarum polycephalum* is particularly interesting because of its fast growth rate, high biomass production, and ease of culturing. Therefore, it has been used as the model organism in

Myxomycetes: Biology, Systematics, Biogeography, and Ecology. http://dx.doi.org/10.1016/B978-0-12-805089-7.00011-1

numerous studies in several different fields. In terms of industrial applications, Brewer et al. (1964) successfully grew some strains of this species at a pilot scale for biomass production. This type of success, combined with the fact that the plasmodium has no cell walls, inspired Tran et al. (2012) to investigate the feasibility of using plasmodia as a source of lipids for biodiesel production. Their initial study suggested that myxomycetes can be used as a new source of lipids for biodiesel production, and the feasibility of direct conversion of the plasmodium into biodiesel was demonstrated to be possible.

In brief, the purpose of the first part of this chapter is to summarize significant achievements in research into myxomycete bioactive compounds and initial results in other fields using these organisms. In addition, some potential research applications that could be carried out with myxomycetes are discussed.

MYXOMYCETE ANTIMICROBIAL COMPOUNDS

Antimicrobial compounds have been isolated from both fruiting bodies and plasmodia of various species of myxomycetes. In general, most bioactive compounds from myxomycetes have been shown to display their strongest inhibition effects on Gram-positive bacteria (e.g., *Bacillus subtilis*), followed by fungi (e.g., *Candida albicans*), but with low or no inhibition effects on Gram-negative bacteria (Sawada et al., 2000). However, the interactions that take place between the bioactive compounds and the microbial cells have not yet been studied. Some selected compounds that have been isolated from myxomycetes are summarized in Table 11.1.

As mentioned by Considine and Mallette (1965), early research directed toward the antimicrobial compounds of myxomycetes was carried out using *F. septica*, as reported by Locquin and Prevot (1948). Two pigmented antibiotic materials thought to be anthraquinoic acids were extracted from *F. septica*. These authors also mentioned the research of Sobels (1950), who reported that the slimy substances produced by the plasmodia of myxomycetes inhibited the growth of certain bacteria and yeasts. In similar research carried out by Asgari and Henney (1977), the slimy substances obtained from of the plasmodium of *Physarum flavicomum* were found to be composed mostly of glucoprotein and exhibited inhibition effects against *B. subtilis* by degrading the cell wall of the latter. However, this bacterium was soon recovered from this inhibition.

Several studies have been carried out on *Physarum gyrosum* (Considine and Mallette, 1965; Mayberry et al., 1962; Schroeder and Mallette, 1973; Taylor and Mallette, 1976). Fergus and Mallette (1962) found that extracts from *P. gyrosum* display inhibitory activity against the bacteria *Staphylococcus aureus* and *Pseudomonas aeruginosa*. Considine and Mallette (1965) then found that diffusible materials from a culture of *P. gyrosum* could inhibit the growth of *B. subtilis*, *Bacillus cereus*, *Escherichia coli*, *S. aureus*, and *P. aeruginosa*; these compounds were partially purified. Schroeder and Mallette (1973) succeeded in purifying the antibiotic compound obtained from the culture of

TABLE 11.1 Myxomycete Antimicrobial Compounds

Species	Sample source	Compounds	Pathogens tested	References
Cribraria purpurea	Fruiting body	Cribrarione A	*Bacillus subtilis*	Naoe et al. (2003)
Didymium bahiense	Plasmodium	Bahiensol	*B. subtilis*	Misono et al. (2003a)
Enteridium lycoperdon	Fruiting body	Enteridinine A Enteridinine B	*B. subtilis, Escherichia coli, Candida albicans, Saccharomyces cerevisiae, Staphylococcus aureus*	Řezanka et al. (2004)
Lycogala epidendrum	Fruiting body	Lycogalinoside A Lycogalinoside B	*B. subtilis, E. coli, C. albicans, S. cerevisiae, S. aureus*	Řezanka and Dvořáková (2003)
Physarum melleum	Plasmodium	Crude extract	*B. subtilis*	Nakatani et al. (2005b)
Physarella oblonga	Plasmodium	Stigmasterol and fatty acids	*B. subtilis, Botrytis cinerea, E. coli, Fusarium oxysporum, S. aureus, Rhizoctonia solani, Trypanosoma cruzi*	Herrera et al. (2011)
P. melleum		Undetermined		
Unidentified species		Undetermined		

this myxomycete. The results they obtained showed that the antibiotic involved was different from other common antibiotics, such as pyromicin-anthracycline and tetracyclines. Subsequent research found that this myxomycete produced the same antibiotic materials when grown in a liquid medium as it did when grown on agar plates (Taylor and Mallette, 1976), and the antibiotic material could inhibit the growth of *B. cereus*. However, the chemical structure of the antibiotic was not reported.

More recently, two novel polypropionate lactone glycosides, which were given the names lycogalinosides A and B, were isolated from fruiting bodies of *L. epidendrum*. Lycogalinoside A showed significant inhibitory activity against the growth of *S. aureus* and *B. subtilis*, with a minimal inhibitory concentration (MIC) of 52 and 12 mg/mL, respectively. Lycogalinoside B showed a weaker activity but higher antifungal activity than lycogalinoside A. The structures of these two compounds were determined by using spectral data combined with their degradation products. The results obtained revealed that the chemical structure of the two compounds is similar to those usually produced by *Streptomyces*. Therefore, there is at least a chance that some species of *Streptomyces*, which had been consumed by this myxomycete at an early stage of the life cycle (probably the amoeboflagellate stage), actually produced these compounds (Řezanka and Dvořáková, 2003). In a similar study, two novel deoxysugar esters (enteridinines A and B) were isolated from fruiting bodies of *Enteridium lycoperdon*. Enteridinine A was found to significantly inhibit growth of Gram-positive bacteria (*S. aureus* and *B. subtilis*), along with demonstrating a mild growth inhibition of Gram-negative bacteria and some yeasts (*C. albicans* and *Saccharomyces cerevisiae*). On the other hand, enteridinine B showed a higher antifungal activity but a weak antibacterial activity (Řezanka et al., 2004). Early studies of *L. epidendrum* identified three novel dimethyl pyrrole dicarboxylates, which were given the names lycogarubins A–C, from fruiting bodies of this myxomycete. These compounds were closely related to arcyriarubins and arycyriaflavins isolated from the fruiting bodies of *Arcyria denudata*. Lycogarubin C showed anti-HSV-I virus activity (IC_{50} of 17.2 μg/mL) in vitro (Hashimoto et al., 1994).

In another study carried out by Řezanka et al. (2004), a fulicineroside, glycosidic dibenzofuran metabolite, was isolated from the fruiting bodies of *Fuligo cinerea*. The compound also displayed antimicrobial activities similar to those of enteridinines A and B isolated from *E. lycoperdon*. Remarkably, the compound displayed significant inhibition of the growth (about 83%) of crown gall tumors on potato disks inoculated with *Agrobacterium tumefaciens* carrying a tumor-inducing plasmid (Řezanka et al., 2005).

In addition to those myxomycetes that form large fruiting bodies, some other species (e.g., *Cribraria purpurea*) with small fruiting bodies have also been investigated. A new dihydrofuranonaphthoquinone pigment (Cribrarione A) isolated from the fruiting bodies of *C. purpurea* was found to possess antimicrobial activity against *B. subtilis* (Naoe et al., 2003). In one of the relatively rare studies that have used plasmodia as the source of the samples that were investigated, Misono et al. (2003a) successfully isolated and structurally characterized a new antimicrobial glycerolipid (bahiensol) from a plasmodial culture of *Didymium bahiense*.

It should be noted that plasmodia used for studies of bioactive compounds in myxomycetes are derived mostly from members of the Physarales due to the fact that they are easily cultured and many species normally form quite large plasmodia. Another study, which used a plasmodium as the sample source (Nakatani et al., 2005b), a crude extract obtained from the plasmodium of *Physarum melleum* was found to display antimicrobial activity against *B. subtilis*. However, the activity was lost when the extract was purified. It was proposed that the unstable yellow pigments were responsible for the antimicrobial activity. In a more recent study, plasmodial cultures of three different myxomycetes (*Physarella oblonga*, *P. melleum*, and an unidentified species) were tested for their antimicrobial activities against *B. subtilis*, *S. aureus*, *E. coli*, *Fusarium oxysporum*, *Botrytis cinerea*, and *Rhizoctonia solani* by measuring the diameters of inhibition zones. Plasmodial extracts of *P. oblonga* showed a low level of toxicity against epymastigote forms of *Trypanosoma cruzi* but displayed no evidence of any effect on the growth of other microorganisms. Similarly, in the same research, extracts from an unidentified species of myxomycete showed strong antimicrobial and antifungal activities against isolated strains of *B. cereus*, *F. oxysporum*, and *Rhizoctonia solani*, whereas those from *P. melleum* displayed a growth inhibition toward *F. oxysporum.* Crude extract from *P. melleum* exhibited a 10–20 mm inhibition zone on *F. oxysporum* and a more than 20 mm inhibition zone on *R. solani*. A crude extract from an unidentified myxomycete strongly inhibited the growth of *B. subtilis*, *F. oxysporum*, and *R. solani* (Herrera et al., 2011). Crude extracts of *P. oblonga* were found to contain stigmasterol and fatty acids, but the composition of the plasmodial extracts of two other species was not determined.

MYXOMYCETE ANTICANCER COMPOUNDS

Anticancer compounds, including alkaloids (arcyriaflavin C, makaluvamine A, and makaluvamine C), cycloanthranilylproline (fuligocandin B), and the polymer polycefin, have been isolated from the fruiting bodies and plasmodia of myxomycetes (Table 11.2). Among these compounds, there have been several novel alkaloids with a high potential for cancer treatment. For example, arcyriaflavin analogues have been considered to have great potential as anticancer drugs, and the pigments of *A. denudata* obtained from the fruiting bodies of these species were found to contain arcyriarubins A–C as well as arcyriaflavins A–C and D. In addition, a new bisindole alkaloid (dihydroarcyriarubin C) isolated from the fruiting bodies of *Arcyria ferruginea* showed cytotoxic activity against HeLa cells at a concentration of 1 mg/mL. Studies of the effect of this compound on the HeLa cell cycle found that arcyriaflavin C inhibits G2/M

TABLE 11.2 Myxomycete Antitumor Compounds

Species	Sample sources	Compounds	Cancer cells tested	References
A. denudate	Fruiting body	Arcyroxocin B	Jurkat cells	Kamata et al. (2006)
A. obvelata		Dihydroarcyriacyanin A		
A. ferruginea and *T. casparyi*	Fruiting body	Arcyriaflavin C	P388 leukemia cells, Hela cells, Wnt signal inhibitory activity	Nakatani et al. (2003)
D. bahiense	Fruiting body	Makaluvamine A Makaluvamine B	Tumor cell-line HCT 116 Human ovarian carcinoma	Barrows et al. (1993); Ishibashi et al. (2001); Dembitsky et al. (2005); Nakatani et al. (2005a)
D. iridis		Makaluvamine I		
L. epidendrum	Fruiting body	Two new bisindole alkaloids	Cytotoxic against HeLa, Jurkat, and vincristine resistant KB/VJ300 cells; one of them inhibited protein tyrosine kinase activity	Hosoya et al. (2005)
P. polycephalum	Plasmodium	Polycefin	Tumor brain cells MDA–MB 468 breast tumor cells	Lee et al. (2006) Ljubimova et al. (2008)
T. favoginea var. *persimilis*	Fruiting body	Kehokorins A–C	HeLa cells	Kaniwa et al. (2006)
T. dimorphotheca	Fruiting body	Tubiferal A Tubiferal B	VCR-resistant KB cell lines	Kamata et al. (2006)

and G1 phases at low concentrations (100 and 10 ng/mL, respectively) (Nakatani et al., 2003). In the follow-up research, *cis*-dihydroarcyriarubin C and arcyriaflavin C were found to display Wnt (Wingless-related integration site) signal inhibitory activity (Kaniwa et al., 2006).

Previously, arcyriaflavin C was isolated from *Metatrichia vesparium* (Dembitsky et al., 2005). This compound was thought to cause topoisomerase I-mediated DNA cleavage, potent inhibition of protein kinase C and cell-cycle-regulating cyclin-dependent kinase (CDK), and cell-cycle checkpoint inhibition (Nakatani et al., 2003). Other alkaloids were isolated from fruiting bodies of *A. denudata* (arcyroxocin B) and *Arcyria obvelata* (dihydroarcyriacyanin A). Arcyroxocin B and dihydroarcyriacyanin A exhibited cytotoxic activity against Jurkat cells with IC_{50} values of 7 and 10 μg/mL, respectively (Kamata et al., 2006). Another group of alkaloids, the makaluvamines, have been shown to exhibit in vitro cytotoxicity against the human colon tumor cell line HCT 116 and ovarian cancer (Dembitsky et al., 2005). Makaluvamines A and B were isolated from the fruiting bodies of *D. bahiense* (Ishibashi et al., 2001), and makaluvamine I was produced from the fruiting bodies of *Didymium iridis* (Nakatani et al., 2005a).

In addition to these alkaloids, two novel rearranged triterpenoid lactones, tubiferal A and tubiferal B, were extracted and purified from fruiting bodies of *Tubifera dimorphotheca*. Tubiferal A showed antitumor activity against vincristine-resistant KB cell lines (Kamata et al., 2004). Moreover, three new dibenzofurans, kehokorins A–C, were isolated from the fruiting bodies of *Trichia favoginea* var. *persimilis*. Among these compounds, kehokorin A was found to exhibit cytotoxicity against HeLa cells, with an IC_{50} value of 1.5 μg/mL. Two new sterols, trichiol and 3-epitrichiol acetate, were isolated from the fruiting bodies of *T. favoginea* var. *persimilis* and their structures elucidated by spectral data. Trichiol was cytotoxic against HeLa cells, while 3-epitrichiol acetate was shown to exhibit reversal effects against TNF-related apoptosis, including the ligand (TRAIL)-resistant Jurkat cell lines (Kaniwa et al., 2006). In addition, two new bisindole alkaloids, 6-hydroxystaurosporinone and 5,6-dihydroxyarcyriaflavin A, were isolated from field-collected fruiting bodies *L. epidendrum*. These compounds showed significant cytotoxicity against HeLa and Jurkat cells as well as slightly weak activity against vincristine-resistant KB/VJ300 cells. Remarkably, 6-hydroxystaurosporinone inhibited protein tyrosine kinase activity (Hosoya et al., 2005).

In a subsequent study carried by Hasegawa et al. (2007), fuligocandin B (FCB), a cycloanthranilylproline derivative, was extracted from *Fuligo candida*. The compound exhibited significant synergism with TRAIL (tumor necrosis factor–related apoptosis-inducing ligand), a compound used in cancer treatment. Apoptosis could be recorded for TRAIL-resistant cell lines when these cell lines were treated with FCB and TRAIL. Single treatment of these compounds did not lead to apoptosis. Therefore, fuligocandin B would seem to have a considerable potential in cancer therapy (Hasegawa et al., 2007).

The compound with the most potential isolated thus far from myxomycetes is polycefin (Figs 11.1 and 11.2). Polycefin is a polymer that was purified from extracts obtained from the plasmodium of *P. polycephalum*. Polycefin was synthesized for targeted delivery of morpholino antisense oligonucleotides into certain tumors (e.g., breast and brain tumors). Polycefin was found to be nontoxic to normal and tumor astrocytes in a wide range of concentrations (Lee et al., 2006; Ljubimova et al., 2008). Polycifin has been considered as the most remarkable nontoxic, nonimmunogenic, and biodegradable nanoconjugate drug delivery system currently known (Keller and Everhart, 2010).

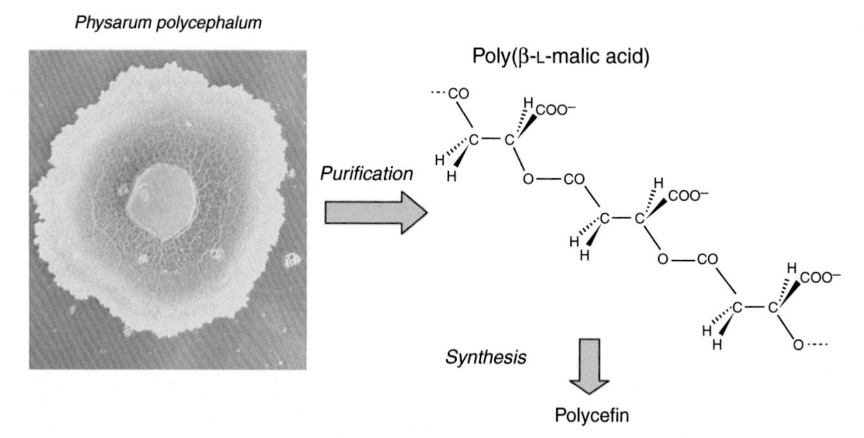

FIGURE 11.1 ***Polycefin obtained from Physarum polycephalum.***

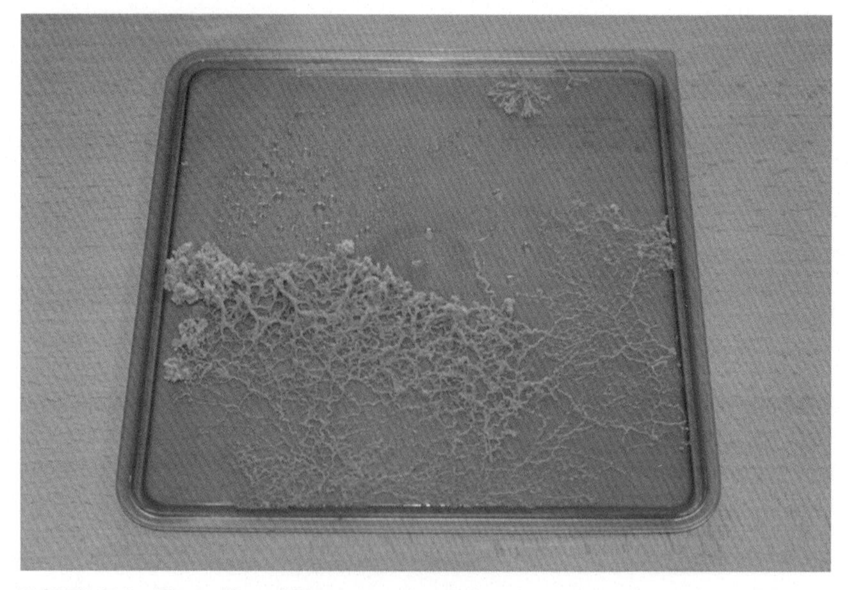

FIGURE 11.2 **Plasmodium of *Physarum polycephalum* as grown in laboratory condition.**

OTHER BIOACTIVE COMPOUNDS OBTAINED FROM MYXOMYCETES

One of the remarkable results obtained thus far from research on myxomycete bioactive compounds was reported by Loganathan (1998), who found that 3,4-dihydroxyphenylalanine (L-DOPA), a neurotransmitter precursor used in the treatment of Parkinson's disease, could be commercially produced from fruiting bodies of *Stemonitis herbatica*.

As already mentioned, numerous novel compounds have been isolated from myxomycetes, but only a limited number of those have been investigated for their biological properties. Therefore, future research to exploit their medical applications is crucial. Examples of some of the novel compounds isolated from myxomycetes that have not yet had their functions and/or biological activities investigated include bisindole sulfate from *A. denudata* (Kamata et al., 2006), arcyxocin A from *A. obvelata* (Steglich, 1989), cribrarione B from *Cribraria cancellata* (Iwata et al., 2003), cribrarione C from *Cribraria meylanii* and fuligoic acid from *Fuligo septica* f. *flava* (Shintani et al., 2009), clionasterol from *Didymium squamulosum* (Ishibashi et al., 1999), cycloanthranilylproline from *F. septica* (Nakatani et al., 2003), lycogalic acid A and B and lycogarides A–G from *L. epidendrum* (Buchanan et al., 1996; Fröde et al., 1994), melleum A and B from *P. melleum* (Nakatani et al., 2005a,b), and physarigins A–C from *Physarum rigidum* (Misono et al., 2003b).

INITIAL STUDIES ON THE USE OF MYXOMYCETE LIPIDS FOR BIODIESEL PRODUCTION

Biodiesel production from microbial biomass has included two steps, which are lipid extraction and then the conversion of the extracted lipids into biodiesel through a process called transesterification. Microorganisms, including microalgae and various oleaginous microorganisms (yeasts and bacteria), have been considered the best alternatives for the production of oils and fats to replace the use of agricultural and animal resources. The disadvantages of using bacteria and fungi are related to the limited amount of material that can be produced. Algae do not have this problem, but they require a larger acreage to culture and a longer fermentation period than most other microorganisms (Chinnasamy et al., 2010). Lipid extraction from algae and other oleaginous microorganisms (yeasts and bacteria) involves cell disruption, and this process is often challenging to carry out because of the rigid cell walls of these microorganisms. The lipid extraction yield is negatively influenced as a result. To save time and energy, direct conversion of biomass into biodiesel (direct transesterification or in situ transesterification) has been investigated as a replacement of the conventional method.

As described in detail in Chapter 1 of this book and already referred to in the current chapter, the life cycle of a myxomycete is characterized by a distinctive multinucleate trophic (feeding) stage called a plasmodium, which has no cell walls. Among the myxomycetes, the plasmodium of *P. polycephalum*, a member of the order Physarales, is particularly interesting because of its rapid growth rate and ease of culturing. Tran et al. (2012) found that *P. polycephalum* produced remarkably high amounts of biomass. Analysis of the lipids present in the plasmodium of *P. polycephalum* by thin layer chromatography (TLC) and fatty acid methyl esters (FAMEs) by gas chromatography–mass spectrometry (GC–MS) techniques showed that the major lipid type present is a triglyceride (95.5%), followed by phospholipids (2.6%), diglyceride (0.92%), and monoglyceride (0.92%). Myxomycete lipids consist of three dominant fatty acids. These are oleic acid (20%), linoleic acid (33%), and palmitoleic acid (17%). These results suggested that *P. polycephalum* has considerable potential as a source of lipids for biodiesel production. In another study, the same authors also reported that this myxomycete could produce a relatively high amount of biomass when grown on defatted rice bran (Tran et al., 2015). Since a plasmodium has no cell walls, the direct conversion into biodiesel is currently under investigation, with the initial results being positive.

In the research just described, a solid culture was used for biomass production. This presents a major challenge for large-scale production. However, some strains of *P. polycephalum* have been reported to grow in liquid cultures, even in 30-L and 50-gal conventional baffled fermenters in pilot studies (Brewer et al., 1964). The yields obtained from liquid cultures were 6–10 g DCW/L, but it should be possible to increase the yield by adjusting the cultivation conditions as well as the composition of the medium used. In the research mentioned earlier, the culture of this myxomycete in a fermentor was scaled up to a working volume of 100 L, which resulted in 1 kg dry cell weight (Brewer et al., 1964).

It should be noted here that when cultured in liquid medium, microplasmodia are formed instead of a typical plasmodium (i.e., which is what appears on a solid culture). However, microplasmodia do not differ from typical plasmodia in terms of their basic properties. Therefore, the results obtained from initial studies of biodiesel production from plasmodia collected from solid cultures do provide a relevant foundation for future biodiesel research on the use of submerged fermentation.

PART B: MYXOMYCETES IN UNCONVENTIONAL COMPUTING AND SENSING

Andrew Adamatzky

The myxomycete *P. polycephalum* has a rather sophisticated life cycle (Stephenson and Stempen, 1994), which includes fruiting bodies, spores, single-celled myxamoebae, and a multinucleate plasmodium. The plasmodium is a coenocyte, and nuclear divisions occur without cytokinesis. As such, it is a single cell with thousands of nuclei. The plasmodium can reach a considerable size, up to tens of centimeters when conditions are good. The plasmodium consumes microscopic particles and bacteria. During its foraging behavior, the plasmodium spans scattered sources of nutrients with a network of protoplasmic tubes. The plasmodium optimizes its protoplasmic network to cover all sources of nutrients, to stay away from repellents, and to minimize transportation of metabolites inside its body. The ability of the plasmodium to optimize its shape (Nakagaki et al., 2001) first attracted the attention of biologists, then computer scientists (Adamatzky, 2010a), and finally engineers. As a result, the field of myxomycete computing was born.

Thus far, the plasmodium is the only useful computation stage of the life cycle of *P. polycephalum*. Therefore, from this point onward the word *Physarum* will be used when referring to the plasmodium. Most computing and sensing devices make use of one or more key features of the physiology and behavior of *Physarum*. For example, *Physarum* senses gradients of chemoattractants and repellents (Durham and Ridgway, 1976; Rakoczy, 1963; Ueda et al., 1976); it responds to chemical or physical stimulation by changing patterns of electrical potential oscillations (Durham and Ridgway, 1976; Kishimoto, 1958) and protoplasmic tube contractions (Teplov et al., 1991; Wohlfarth-Bottermann, 1979). *Physarum* optimizes its body to maximize its protoplasm streaming (Dietrich, 2015), and it is made up of hundreds, if not thousands, of biochemical oscillators (Kauffman and Wille, 1975), with varied modes of coupling (Grebecki and Cieslawska, 1978).

Herein we offer very short descriptions of actual working prototypes of *Physarum*-based sensors, computers, actuators, and controllers. Details can be found in the pioneer book on *Physarum* machines (Adamatzky, 2010a) and the "bible" of myxomycete computing (Adamatzky, 2016).

SHORTEST PATH

When presented with a maze, the objective is to find the shortest path between the central chamber and an exit. This was the first problem ever solved by *Physarum*. There are two *Physarum* processors that can solve the maze. The first prototype (Nakagaki et al., 2001) works as follows. The myxomycete is inoculated everywhere in a maze. *Physarum* then develops a network of protoplasmic tubes spanning all channels of the maze. This network represents all possible solutions. Then, oat flakes are placed at a source and a destination site. Tubes lying along the shortest (or near shortest) path between the two sources of nutrients develop an increased flow of cytoplasm. These tubes become thicker. Tubes branching to sites without nutrients become smaller due to the lack of cytoplasm flow and eventually collapse. The "thickest" tube represents the shortest path between the sources of nutrients. The selection of the shortest protoplasmic tube is implemented via interaction of propagating biochemical, electrical potential, and contractile waves within the mass of the plasmodium (see mathematical model in Tero et al., 2006). The approach is not efficient, though because we must literally distribute the computing substrates everywhere in the physical representation of the problem. The number of computing elements would be proportional to a sum of lengths of the maze's channels.

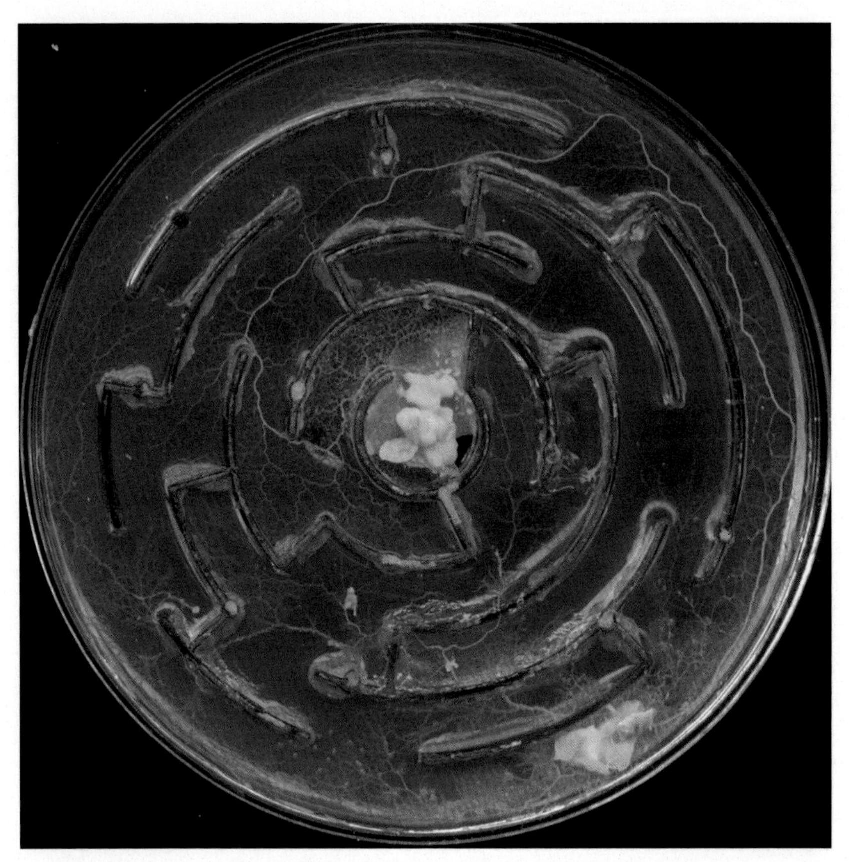

FIGURE 11.3 Plasmodium of *Physarum polycephalum* "solving" a maze. *(Originally published in Adamatzky, A., 2012. Slime mold solves maze in one pass, assisted by gradient of chemoattractants. IEEE Trans. Nanobiosci.11(2), 131–134.)*

The second prototype of the *Physarum* maze solver is based on its chemoattraction. An oat flake is placed in the central chamber. The myxomycete is inoculated somewhere in a peripheral channel. The oat flake releases chemoattractants, and these chemoattractants diffuse along the channels of the maze. The *Physarum* explores its vicinity by branching out and thus extending protoplasmic tubes into the opening of nearby channels. When a wave front of diffusing attractants reaches *Physarum*, it halts lateral exploration. Instead, *Physarum* develops an active growing zone propagating along the gradient of the attractants' diffusion. The "thickest" tube represents the shortest path between the sources of nutrients (Fig. 11.3). This approach is efficient because the number of computing elements would be proportional to the length of the shortest path.

SPANNING TREE

A spanning tree of a finite planar set is a connected, undirected, acyclic planar graph whose vertices are points of the planar set. The tree is a minimal spanning tree where the sum of edge lengths is minimal (Nesetril et al., 2001). An algorithm for computing a spanning tree of a finite planar set based on morphogenesis of a neuron's axonal tree was initially proposed by Adamatzky (1991), for which planar data points are marked by attractants (e.g., neurotrophins), and a neuroblast is placed at some site. Growth cones sprout new filopodia in the direction of maximal concentration of attractants. If two growth cones compete for the same site of attractants, then a cone with highest energy (closest to previous site or branching point) wins. Fifteen years later, we implemented the algorithm with *Physarum* (Adamatzky, 2008).

The degree of *Physarum* branching is inversely proportional to a quality of its substrate. Therefore, to reduce a number of random branches we cultivated *Physarum* not on agar but on just humid filter paper. The planar data set is represented by a configuration of oat flakes. *Physarum* is inoculated at one of the data sites, and *Physarum* then propagates to the virgin oat flake closest to the inoculation site (it branches if there are several virgin flakes nearby). It colonizes the next set of flakes. The propagation goes on until all data sites are spanned by a protoplasmic network. The protoplasmic network approximates the spanning tree. However, the resultant tree does not remain static. Later cycles can be formed and the tree is transformed to one of proximity graphs (e.g., a relative neighborhood graph or a Gabriel graph) (Adamatzky, 2009).

APPROXIMATION OF TRANSPORT NETWORKS

Motorway networks are designed with an aim of efficient vehicular transportation of goods and passengers. *Physarum* protoplasmic networks evolved for efficient intracellular transportation of nutrients and metabolites. To uncover similarities between biological and human-made transport networks and to project behavioral traits of biological networks onto development of vehicular transport networks, we conducted an evaluation and approximation of motorway networks by *Physarum* in 14 geographical regions. These were Africa, Australia, Belgium, Brazil, Canada, China, Germany, Iberia, Italy, Malaysia, Mexico, the Netherlands, the United Kingdom, and the United States (Adamatzky, 2012a).

We represented each region with an agar plate, imitated major urban areas with oat flakes, inoculated *Physarum* in a capital, and analyzed the structures of protoplasmic networks developed. We found that the networks of protoplasmic tubes grown by *Physarum* match, at least partly, the networks of human-made transport arteries. The shape of a country and the exact spatial distribution of urban areas, represented by sources of nutrients, play a key role in determining the exact structure of the plasmodium network. In terms of absolute matching between *Physarum* networks and motorway networks, the regions studied can be arranged in the following order of decreasing matching: Malaysia, Italy, Canada,

Belgium, China, Africa, the Netherlands, Germany, the United Kingdom, Australia, Iberia, Mexico, Brazil, and the United States. We compared the *Physarum* and the motorway graphs using such measures as average, longest, and shortest paths, average degrees, number of independent cycles, the Harary index, the P-index, and the Randic index. Using these measures, we find that motorway networks in Belgium, Canada, and China are most similar to *Physarum* networks.

With regard to measures and topological indices, we demonstrated that the Randic index could be considered the most biocompatible measure of transport networks because it matches very well the myxomycete and human-made transport networks, yet efficiently discriminates between transport networks of different regions (Adamatzky et al., 2013a). Many curious discoveries have been made. Just a few of these are listed next. All segments of the trans-African highways not represented by *Physarum* have components of nonpaved roads (Adamatzky and Kayem, 2013) (Fig. 11.4). For Australia, the east coast transport chain from the Melbourne urban area in the south to the Mackay area in the north, and the highways linking Alice Springs and Mount Isa and Cloncurry, are represented by the protoplasmic tubes formed by the myxomycete in almost all experiments in approximation of Australian highways (Adamatzky and Prokopenko, 2012). If the two parts of Belgium were separated with Brussels in Flanders, the Walloon region of the Belgian transport network would be represented by a single chain from Tournai in the northwest to the Liege area in the northeast and down to southernmost Arlon; motorway links connecting Brussels with Antwerp, Tournai, Mons, Charleroi, and Namur, and links connecting Leuven with Liege and connecting Antwerp with Genk and Turnhout are redundant from the point of view of *Physarum* (Adamatzky et al., 2013b). The protoplasmic network forms a subnetwork of the human-made motorway network in the Netherlands; a flooding of a large area around Amsterdam would lead to a substantial increase in traffic at the boundary between the flooded and nonflooded areas, paralysis and abandonment of the transport network, and migration of the population from the Netherlands to Germany, France, and Belgium (Adamatzky et al., 2013b). *Physarum* imitates the separation of Germany into East Germany and West Germany in 1947 (Adamatzky and Schubert, 2012).

People migrate toward sources of a safe life and higher income. *Physarum* migrates into environmentally conformable areas and toward sources of nutrients. Adamatzky and Martinez (2013) explored this analogy to imitate Mexican migration to the United States. We have made a three-dimensional nylon terrain of the United States and placed oat flakes (Fig. 11.5), to act as sources of attractants and nutrients, at 10 areas with the highest concentrations of migrants—New York, Jacksonville, Chicago, Dallas, Houston, Denver, Albuquerque, Phoenix, Los Angeles, and San Jose. We inoculated *Physarum* in a locus between Ciudad Juarez and Nuevo Laredo, allowed it to colonize the template for 5–10 days, and analyzed the routes of migration. From results of laboratory experiments, we extracted topologies of migratory routes, and highlighted the role of elevations in shaping the human movement networks.

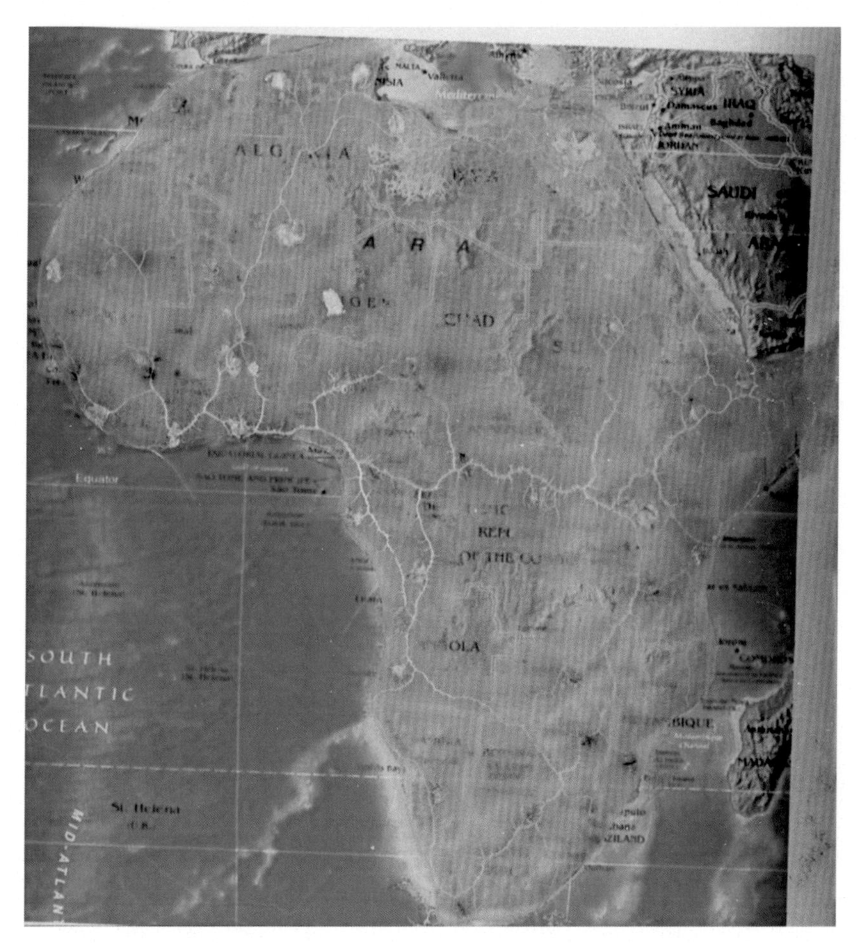

FIGURE 11.4 Trans-African transport formed by the plasmodium of *Physarum polycephalum*. *(Originally published in Adamatzky, A., Kayem, A.V.D.M., 2013. Biological evaluation of Trans-African highways. Eur. Phys. J. Spec. Top 215. (1), 49–59.)*

SPACE EXPLORATION

We employed the foraging behavior of *Physarum* to explore scenarios of future colonization of Earth's moon and Mars (Adamatzky et al., 2014). We have grown *Physarum* on three-dimensional templates of these celestial bodies, and analyzed formation of the exploration routes, including dynamic reconfiguration of the transportation networks as a response to the addition of hubs. The developed infrastructures were explored using proximity graphs and *Physarum*-inspired algorithms of supply chain designs. Interesting insights about how various lunar missions may develop and how interactions between hubs and landing sites can be established are given in Adamatzky et al. (2014).

FIGURE 11.5 Plasmodium of *Physarum polycephalum* on a terrain model of the United States. *(Originally published in Adamatzky, A., Martinez, G.J., 2013. Bio-imitation of Mexican migration routes to the USA with myxomycete on 3D terrains. J. Bionic Eng. 10 (2), 242–250.)*

TACTILE SENSOR

A tactile sensor is a device that responds to a physical contact between the device and an object. When a segment of a glass capillary tube is placed across a protoplasmic tube, which spans reference and recording electrodes, *Physarum* demonstrates two types of responses to application of this load: an immediate response with a high-amplitude impulse and a prolonged response with changes in its oscillation pattern (Adamatzky, 2013a). The immediate response is a high-amplitude spike, for which the amplitude is 12.33 mV and its duration is 150 s. The prolonged response is an envelope of increased amplitude oscillations. For example, an average amplitude of oscillations before stimulation is 2.3 mV and duration of each wave is 120 s. The amplitude of waves in the prolonged response to stimulation is 5.29 mV with a period of a wave increased to 124 s.

The tactile sensor developed is nonreusable. While the load rests on the protoplasmic tube, *Physarum* starts colonizing the load. Removal of the load damages the protoplasmic tube. To rectify this deficiency, we designed a *Physarum* tactile bristle (Adamatzky, 2014).

To make a tactile bristle with *Physarum*, we stuck a bristle in the agar blob on the recording electrode. In a couple of days after inoculation of *Physarum* to an agar blob on a references electrode, the *Physarum* propagates to and colonizes the agar blob on a recording electrode. *Physarum* climbs up the bristle and occupies one-third to half of the bristle's length. The sensor works by deflecting the bristle. A sensed object does not come into direct contact with *Physarum* but only with the tip of the bristle not colonized by *Physarum*. A typical response of *Physarum* to deflection of the bristle consists of an immediate response—a high-amplitude impulse—and a prolonged response. The high-amplitude impulse is always well pronounced, but the prolonged response oscillations can sometimes be distorted by other factors (e.g., growing branches of a protoplasmic tube or additional strands of plasmodium propagating between the agar blobs). Responses are repeatable not only in different experiments but also during several rounds of stimulation in the same experiment.

COLOR SENSOR

A color sensor is a device that gives a wavelength-dependent response when illuminated. *Physarum* is photo-sensitive. It changes its pattern of electrical potential oscillatory activity when illuminated (Block and Wohlfarth-Bottermann, 1981; Wohlfarth-Bottermann and Block, 1981). Moreover, *Physarum* distinguishes the color of the illumination (Adamatzky, 2013b). We placed *Physarum* between two electrodes and illuminated it with red, green, blue, or white light. We also illuminated *Physarum* with white light via a transparent lens. The amount of light on the blob was 80–120 lux for each color. We say the *Physarum* recognizes a color of the light if it reacts to illumination with the color by a unique change in amplitude and periods of oscillatory activity. We found that *Physarum* recognizes when red and blue light are switched on and when red light is switched off. Red and blue illuminations decrease the frequency of oscillations. Red light increases the amplitude of oscillations but blue light decreases the amplitude. *Physarum* does not differentiate between green and white lights. Switching off red light leads to increase of period and decrease of amplitude of oscillations.

CHEMICAL SENSOR

A chemical sensor is a device that gives a selective response when exposed to a target chemical substance. *Physarum* senses and responds to volatile aromatic substances (Adamatzky, 2011, 2012b; Adamatzky and De Lacy Costello, 2012). We studied the binary preferences of *Physarum* with regard to various volatile chemicals (De Lacy Costello and Adamatzky, 2013) and derived an experimental mapping between subsets of chemoattractants and chemorepellents, including arnesene, tridecane, *S*-(−)-limonene, *cis*-3-hexenylacetate, geraniol, benzyl alcohol, linalool, nonanal, and the amplitude and frequency of electrical potential oscillation of *Physarum* (Whiting et al., 2014). *Physarum* increases the frequency of electrical potential oscillations when exposed to the strongest attractants—farnesene, tridecane, *S*-(−)-limonene, and *cis*-3-hexenylacetate. Exposure to repellents—linalool, benzyl alcohol, nonanal—leads to a decrease of oscillation frequency, and, for linalool and benzyl alcohol, an increase of the oscillation amplitude. The *Physarum* chemical sensor discriminates between individual chemicals by changing amplitude and frequency of its electrical potential oscillations; it can detect a chemical from a distance of several centimeters.

ROBOT CONTROLLERS

Physarum responds to stimuli by changing the pattern of its electrical potential oscillations and cytoplasm shuttling. By interfacing the *Physarum* with actuators, we can make the myxomycete a controller for robots. Two prototypes of such robotic controllers have been developed. The first is a controller for a hexapod robot (Tsuda et al., 2006), and the second is a controller for a robotic android head (Gale and Adamatzky, 2016). A *Physarum* controller for the hexapod robot is made of a star-shaped template (Tsuda et al., 2006). It has six circular wells connected by channels that meet at a single point. The *Physarum* grows inside the template. *Physarum* in each well acts as a cytoplasm shuttle streaming oscillator. The *Physarum* oscillators in the wells are coupled via the *Physarum* body colonizing the channels between the wells. Blue light is used as a stimulus. The shuttle streaming of cytoplasm is measured via light absorbance. Oscillations of shuttle streaming in the wells, as a response to a stimulation with light, modulate phase and frequency of the movement of the robot's legs and thus cause the robot to change its direction of movement.

An electrical activity of *Physarum* in response to stimulations is converted to an affective state in the design of a *Physarum* emotional controller (Gale and Adamatzky, 2016). A *Physarum* is inoculated on a multiple-electrode array and stimulated with nutrients (attractant) and light (repellent). Extracellular electrical potential is recorded. The recorded data are split into chunks. We employed a circumplex model of affect, where emotions are plotted in two dimensions determined by the polarity and arousal level. The chunks are assigned polarity. Potential recorded during stimulation with attractant is given positive polarity. Data obtained during illumination of *Physarum* are assigned negative polarity. A level of arousal is proportional to the amplitude of the electrical potential. Emotions are assigned to the data chunks based on the polarity and the arousal of chunks and are fed into an android robot. The data activate motors placed in positions matching sites of real muscles in a human face. Actuation of the motors causes movements of an artificial skin. The movements are expressed as affective facial expression of the android.

MUSIC GENERATION

Physarum expresses its physiological states in patterns of its electrical potential oscillators (Adamatzky and Jones, 2011). In Adamatzky (2010c), 5–10 days' recordings were processed and converted to a sound track by mapping parameters of electrical potential oscillation to pitch, attack, and duration of tones. The first-ever sound track was produced from electrical activity in *Physarum* as reported by Adamatzky (2010b). The music reflected a physiological transition of *Physarum* from a comfortable foraging state to a state of active search for disappearing nutrients to a decision-making state to transformation to a sclerotium. When the track was played to an auditorium at various presentations, the majority of people had the feeling of a dramatic development in the "life of *Physarum*." Subsequently, the transformation of *Physarum* activity recording into sounds has been taken to a more professional level, from a musical point of view, in Miranda et al. (2011). There, electrical activity of *Physarum* was converted to parameters of sinusoidal oscillators; the rhythmic behavior of *Physarum* was shown to produce different timbres.

Memristive properties of *Physarum* were used to generate musical responses, as reported by Braund et al. (2016). Vocabulary notes were assigned voltage values, and the *Physarum* current–voltage response to electrical stimulation was recorded. Discrete voltages were converted to notes. The notes were fed into a MIDI keyboard. During interactive music performance between a human composer and *Physarum*, a feedback to the *Physarum* was implemented. Parts of a well-known melody (Elgar's "Nimrod" and Beethoven's "Für Elise") were generated for live performances with *Physarum* (Braund et al., 2016).

REFERENCES

Adamatzky, A., 1991. Neural algorithm for constructing minimal spanning tree. Neural Network World 6, 335–339.

Adamatzky, A., 2008. Growing spanning trees in plasmodium machines. Kybernetes 37 (2), 258–264.

Adamatzky, A., 2009. Developing proximity graphs by *Physarum polycephalum*: does the plasmodium follow the Toussaint hierarchy? Parallel Process. Lett. 19 (1), 105–127.

Adamatzky, A., 2010a. *Physarum* Machines: Computers From Slime Mould (World Scientific Series on Non Linear Science Series A: Vol. 74). World Scientific, London, England.

Adamatzky, A., 2010b. *Physarum* music. Available from: https://youtu.be/F79D_YWXycI

Adamatzky, A., 2010c. Myxomycete logical gates: exploring ballistic approach. Applications, Tools and Techniques on the Road to Exascale Computing 41–56 (arXiv:1005.2301).

Adamatzky, A., 2011. On attraction of myxomycete Physarum polycephalum to plants with sedative properties. Nature Precedings 10.

Adamatzky, A., 2012a. Bioevaluation of World Transport Networks. World Scientific, London, England.

Adamatzky, A., 2012b. Simulating strange attraction of acellular myxomycete *Physarum polycephalum* to herbal tablets. Math. Comput. Model. 55 (3), 884–900.

Adamatzky, A., 2013a. Myxomycete tactile sensor. Sens. Actuators B Chem. 188, 38–44.

Adamatzky, A., 2013b. Towards myxomycete colour sensor: recognition of colours by *Physarum polycephalum*. Org. Electron. 14 (12), 3355–3361.

Adamatzky, A., 2014. Tactile bristle sensors made with slime mold. Sens. J. IEEE 14 (2), 324–332.

Adamatzky, A., 2016. Advances in *Physarum* Machines: Sensing and Computing With Myxomycete. Springer, Heidelberg, Germany.

Adamatzky, A., Akl, S., Alonso-Sanz, R., Van Dessel, W., Ibrahim, Z., Ilachinski, A., Jones, J., Kayem, A.V.D.M., Martinez, G.J., De Oliveira, P., 2013a. Are motorways rational from myxomycete's point of view? Int. J. Parallel Emergent. Distrib. Syst. 28 (3), 230–248.

Adamatzky, A., De Lacy Costello, B., 2012. *Physarum* attraction: why slime mold behaves as cats do? Commun. Integr. Biol. 5 (3), 297–299.

Adamatzky, A., Jones, J., 2011. On electrical correlates of *Physarum polycephalum* spatial activity: can we see *Physarum* machine in the dark? Biophys. Rev. Lett. 6 (01–02), 29–57.

Adamatzky, A., Kayem, A.V.D.M., 2013. Biological evaluation of trans-African highways. Eur. Phys. J. Spec. Top. 215 (1), 49–59.

Adamatzky, A., Lees, M., Sloot, P., 2013b. Bio-development of motorway network in the Netherlands: a myxomycete approach. Adv. Complex Syst. 16 (02–03), 1250034.

Adamatzky, A., Martinez, G.J., 2013. Bio-imitation of Mexican migration routes to the USA with myxomycete on 3D terrains. J. Bionic Eng. 10 (2), 242–250.

Adamatzky, A., Prokopenko, M., 2012. Myxomycete evaluation of Australian motorways. Int. J. Parallel Emergent Distrib. Syst. 27 (4), 275–295.

Adamatzky, A., Schubert, T., 2012. Schlauschleimer in Reichsautobahnen: myxomycete imitates motorway network in Germany. Kybernetes 41 (7–8), 1050–1071.

Adamatzky, A., Armstrong, R., De Lacy Costello, B., Deng, Y., Jones, J., Mayne, R., Schubert, T., Sirakoulis, G.Ch., Zhang, X., 2014. Myxomycete analogue models of space exploration and planet colonisation. J. Br. Interplanet. Soc. 67, 290–304.

Asgari, M., Henney, H.R., 1977. Inhibition of growth and cell wall morphogenesis of *Bacillus subtilis* by extracellular slime produced by *Physarum flavicomum*. Cytobios 20 (79–80), 163–177.

Barrows, L.R., Radisky, D.C., Copp, B.R., Swaffar, D.S., Kramer, R.A., Warters, R.L., Ireland, C.M., 1993. Makaluvamines, marine natural products, are active anti-cancer agents and DNA topo II inhibitors. Anticancer Drug Des. 8 (5), 333–347.

Block, I., Wohlfarth-Bottermann, K.E., 1981. Blue light as a medium to influence oscillatory contraction frequency in *Physarum*. Cell Biol. Int. Rep. 5 (1), 73–81.

Braund, E., Sparrow, R., Miranda, M., 2016. *Physarum*-based memristors for computer music. In: Adamatzky, A. (Ed.), Advances in *Physarum* Machines. Springer, Heidelberg, Germany.

Brewer, E.N., Kutraishi, S., Garver, J.C., Strong, F.M., 1964. Mass culture of a slime mold, *Physarum polycephalum*. Appl. Microbiol. 12, 161–164.

Buchanan, M.S., Hashimoto, T., Asakawa, Y., 1996. Acylglycerols from the slime mould, *Lycogala epidendrum*. Phytochemistry 41, 791–794.

Chinnasamy, S., Bhatnagar, A., Hunt, R., Das, K.C., 2010. Microalgae cultivation in a wastewater dominated by carpet mill effluents. Bioresour. Technol. 101, 3097–3105.

Considine, J.M., Mallette, M.F., 1965. Production and partial purification of antibiotic materials formed by *Physarum gyrosum*. Appl. Microbiol. 13, 464–468.

De Lacy Costello, B., Adamatzky, A., 2013. Assessing the chemotaxis behavior of *Physarum polycephalum* to a range of simple volatile organic chemicals. Commun. Integr. Biol. 6 (5), e25030.

Dembitsky, V.M., Rezanka, T., Spizek, J., Hanus, L.O., 2005. Secondary metabolites of slime molds (myxomycetes). Phytochemistry 66 (7), 747–769.

Dietrich, M.R., 2015. Explaining the pulse of protoplasm: the search for molecular mechanisms of protoplasmic streaming. J. Integr. Plant Biol. 57 (1), 14–22.

Durham, A.C., Ridgway, R.B., 1976. Control of chemotaxis in *Physarum polycephalum*. J. Cell Biol. 69 (1), 218–223.

Fergus, C.L., Mallette, M.F., 1962. Pure culture of the slime mold *Physarum gyrosum*. Mycologia 54, 580–581.

Fröde, R., Hinze, C., Josten, I., Schmidt, B., Steffian, B., Steglich, W., 1994. Isolation and synthesis of 3,4-*bis*(indol-3-yl)pyrrole-2,5-dicarboxylic acid derivatives from the slime mould *Lycogala epidendrum*. Tetrahedron Lett. 35, 1689–1690.

Gale, E., Adamatzky, A., 2016. Translating myxomycete responses: a novel way to present data to the public. In: Adamatzky, A. (Ed.), Advances in *Physarum* Machines. Springer, Heidelberg, Germany.

Grebecki, A., Cieslawska, M., 1978. Plasmodium of *Physarum polycephalum* as a synchronous contractile system. Cytobiologie 17 (2), 335–342.

Hasegawa, H., Yamada, Y., Komiyama, K., Hayashi, M., Ishibashi, M., Sunazuka, T., Izuhara, T., Sugahara, K., Tsuruda, K., Masuda, M., Takasu, N., Tsukasaki, K., Tomonaga, M., Kamihira, S., 2007. A novel natural compound, a cycloanthranilylproline derivative (Fuligocandin B), sensitizes leukemia cells to apoptosis induced by tumor necrosis factor related apoptosis-inducing ligand (TRAIL) through 15-deoxy-$\Delta^{12,14}$ prostaglandin J_2 production. Blood 110 (5), 1664–1674.

Hashimoto, T., Yasuda, A., Akazawa, K., Takaoka, S., Tori, M., Asakawa, Y., 1994. Three novel dimethyl pyrroledicarboxylate, lycogarubins A-C, from the myxomycetes *Lycogala epidendrum*. Tetrahedron Lett. 35, 2559–2560.

Herrera, N.A., Rojas, C., Franco-Molano, A.E., Stephenson, S.L., Echeverri, F., 2011. *Physarella oblonga*-centered bioassays for testing the biological activity of myxomycetes. Mycosphere 6, 637–644.

Hosoya, T., Yamamoto, Y., Uehara, Y., Hayashi, M., Komiyama, K., Ishibashi, M., 2005. New cytotoxic bisindole alkaloids with protein tyrosine kinase inhibitory activity from a myxomycete *Lycogala epidendrum*. Bioorg. Med. Chem. Lett. 15, 2776–2780.

Ishibashi, M., Iwasaki, T., Imai, S., Sakamoto, S., Yamaguchi, K., Ito, A., 2001. Laboratory culture of the myxomycetes: formation of fruiting bodies of *Didymium bahiense* and its plasmodial production of makaluvamine A. J. Nat. Pro. 64, 108–110.

Ishibashi, M., Mitamura, M., Ito, A., 1999. Laboratory culture of the myxomycete *Didymium squamulosum* and its production of clionasterol. Nat. Med. 53, 316–318.

Iwata, D., Ishibashi, M., Yamamoto, Y., 2003. Cribrarione B, a new naphthoquinone pigment from the myxomycete *Cribraria cancellata*. J. Nat. Prod. 66, 1611–1612.

Kamata, K., Onuki, H., Hirota, H., Yamamoto, Y., Hayashi, M., Komiyama, K., Sato, M., Ishibashi, M., 2004. Tubiferal, A, a backbone-rearranged triterpenoid lactone isolated from the myxomycete *Tubifera dimorphotheca*, possessing reversal of drug resistance activity. Tetrahedron Lett. 60 (44), 9835–9839.

Kamata, K., Suetsugu, T., Yamamoto, Y., Hayashi, M., Komiyama, K., Ishibashi, M., 2006. Bisindole alkaloids from myxomycetes *Arcyria denudata* and *Arcyria obvelata*. J. Nat. Prod. 69 (8), 1252–1254.

Kaniwa, K., Arai, M.A., Li, X., Ishibashi, M., 2006. Synthesis, determination of stereochemistry, and evaluation of new bisindole alkaloids from the myxomycete *Arcyria ferruginea*: an approach for Wnt signal inhibitor. Bioorg. Med. Chem. Lett. 17, 4254–4257.

Kauffman, S., Wille, J.J., 1975. The mitotic oscillator in *Physarum polycephalum*. J. Theor. Biol. 55 (1), 47–93.

Keller, H.W., Everhart, S.E., 2010. Importance of myxomycetes in biological research and teaching. Fungi 3 (1), 13–27.

Kishimoto, U., 1958. Rhythmicity in the protoplasmic streaming of a slime mood, *Physarum polycephalum*. i. A statistical analysis of the electrical potential rhythm. J. Gen. Physiol. 41 (6), 1205–1222.

Lee, B.S., Fujita, M., Khazenzon, N.M., Wawrowsky, K.A., Wachsmann-Hogiu, S., Farkas, D.L., Black, K.L., Ljubimova, J.Y., Holler, E., 2006. Polycefin, a new prototype of a multifunctional nanoconjugate based on poly(β-L-malic acid) for drug delivery. Bioconjug. Chem. 17 (2), 317–326.

Ljubimova, J.Y., Fujita, M., Khazenzon, N.M., Lee, B.S., Wachsmann-Hogiu, S., Farkas, D.L., Black, K.L., Holler, E., 2008. Nanoconjugate based on polymalic acid for tumor targeting. Chem. Biol. Interact. 171 (2), 195–203.

Locquin, M., Prevot, A.R., 1948. Etude de quelques antibiotiques produits par les myxomycetes. Ann. Inst. Pasteur Paris 75, 8–13.

Loganathan, P., 1998. Production of DL-DOPA from acellular slime-mould *Stemonitis herbatica*. Bioprocess Eng. 18, 307–308.

Mayberry, J.M., Fergus, C.L., Mallette, M.F., 1962. Pure culture of the slime mold *Physarum gyrosum*. Mycologia 54, 580–581.

Miranda, E.R., Adamatzky, A., Jones, J., 2011. Sounds synthesis with myxomycete of *Physarum polycephalum*. J. Bionic Eng. 8 (2), 107–113.

Misono, Y., Ishibashi, M., Ito, A., 2003a. Bahiensol, a new glycerolipid from a cultured myxomycete *Didymium bahiense* var. *bahiense*. Chem. Pharm. Bull. (Tokyo) 51 (5), 612–613.

Misono, Y., Ito, A., Matsumoto, J., Sakamoto, S., Yamaguchi, K., Ishibashi, M., 2003b. Physarigins A–C, three new yellow pigments from cultured myxomycete *Physarum rigidum*. Tetrahedron Lett. 44, 4479–4481.

Nakagaki, T., Yamada, H., Toth, A., 2001. Path finding by tube morphogenesis in an amoeboid organism. Biophys. Chem. 92 (1), 47–52.

Nakatani, S., Kamata, K., Sato, M., Onuki, H., Hirota, H., Matsumoto, J., Ishibashi, M., 2005b. Melleumin A, a novel peptide lactone isolated from the cultured myxomycete *Physarum melleum*. Tetrahedron Lett. 46, 267–271.

Nakatani, S., Kiyota, M., Matsumoto, J., Ishibashi, M., 2005a. Pyrroloiminoquinone pigments from *Didymium iridis*. Biochem. Syst. Ecol. 33, 323–325.

Nakatani, S., Naoe, A., Yamamoto, Y., Yamauchi, T., Yamaguchi, N., Ishibashi, M., 2003. Isolation of bisindole alkaloids that inhibit the cell cycle from myxomycetes *Arcyria ferruginea* and *Tubifera casparyi*. Bioorg. Med. Chem. Lett. 13 (17), 2879–2881.

Naoe, A., Ishibashi, M., Yamamoto, Y., 2003. Cribrarione A, a new antimicrobial naphthoquinone pigment from a myxomycete *Cribraria purpurea*. Tetrahedron 59, 3433–3435.

Nesetril, J., Milkova, E., Nesetrilova, H., 2001. Otakar Boruvka on minimum spanning tree problem: translation of both the 1926 papers, comments, history. Discrete Math. 233 (1), 3–36.

Rakoczy, L., 1963. Application of crossed light and humidity gradients for the investigation of slime-molds. Acta Soc. Bot. Pol. 32 (2), 393–403.

Řezanka, T., Dvořáková, R., 2003. Polypropionate lactones of deoxysugars glycosides from slime mold *Lycogala epidendrum*. Phytochemistry 63 (8), 945–952.

Řezanka, T., Dvořáková, R., Hanuš, L.r.O., 2004. Enteridinines A and B from slime mold *Enteridium lycoperdon*. Phytochemistry 65 (4), 455–462.

Řezanka, T., Hanus, L.O., Kujan, P., Dembitsky, V.M., 2005. The fulicineroside, a new unusual glycosidic dibenzofuran metabolite from the slime mold *Fuligo cinerea* (L.) Wiggers. Eur. J. Org. Chem. 13, 2708–2714.

Sawada, T., Aono, M., Asakawa, S., Ito, A., Awano, K., 2000. Structure determination and total synthesis of a novel antibacterial substance, AB0022A, produced by a cellular slime mold. J. Antibiot. (Tokyo) 53 (9), 959–966.

Schroeder, H.R., Mallette, M.F., 1973. Isolation and purification of antibiotic material from *Physarum gyrosum*. Antimicrob. Agents Chemother. 4, 160–166.

Shintani, A., Ohtsuki, T., Yamamoto, Y., Hakamatsuka, T., Kawahara, N., Goda, Y., Ishibashi, M., 2009. Fuligoic acid, a new yellow pigment with a chloronated polyene-pyrone acid structure isolated from the myxomycete *Fuligo septica* f. *flava*. Tetrahedron Lett. 50, 3189–3190.

Sobels, J.C., 1950. Nutrition de quelque myxomycetes en cultures et associees et leurs proprietes antibiotiques. Antonie Van Leeuwenhoek J. Microbiol. Serol. 16, 123–243.

Steglich, W., 1989. Slime molds (myxomycetes) as a source of new biologically active metabolites. Pure Appl. Chem. 61, 281–288.

Stephenson, S.L., Stempen, H., 1994. Myxomycetes: A Handbook of Slime Molds. Timber Press, Portland, OR.

Taylor, R.L., Mallette, M.F., 1976. Growth of *Physarum gyrosum* on agar plates and in liquid culture. Antimicrob. Agents Chemother., 613–617.

Teplov, V.A., Romanovsky, Y.R., Latushkin, O.A., 1991. A continuum model of contraction waves and protoplasm streaming in strands of *Physarum plasmodium*. Biosystems 24 (4), 269–289.

Tero, A., Kobayashi, R., Nakagaki, T., 2006. *Physarum* solver: a biologically inspired method of road-network navigation. Phys. A 363 (1), 115–119.

Tran, H.T.M., Stephenson, S.L., Chen, J., Pollock, E.D., Goggin, F.L., 2012. Evaluating the potential use of myxomycetes as a source of lipids for biodiesel production. Bioresour. Technol. 123, 386–389.

Tran, H.T.M., Stephenson, S.L., Pollock, E.D., 2015. Evaluation of *Physarum polycephalum* plasmodial growth and lipid production using rice bran as a carbon source. BMC Biotechnol. 15, 67.

Tsuda, S., Zauner, K.P., Gunji, Y.P., 2006. Robot control: from silicon circuitry to cells. Biologically Inspired Approaches to Advanced Information Technology, vol. 3853. Springer, Berlin, Heidelberg, pp. 20–32.

Ueda, T., Muratsugu, M., Kurihara, K., Kobatake, Y., 1976. Chemotaxis in *Physarum polycephalum*: effects of chemicals on isometric tension of the plasmodial strand in relation to chemotactic movement. Exp. Cell Res. 100 (2), 337–344.

Whiting, J.G.H., De Lacy Costello, B., Adamatzky, A., 2014. Towards myxomycete chemical sensor: mapping chemical inputs onto electrical potential dynamics of *Physarum polycephalum*. Sens. Actuators B Chem. 191, 844–853.

Wohlfarth-Bottermann, K.E., 1979. Oscillatory contraction activity in *Physarum*. J. Exp. Biol. 81 (1), 15–32.

Wohlfarth-Bottermann, K.E., Block, I., 1981. The pathway of photosensory transduction in *Physarum polycephalum*. Cell Biol. Int. Rep. 5 (4), 365–373.

FURTHER READING

Adamatzky, A., 2013. *Physarum* wires: self-growing self-repairing smart wires made from slime mould. Biomed. Eng. Lett. 3 (4), 232–241.

Chapter 12

Myxomycetes in Education: The Use of These Organisms in Promoting Active and Engaged Learning

Katherine E. Winsett*, Thomas Edison E. dela Cruz**, Diana Wrigley de Basanta†
*Wake Technical Community College, Raleigh, NC, United States; **University of Santo Tomas, Manila, Philippines; †Royal Botanic Garden (CSIC), Madrid, Spain

INTRODUCTION

Myxomycetes are a potentially important educational tool and represent an excellent component of classroom activities for students of almost any age. They provide an opportunity to work with living organisms that pose little or no risk to students or teachers. Myxomycetes are nonpathogenic, nontoxic, and hypoallergenic, and will not affect other plants or animals. In addition, they are an inexpensive, locally available natural resource, irrespective of where a school might be located. Myxomycetes are commonly used as examples of protists in biology textbooks in the United States and other countries, primarily because of their extraordinary life cycle, and they are excellent model organisms for a variety of educational uses.

The biology of myxomycetes is such that their uses in an educational setting are both diverse and appropriate for all levels of education (Tables 12.1 and 12.2). The relative ease with which they are cultured from decaying plant material obtained from virtually any source (even including houseplants), their maintenance at any life stage in the classroom laboratory, and the opportunity to demonstrate standard concepts of biology as they relate to organisms outside of the animal and plant kingdoms make myxomycetes a unique addition to any classroom or educational program. This chapter provides a discussion of how myxomycetes have been used in formal and informal classroom environments involving students of various ages and skill levels. Elsewhere in this volume (Chapter 1) there are detailed descriptions of the biology of these organisms.

Myxomycetes: Biology, Systematics, Biogeography, and Ecology. http://dx.doi.org/10.1016/B978-0-12-805089-7.00012-3

TABLE 12.1 Examples of Learning Outcomes and Suggested Learning Activities With Myxomycetes

Learning activities	Levels				Expected student learning outcomes
	Kinder-garten	Elementary school	Junior and senior high school	Under-graduate	
Collecting myxomycetes in the field	√	√	√	√	Remembering
Setting up moist chamber culture	√	√	√	√	Remembering Understanding Applying
Storing fruiting bodies and maintaining a collection		√	√	√	Remembering Understanding Applying
Culturing myxomcyetes on agar			√	√	Understanding Applying Creating
Observing fruit-ing bodies	√	√	√	√	Applying Evaluating
Observing life cycle of myxo-mycetes			√	√	Applying Evaluating
Identifying myxomycetes using online resources			√	√	Applying Analyzing Evaluating
Classifying myxomycetes			√	√	Applying Analyzing Evaluating
Assessing ecological patterns of myxomycetes			√	√	Analyzing Evaluating
Observing feed-ing behavior of myxomycetes			√	√	Applying Evaluating
Developing infomaterials for myxomycetes			√	√	Creating

TABLE 12.2 Some Examples of the Use of Myxomycetes in High School Biology Courses

Topic	Examples
Cell biology	Characteristics of living things, cell structures, cell reproduction, synchronous mitosis, cyclosis
Organismal biology	Phagocytosis, food vacuoles, digestion, egestion, response to stimuli
Behavior	Food preferences, feeding rates, simple mazes
Classification	Taxonomic characters, taxonomic problems, use of dichotomous keys, making permanent slide mounts
Ecology	Microbial predation, nutrient recycling, importance of microorganisms in ecosystems, ecological models and indices, comparisons of ecological parameters
Independent projects	Most of the aforementioned, also cooperation in science with group projects, formulation of hypotheses, experimental design, writing a scientific paper

Myxomycetes can be used to teach any number of scientific concepts ranging from observation to classification, and they can even serve as a subject to introduce the concepts of molecular biology. It would be futile for us to attempt to describe all of the ways in which myxomycetes can be used in the teaching of science, but the following examples are derived from our combined experience teaching in classrooms, workshops, and laboratories, as well as in the field. There are references to additional work appearing in the literature, and some of the examples are organized into informal lesson plans.

What we have learned from using myxomycetes in the teaching of science is that they are outside the experience of many people. Their use for teaching concepts is valuable, but these organisms are unique. They are different from the more familiar organisms (plants, animals, and fungi normally used in the classroom setting), but they offer both a valuable tool in the classroom and a totally new experience for students. The techniques used for isolating myxomycetes from nature are quite simple. Once they are mastered, a whole world of opportunity opens up for using these organisms to educate any audience about any number of scientific concepts. The techniques described in the next section are only ways in which we have used myxomycetes, along with a general discussion of alternative opportunities. Consider these descriptions a jumping-off point. They were and are successful in our experience, but we have every confidence that with the information presented in the rest of this volume, as well as experience in the techniques described in this chapter, any and all teachers of science can find a number of other uses for these organisms in their own classrooms.

TECHNIQUES

To acquire sample specimens for use or to develop other inquiry-based teaching opportunities requires a handful of relatively simple techniques. Each of these is discussed here.

Collecting Myxomycetes in the Field

Myxomycetes are found everywhere, including every known terrestrial habitat examined to date. Wherever decaying plant material (e.g., decaying wood and dead leaves) is present, myxomycetes will thrive, feeding on microorganisms, such as bacteria and fungi, that are associated with the decomposition of such material. In the field, myxomycetes form tiny fruiting bodies. For most species, their minute fruiting bodies are no taller than a millimeter or two; hence close observation of decaying plant material is needed to see them. Other species may form conspicuous fruiting structures easily visible to the naked eye. Some of these are 20 cm or more in maximum extent. To observe myxomycetes directly during field collecting, it is usually necessary to use a hand lens or an eye lens similar to what watchmakers use. With this tool, you can observe specimens directly in the field and bring back to the laboratory the fruiting bodies of only myxomycetes and not something else. There are many small fruiting bodies of fungi also present in the field that can easily be mistaken for those of a myxomycete.

But before a teacher goes on a field trip with students, it is important to secure any permits necessary to visit the collecting area. Securing such permits early saves much valuable time. It is also important to learn as much as possible, in advance, about the collecting area. Moreover, it is useful to take a map along during the collecting trips to identify the different habitat types where myxomycetes and the substrates with which they are associated can be collected. If the students are to assess ecological patterns, it is important to describe each of the collecting areas or sampling points. Important parameters that should be noted include the type of habitat, GPS coordinates, common plant genera within the sampling area, elevation, and canopy coverage. A checklist of relevant environmental parameters (e.g., relative humidity and temperature) can also be included.

Collecting myxomycetes in the field requires a certain amount of timing. Fruiting bodies often appear a week or so after a heavy rain followed by sunny days. When the ideal climatic conditions occur, students can be taken to the chosen sampling areas. Piles of leaf litter on the ground or decaying twigs should be examined. These represent ideal places for myxomycetes to thrive and later form fruiting bodies. In wooded areas, decaying logs and stumps also should be checked for myxomycetes. Ideal logs for myxomycetes are those that retain moisture and show obvious evidence of decay. Fruiting bodies on a piece of substrate observed directly in the field and then collected can be glued to a small piece of paper cut to fit in a small cardboard "pill box." Alternatively, a compartmentalized plastic or tin container lined with corkboard can be used to pin pieces of substrate that have fruiting bodies of myxomycetes present. While in the field, students can also collect samples to use for preparing moist chamber cultures. Leaf litter on the ground and dead but still attached plant parts above the ground, twigs, pieces of the dead outer bark from living trees, dead grass, inflorescences, decayed fruits, or soil can be collected and placed in paper bags, brought to the laboratory, and placed in moist chamber cultures. If these materials are wet when collected, it is ideal to air-dry these samples before setting up moist chamber cultures. More detail relating to the procedures used for isolating myxomycetes can be found in Chapter 10.

Setting Up Moist Chamber Cultures

While many of the more common species of myxomycetes are found regularly and relatively easily in nature, the most efficient way to acquire specimens for study is through the culturing of dead plant material. A thorough description of this technique is found in Stephenson and Stempen (1994) and in Chapter 10 of this volume, but a brief outline of the technique is supplied herein. Fortunately for educators, this is a simple technique that can be accomplished using basic laboratory materials or readily available alternatives. In its simplest description, a moist chamber culture is the establishment of a microcosm of the environment in which these organisms are found in nature. Decaying plant material (including dead leaves or twigs) is kept moist over a period of weeks, providing the microhabitat necessary for myxomycetes to go through their life cycle. In the laboratory, a Petri dish is lined with filter paper on which a single layer of the decaying plant material is placed, and then the material is covered with distilled water. The dish and its contents are kept covered, and after 24 h the water is poured off. Every few days, a little distilled water is added to keep the filter paper wet but with no standing water. The covered Petri dish is placed in indirect light and is now a simple version of the type of microhabitat in nature that generally promotes myxomycetes to go through their life cycle. Different species will emerge at different times, but after a few weeks the fruiting bodies of some of the more common species can be found fruiting on the leaves or twigs or even the filter paper and in some cases the sides or top of the Petri dish. The spores of myxomycetes are found everywhere, so it is reasonable to assume that spores will be added to each dish along with the dead plant material. However, it is often the case that myxomycete fruiting bodies do not emerge in some moist chamber cultures. Even when the dead plant material comes from the same collected sample, some cultures are not successful. It is not fully understood why, but to maximize the chances of finding fruiting bodies in culture, one should set up replicates from each sample. A handful of leaves from the ground can be used to set up a number of moist chamber cultures.

The materials used for the moist chamber culture technique are available in any catalog of scientific or laboratory supplies, and they are reasonably inexpensive. However, because the purpose is just to create a microhabitat to promote the life cycle of myxomycetes, other materials can be used. For example, the filter paper used to maintain moisture levels in the dish can be replaced with a circle of paper towel. If plastic or glass Petri dishes are not available, clear plastic containers like those in which food is stored, lined on the bottom with paper towels, can serve the same purpose.

Once a week, the plates should be examined using a dissecting microscope (ideally) or a magnifying glass to look for plasmodium and fruiting bodies. Some species are visible with the naked eye, but some magnification is ideal to confirm that the structure observed is indeed the fruiting body of a myxomycete.

Storing Fruiting Bodies and Maintaining a Collection

Once fruiting bodies are found in nature or obtained in moist chamber cultures, it is necessary to allow them to dry to prevent colonization by fungi or the loss of spores. The myxomycete fruiting body will be found attached to some substrate, usually a piece of a dead leaf or a twig. This substrate material is removed along with the attached fruiting bodies for storage.

Like plant specimens found in a herbarium or pressed between the pages of a book, dried specimens of myxomycetes can maintain the structures necessary for identification. In the case of myxomycetes, the substrate—the leaf or piece of wood to which the fruiting body is attached—can be glued using simple white glue to a strip of card-stock paper. These strips can then be stored in matchboxes or other similarly sized cardboard boxes (Fig. 12.1). Cardboard "pill boxes" can be ordered from scientific supply catalogs, but a collection of matchboxes can serve the same purpose quite well. To examine any specimen, the card-stock paper with the specimen attached can be removed, thus providing easier access to the sample. Once dried, these specimens can be stored for many years.

Culture of Myxomycetes on Agar

It is also possible to see the stages of the life cycle in agar culture as has been discussed in Chapter 10. This technique is quite beneficial for creating a way in which to see the microscopic components of the life cycle. For example, the plasmodia of species in the order Physarales are visible to the naked eye because of their size and remarkable colors. Developing the same plasmodium in agar culture allows for viewing the cytoplasmic streaming within the plasmodium under a microscope. The study of agar culture of myxomycetes is quite interesting. It has not been possible to get all species to go from spore to spore in agar culture (Haskins and Wrigley de Basanta, 2008), and even within species that can easily be induced to go from spore to spore in culture, different samples do not have equal success. There is no clear indication of why this is the case, but it has been observed regularly. Fortunately, for those specimens that do germinate

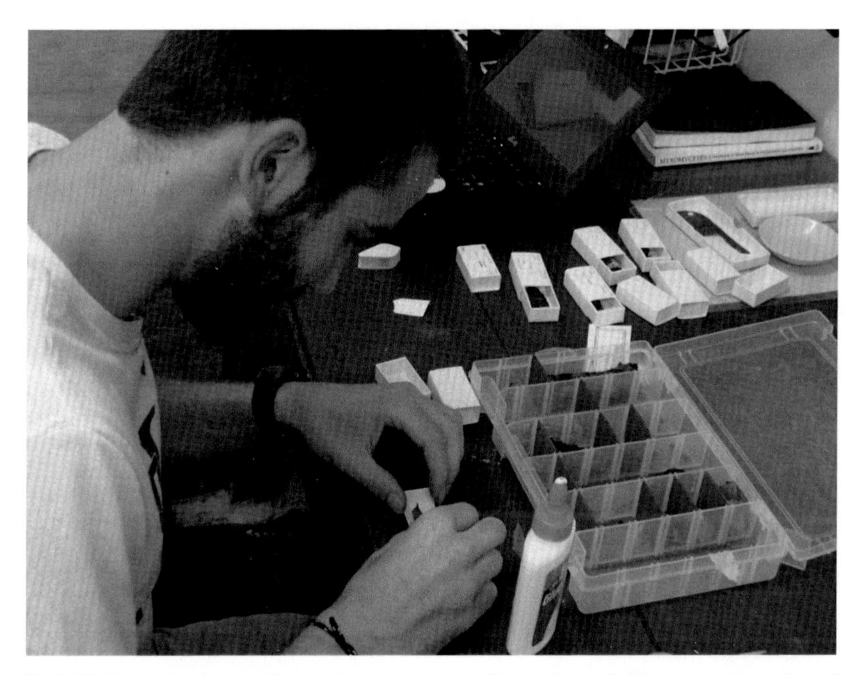

FIGURE 12.1 Placing specimens of myxomycetes collected in the field into small pasteboard boxes.

and go through the life cycle in culture, many will do so quite easily. The easiest way to get started culturing myxomycetes on agar is to use the strains available from biological supply companies. Carolina Biological Supply (Carolina.com) has a strain of *Physarum polycephalum* that is an excellent example of a readily available sample that will grow very well, and do so repeatedly.

Haskins and Wrigley de Basanta (2008) provide a good review of the techniques for establishing agar cultures of various species of myxomycetes. The key for establishing successful cultures is to provide food organisms without the plate becoming overwhelmed with the hyphae of fungi. While we direct you to the aforementioned paper for a detailed explanation of techniques, it is worth noting that for high school and undergraduate students, the value of the elements of the life cycle that can be revealed through agar culture should not be dismissed. An agar plate can be placed on the stage of a compound microscope, allowing an unprecedented view of the amoeboflagellate cells, the plasmodium, and cytoplasmic streaming within the plasmodium. If a specimen in your collection is identified as one that will germinate easily, another technique is to make an agar slide (F. Spiegel, personal communication). A small cube of agar (water agar is appropriate) can be melted onto the slide by placing the cube on a glass slide, setting a cover slip over top, and passing the slide gently across a flame. Allow the agar to cool, remove the cover slip, place some spores onto the surface of the agar, and then replace the cover slip. The spores can germinate on the agar, and the technique allows students to see germination and the movement of amoeboflagellates for themselves with unprecedented clarity through the use of a compound microscope.

Identifying Myxomycetes

As is the case with other organisms, such as plants and animals, identification of myxomycetes relies heavily on morphological features. Examples of morphological features are the type, color, and appearance of the fruiting body and its parts (e.g., hypothallus, peridium, and stalk) and the presence or absence of whitish powder or lime. These are often best observed under a dissecting microscope, but a hand lens will work, too. Detailed observations of spores, capillitium, and other microscopic features are also observed under a compound light microscope. All morphometric data are then compared with species descriptions in the published literature and/or online identification guides for species identification. The monograph (*The Myxomycetes*) by Martin and Alexopoulos (1969) is one of the most comprehensive books, with detailed descriptions of a large number of species of myxomycetes. However, this book is no longer in print, and securing a copy is often very difficult or almost impossible. Additional references on myxomycete identification that have been published more recently and contain detailed morphological features and photographs of specimens include Neubert et al. (1993, 1995, 2000), Stephenson and Stempen (1994), and Poulain et al. (2011). Online identification guides are also easily accessible for species identification. Websites, such as those constructed as a result of the Eumycetozoan Research Project based at the University of Arkansas (http://slimemold.uark.edu), contain searchable databases and images of myxomycetes for easy comparison, as well as other educational materials relating to these organisms. The website of Discover Life in America (http://www.discoverlife.org) has many images and other useful information, and many other websites developed by various individuals who work with myxomycetes provide images (often rather spectacular) that have been identified. A Google search for myxomycete images will easily reveal many of these. In a study carried out by dela Cruz et al. (2012), students were found to be highly visual and thus preferred a tool with images more than the plainly written keys available for species identification.

One of the great benefits of teaching with myxomycetes is the relative simplicity of the techniques used to grow them. With very few differences, the techniques outlined earlier are standard for those individuals who are involved in carrying out research directed toward studying many aspects of the biology and ecology of these organisms. For this reason, the study of myxomycetes is accessible to students of biology at all levels. What follows are some examples of concepts that can be studied using myxomycetes and some examples from the collective experiences of the coauthors. These are summarized in Table 12.1. It is worth noting that we recognize that there are very few courses that have myxomycetes as an explicit topic of study, so the examples presented herein are descriptions of how we have used these organisms in various ways to illustrate or teach general science or biology topics and not just to teach about the myxomycetes themselves.

Collecting and Classifying Myxomycetes

Using the basic technique for moist chamber cultures described herein, set up a series of cultures using available ground litter. This in itself can be an activity for students (from elementary through high school) allowing for opportunities to observe the litter using a magnifying glass, or the cultures can be set up ahead of time and then used when myxomycetes have developed. An identification guide of common species (Fig. 12.2) can be used as an initial tool for students to match the myxomycetes developed in a moist chamber as an exercise in classification. Less important in this exercise are the names of the species, which are all in Latin since, unlike most plants and animals, there are very few common names for particular species of myxomycetes. Instead, students can use tools, such as a magnifying glass, to observe these nearly microscopic specimens and to recognize the differences that exist among different species, as well as the detail that can be observed even in such small organisms. Myxomycetes can be easily saved in matchboxes for use in the future, ultimately building up a collection that can be used for this exercise year after year. The materials and method for doing this are described in Chapter 10.

Observations

What may seem a basic concept, using myxomycetes to develop observational skills, is an opportunity that spans all age levels. There is a great deal to be observed when examining the fruiting bodies of myxomycetes. There are many colorful examples that can be seen with the naked eye, but with every increase in magnification there is even more to see. The fruiting bodies exhibit different shapes and sizes, the stalks can be smooth or furrowed, and the spores mass itself can show striking differences in color. These characteristics can be seen quite easily with something as simple as a magnifying glass or jeweler's loupe. If a dissecting microscope is available, students can get a clear view of the major structures of the myxomycete fruiting body. Further magnification, which is possible by making a slide and observing the spores and capillitium through a compound microscope, offers even further opportunities for observing the characteristics used in identification. The spores and capillitium often have a distinctive or even unique ornamentation, which is important in the classification of myxomycetes.

The life stages of a myxomycete are significantly different from those of plants, animals, or fungi. Some stages of their life history seem to counter the accepted norms taught in general biology. For example, the plasmodium is a

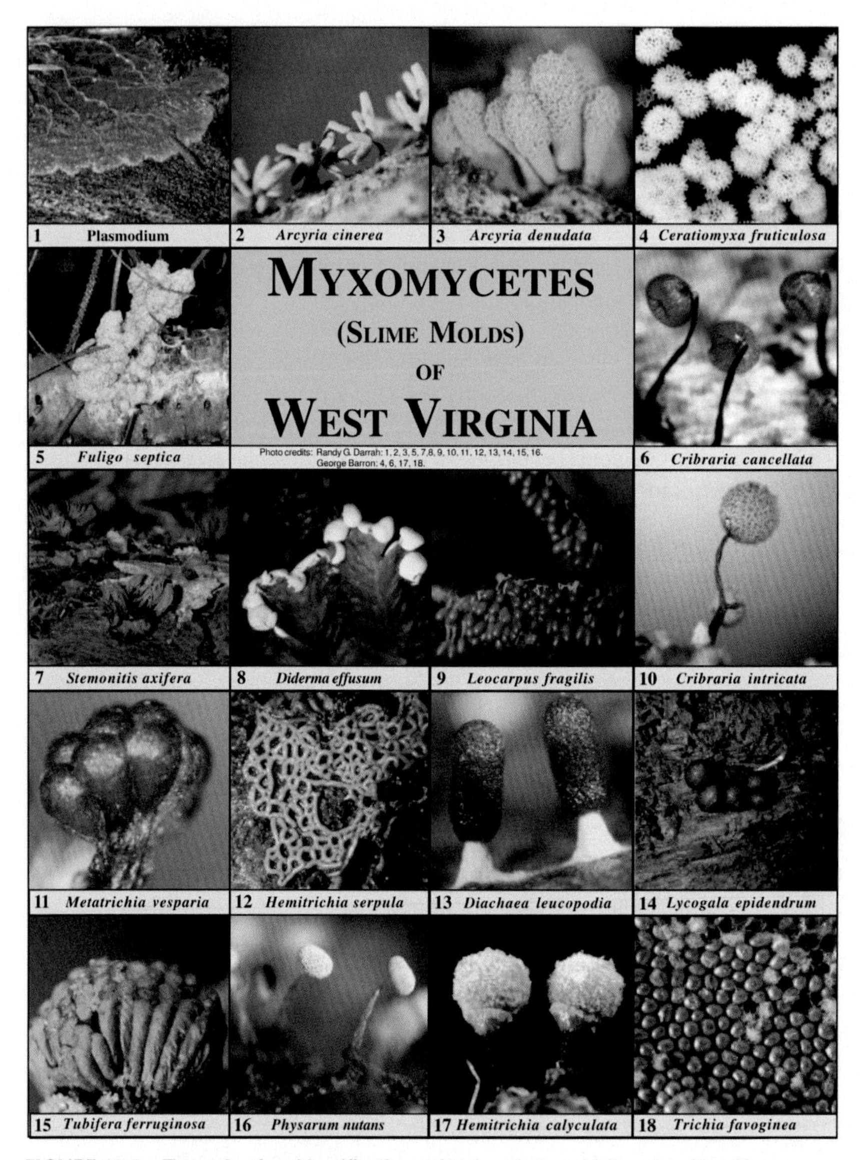

FIGURE 12.2 Example of an identification guide (or photo guide) used to identify common species of myxomycetes.

single cell, but it is often relatively large and easily visible to the naked eye. Its irregular shape and movement strategy belie the usual assumption of the idealized cell and its organelles as depicted in textbooks and models. The plasmodium is an excellent example of a multinucleate cell—a single cell that contains many nuclei. The nuclei undergoes mitosis, but cytokinesis (cell division) does not occur. This allows the cell to become larger and larger. While this makes for an interesting example when teaching about cell division, the ease with which a plasmodium can be grown in culture enhances the opportunity to observe the details. A simple technique is to place a sclerotium of *P. polycephalum* obtained from a biological supply house on a thin layer of water agar in a Petri dish with a few rolled oats present. The bright yellow plasmodium will activate in a few hours, and students can observe cytoplasmic streaming and see the multiple nuclei under a basic dissecting microscope. While there are images of the life stages to be found in the literature and on the web, the films produced at the University of Iowa in 1961 represent one of the best opportunities to see and hear a description of the life cycle and see the life stages in action. There are three of these films—entitled Slime Molds I, II, and III—that feature excellent time-lapse photography and photomicrographs (Koevenig et al., 1961). Fortunately, copies of these films can still be found, but at present the easiest access to high-quality versions of their content is in the Internet Archive (e.g., https://archive.org/details/slimemolds3identification). The first film is an excellent live-action description of the life cycle. It is highly recommended to accompany any laboratory investigations of the life cycle to help students prepare a mental image and acquire some background information for what they will see in their own cultures. Another excellent trio of films (Haskins, 1973a,b, 1974) on the development of *Echinostelium minutum* and *Stemonitis flavogenita* is also available on the Internet (https://av.tib.eu/search?q=Haskins). Several other time-lapse videos are available on the YouTube website (www.youtube.com). Two examples are "Amazing Slime Mold" by francischeefilms (https://www.youtube.com/watch?v=VJkJbM3y5R4) and "Are You Smarter Than a Slime Mold?" by It's Okay to Be Smart (https://www.youtube.com/watch?v=K8HEDqoTPgk).

Like the plasmodium, it is possible to observe other stages of the myxomycete life cycle. Direct observation of spore germination is an experience most students will never forget. In a paper by Keller and Everhart (2010), there is a detailed description of various techniques for achieving this in a classroom. In this same paper, there is also a review of literature describing other uses of myxomycetes in teaching and research. The cells produced from germinating spores divide, so a significant population of amoeboflagellates can be found on an agar plate just a few days after spore germination. Some amoeboflagellates emerge from spores more quickly than others, but spores from *Fuligo septica*, a large and very common myxomycete often found on mulch in flower gardens, can emerge rather quickly, in as little as 60 min (Keller and Everhart, 2010). Given additional time, other species can emerge over the course of several hours, a day, or a few days. Myxomycetes also offer the opportunity to examine fruiting bodies in detail so that older students can have a significant challenge.

There are many lesson plans that can be derived from the observation of myxomycetes. For younger students, they provide another focus for their abounding curiosity, in addition to the means of training children to recognize details using simple scientific tools. What follows is an example of an activity used in an international school setting in Spain with second graders

(7-year-olds). The objective was to explore the history and nature of science as part of science standards for the age group and was designed by a second-grade teacher. It was decided that the second grade would go on an expedition "in search of myxomycetes." An expedition provides a wonderful opportunity to use the scientific method and observation skills, to ask questions, and to investigate the environment. It allowed the children to experience what it is like to work as a scientist in the field. Language arts and math were easily integrated, as scientists go about measuring, mapping, graphing data, and writing about their ideas and conclusions. The activity involved several stages briefly listed herein.

Children made a hypothesis by posing a question: Are there any slime molds at our school? They made a prediction yes or no and drew a bar graph of the class results. They wrote a letter to a scientist to find out what was needed to go on an expedition, and a date was set for a visit by the high school science chair and biologist (who happened to be the third coauthor) to give them more information. Lists were made of what would be needed for the day outside, and letters were sent home informing parents and inviting them to participate. On the day of the expedition, there were five color-coded sites on campus, and the children, limited to collecting one sample from each area, set off in small groups with hand lenses to search the areas. When they found samples, they took them to the adults for confirmation and boxed them. In the absence of a myxomycete fruiting body, they collected samples of dead plant material (e.g., leaves, stems, and bark) to take back to the classroom in an envelope. All samples were labeled with the color code of where they were found. The following day, the children set up moist chamber cultures in plastic dishes and observed them during a 15-min period every week. There was a continued support from the biology teacher and a subsequent visit to the science laboratory to see their resulting collections using binocular stereoscopes. The activity was repeated with different groups over several years, and in total these young children found 17 different species on the school grounds, a new record for Spain. It was a successful activity on many levels, as the children were filled with curiosity and excitement over the myxomycetes they found and the simplified life cycle that was explained to them. The use of such a small model organism encouraged them to observe very closely and carefully. They were keen observers, thought nothing of scrambling beneath bushes or over obstacles to turn over leaf litter in their search, and did some detailed and precise sketches of what they saw through the stereoscopes.

A focus of interest in any classroom on a nature or science table is a terrarium. The third coauthor has had considerable success with a myxomycete terrarium. A large plastic aquarium was set up with a little soil, small pieces of decaying wood and twigs, mosses, leaf litter, or any other type of plant material. Initially, the contents were watered well, and then they were periodically spray-moistened and left near a window but not in direct sunlight. Many plasmodia appeared and climbed up the transparent walls of the terrarium, allowing students to experience the wonder of their movement, their occasional disappearance into the plant material in the terrarium, and their color. Results included some small white phaneroplasmodia, some greenish veins and fans, and bright yellow plasmodia, especially if the terrarium had been "primed" with sclerotia of *P. polycephalum*. In time, some produced fruiting bodies on twigs, dead leaves, or the small pieces of decaying wood, delighting the students with the (usually) overnight appearance of such a different-looking stage in the life cycle and the fact that the plasmodia "vanished." The terrarium can be used as a source of specimens for many activities, and naturally supported various other living organisms as well. Mini field collecting could take place within the classroom, since samples can be carefully taken out for examination with the stereoscope, including smaller fruiting bodies that were invisible to the naked eye. Even small plasmodia can be carefully lifted from dead leaves and placed on thin-layer water agar plates as described later, to observe cell structures and cyclosis. The great discussion that ensues from carefully led questioning, either on work cards beside the terrarium or in class or group discussions, is enormously valuable and very long-lasting for very little effort. An extension to this tried-and-tested general activity would be to set up several terraria representing different ecosystems that can be used to generate comparative observations and discussion on the differences that exist in the substrates present and the different assemblages of myxomycetes that appear.

Observing Feeding Behavior of Myxomycetes

Myxomycetes are members of nature's clean-up crew. As "micropredators," they feed upon bacteria and fungi that come in contact with the amoeboflagellates or the larger feeding form, the plasmodium. Feeding behavior can therefore be observed and illustrated with myxomycetes in agar culture. A plasmodium can be initially maintained at one side of a water agar plate. To prepare a water agar, simply dissolve 15–20 g of agar in 1 L of tap or distilled water. This is sterilized by autoclaving and allowed to solidify in Petri dishes. Alternatively, water agar can be dispensed and solidified in clean, microwave-safe plastic containers. Suspensions of different species of bacteria, yeasts, or fungal spores are prepared in sterile distilled water and streaked on the surface of the agar perpendicular to the growing plasmodium. Students can observe over a 12–24-h period the plasmodium as it feeds on the different microorganisms. The feeding rate can be measured as the time required for the plasmodium to completely consume its "food" microorganism. Stained yeast cells can be used to better show food vacuoles.

Assessing Ecological Patterns of Myxomycetes

Basic ecological concepts can also be applied in the study of myxomycetes. A species listing can be generated from any survey of myxomycetes. Students count the number of species and the number of genera, and then compute the taxonomic diversity expressed as the species:genus (S:G) ratio or the species richness as the number of species in a given habitat or substrate. The occurrence of species can also be expressed as the number of records for a particular species and expressed based on abundance indices as rare, occasionally occurring, common, or abundant. A higher ecological analysis (i.e., alpha or beta species diversity) can then be generated from these data sets. A detailed review on the different diversity indices used for myxomycetes is reported in another chapter within this book.

High School and Beyond

The activities with myxomycetes discussed previously can be adapted to almost any age of student; however, high school students enrolled in advanced courses in preparation for college and undergraduate college students have to adhere to specific course outlines and expectations. Most high school biology courses demand an inquiry approach to learning, and aim to introduce students not only to the principal processes governing living organisms, but also to the process of scientific investigation through laboratory investigation. In our experience, work with myxomycetes as model organisms has been demonstrated to be successful in biology classes leading to the US College Board Advanced Placement (AP), United Kingdom Advanced Level Biology (A-level), and International Baccalaureate Higher Level Biology (IB) examinations. The last curriculum includes an obligatory Group 4 project that involves a collaborative and interdisciplinary group activity to which myxomycete studies lend themselves perfectly. Some ideas given to students for study projects include a comparison of the assemblages of myxomycetes associated with different species of plants, myxomycetes found on the bark of living trees versus leaf litter, the effect of tree age on the myxomycetes present, the effect of different orientation or height of the tree bark, a comparison of bare bark and bark with various epiphytes (mosses and lichens) present, the effects of abiotic parameters, such as pH, temperature, or water availability, and many more. Some of the topics in which myxomycetes have been used at the high school level, either as example organisms or for practical activities, are listed in Table 12.2. However, the possibilities are limited only by the imaginations of the students and their teachers.

Many of these topics are also suitable for college-level research projects. A comprehensive review of ways to use myxomycetes in education and research is provided by Keller and Braun (1999). The authors not only give a list of possible laboratory exercises suitable for study at the high school level but also many research suggestions for college level and beyond. Among other literature reviews and useful suggestions for studies at these higher levels are those given by Ashworth and Dee (1975), Ing (1984), and Stephenson (1985).

There have been many doctoral dissertations based on studies of myxomycetes. Some prominent examples include those by Gray (1938), Wollman (1966), Keller (1971), and Lado (1984). More recently, many others have completed their PhD studies on the myxomycetes, including the first two coauthors (e.g., dela Cruz, 2006; Pando, 1994; Rojas, 2010; Winsett, 2010). Most research grants now either require or are favored by the inclusion of an educational component. For example, several PhD students are or have been included on the team of scientists involved in the Myxotropic project (http://www.myxotropic.org/home) funded by the Spanish government. In addition, students from the host countries have been involved in every collecting trip of this and previous projects, and have been exposed to both instruction in and the practical experience of collecting strategies and techniques, recognizing species, curation of specimens, and recording protocols used by the members of team. This type of outreach has also been a major focus of US National Science Foundation (NSF)-funded collecting expeditions to countries around the world (http://slimemold.uark.edu), and has frequently involved formal lectures, presentations, and workshops to local students in the host country.

Workshops given by specialists are an important and cost-effective way to train parataxonomists. This training was a major focus of the Planetary Biodiversity Inventory project based at the University of Arkansas and supported by the NSF (http://slimemold.uark.edu), in which graduate students participated in collecting trips to localities throughout the world. Parataxonomists play a very important role in building the knowledge base on which data sets and analyses are ultimately based. Moreover, although the primary focus of the Planetary Biodiversity Inventory project was to collect and record myxomycetes and other eumycetozoans to elucidate their ecology and biogeographical distribution, the project generated excellent educational materials that are still available on the website listed earlier. The two editors of the present volume have been responsible for a number of myxomycete workshops in regions of the world ranging from several countries in Central America to Australia, Thailand, Kenya, Namibia, and Vietnam (Fig. 12.3).

These workshops have involved several different types of participants. In some instances, only students were involved, but other workshops have involved both students and their teachers, as has been the case at the University of Arkansas (Fig. 12.4). Workshops can be carried out in a laboratory or classroom setting, and sometimes circumstances (i.e., limited facilities and the time slot available) dictate that this has to be the case. However, the educational experience of the students is considerably enhanced if a field component is possible (Fig. 12.5). On two occasions (2003 and 2016), a "myxoblitz" has been held in the Great Smoky Mountains National Park. The purpose of the "myxoblitz" was to see how many species of myxomycetes could be recorded from several localities during a period of time that ranged from a few hours to several days. The "myxoblitz" was advertised, and the only requirement for participating was an interest in doing so. This type of activity represents an exceedingly worthwhile way of introducing myxomycetes to a diverse range of participants of all ages and backgrounds. The individual who might never have seen a myxomycete in the field before may have the opportunity to work alongside a leading authority in the group (Fig. 12.6).

FIGURE 12.3 Participants preparing moist chamber cultures as part of a workshop in Kenya.

FIGURE 12.4 Myxomycete workshop involving teachers and students held at the University of Arkansas.

FIGURE 12.5 Fruiting bodies and plasmodium of a myxomycete observed in the field during a workshop held in northern Thailand.

FIGURE 12.6 Participants in a "myxoblitz" held in the Great Smoky Mountains National Park.

Apart from their use in formal education activities carried out in schools and universities, myxomycetes are important model organisms for education in a much broader sense, for enjoyment and for conservation. The more local groups and communities understand the interdependence of all life, especially organisms, such as myxomycetes, located at the base of food chains and so vitally important in nutrient recycling, the less global problems will deepen. The health and resilience of natural habitats are dependent on microorganisms, and acquiring an appreciation of their vital roles should help to protect them. Training local people in areas of the world where information is scarce or lacking, and training wardens and resource managers at national parks and forested areas to recognize and appreciate these essential members of every ecosystem, is of fundamental importance for conservation.

Another area of education involves groups of amateur enthusiasts who make up nature societies and clubs all over the world. They often maintain archives and curated collections, and generate publications of considerable importance. Their work, expeditions (even if local), and meetings have brought about landmark discoveries in many areas of science, including in the study of myxomycetes. One such club is the Essex Field Club in the United Kingdom, of which Charles Darwin and Alfred Russel Wallace were honorary founder members (http://www.essexfieldclub.org.uk/). Both Arthur Lister and his daughter Gulielma (Chapter 2) were active early members of this club and developed a global network of researchers and enthusiasts with whom they exchanged material and information over a period of many years. Another example of a large association of such clubs that has greatly increased knowledge and education of myxomycetes is the Fédération Mycologique et Botanique Dauphiné-Savoie (http://fmbds.org/) in France, which is dedicated to raising awareness of mycological and botanical knowledge. Its annual international meetings on nivicolous myxomycetes, those species that develop near the melting snow, are legendary. This group also generates publications, such as the "Bulletin Mycologique et Botanique Dauphiné-Savoie," and hosts exhibitions, meetings, and workshops.

Developing Myxomycete Infomaterials for Promoting Biodiversity Conservation

"Charismatic" species of plants and animals often become poster organisms for informational material directed toward promoting or raising awareness of the need for biodiversity conservation. Species that are not so popular with the public have rarely been used, and this is certainly the case for those that are minute and invisible to the naked eye. However, the intricate fruiting body structures of myxomycetes when magnified and photographed offer a remarkable "subject" for postcards, posters, and book covers. When developed, photo guides can be used as an informal tool for identifying common species of myxomycetes in a given park or other locality. For example, Macabago and dela Cruz (2012) developed a myxomycete photo guide as a teaching tool in the Philippines. To prepare this, common species of myxomycetes were chosen from those recorded in a given study area. The specimens were photographed and arranged in a single PowerPoint slide. In addition, a scale bar was included for easy reference to the size of the specimen. Some basic taxonomic description of the species was provided directly opposite the photograph on a second PowerPoint slide such that on a printed card the accompanying information would be directly behind the image of the species of myxomcyete. The two slides could be printed as a single, back-to-back miniposter or postcard, which could be used as a quick visual guide, as shown earlier in Fig. 12.2.

An interesting visual educational tool to introduce younger children to myxomycetes was devised by de Haan (2005) as a comic strip. The adventures of "Mike the Myxo" (Fig. 12.7) related much of the life history and characteristics of a myxomycete in the context of an amusing illustrated story that can be used to raise interest and to teach the main characteristics of the group. The fruiting bodies of myxomycetes have been referred to as "miniature works of art" found in nature, and many artists have been attracted to them. Some of the sketches and paintings that these artists have produced reveal the inherent beauty of the fruiting body, and this catches the attention of even the layperson with no previous exposure to these organisms. An example of such a sketch done by the artist Angela Mele can be seen in Fig. 12.8.

Photographing Myxomycetes

At any age, from elementary school to adult, interest in this fascinating group of organisms can be captured by using images of different stages of their life cycle. The methods used to obtain images of the unicellular stages have been briefly discussed in an earlier chapter. However, the elegant beauty and design of spores, capillitium, or even the entire fruiting body of smaller myxomycetes can be achieved using scanning or transmission electron micrographs. These images are very costly and more frequently used in research to resolve complex taxonomic questions and problems, but they could be used more generally to raise awareness and interest. In addition, digital technology can now enable anyone to produce high-quality images (and even video) with relatively simple and inexpensive equipment.

The small size of the fruiting bodies of most myxomycetes can make photographing them quite challenging. However, with the acquisition of a few easily purchased items of equipment and/or the use of several commonly available, inexpensive items, good images can be obtained. Since the fruiting bodies are immobile, this allows images to be taken without fear of the subject moving. Images can easily be taken with a simple digital camera without motion blur, as long as the substrate and camera are stable (Fig. 12.9). A cable release, or using the camera's timer, eliminates the possibility of camera movement due to pressing the shutter button. Expensive light-gathering lenses are not necessary, thus allowing any lens to be used. Even cell phones can and have been used to take very good images. A sturdy tripod is optimal for stability, but anything that

FIGURE 12.7 "Mike the Myxo" as drawn by Myriam de Haan.

allows the camera to remain stationary will work. A backpack makes a satisfactory replacement, as does a folded jacket or a sock filled with rice, beans, or sand. A macro lens will allow one to get much closer to the subject, but there are also other lens attachments that allow closer focusing (Fig. 12.10).

Images should be taken with the lowest ISO setting to ensure there will be as little grain (noise) as possible. The aperture should be set to give the most satisfactory amount of subject focus. This setting can be adjusted to obtain results that are the most appealing. Shutter speed can be increased or decreased to

FIGURE 12.8 **Drawing of the myxomycete *Collaria arcyrionema* as done by the artist Angela Mele.**

FIGURE 12.9 **Photographing myxomycetes with a relatively simple digital camera.**

FIGURE 12.10 **Randy Darrah using a more sophisticated camera to photograph a myxomycete.**

give the best subject exposure. A light source, such as a flash, sunlight, or small flashlight, can be used to provide additional illumination and emphasis of the subject.

Samples that are collected, brought back from the field, and worked with indoors provide even more control of the variables involved in photography, allowing the opportunity for even better results. A stereoscope or microscope with a built-in camera or an adapter to attach a camera or video camera can be used to capture small details of the myxomycete fruiting body and its structures. Moreover, many of today's digital cameras can synchronize with the Internet, allowing easy and rapid sharing of images with others.

CONCLUSIONS

The potential for using myxomycetes in education is enormous. As a safe, nontoxic, hypoallergenic living organism, its place in the classroom of even the very young student is virtually unrivaled. The myxomycete as an example of a true microorganism that is sometimes visible to the naked eye, with observable cyclosis, feeding behavior, a complex life cycle consisting of several stages, and a model of synchronous mitosis, is unique. The relative ease of laboratory culture in large quantities of some members of this group makes it an ideal candidate for many higher research possibilities as discussed herein and in earlier chapters. The sheer beauty and elegance of some myxomycete fruiting bodies make them a natural focus of interest and wonder, and their indisputable importance in the correct functioning of ecosystems only underlines the need to teach with them and about them whenever possible. Some of the educational opportunities afforded by these characteristics, particularly those used by the coauthors, have been described in this chapter. It is not intended to be more than an example of our experiences and of the didactic potential to encourage other educators to consider the advantages of using these organisms. We have also given examples of literature pertaining to educational uses, but it is in no way intended to be an exclusive list, and there are probably many other examples of which we are unaware.

ACKNOWLEDGMENTS

We would like to thank Mrs. Susan Allen of the American School of Madrid, Spain, who designed and carried out the second-grade project described herein. Appreciation is extended to Myriam de Haan for allowing us to include an image of "Mike the Myxo" and to Angela Mele for our use of one of her sketches of myxomycetes. Gratitude is expressed to Randy Darrah for providing the information on photography included herein.

REFERENCES

Ashworth, J., Dee, J., 1975. The Biology of Slime Molds. In: Arnold, E. (Ed.), Institute of Biology's Studies in Biology. London.

de Haan, M., 2005. The Adventures of Mike the Myxo. Royal Antwerp Mycological Society, Antwerp, Belgium, English translation aided by Henry Beker.

dela Cruz, T.E.E., 2006. Marine *Dendryphiella* species from different geographical locations: an integrated, polyphasic approach to its taxonomy and physioecology. Dissertation, Technical University, Braunschweig, Germany, 195 p.

dela Cruz, T.E., Pangilinan, M.V., Litao, R.A., 2012. Printed identification key or web-based identification guide: an effective tool for species identification? J. Microbiol. Biol. Educ. 13, 180–182.

Gray, W.D., 1938. The effect of light on the fruiting of myxomycetes. Am. J. Bot. 25, 511–522.

Haskins, E.F., 1973a. *Echinostelium minutum* (Myxomycetes). Amoebal Phase. Encyclopaedia Cinematographica. Film E 1816 des Institut für den Wissenschaftlichen Film. University of Göttingen, Göttingen, Germany.

Haskins, E.F., 1973b. *Echinostelium minutum* (Myxomycetes). Plasmodial Phase (Protoplasmodium). Encyclopaedia Cinematographica. Film E 1817 des Institut für den Wissenschaftlichen Film. University of Göttingen, Göttingen, Germany.

Haskins, E.F., 1974. *Stemonitis flavogenita* (Myxomycetes)—Plasmodial Phase (Protoplasmodium). Encyclopaedia Cinematographica. Film E 2000 des Institut für den Wissenschaftlichen Film. University of Göttingen, Göttingen, Germany.

Haskins, E.F., Wrigley de Basanta, D., 2008. Methods of agar culture of myxomycetes: an overview. Rev. Mex. Micol. 27, 1–7.

Ing, B., 1984. Myxomycetes in biology teaching. J. Biol. Educ. 18, 277–285.

Keller, H.W., 1971. The genus *Perichaena*: a taxonomic and cultural study. PhD thesis, University of Iowa, Iowa City.

Keller, H.W., Braun, K.L., 1999. Myxomycetes of Ohio: their systematics, biology, and use in teaching. Ohio Biological Survey Bulletin XIII NS, Columbus, OH.

Keller, H.W., Everhart, S.E., 2010. The importance of myxomycetes in biological research and teaching. Fungi 3, 13–27.

Koevenig, J.L., Alexopoulos, C.J., Martin, G.W., Porter, T.R., 1961. Slime Molds. Iowa City: State University of Iowa. 8 p. Slime Molds I: Life Cycle. U-5518. Slime Molds II: Collection, Cultivation, and Use. U-5519. Slime Molds III: Identification. U-5520.

Lado, C., 1984. Estudio taxonómico, florístico y corológico de la clase myxomycetes en las provincias de Ávila, Madrid y Segovia (España peninsular). PhD thesis, Universidad de Alcalá de Henares.

Macabago, S.A.B., dela Cruz, T.E.E., 2012. Development of a myxomycete photoguide as a teaching tool for microbial taxonomy. J. Microbiol. Biol. Educ. 13, 67–69.

Martin, G.W., Alexopoulos, C.J., 1969. The Myxomycetes. University of Iowa Press, Iowa City.

Neubert, H., Nowotny, W., Baumann, K., 1993. Die Myxomyceten Deutschlands und des angrenzenden Alpenraumes unter besonderer Berücksichtigung Österreichs. Band 1: Ceratiomyzales, Echinosteliales, Liceales, Trichiales. Karlheinz Baumann Verlag, Gomaringen, Germany.

Neubert, H., Nowotny, W., Baumann, K., 1995. Die Myxomyceten Deutschlands und des angrenzenden Alpenraumes unter besonderer Berücksichtigung Österreichs. Band 2: Physarales. Karlheinz Baumann Verlag, Gomaringen, Germany.

Neubert, H., Nowotny, W., Baumann, K., 2000. Die Myxomyceten Deutschlands und des angrenzenden Alpenraumes unter besonderer Berücksichtigung Österreichs. Band 3: Stemonitales. Karlheinz Baumann Verlag, Gomaringen, Germany.

Pando, F., 1994. Estudio de los mixomicetes corticícolas de la España peninsular e islas Baleares. PhD thesis, Universidad Complutense de Madrid.

Poulain, M., Meyer, M., Bozonnet, J., 2011. Les Myxomycètes. 2 vols. Fédération mycologique et botanique Dauphiné-Savoie, Sevrier, France.

Rojas, C.A., 2010. Biogeography and microhabitat distribution of myxomycetes in high-elevation areas of the Neotropics. PhD thesis, University of Arkansas, Fayetteville, AR.

Stephenson, S.L., 1985. Slime molds in the laboratory II: moist chamber cultures. Am. Biol. Teach. 47, 487–489.

Stephenson, S.L., Stempen, H., 1994. Myxomycetes: A Handbook of Slime Molds. Timber Press, Portland, OR.

Winsett, K., 2010. Intraspecific variation in two cosmopolitan species of myxomycetes, *Didymium squamulosum* and *Didymium difforme* (Physarales: Didymiaceae). PhD dissertation, University of Arkansas, Fayetteville, AR.

Wollman, C., 1966. Laboratory culture of selected species of myxomycetes with special reference to the gross morphology of their plasmodia. PhD dissertation, University of Texas, Austin, TX.

FURTHER READING

Farr, M.L., 1981. How to Know the True Slime Molds. Picture Key Nature Series. William C. Brown Co., Dubuque, IA.

Gray, W.D., Alexopoulos, C.J., 1968. Biology of Myxomycetes. Ronald Press, New York, NY.

Chapter 13

Myxomycetes in the 21st Century

Carlos Rojas*, Tetiana Kryvomaz**
**Engineering Research Institute, University of Costa Rica, San Pedro de Montes de Oca, Costa Rica; **Kyiv National Construction and Architecture University, Kyiv, Ukraine*

CONNECTION WITH THE WORLD

The myxomycetes have been formally studied since the middle part of the 17th century (Stephenson et al., 2008). Although the biology of the group has not been documented to the extent that is the case for most groups of vertebrates and plants, there is a good baseline of information relating to the fruiting body stage of myxomycetes. As noted in other chapters of this volume, this is more easily observed for taxonomic and basic ecological aspects. However, the communication of the accumulated information on myxomycetes has been conducted primarily through formal scientific and academic channels, such as articles, volumes, scientific notes, and university courses. In fact, there is a good chance that most of the readers of this volume first learned about myxomycetes directly from any of these strategies of knowledge transmission.

Recently, with the establishment of the Internet as the most important platform for global communication (Gubbia et al., 2013), there has been a fantastic development of high quality websites, blogs, social media platforms, and mobile apps for different groups of living organisms. This natural expansion of the technology-based modern lifestyle to the scientific world has created new paths for scientific communication and has changed the way our societies transfer knowledge (Cummings and Teng, 2003). Some authors have even stated that the Internet has modified the neurobiological and cognitive pathways of the human learning process (Roth and Dicke, 2005). These changes have permeated science and allowed for a rapid growth of nonscientists performing basic scientific assessments. For instance, the more prominent role of citizen science projects, most of which are based on emotional connections with nature promoted by multiscale information transfer, is highly relevant in the current structure of scientific research (Boney et al., 2014).

Myxomycetes: Biology, Systematics, Biogeography, and Ecology. http://dx.doi.org/10.1016/B978-0-12-805089-7.00013-5

In this modern context, it is understandable that effective communication seems to be a partial function of the popularity-based commercial background associated with most of the "free Internet" available to the public at large, since companies rapidly recognized the power of such a communication channel. In this manner, the more individuals promoting a specific topic, usually due to its commercial value, the more likely that such a topic would survive for longer on the Internet. We all have seen examples of the latter in the form of straightforward, short, "viral" pieces of information. Scholars have documented that most individuals do not remain focused on one webpage when "surfing the Internet" for longer than 10–20 s, unless they find something interesting (Gong et al., 2012), and this pattern has modified the way messages are communicated. However, just how easy is it to create such effective messages with information on living organisms? The average person is already very much aware of many different kinds of macroorganisms, ranging from trees to household pets. What about microorganisms to which most individuals cannot relate? Most academic communication strategists are still debating on the effectiveness of the different styles of scientific literacy approaches (MacDermott and Hand, 2015). However, there is a large consensus on the fact that Internet-based strategies should be a key element in any type of communication plan that is developed. That does not imply that communication should only take place in this platform because written works and classical human interaction would always be imperative (Raber and Richter, 2008), but it demonstrates that the Internet should be present in the desired plan of communication. Nevertheless, it is important to keep in mind that modern online-based communication is designed in short, straightforward, highly stimulating pieces. In this sense, myxomycetes have the advantage of forming beautiful fruiting bodies that when photographed or recorded with special techniques can communicate strong biophilic messages about the microscopic world around us.

The coauthors of this volume and myxomycete specialists worldwide have used their own ways to transfer accumulated knowledge in spite of their own limitations in the field of human communications. That is simply one of the shortcomings of the academic training in science. Despite that, some have even developed websites and online resources devoted to the myxomycetes and the dissemination of information around and beyond their scientific projects. In fact, most of the myxomycete specialists worldwide at some point in their lives have become teachers, either in formal academic venues or in informal ones. Also, the present volume was developed as an effort to maintain the natural process of communicating condensed information on a group of microorganisms for which a comprehensive text had not been published for almost a half century. What all this means is that very likely there has never been any point in history when such an active attempt to communicate information on myxomycetes has existed as at present. However, it is the responsibility of all science communicators, teachers, naturalists, and myxomycete professionals to understand that integrated messages are also much more effective than simple monothematic ones (Thorson and Moore, 1996). In this manner, it is not the same to provide a message on the taxonomy of myxomycetes, whether it is classical or molecular-based, by itself as opposed to integrating such a message in the context of technique development, the relevance of scientific consensus through debate, and why, in the end, for a nonscientist, taxonomy and myxomycetes ultimately matter as well.

A relevant point to consider in this discussion is that myxomycetes are an excellent gateway for individuals to enter the world of microscopic organisms (Keller and Everhart, 2010). The fruiting bodies of many species are simply beautiful, their life history intriguing, and their potential still understudied. Since research is driven by curiosity, these organisms are fantastic for integrating such elements of scholarly dissemination of information along with modern multilevel (i.e., political, social, and ecological) trends.

During the past few decades, myxomycete specialists, as well as many other biologists specializing in various other taxonomic groups, have faced a common problem. The number of individuals, particularly students, interested in classical aspects of biology has decreased, or at least that seems to be the general consensus. The traditional strategy to counterbalance such a phenomenon has been to reach out and develop different types of workshops, hands-on activities, and field-based forays. The latter has resulted in an increasing number of "amateur" groups of individuals, some of whom possess very specialized knowledge of taxonomic aspects of the more prominent groups of organisms. However, this trend is highly dependent on many socioeconomic variables that allow individuals in different regions of the world to show those levels of organization. In this manner, the relevant question is how efforts can be socioeconomically balanced out when most of the world does not have the much-needed, respective specialists or communication facilitators to "plant the seed" and mobilize such initiatives in the first place? This core issue seems to be a good stage for integrated global communication approaches, such as online courses and video-based online strategies. However, for a section of the world where the Internet is a simple fact of life, it is hard to remember that there is still a large section of the human population for which these strategies are technologically limited. These are the cases in which classical approaches still matter and the reason why educators should not give up on using traditional techniques. As we have argued, the biophilic relationships that some individuals can establish with organisms, such as the myxomycetes, may be relevant in the process of communication.

Along with the problems associated with communication channels, language barriers can impose an obstacle to the strategies of communication used to disseminate information on myxomycetes. As is the case for most groups of living organisms, English is the language most often used for scientific publications. Even though some popular works and several scientific works on myxomycetes have also been published in other languages, most of the important literature on myxomycetes is still in English. This shortcoming implies that local researchers in different parts of the world also have some responsibility in translating and communicating the mainstream information on myxomycetes in their local languages. Only with a multilanguage effort effective communication can be reached at larger geographical scales. In this regard, noteworthy achievements are represented in the works of Neubert et al. (1993, 1995, 2000), Hagiwara and Yamamoto (1995), Lado and Pando (1997), Poulain et al. (2011), and others, which in addition to being based on many decades of field research on myxomycetes, have attempted to communicate myxomycete information in languages other than English for regional populations of readers.

Despite the different strategies of message communication, it seems that the most effective way to transfer information is with mixed approaches (Jewitt, 2012) coming from committed groups of individuals, specialists or not, and directed toward one group of organisms. In this sense, citizen-science approaches could be an effective way to develop complex networks to share information on myxomycetes, given their multiple characteristics for biophilia-based (Wilson, 1984) integration with several aspects of the human lifestyle (i.e., they are simply beautiful and we want them around). After all, myxomycetes can be observed and detected in the field by amateurs and specialists, and basic training could be accomplished with committed individuals at a comparatively low cost. Such strategies should be one of the relevant topics of conversation in professional meetings and nonacademic forums for common strategies and consensus to be reached. This is the way to capture the attention of groups of policy makers who can match the scientific and popular developments relating to one group of organisms, such as the myxomycetes, with integrated approaches to manage nature that would consider other microorganisms, as well. In this end, when popular pressure exists toward the conservation of biological resources, particularly in social media, political groups seem to respond. The latter has effectively worked for common global issues, such as climate change and management of natural resources, and it has been effective for NGO-based efforts directed toward flagship species. However, there is still much work to do to make individuals understand that nonvisible microorganisms that may not have the marketing capability of larger organisms do perform an important role and even have a high value for the dynamics of biological systems. In fact, there is still much work to do to make the public generate hedonic relationships with noneconomically important organisms. After all, not everything has to have a visible value.

In this sense, outside of classrooms and laboratories not much is known about myxomycetes. Despite this, individual efforts from professionals, naturalists, academic groups, and organized individuals have still attracted attention toward myxomycetes. Their aesthetic beauty has been an aspect of fascination for the nonspecialized public and several volumes on groups historically linked with myxomycetes have been illustrated with images of these organisms. As an example, Fig. 13.1 shows the alternative option for the cover of the present volume and demonstrates the beauty and versatility of myxomycete images, as well as their associated biophilic message with respect to the world of microscopic

FIGURE 13.1 Alternative cover volume design evaluated for the present volume. The image corresponds to a *Diderma miniatum* collected by Dr. Carlos Lado in Peru and photographed by Carlos de Mier in the Royal Botanical Gardens of Madrid, Spain.

organisms. There is no doubt that nonspecialized eyes seeing such an image (or the selected cover of this volume) would still find it appealing and interesting.

However, when talking to a layperson about locally known plants or animals, it is still much easier to relate to what they already know than with such cryptic topics as microorganisms. In fact, personal accumulated knowledge on microscopic organisms is likely very low for most the world population. This is ironic, since microbial systems are essential for the existence of specific ecosystems (Pointing and Belnap, 2012), commercial products (Velivelli et al., 2014), and lifestyles (Kramer et al., 2013). For example, myxomycetes have been recently reported to absorb heavy metals (Rea-Maminta et al., 2015) and thus have been proposed as bioremediators, but most individuals ignore this information. The widespread low level of microbial literacy is a constraint for the true integrated development of applied strategies, and in developing countries the valuation of microbial-based ecosystem services seems to be a black box still waiting for attention. We are not proposing a commercially based approach to address myxomycetes since their real value, as with several other groups of organisms on the planet, may simply be related to their very existence. However, such an applied field of knowledge is still in a very basic stage.

Such a constraint has been detected by both economists and biologists and even though some individuals would not agree on a quantified economic strategy based on the concept of natural capital, this concept has allowed economic theory to enrich the models used by natural science and particularly conservation specialists. In this way, biological systems are conceptualized as organizations of sinks, sources, and flows or quantifiable energy that permit human social systems to utilize both goods and services from nature. Of course, the key element is the maximization of the human-nature interaction in terms of sustainability and intelligent, responsible use of resources. However, how accurate such models can be in systems where microbial ecosystem functions are poorly documented? What about myxomycetes, whose trophic stages are extremely poorly studied (Chapter 9)? To cope with such limitations, the elaboration of interesting ideas for human development toward balancing conservation and utilization or resources has been an active task of policy makers in the past few decades.

For instance, the economic strategy for more equal development known as the triple helix (Leydesdorff, 2013), which calls for three sectors of society—academia, government and industry—to be engines of development has gained recent popularity in some countries. Most developed countries in the northern hemisphere have understood such a concept for decades and even though the system has weakened due to globalization and free-market dynamics, it still somehow works in these areas. The problem in developing regions is that the lack of long-term political, economic, and social stability has not permitted a true development of academy-based research, social outreach, and scientific empowerment in political spheres. In these areas, more than in developed countries, private industries are disconnected with universities and local governments. The latter are expected to perform an auditing role on the former while universities are expected to do the same on both governments and industries.

Of course, the true development of a scientific path moves forward more rapidly when teams of organized individuals, independent of their origin and approaches, share common interests and work together toward achieving a goal. The accumulation of information presented in this volume is an example of the latter, since during the last 50 years or so, many individuals from different parts of the world have permitted the fastest rate of accumulation ever documented. However, as mentioned before, such a community of active researchers and individuals interested in myxomycetes still lacks the political leverage to introduce this taxonomic group of microorganisms in agendas outside the scientific environment. This is not surprising, taking into consideration that a larger and much more involved group of organisms in human development, such as the fungi, are also ignored for the most part. In this sense, it is imperative for everybody dealing with myxomycetes to use as many communication channels as possible and incorporate the group into as many different agendas as possible. For a true integrated strategy of natural resources management, sustainability and economic development, it is necessary to understand that microbial communities perform a unique role on the earth and that even though undocumented roles such as the ones performed by myxomycetes do exist, that does not imply that those organisms lack importance for ecosystem functioning.

RESEARCH PERSPECTIVES

Most of the research directed toward myxomycetes has been carried out in universities and research centers associated with academic institutions. However, extremely valuable input has been provided for a number of years by individuals who have not been affiliated with academia and yet have devoted their time to collecting, identifying, and understanding myxomycete relationships with the environment. Over time, very important contributions have been provided by both groups, and most myxomycete collections contain several records from both, as well. In this sense, it is important to acknowledge the role of nonacademically affiliated myxomycete researchers for the construction of hypotheses and generation of information about the group.

The modern constraint associated with the prominent role of researchers who are not affiliated with academic institutions is that they usually cannot apply for public funds and thus are limited to generating smaller scale datasets; they are more inclined to continue their research within a specific field of action (i.e., taxonomy) and they are usually not encouraged to train younger researchers. This is particularly relevant in areas of the world where myxomycete research is absent or slow since the process of generational change would reach only those younger researchers geographically close to these potential mentors. In contrast, when researchers from academic institutions obtain funds, usually through a network of collaborators, some money can be invested in training newer generations of researchers from different parts of the world and in connecting several disciplines outside the biological sciences. Given the reduced number of funding opportunities in the current global context, most academic myxomycete researchers tend to form strong international networks that extend by default their scale of action and influence. One of the most prominent examples of the latter has taken place in the past decade at the University of Arkansas in the United States, where a generation of researchers from North and South America, Africa, Southeast Asia, and Europe has been completely or partially trained. In this sense, the role of academically affiliated researchers may be more important for developing standard methodologies and continuing the existence of international networks. Either way, all valuable inputs for the generation of myxomycete information should be recognized, understanding the limitations of contextual elements.

What is interesting in the historical development of the generation of information on myxomycetes is that the internationalization of research seems to have been present for a long time. As has been the case for many groups of living organisms, the first documented processes of myxomycete study took place in Europe and rapidly transferred to North America (Lado and Wrigley de Basanta, 2008). After that, both European and American researchers started carrying projects in areas of the world that had been undocumented at the time; these included Africa and Latin America. Later, the Far East, particularly Japan, became an important source of information on myxomycetes and more recently, individuals in Southeast Asia have activated research in that part of the world. At this point, most of the ecological systems on Earth have been documented in one way or another and even though there are still geographical gaps to fill in, there is a "good" idea of the distribution of myxomycetes, based on the occurrence of fruiting bodies throughout the world (Fig. 13.2).

The internationalization of myxomycete research has taken a different shape in the last decades and turned into a collaborative effort directed toward the goal of understanding the phylogenetic position and molecular relationships among groups within the myxomycetes. This has been very important for the development of new hypotheses relating to the group but has created a modern technological and infrastructural gap between researchers based in developed and developing areas of the world. Since this phenomenon does not apply only to myxomycetes but to many other groups of organisms, some governments have created legislation to protect their own intracellular diversity, the potential use of the information contained at that level, and the intellectual property rights of local researchers.

Developing countries have been particularly active in regulating "access to biodiversity." This is a concept understood in an international legal framework as "access to biochemical, molecular, and genetic information," based on claiming moral rights associated with geographical location. The Nagoya protocol, sponsored by the Convention on Biological Diversity (UNEP, 2011),

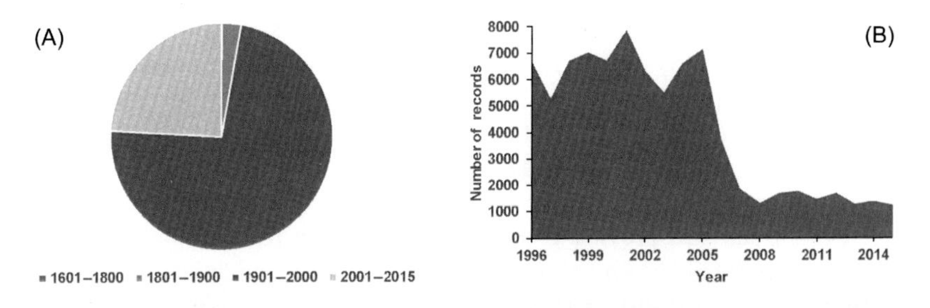

FIGURE 13.2 Myxomycetes have been studied since the middle of the 17th century, but most of the records included in the GBIF platform (A; www.gbif.org) were collected in the 1901–2000 period. During the most recent years, there has been a decline in the number of records in GBIF (B) even though the research activity has not declined. The distribution of genera, such as *Physarum* (C), clearly shows a cosmopolitan range even though several species, such as *P. echinosporum* (Chapter 8), have limited distribution (www.discoverlife.org).

which started in 2014, is one example of the latter. Up to now, this protocol has been signed by over 80 countries in the world, mainly in Latin America, Africa, Europe, Australia, and the Asian continent, and it has created a legal framework for the sharing of economic benefits product of bioprospecting and marketing of biodiversity. The issue observed by some individuals (Richerzhagen, 2014) is that this protocol also limits the capacity of foreign researchers to operate projects in signatory countries, since it does establish that local counterparts should be part of the project.

Modern myxomycete research should take place in this specific environment, which differs dramatically from the setting of any other time in history. Only a few years ago, the regulatory framework for biological research, particularly examples focusing on the generation of molecular information, was more loosely established in most countries. However, even for basic routine tasks, such as movement of collections from one country to another, most internal regulations establish the signing of bi- or multiparty research agreements, material transfer agreements, and the previous signing of informed consents between researchers and the owners, public or private, of the areas where research has taken place. The reality is that, in a similar manner to other types of researchers for other groups or organisms, most myxomycete scientists operate behind legal operative international regulations. This is one aspect that obviously must change in the development of research projects on myxomycetes worldwide, not only for moral reasons, but more importantly because the limited funding agencies that can still provide grants for myxomycete research have already complied, or will likely do so, with such international guidelines. However, it is important to critically analyze the role of such international agreements in terms of hindering or promoting research in the first place. In this manner, the responsibility of the younger generations of myxomycete researchers is to understand the current stage of international research to practice and transmit to their students/apprentices good practices for the implementation of international team–based research and the development of critical thinking skills.

In addition to such legal international limitations, one aspect that has been prominent in the history of myxomycete research is the classical monothematic approach taken by most researchers. There are limited numbers of scientific articles and notes on myxomycete research from any other perspective than the biological one. With the notable exception of some modern investigators who have developed applied research using myxomycetes in the fields of civil engineering, computer sciences, medicine, pharmacy, and alternative energy strategies, there are limited historical examples of myxomycete applications. This volume contains a complete chapter (Chapter 11) devoted to such topics, but alternative approaches on myxomycetes are traditionally not considered by most researchers due to their classical biological background. In this sense, there is a whole field of work for the future in these directions if different groups of individuals receive information on myxomycetes and learn about their fascinating life cycles, system relationships, and potential for various applications. At least for some parts of the world, the basic surveying and accumulated knowledge on the myxomycete biota present within political or natural boundaries already exists, and more applied projects for human development using myxomycetes could be designed. Even though this is not the case for most of the world, given the technological gaps between developed and developing countries already discussed, the latter research path may represent an alternative for some groups of researchers based in developing countries and for multidisciplinary approaches to be executed in these areas. It is important to mention that all or most of the currently integrated ecological methodologies can be applied to the study of myxomycetes.

However, the development of intraacademic merging and collaboration should be promoted in the first place for this type of proposal to be maturely designed and academically supported. The reality is that within academic institutions, particularly those of a comparatively large size, most scholars are disconnected from the research efforts of other individuals within the same institution. Similarly, it is very hard for most individuals to keep up with scientific developments with all of the multiangled social stimulation provided by the modern lifestyle. It would be very difficult to produce a general communication channel that could keep individuals interested in myxomycetes for long periods of time, but it is probably not so difficult to generate such channels for specific social subgroups. In this sense, a multidisciplinary approach integrating specialists in other disciplines of human knowledge could help the more classical myxomycete scientists disseminate their work in more creative and effective ways. Whether or not such a strategy could create a local "snowball" effect would be determined by the context, but most myxomycete specialists, academic or not, are recognized in their particular workplace, regardless of the general interest that other colleagues may have on myxomycetes.

Whatever strategies for the development of future research lines may be generated, in the modern context myxomycete specialists have recently been and will likely continue to reach out to colleagues in other disciplines to design more creative projects as well. This type of skill for a multidisciplinary approach is important to transmit to the younger generation of myxomycete specialists, in a manner that would give equal importance to the transfer of knowledge at both the information and the disciplinary formation level. For a group of organisms, such as the myxomycetes, which will likely continue attracting much less attention than their vertebrate or botanical counterparts, it is imperative to provide a much more integrated conceptual framework for research to take place. For example, it may be very attractive for younger researchers to begin using modern techniques to generate data and produce popular articles in spite the fact that there are still many basic questions on myxomycetes that have not been answered. In this way, a balance between the two fields of action is necessary, since the key to biological comprehensiveness is the integration of results through theoretical elements. In the end, science is the activity responsible for the identification of patterns in nature, and this can be substantiated only from many different angles when researchers work from several different perspectives.

In this sense, it is important to understand that the advancements, both theoretical and experimental, generated for a group of organisms are a byproduct of the intensity and effort of work over time. Since myxomycetes have not been studied with the intensity of other groups of organisms, there are still several basic questions to address, and there are still many possibilities to readdress questions with modern techniques and higher resolution datasets, particularly with the extremely understudied tropic stages. This is currently an active approach in the taxonomic and phylogenetic areas of myxomycete research, but it could easily be applied to ecological and biogeographical fields, for example. For some very well studied parts of the world, perhaps this approach is the next logical step.

The effective execution of myxomycete research projects depends heavily upon two main aspects: namely, human expertise and infrastructural/equipment capabilities. We have addressed the fact that a large gap between developed and developing areas may exist due to intrinsic modern characteristics of both areas. However, for those developing regions, countries, for example, that may lack a high level of human expertise on myxomycetes, it is crucial for researchers to develop plans for the improvement of conditions and learn from the lessons of colleagues. It is understandable that not all developing regions have the same capacities, but there are always opportunities to address issues and improve research conditions. This does not mean that everybody should carry out the same tasks or do the same work in different geographical regions, and points more toward understanding in which directions the local capability may grow and improve and which directions may not be feasible for future growth. For the latter, it is the establishment of relationships with other complementary groups that could determine the success of the global research endeavors.

In either case, one aspect that is important to remember is that, due to the modern constraints of the research environments, most individual efforts should become local team strategies. Contrary to past research, the modern paradigm of research calls for teams rather than for individuals. In this sense, research networks are essential, and whether they are national or regional, they are very important. However, local-based teams, within the institutions or the organized groups where the effort to study myxomycetes arise, are a key element for the execution of modern comprehensive projects. These teams not only provide an increased effort to generate more robust datasets but an internal leverage that can boost the support to the research initiatives. In this sense, some scholars have pointed out that the formation of a critical intellectual mass on one topic of human knowledge is a key for the correct implementation of plans and ideas (Guest, 1997). When this is implemented at the local level, it may provide the necessary boost for the lead local researchers to move on with their plans for research.

An example of the latter is the well-documented advancement of techniques and protocols for molecular analyses of myxomycetes. Only 2 decades ago, when other living organisms already had baseline work in place, most of the work on myxomycetes was in a very early stage of development. However, the modern trend to look at molecular data for different types of biological analyses created a need to move research in that direction, fostered teams to work together and helped these individuals generate the much-needed ground work that determined the protocols used today. Even though this field of action is still ongoing and new developments are generated every year by different teams of researchers, it took a longer time for myxomycetes to be documented at a level equivalent to the levels of many other groups of organisms. These types of technical challenges, better addressed in other chapters of this volume, have created important limitations for the generation of information on the group. Moreover, as already mentioned, it has also indicated that more applied lines of investigation are still in very early stages of implementation.

CONSERVATION AND MANAGEMENT

Most areas in the world have not yet been well documented with respect to the diversity of myxomycetes (as represented by fruiting bodies) present, the functional diversity of the group is still understudied, and survey efforts are currently not normalized using appropriate standards. In this situation, it is very hard to determine the management status of the different species or groups within the myxomycetes and thus design appropriate methods for their management. In recent years, there have been initiatives to create groups of researchers that can propose guidelines for the management and conservation of myxomycetes. An innovative example was the effort of Schnittler et al. (2011), who put together a red list analysis using a threat and conservation approach for German myxomycetes. In this work, the authors tried to cover the dissemination of knowledge about myxomycete ecology and natural history in Germany, they compiled a checklist of myxomycetes for that country, and attempted to assess multiorigin threats to the existence of the group. However, the current baseline information associated with most biological systems in the world is so basic that it seems impossible to accomplish a similar task for most other regions at the present time. When only the biodiversity aspect is considered, it is still hard to determine for most species of myxomycete whether they are becoming rare, are simply uncommon or are strongly associated with specific biological systems (or geographical zones) and thus not present in most others.

Although some species of myxomycete are currently included in several regional Red Lists, there are several methodological and terminological problems in the evaluation of their conservation status with respect to IUCN criteria (International Union for Conservation of Nature and Natural Resources). The IUCN Red List criteria assign species to categories of extinction risk, using quantitative rules based on population sizes, population decline rates, range areas, and range declines. The IUCN quantitative analyses estimate the extinction probability of a taxon based on its known life history, habitat requirements, threats, and any specified management options. The analysis of population viability is an example of such approach, and it is expected that quantitative analyses make full use of all relevant available data. In a situation in which there is limited information, whatever data are available can be used to provide an estimate of extinction risk. However, the application of the IUCN criteria to myxomycetes in fact generates more questions that the actual applicability of the current methodology, due to the life history and characteristics of the group. For example, how is an "individual" defined for myxomycetes? Can changes in population size be assessed in myxomycetes? How can population growth rates be determined for myxomycetes? What is a "population" in myxomycetes? Is it possible to determine the dispersal ability of species within the group? What is "endemic" and "rare" for myxomycetes?

As we can observe, it is not easy to identify what constitutes an "individual organism" for myxomycetes because their life cycle includes partially documented life cycles, one reproductive stage and two vegetative ones, several morphologies at all different levels and limited documentation of population-based parameters. Should single fruiting bodies or groups of crowded fruiting bodies be counted as one individual? Despite the obvious constraints of such methodology, IUCN approaches require practical protocols as well. Also, spores, plasmodia (except perhaps for protoplasmodia and those that give rise to aethalia and pseudoaethalia), and amoeboflagellates provide evidence of the occurrence of myxomycetes in a specific microsite or locality, but they cannot be counted as individuals. There is no feasible way of assessing either the absolute or relative abundance of each of these stages in myxomycetes, even on a limited spatial scale. The absence of such data makes it impossible to evaluate population sizes following IUCN criteria (Kryvomaz and Stephenson, 2016).

Moreover, appreciable changes in species composition, overall abundance, and diversity are observed for myxomycetes from 1 year to the next at the locality level. In theory, it might well be possible to assess changes in at least some microhabitats (e.g., soil) with the use of direct environmental sampling with some of the molecular techniques now in use, but such sampling would be limited to relatively small microsites, and it is highly questionable that these would be of any value in assessing population changes for the entire habitat. A population, as defined by IUCN, is the total number of mature individuals of the taxon. It is the fundamental basis for the determination of the critically endangered (CR), endangered (EN), or vulnerable (VU) categories. Such determination is performed by calculating the reduction in population size in the past and predicting what the size of the group of individuals will be in the future. However, quantitative studies of myxomycetes have not yet produced data of this type. Records of myxomycetes are based almost exclusively upon fruiting bodies collected in nature or recovered in moist chamber cultures, but there is no way of knowing what the numbers generated in this approach represent. The typical number of functional individuals at an average site could be interpreted for myxomycetes as the total number of functional individuals over multiple years. It is important to consider both known localities and an estimate of the likely number of unrecorded localities to obtain an estimate of the total number of localities and hence the total population size. There are relatively little data on the variation in abundance displayed by species of myxomycetes during a single season or over several years. Most studies are carried out during a single expedition, and this type of effort does not yield data about phenology. As we can see, the knowledge of the spatial and temporal distribution of myxomycetes in the various microhabitats in which they occur is still far too limited.

Given that scenario, an interesting proposal to consider for the conservation of myxomycetes is the "microhabitat approach" proposed by Schnittler et al. (2011), in which the target of conservation is the microhabitat used by microorganisms rather than the species. This is similar to the idea of targeting ecosystems rather than individual species of macroscopic organisms due to the added value of ecological networks, ecosystem functioning, and the maintenance of important biological processes (Cadotte et al., 2011). Using this approach, the determination of the species of myxomycetes involved could be carried out by consolidated specialists, the modern research approaches could be the responsibility of highly trained students, and the basic ecological quantification could easily be determined by conservationists and/or laypersons. Such a division of labor has the potential of increasing the quality and quantity of data in several important selected biological systems worldwide.

Despite the lack of a consolidated strategy for the conservation of myxomycetes, the role of different actors in the successful inclusion of myxomycetes in the lists of managed groups is a key for the development of standard guidelines. Since several other groups of living organisms have already been included in regional and international management systems, the different stakeholders in the field of the conservation have already accumulated experience in the creation, implementation, and control of such systems (i.e., NGOs, government branches dealing with conservation, universities, and communities of individuals surrounding protected areas). In this way, there is no reason to think that most of these management models would not work on myxomycetes, particularly those created to apply to microscopic groups (like the "microhabitat approach"). In fact, such consideration might even represent an interesting option for some myxomycete specialists to engage efforts in an integrated field that makes use, but not exclusively, of biological information.

The concepts of management and conservation require more integrated, ideally holistic strategies and multidisciplinary teams than most classical biological topics. However, it still seems that most myxomycete specialists are disconnected with these types of initiatives and, as mentioned earlier, most projects designed in the past decade target intrinsic aspects of the biology of myxomycetes, such as their evolutionary path, phylogeny, and molecular taxonomy. The latter does make sense considering the limited available information about myxomycetes using modern techniques during the period when those protocols have been used. However, it still points out the fact that most myxomycete research is limited in its applicability for other disciplines, multidisciplinary design, and conceptualization in the context of management and conservation.

Most individuals involved in wildlife or nature management understand that when it comes to nature there are two main discursive approaches in the field. The first one is the classical scientific path that attempts to manage nature based on empirical data collected independently by groups of researchers. Advocates of this approach tend to use primarily "hard" data to make conclusions that can be used for the rational management of the biological group or system under scrutiny. The second approach is the politically supported path in which decisions are made based on contextual elements taking place at a larger scale within national or international political agendas. With this strategy, the management of biological groups or systems is primarily seen as a function of current international pressures, legislation in place, and even popularity of jurisdiction-level decisions. In practice, most universities and research institutions tend to favor the first approach, whereas NGOs and governments primarily use the second one.

The goal of any stakeholder in the management and conservation field is the implementation of sustainable practices associated with the preservation of either a group of organisms or the biological systems sustaining such a group.

In this manner, since such preservation should theoretically be as strong as possible, it seems illogical to believe that only one of the approaches mentioned earlier would be enough to provide the requirements for a realistic implementation of a management program. Most successful conservation stakeholders have learned over time that a series of social, cultural, economic, political, historical, and spiritual variables play an important role in the successful implementation of the initiatives and that strongly biased approaches usually fail.

In this sense, most individuals would not understand why some conservation teams are performing "perceptional" or "conceptual" studies in relation to the management of nature. However, it is easy to understand that one of the keys of sustainability is the promotion of local actions in favor of local communities. As such, it seems highly unrealistic to determine large geographical-scale guidelines for the management of biological systems without considering the homogeneity of perceptional social systems within the same geographical unit. Such a mistake has been documented to undermine the value of some visions of life, usually practiced by small, nonpowerful social groups, and conceptually empower those already with power. For myxomycetes, for example, it is very interesting to see the historical empowerment and current semiindependence of some geographical areas where myxomycete specialists have worked. This is not surprising given the local natural extensionist action of researchers, but it shows that there are simply not enough researchers of myxomycetes worldwide to cover a large extent of the planet. However, the observed normal evolution of myxomycete research has not only enriched the field through healthy competition and different project designs but also favored much needed research synergies, training of young researchers, and institutional collaborations. Such a legacy should be maintained by the current and the next generation of myxomycete specialists and even boosted through more integrated approaches that also consider the multiperspective considerations mentioned earlier in this chapter.

Given all of these considerations, does it make sense to generate strategies for the conservation of myxomycetes? To what extent are these organisms really threatened (certainly a difficult question to address)? Or should they be simply included in conservation agendas targeting microbial diversity? It would have been easy to justify either path if myxomycetes were documented to affect in any way our human lifestyle, but it probably does makes more sense to include them in general microbial conservation programs given their relative innocuity. The reasons may come from two different directions. On one hand, most microbial communities beyond the myxomycetes are also poorly documented in most biological systems worldwide, and on the other hand, the case of most discrete nonpathogenic microbial communities is hard to make due to the cognitive distance between mildly innocuous microbes and human societies.

For instance, it is reasonable to think that only some groups of microscopic organisms have a positive health effect on societies; however, due to the poor documentation of across-group microbial relationships, there is a current trend to generalize the concept of microbial conservation for collective health (Berg et al., 2014; O'Doherty et al., 2016; Panizzon et al., 2015). In that scenario, it seems reasonable to include myxomycetes in the groups of organisms to be managed, but such a trend also shows one more fields of work where myxomycete specialists could potentially engage with public health researchers. There has been some research directed toward the potential health interactions of myxomycetes and humans, particularly with respect to allergies (Lierl, 2013), but the documentation of a more complete scenario has not as yet been carried out in a comprehensive manner.

In terms of ecological relevance, the topic of management and conservation of myxomycetes is also limited in extent, scope, and design. As mentioned earlier in this chapter and more thoroughly in other chapters of this volume, most of the historical research on myxomycetes has been primarily taxonomic and most of the ecologically meaningful relationships studied by researchers comprise only a basic glimpse of the interactions of the group and the surrounding biotic and abiotic elements. It is perhaps safe to say that most of the conceptual advancements in the ecological study of myxomycetes were developed in the first part of the 20th century and that most modern researchers have supported the "gap filling" problem at this point. Most ecological research has either taken a basic approach or considered the ecological relationships as a byproduct of the taxonomic goals pursued. In that way, the ecology of myxomycetes as we know it today is a compendium of natural history and biology of the particular species or assemblages of species and not necessarily a more complex depiction of network interactions, energy flows, and life cycle–integrated analyses entangled in theoretical biological elements.

Interestingly, myxomycetes have several particularities that can be explored in depth from different perspectives in the future to develop such theoretical connections and strengthen the working vision of research teams. For this reason, and also given the recent advancements in the techniques and approaches followed by researchers, it seems logical that the near future may witness many interesting ecological elements relating to myxomycetes that are as yet undocumented in the published literature.

Such potential advances would be more relevant if conceptual elements of the management and conservation approaches were considered. As mentioned earlier, integrated stories tend to be favored, particularly at the granting agency level, over simple ones. For instance, the topics of microbial trophic networks, gene flow, forest fragmentation, and ecosystem services could be adapted to construct an integrated project intended to document the landscape-mediated gene flow capacity of microbes in the framework of management and conservation of ecosystem services. Such a comprehensive project sounds different from simply focusing on the different parts, and it may provide more data on the contextual interactions that affect the biology of the group with the advantage that it can also generate information for those researchers working on the conservation strategies of myxomycetes.

It is understandable that at the local level there should still be efforts to document taxonomic assemblages associated with different forest types, substrates, and other ecological compartments. However, at the regional and higher levels, efforts should target the integrated approach recently mentioned while also including elements for the development of new lines of research. This is the reason why teams of researchers are necessary for larger geographical scales and the reason why extensionism and training, either through technology-based strategies or other techniques of communication, should still take place.

Despite the latter, it is important to remember that myxomycetes, as is the case for any other group of organisms living on the planet, do interact with our species and that there are several other issues for which there may only be limited documentation that could be studied in the future. In particular, the negative effects on myxomycete populations facilitated by global phenomena, such as climate change, land use change, and human-mediated dispersal (i.e., myxomycetes as invasive species) are good areas for future study within the field of conservation. For instance, since temperature and moisture are thought to be the main factors limiting the occurrence of myxomycetes, the anticipated changes in climate regimes are almost certainly going to have a significant impact upon their distribution and ecology. This will likely be true for those species of myxomycetes restricted to specific types of microhabitats (e.g., alpine snowbanks), for those which are confined to geographical areas that are limited in extent (e.g., small oceanic islands), and for assemblages of species associated with deserts and the polar regions of the world, which are highly sensitive ecological systems. But any evaluation of the impact of climate change on myxomycetes is extremely difficult. The main question is whether myxomycetes react to climate change at all? If they do react, in what way is such effect expressed? Are there any species of myxomycetes that might be threatened by climate change? Is it even possible to assess a cause–effect relationship for organisms that are ephemeral and usually so hidden in nature? In future research, at least two possible effects of climate change should be clearly distinguished.

First, the effect of climate change on the composition of myxomycete assemblages, which does not necessarily consider the threat to individual species, must be assessed. Second, the effects of climate change on a particular species of myxomycete, which may well be threatened and thus would warrant inclusion on Red Lists, needs to be evaluated. Using species distribution models, modeling metapopulation dynamics and various other methods are difficult to apply to myxomycetes, in part because there are limited (and rather fragmentary) data available from studies of their distribution around the world. Moreover, distribution data are based on the visible stage in the life cycle—the fruiting body—which is possible to find and identify but only reveals limited information regarding the abundance of the species. Truly definitive confirmation of species occurrences at a specific locality would require a DNA-based analysis of soil, organic matter, living plants, atmosphere, and water—essentially any place where it is possible to find myxomycetes in one of the cryptic but active stages of their life cycle (i.e., amoeboflagellates and plasmodia). Most of the data about the occurrence of myxomycetes in many areas of the world is based on the results obtained in single expeditions. To truly determine the impact of climate change on the distribution of myxomycetes, there needs to be regular monitoring of myxomycete habitats in several study sites, where all possible climatic parameters (e.g., amount and type of precipitation, humidity, and temperature) would be recorded daily, weekly, monthly, seasonally, and on a yearly basis. Only then would it be possible in that way to determine the ecological resilience of myxomycete populations and the role of biological systems to modulate such resilience. All these future directions of study around the concept of climate change may provide much needed information on myxomycetes, contribute to the multidisciplinary approach and collaborative efforts to study the group, and construct the basis for more complex research projects involving the myxomycetes as a group, or of a particular species of interest within the group.

The management and conservation of myxomycetes should not be a discrete field of action within the community of myxomycete researchers and specialists, but a common horizontal element embedded in any project on myxomycetes. The likelihood of microbial groups to be integrated in more formal conservation efforts depends heavily on the activity of the research community and their interaction with other individuals, agencies, government offices, NGOs and the public. The strategy should probably look to integrate myxomycetes in more general efforts to manage and conserve microbial communities, taking into consideration their important participation toward ecosystem functioning and, by default, the goods and services provided. In this sense, it is also imperative to begin integrating this element in professional meetings and in extension programs more strongly than in the past, to generate other types of support from the academic and the general public communities as well.

PUTTING THE PIECES TOGETHER

Modern researchers, specialists, and individuals interested in myxomycetes face a common shortcoming, the poor geographical documentation of the group (although better than for most other groups of protists) and the limited availability of integrated information have not moved myxomycetes higher up in social agendas. Except for some myxomycetes that have been studied as model organisms, the biology of most species and the two tropic stages in the life cycle are poorly documented. Most of the work on the group has been strictly scientific, and specialists have not looked to include this in their environmental plans. The range of action by myxomycete specialists has been limited by their number and their monothematic vision of work. Language barriers, training centers for newer generations of specialists, and a large infrastructural/technical gap between developed and developing institutions are still present worldwide.

Despite these challenges and obstacles for the generation of information in myxomycetes around the world, this is the most active moment in history for the documentation of the biology or myxomycetes. A very tangible proof of the latter is the present volume, which summarizes through the eyes of world-class researchers and specialists, the accumulated knowledge on myxomycetes during the last half century or so. Since a multidisciplinary approach, including elements from applied sciences, education, and communication has been pursued herein, it is expected that this volume will represent a major contribution to the development of myxomycete activity in the foreseeable future.

However, it is still the responsibility of all the individuals involved one or way or another with myxomycetes to promote the inclusion of the group in other fields of research. Topics, such as management and conservation, climate change monitoring, and applicability of myxomycetes in new areas of science, should be transversal elements in the development of research in the future. A mixed-channel, multilanguage approach to disseminating data and more modern communication strategies should be used by teams of researchers looking to sustain integrated lines of research (Fig. 13.3).

Given the accumulation of information on myxomycetes facilitated by researchers in the past and the newer developments generated by modern teams, it is necessary to give credit to the comparatively smaller number of individuals that have engaged in the generation of information about the group. It is understandable that a deeper disconnection between the nonspecialized public

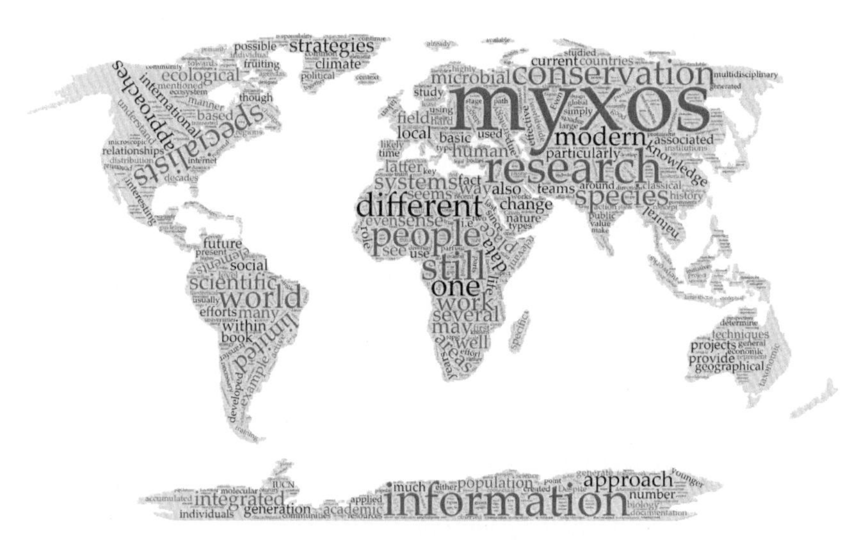

FIGURE 13.3 Word map showing the concept relationships among communication strategies, research, and conservation/management of myxomycetes proposed in the present chapter. This is intended to serve a basis for debate in different spaces and to promote the discussion of nonbiological elements in biological venues. As observed in other chapters of this volume, the community of myxomycete researchers also use the word "myxos" to refer to them.

and myxomycetes has always existed than is the case for other groups of living organisms, such as vertebrates and plants. However, it seems about time for myxomycetes to be included in general microbial conservation agendas and other types of social efforts with the objective of managing more responsibly the natural resources of the planet. Biological systems and important energy transfer networks relying on bottom-up dynamics could use integrated information on myxomycetes to quantify more accurately their own organization and, with that, provide policy makers with arguments to create guidelines for microbial conservation.

The natural paths taken by researchers at different moments in history are logical and contextual, and they should be embraced by the community of myxomycete specialists rather than criticized. At least in recent decades, the former has been the trend observed by the research community, and such a legacy of professional respect and tolerance should be maintained by the younger generations of specialists. Given the fact that more complex research designs, including transversal elements and deeper theoretical considerations, should be expected in the future, it is important for specialists to address their contextual limiting factors and understand their real capacities. This relevant task is key for the establishment of research groups that can work together, share capacities, and push the research goals ahead, particularly for a group of microorganisms for which there is still only a very limited number of specialists around the world.

Overall, myxomycetes represent a very interesting assemblage of microorganisms with many characteristics that are unique to the group. Although their formal study, as documented in Chapter 2 of the present volume, has accumulated over 350 years of work and a substantial body of information on the biology of myxomycetes is currently available, there is still plenty of room for new research to be carried out. The higher resolution capacity of modern techniques and the possibilities of multidisciplinary integration provided by the current research pressures, represent both challenges and opportunities for the community of specialists on the group. In the 21st century, myxomycetes are far from being well documented from most of the different perspectives from which the group can be studied, and creative team-based work is necessary to move forward common goals intended to continue the documentation of the group. Several current researchers have understood this, but it is the task of the younger generation of individuals interested in the group to continue promoting the study of myxomycetes all over the world.

REFERENCES

Berg, G., Grube, M., Schloter, M., Smalla, K., 2014. The plant microbiome and its importance for plant and human health. Front. Microbiol. 5, 491.

Boney, R., Shirk, J.L., Phillips, T.B., Wiggins, A., Ballard, H.L., Miller-Rushing, A.J., Parrish, J.K., 2014. Next steps for citizen science. Science 343 (6178), 1436–1437.

Cadotte, M.W., Carscadden, K., Mirotchnick, N., 2011. Beyond species: functional diversity and the maintenance of ecological processes and services. J. Appl. Ecol. 48 (5), 1079–1087.

Cummings, J.L., Teng, B.S., 2003. Transferring R&D knowledge: the key factors affecting knowledge transfer success. J. Eng. Technol. Manage. 20 (1–2), 39–68.

Gong, X., Borisov, N., Kiyavash, N., Schear, N., 2012. Website detection using remote traffic analysis. In: Fishner-Hubner, S., Wright, M. (Eds.), Privacy Enhancing Technologies. Springer Verlag, Berlin, pp. 58–78.

Gubbia, J., Buyyab, R., Marusica, S., Palaniswamia, M., 2013. Internet of Things (IoT): a vision, architectural elements, and future directions. Fut. Gener. Comput. Syst. 29 (7), 1645–1660.

Guest, D.E., 1997. Human resource management and performance: a review and research agenda. Int. J. Hum. Resour. Manage. 8 (3), 263–276.

Hagiwara, H., Yamamoto, Y., 1995. Myxomycetes of Japan. Heibonsha, Tokyo, Japan.

Jewitt, C., 2012. Multimodal Teaching and Learning. Blackwell, Hoboken, NJ.

Keller, H.W., Everhart, S.E., 2010. Importance of myxomycetes in biological research and teaching. In: Papers in Plant Pathology, University of Nebraska (Paper 366).

Kramer, A., Bekeschus, S., Bröker, B.M., Schleibinger, S., Razavi, B., Assadian, O., 2013. Maintaining health by balancing microbial exposure and prevention of infection: the hygiene hypothesis versus the hypothesis of early immune challenge. J. Hosp. Infect. 83, S29–S34.

Kryvomaz, T., Stephenson, S.L., 2016. Preliminary evaluation of the possible impact of climate change on myxomycetes. Nova Hedwigia 104, 5–30.

Lado, C., Pando, F., 1997. Myxomycetes: I. Ceratiomyxales, Echinosteliales, Liceales, Trichiales. Vol. 2. Flora Micológica Ibérica. CSIC, Madrid, Spain.

Lado, C., Wrigley de Basanta, D., 2008. A review of neotropical myxomycetes (1828–2008). Anales Jard. Bot. Madrid 65 (2), 211–254.

Leydesdorff, L., 2013. Triple helix of university-industry-government relations. In: Carayaniss, E.G. (Ed.), Encyclopedia of Creativity, Invention, Innovation and Entrepreneurship. Springer, New York, NY, pp. 1844–1851.

Lierl, M.B., 2013. Myxomycete (slime mold) spores: unrecognized aeroallergens? Ann. Allergy Asthma Immunol. 111 (6), 537–541.

MacDermott, M.A., Hand, B., 2015. Improving scientific literacy through multimodal communication: strategies, benefits and challenges. School Sci. Rev. 97, 15–20.

Neubert, H., Nowotny, W., Baumann, K., 1993. Die Myxomyceten, Band 1: Ceratiomyxales, Echinosteliales, Liceales, Trichiales. Karlheinz Baumann Verlag, Germany.

Neubert, H., Nowotny, W., Baumann, K., 1995. Die Myxomyceten, Band 2: Physarales. Karlheinz Baumann Verlag, Germany.

Neubert, H., Nowotny, W., Baumann, K., 2000. Die Myxomyceten, Band 3: Stemonitales. Karlheinz Baumann Verlag, Germany.

O'Doherty, K.C., Virani, A., Wilcox, E.A., 2016. The human microbiome and public health: social and ethical considerations. Am. J. Public Health 106 (3), 414–420.

Panizzon, J.P., Pilz, J.H.L., Knaak, N., Ramos, R.C., Ziegler, D.R., Fiuza, L.M., 2015. Microbial diversity: relevance and relationship between environmental conservation and human health. Braz. Arch. Biol. Technol. 58 (1), 137–145.

Pointing, S.B., Belnap, J., 2012. Microbial colonization and controls in dryland systems. Nat. Rev. Microbiol. 10, 551–562.

Poulain, M., Meyer, M., Bozonnet, J., 2011. Les Myxomycètes. FMBDS, Delémont, France.

Raber, M., Richter, J., 2008. Bringing social action back into the social work curriculum: a model for "hands-on" learning. J. Teach. Soc. Work 19, 77–91.

Rea-Maminta, M.A.D., Dagamac, N.H.A., Huyop, F.Z., Wahab, R.A., dela Cruz, T.E.E., 2015. Comparative diversity and heavy metal biosorption of myxomycetes from forest patches on ultramafic and volcanic soils. Chem. Ecol. 31, 741–753.

Richerzhagen, C., 2014. The Nagoya protocol: fragmentation or consolidation? Resources 3, 135–151.

Roth, G., Dicke, U., 2005. Evolution of the brain and intelligence. Trends. Cogn. Sci. 9 (5), 250–257.

Schnittler, M., Kummer, V., Kuhnt, A., Krieglsteiner, L., Flatau, L., Müller, H., Täglich, U., 2011. Rote Liste und Gesamtartenliste der Schleimpilze (Myxomycetes) Deutschlands. Vegetationskunde 70 (6), 125–234.

Stephenson, S.L., Schnittler, M., Novozhilov, M., 2008. Myxomycete diversity and distribution from the fossil record to the present. Biodivers. Conserv. 17 (2), 285–301.

Thorson, E., Moore, J. (Eds.), 1996. Integrated Communication: Synergy of Persuasive Voices. Lawrence Erlbaum, Mahwah, NJ.

UNEP, 2011. Nagoya Protocol on Access to Genetic Resources and the Fair and Equitable Sharing of Benefits Arising From Their Utilization to the Convention on Biological Diversity. Secretariat of the Convention on Biological Diversity, Montreal, Canada.

Velivelli, S.L.S., Sessitsch, A., Prestwich, B.D., 2014. The role of microbial inoculants in integrated crop management systems. Potato Res. 57, 291–309.

Wilson, E.O., 1984. Biophilia. Harvard University Press, Cambridge, MA.

Index

D

N

O

P

S

T

U

V

W

Y

Z

Printed in the United States
By Bookmasters